FUNDAMENTAL MATH AND PHYSICS FOR SCIENTISTS AND ENGINEERS

FUNDAMENTAL MATH AND PHYSICS FOR SCIENTISTS AND ENGINEERS

DAVID YEVICK

HANNAH YEVICK

Published by John Wiley & Sons, Inc., Hoboken, New Jersey
Published simultaneously in Canada

Library of Congress Cataloging-in-Publication Data:

Yevick, David, author.
Fundamental math and physics for scientists and engineers / David Yevick, Hannah Yevick.
pages cm
"Published simultaneously in Canada"–Title page verso.
Includes index.
ISBN 978-0-470-40784-4 (pbk.)
1. Mathematics. 2. Physics. I. Yevick, Hannah, author. II. Title.
QA39.3.Y48 2015
510–dc23

2014032265

Cover Image: iStockphoto©nadla

Printed in the United States of America

To George

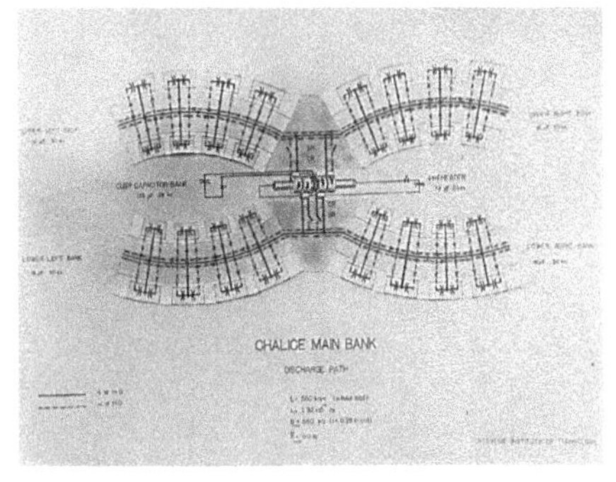

CONTENTS

1

INTRODUCTION

Unique among disciplines, physics condenses the limitlessly complex behavior of nature into a small set of underlying principles. Once these are clearly understood and supplemented with often superficial domain knowledge, any scientific or engineering problem can be succinctly analyzed and solved. Accordingly, the study of physics leads to unsurpassed satisfaction and fulfillment.

This book summarizes intermediate-, college-, and university-level physics and its associated mathematics, identifying basic formulas and concepts that should be understood and memorized. It can be employed to supplement courses, as a reference text or as review material for the GRE and graduate comprehensive exams.

Since physics incorporates broad areas of science and engineering, many treatments overemphasize technical details and problems that require time-consuming mathematical manipulations. The reader then often loses sight of fundamental issues, leading to gaps in comprehension that widen as more advanced material is introduced. This book accordingly focuses exclusively on core material relevant to practical problem solving. Fine details of the subject can later be assimilated rapidly, effectively placing leaves on the branches formed by the underlying concepts.

Mathematics and physics constitute the language of science. Hence, as with any spoken language, they must be learned through repetition and memorization. The central results and equations indicated in this book are therefore indicated by shaded text. These should be rederived, transcribed into a notebook or review cards with a summary of their derivation and memorized. Problems from any source should be solved in conjunction with this book; however, undertaking time-consuming problems

Fundamental Math and Physics for Scientists and Engineers, First Edition.
David Yevick and Hannah Yevick.

without recourse to worked solutions that indicate optimal calculational procedures is not recommended.

Finally, we wish to thank our many inspiring teachers, whose numerous insights guided our approach, in particular Paul Bamberg, Alan Blair, and Sam Treiman, and, above all, our father and grandfather, George Yevick, whose boundless love of physics inspired generations of students.

2

PROBLEM SOLVING

Problem solving, especially on examinations, should habitually follow the procedures below.

2.1 ANALYSIS

1. Problems are very often misread or answered incompletely. Accordingly, circle the words in the problem that describe the required results and underline the specified input data. After completing the calculation, insure that the quantities evaluated in fact correspond to those circled.
2. Write down a summary of the problem in your own words as concisely as possible.
3. Draw a diagram of the physical situation that suggests the general properties of the solution. Annotate the diagram as the solution progresses. Always draw diagrams that accentuate the difference between variables, e.g., when drawing triangles, be sure that its angles are markedly unequal.
4. Briefly contrast different solution methods and summarize on the examination paper the simplest of these (especially if partial credit is given).
5. Solve the problem, proceeding in small steps. Do not perform two mathematical manipulations in a single line. Align equal signs on subsequent lines and check

Fundamental Math and Physics for Scientists and Engineers, First Edition.
David Yevick and Hannah Yevick.

each line of the calculation against the previous line immediately after writing it down. Being careful and organized inevitably saves time.

6. Reconsider periodically if you are employing the simplest solution method. If mathematics becomes involved, backtrack and search for an error or a different approach.
7. Verify the dimensions of your answer and that its magnitude is physically reasonable.
8. Insert your answer into the initial equations that define the problem and check that it yields the correct solution.
9. If necessary and time permits, solve the problem a second time with a different method.

2.2 TEST-TAKING TECHNIQUES

Strategies for improving examination performance include:

1. For morning examinations, 1–3 weeks before the examination, start the day two or more hours before the examination time.
2. Devise a plan of studying well before the examination that includes several review cycles.
3. Outline on paper and review cards in your own words the required material. Carry the cards with you and read them throughout the day when unoccupied.
4. To become aware of optimal solution procedures, solve a variety of problems in books that provide worked solutions and rederive the examples in this or another textbook. Limit the time spent on each problem in accordance with the importance of the topic.
5. Obtain or design your own practice exams and take these under simulated test conditions.
6. In the day preceding a major examination, at most, briefly review notes—studies have demonstrated that last-minute studying does not on average improve grades.
7. Be aware of the examination rules in advance. On multiple choice exams, determining how many answers must be eliminated before selecting one of the remaining choices is statistically beneficial.
8. If allowed, take high-energy food to the exam.
9. Arrive early at the examination location to familiarize yourself with the test environment.
10. First, read the entire examination and then solve the problems in order of difficulty.

11. Maintain awareness of the problem objective; sometimes, a solution can be worked backward from this knowledge.
12. If a calculation proves more complex than expected, either outline your solution method or address a different problem and return to the calculation later, possibly with a different perspective.
13. For multiple choice questions, insure that the solutions are placed correctly on the answer sheet. Write the number of the problem and the answer on a piece of paper and transfer this information onto the answer sheet only at the end of the exam. Retain the paper in case of grading error.
14. On multiple choice tests, examine the possible choices before solving the problem. Eliminate choices with incorrect dimensions and those that lack physical meaning. Those remaining often indicate the important features of the solution and possibly may even reveal the correct answer.
15. Maintain an even composure, possibly through short stretching or controlled breathing exercises.

2.2.1 Dimensional Analysis

Results can be partially verified through dimensional analysis. Dimensions such as those of force, $[MD/T^2]$, are here distinguished by square brackets, where, e.g., D indicates length, T time, M mass, and Q charge. Quantities that are added, subtracted, or equated must possess identical dimensions. For example, $a = v/t$ is potentially valid since the right-hand side dimension of this expression is the product $[D/T][1/T]$, which agrees with that of the left-hand side. Similarly, the argument of a transcendental function (a function that can be expressed as an infinite power series), such as an exponential or harmonic function or of polynomials such as $f(x) = x + x^2$, must be dimensionless; otherwise, different powers would possess different dimensions and could therefore not be summed.

While the dimensions of important physical quantities should be memorized, the dimensions of any quantity can be deduced from an equation expressing this quantity in terms of variables with known dimensions. Thus, e.g., $F = ma$ implies that $[F] = [M][D/T^2] = [MD/T^2]$. Quantities with involved dimensions are often expressed in terms of other standard variables such as voltage.

Example

From $Q = CV$, the units of capacitance can be expressed as $[Q/V]$, with V representing volts. Subsequently, from $V = IR$ with $I = dQ/dt$, the dimensions of, e.g., $t = 1/RC$ can be verified.

3

SCIENTIFIC PROGRAMMING

This text contains basic physics programs written in the Octave scientific programming language that is freely available from http://www.gnu.org/software/octave/index.html with documentation at www.octave.org. Default selections can be chosen during setup. Octave incorporates many features of the commercial MATLAB® language and facilitates rapid and compact coding (for a more extensive introduction, refer to A Short Course in Computational Science and Engineering: C++, Java and Octave Numerical Programming with Free Software Tools, by David Yevick Copyright © 2012 David Yevick). Some of the material in the following text is reprinted with permission from Cambridge University Press.

3.1 LANGUAGE FUNDAMENTALS

A few important general programming concepts as applied to Octave are first summarized below:

1. A program consists primarily of statements that result from terminating a valid expression not followed by the continuation character ... (three lower dots), a carriage return, or a semicolon.
2. An expression can be formed from one or more subexpressions linked by operators such as + or *.

Fundamental Math and Physics for Scientists and Engineers, First Edition.
David Yevick and Hannah Yevick.

3. Operators possess different levels of precedence, e.g., in 2/4 + 3, the division operation possesses a higher precedence and is therefore evaluated before addition. In expressions involving two or more operators with the same precedence level, such as division and multiplication, the operations are typically evaluated from left to right, e.g., 2/4 * 3 equals (2/4) * 3.
4. The parenthesis operator, which evaluates the expression that it encloses, is assigned to the highest precedence level. This eliminates errors generated by incorrect use of precedence or associativity.
5. Certain style conventions, while not required, enhance clarity and readability:
   a. Variables and function names should be composed of one or more descriptive words. The initial letter should be uncapitalized, while the first letter of each subsequent word should be capitalized as in outputVelocity.
   b. Spaces should be placed to the right and left of binary operators, which act on the expressions (operands) to their left and right, as in `3 + 4`, but no space should be employed in unary operator such as the negative sign in `-3 + 4`. Spaces are preferentially be inserted after commas as in `computeVelocity( 3, 4 )` and within parentheses except where these indicate indices.
   c. Indentation should be employed to indicate when a group of inner statements is under the logical control of an outer statement such as in

```
if ( firstVariable == 0 )
   secondVariable = 5;
end
```

   d. Any part of a line located to the right of the symbol % constitutes a comment that typically documents the program. Statements that form a logical unit should be preceded by one or more comment lines and surrounded by blank lines. Statement lines that introduce input variables should end with a comment describing the variables.

3.1.1 Octave Programming

Running Octave: Starting Octave opens a command window into which statements can be entered interactively. Alternatively, a program in the directory `programs` in partition `C:` is created by first entering `cd C:\programs` into the command window, pressing the enter key, and then entering the command `edit`. Statements are then typed into the program editor, the file is saved by selecting Save from the button or menu bar as a MATrix LABoratory file such as `myFile.m` (the .m extension is appended automatically by the editor), and the program is then run by typing `myFile` into the command window. The program can also be activated by including the statement `myFile;` within another program. To list the files in the current directory, enter `dir` into the Octave command window.

Help Commands: Typing `help commandName` yields a description of the command `commandName`. To find all commands related to a word `subject`, type `lookfor subject`. Entering `doc` or `doc topic` brings up, respectively, a complete help document and a description of the language feature `topic`.

Input and Output: A value of a variable `G` can be entered into a program (`.m` file) from the keyboard by including the line `G = input( 'user prompt' )`. The statement `format long e` sets the output style to display all 15 floating-point number significant digits, after which `format short e` reverts to the default 5 output digits.

Constants and Complex Numbers: Some important constants are `i` and `j`, which both equal $\sqrt{-1}$, `e`, and `pi`. However, if a variable assignment such as `i = 3;` is encountered in an Octave program, `i` ceases to be identified with the imaginary unit until the command `clear i` is issued. Imaginary numbers can be manipulated with the functions `real( )`, `imag( )`, `conj( )`, and `norm( )`, and imaginary values are automatically returned by standard functions such as `exp( )`, `sin( )`, and `sinh( )` for imaginary arguments.

Arrays and Matrices: A symbol `A` can represent a scalar, row, or column vector or matrix of any dimension. Row vectors are constructed either by

```
vR = [ 1 2 3 4 ];
```

or

```
vR = [ 1, 2, 3, 4 ];
```

The corresponding column vector can similarly be entered in any of the following three ways:

```
vC = [ 1
2
3
4 ];
```

```
vC = [ 1; 2; 3; 4 ];
vC = [ 1 2 3 4 ].';
```

Here `.'` indicates transpose, while `'` instead implements the Hermitian (complex conjugate) transpose.

A 2×2 matrix

$$\mathrm{mRC} = \begin{pmatrix} 1 & 2 \\ 3 & 4 \end{pmatrix}$$

can be constructed by, e.g., `mRC = [ 1 2; 3 4 ];` after which `mRC(1, 2)` returns $(\mathrm{MRC})_{12}$, here the value 2. Subsequently, `size(mRC)` yields a vector containing the row and column dimensions of `mRC`, while `length( mRC )` returns the maximum of these values. Here, we introduce the convention of appending `R`, `C`, or `RC` to the variable name to respectively identify row vectors, column vectors, and matrices.

Basic Manipulations: A value n is raised to the power m by n^m. The remainder of n/m is denoted rem(n, m) and is positive or zero for $n > 0$ and negative or zero for $n < 0$. The function mod(n, m) returns n modulus m, which is always positive, while ceil( ), floor( ), and fix( ) round floating-point numbers to the next larger integer, smaller integer, and nearest integer closer to zero, respectively.

Vector and Matrix Operations: Two vectors or matrices of the same dimension can be added or subtracted. Multiplying a matrix or vector by a scalar, c, multiplies each element by c. Additionally, eye (n, n) is the $n \times n$ unit or identity matrix with ones along the main diagonal and zeros elsewhere, while ones (n, m) and zeros (n, m) are $n \times m$ matrices with all elements one or zeros so that

$$2 + \mathtt{mRC} = 2 * \mathtt{ones}(2, 2) + \mathtt{mRC} = \begin{pmatrix} 3 & 4 \\ 5 & 6 \end{pmatrix}$$

and

$$2 * \mathtt{eye}(2, 2) + \mathtt{mRC} = \begin{pmatrix} 3 & 2 \\ 3 & 6 \end{pmatrix}$$

Further, mRC * mRC, or equivalently mRC^2, multiplies mRC by itself, while

$$\mathtt{mRC.} * \mathtt{mRC} = \mathtt{mRC.}\hat{}2 = \begin{pmatrix} 1 & 4 \\ 9 & 16 \end{pmatrix}$$

implements component-by-component multiplication. Other arithmetic operations function analogously so that the (i, j) element of M ./ N is M_{ij}/N_{ij}. Functions such as cos(M) return a matrix composed of the cosines of each element in M.

Solving Linear Equation Systems: The solution of the linear equation system xR * mRC = yR is xR = yR / mRC, while mRC * xC = yC is solved by xC = mRC \ yC. The inverse of a matrix mRC is represented by inv(mRC). The eigenvalues of a matrix are obtained through eigenValues = eig(mRC), while both the eigenvalues and eigenvectors are returned through [eigenValues, eigenVectors] = eig(mRC).

Random Number Generation: A single random number between 0 and 1 is generated by rand, while rand(m, n) returns a $m \times n$ matrix with random entries. The same random sequence can be generated each time a program is run by including rand('state', 0) before the first call to rand.

Control Logic and Iteration: The *logical operators* in octave are ==, <, <=, >, >=, ~= (not equal) and the and, or, and not operators—&, |, and ~, respectively. Any nonzero value is taken to represent a logical "true" value, while a zero value corresponds to a logical "false" as can be seen by evaluating, e.g., 3 & 4, which produces the output 1. Thus,

```
if ( S == 2 )
   xxx
elseif ( S == 3 )
   yyy
```

```
else
   zzz
end
```

executes the statements denoted by xxx if the logical statement S == 2 is true, yyy if S == 3, and zzz otherwise. The for loop

```
for loop = 10 : -1 : 0;
        vR(loop) = sin(loop * pi / 10 );
end;
```

yields the array vR = [sin(π) sin(9π / 10) ... sin(π/10) 0], while 1 : 10 yields an array with elements from 1 to 10 in unit increments. *Mistakenly replacing colons by commas or semicolons results in severe and often difficult to detect errors.* If a break statement is encountered within a for loop, control is passed to the statement immediately following the end statement. An alternative to the for loop is the while (logical condition) ... statements ... end construct.

Vectorized Iterators: A *vectorized iterator* such as vR = sin(pi: -pi/10: -1.e-4), which yields, generates, or manipulates a vector far more rapidly than the corresponding for loop. linspace(s1, s2, n) and logspace(s1, s2, n) produce n equally/logarithmically spaced points from s1 to s2. An isolated colon employed as an index iterates through the elements associated with the index so that MRC(:, 1) = V(:); places the elements of the row or column vector V into the first column of MRC.

Files and Function Files: A function that returns variables output1, output2 ... is called [output1, output2, ...] = myFunction(input1, input2, ...) and normally resides in a separate file myFunction.m in the current directory, the first line of which must read function [aOutput1, aOutput2, ...] = myFunction(aInput1, aInput2, ...). Variables defined (created) inside a function are inaccessible in the remainder of the program once the function terminates (unless global statements are present), while only the argument variables and variables local to the function are visible from within the function. A function can accept other functions as an arguments either (for Octave functions) with the syntax fmin('functionname', a, b) or through a function handle (pointer) as fmin(@functionname, a, b).

Built-In Functions: Some common functions are the discrete forward and inverse Fourier transforms, fft( ) and ifft( ) and mean( ), sum( ), min( ), max( ), and sort( ). Data is interpolated by y1 = interp1(x, y, x1, 'method'), where 'method' is 'linear' (the default), 'spline', or 'cubic'; x and y are the input x- and y-coordinate vectors; and x1 contains the x-coordinate(s) of the point(s) at which interpolated values are desired. The function roots([1 3 5]) returns the roots of the polynomial $x^2 + 3x + 5$.

Graphic Operations: plot(vY1) generates a simple line plot of the values in the row or column vector vY1, while plot(vX1, vY1, vX2, vY2, ...) creates a single plot with lines given by the multiple (x, y) data sets. Hence, plot(C, 'g.'), where C is a complex vector, graphs the real against the imaginary part of C in green with point marker style. Logarithmic graphs are plotted with semilogy( ), semilogx( ), or loglog( ) in place of plot( ). Three-dimensional grid and contour plots with nContours contour levels are created with mesh(mRC) or mesh(vX, vY, mRC) and contour(mRC) or contour(vX, vY, mRC, nContours) where vX and vY are row or column vectors that contain the x and y positions of the grid points along the axes. The commands hold on and hold off retain graphs so that additional curves can be overlaid. Subsequently, axis defaults can be overridden with axis([xmin xmax ymin ymax]), while axis labels are implemented with xlabel('xtext') and ylabel('ytext') and the plot title is specified by title('title text'). The command print ('outputFile.eps', '-deps') or, e.g., print('outputFile.pdf', '-dpdf') yields, respectively, encapsulated postscript or .pdf files of the current plot window in the file outputFile.dat or outputFile.pdf (help print displays all options).

Memory Management: User-defined variable or function names hide preexisting or built-in variable and function names, e.g., if the program defines a variable or function length or length( ), the Octave function length( ) becomes inaccessible. Additionally, if the second time a program is executed a smaller array is assigned to an variable, the larger memory space will still be reserved by the variable causing errors when, e.g., its length or magnitude is computed. Accordingly, each program should begin with clear all to remove all preexisting assignments (a single construct M is destroyed through clear M).

Structures: To associate different variables with a single entity (structure) name, a dot is placed after the name as in

```
Spring1.position = 0;
Spring1.velocity = 1;
Spring1.position = Spring1.position + deltaTime * k/m *
                   Spring1.velocity
```

Variables pertaining to one entity can then be segregated from those, such as Spring2.position, describing a different object. The names of structures are conventionally capitalized.

4

ELEMENTARY MATHEMATICS

The following treatment of algebra and geometry focuses on often neglected aspects.

4.1 ALGEBRA

While arithmetic concerns direct problems such as evaluating $y = 2x + 5$ for $x = 3$, algebra addresses arithmetical inverse problems, such as the determination of x given $y = 11$ above. Such generalizations of division can be highly involved depending on the complexity of the direct equation.

4.1.1 Equation Manipulation

Since both sides of an equation evaluate to the same quantity, they can be added to, subtracted from, or multiplied or divided by any number or expression. Therefore,

$$\frac{a}{b} = \frac{c}{d} \tag{4.1.1}$$

can be simplified through *cross multiplication*, e.g., multiplication of both sides by bd to yield

$$ad = bc \tag{4.1.2}$$

Fundamental Math and Physics for Scientists and Engineers, First Edition.
David Yevick and Hannah Yevick.

Similarly, the left hand of one equation can be multiplied or divided by the left-hand side of a second equation if the right-hand sides of the two equations are similarly manipulated (as the right and left sides of each equation by definition represent the same value).

Example

Equating the quotients of the right- and left-hand sides of the following two equations

$$\begin{aligned} 3y(x+1) &= 4 \\ 4(x+1) &= 2 \end{aligned} \tag{4.1.3}$$

results in $3y/4 = 2$.

4.1.2 Linear Equation Systems

An algebraic equation is *linear* if all variables in the equation only enter to first order (e.g., as x and y but not xy). At least N linear equations are required to uniquely determine the values of N variables. The standard procedure for solving such a system first reduces the system to a "tridiagonal form" through repeated implementation of a small number of basic operations.

Example

To solve,

$$\begin{aligned} x+y &= 3 \\ 2x+3y &= 7 \end{aligned} \tag{4.1.4}$$

for x and y, the first equation can be recast as $x = 3 - y$, which yields a single equation for y after substitution into the second equation. Alternatively, multiplying the first equation by two results in

$$\begin{aligned} 2x+2y &= 6 \\ 2x+3y &= 7 \end{aligned} \tag{4.1.5}$$

Subtracting the first equation from the second equation then gives

$$\begin{aligned} 2x+2y &= 6 \\ y &= 1 \end{aligned} \tag{4.1.6}$$

The inverted pyramidal form is termed an *upper triangular linear equation system* and can be solved by *back-substituting* the solution for y from the second equation into the first equation, which then solved for x.

A set of equations can be redundant in that one or more equations of the set can be generated by summing the remaining equations with appropriate coefficients. If the number of independent equations is less or greater than N, infinitely many or zero solutions exist, respectively. Nonlinear equation systems can sometimes be linearized through *substitution* of new variables formed from nonlinear combinations of the original variables. Thus, defining $w = x^2$, $z = y^3$ recasts

$$\begin{aligned} x^2 + 3y^3 &= 4 \\ 2x^2 + y^3 &= 3 \end{aligned} \tag{4.1.7}$$

into the linear equations $w + 3z = 4$, $2w + z = 3$.

4.1.3 Factoring

The inverse problem to polynomial multiplication is termed *factoring*. That is, multiplication and addition yield

$$(ax+b)(cx+d) = acx^2 + (bc+ad)x + bd \tag{4.1.8}$$

which is reversed by factoring the right-hand side into the left-hand product of two lesser degree polynomials. For quadratic (second-order) equations, the quadratic formula states that the roots (solutions) of $ax^2 + bx + c = 0$ are

$$x_{1,2} = \frac{-b \pm \sqrt{b^2 - 4ac}}{2a} \tag{4.1.9}$$

implying that the polynomial $ax^2 + bx + c$ can be factored as $(x - x_1)(x - x_2)$. Equation (4.1.9) is derived by first *completing the square* according to

$$\begin{aligned} ax^2 + bx + c &= a\left(x^2 + \frac{bx}{a}\right) + c \\ &= a\left(x^2 + \frac{bx}{a} + \frac{b^2}{4a^2}\right) - \frac{b^2}{4a} + c \\ &= a\left(x + \frac{b}{2a}\right)^2 - \frac{b^2 - 4ac}{4a} \end{aligned} \tag{4.1.10}$$

Multiplying N terms of the form $(x - \lambda_i)$ yields

$$(x-\lambda_1)(x-\lambda_2)\ldots(x-\lambda_N) = x^N + x^{N-1}\sum_i \lambda_i + x^{N-2}\sum_{\substack{i,j \\ i \neq j}} \lambda_i \lambda_j + \ldots + \prod_i \lambda_i \tag{4.1.11}$$

That, e.g., the coefficient x^{N+1} equals the sum of the roots can aid in factoring polynomials.

Numerous other factorization theorems exist. For example, as can be verified by polynomial division,

$$a^3 \pm b^3 = (a \pm b)\left(a^2 \mp ab + b^2\right) \tag{4.1.12}$$

and in general for odd n,

$$a^n + b^n = (a+b)\left(a^{n-1} - a^{n-2}b + a^{n-3}b^2 - \ldots + a^2b^{n-3} - ab^{n-2} + b^{n-1}\right) \tag{4.1.13}$$

A few less common formulas are

$$\begin{aligned} a^4 + 4b^4 &= (a^2 + 2ab + 2b^2)(a^2 - 2ab + 2b^2) \\ (a+b)(b+c)(a+c) &= a^2(b+c) + b^2(a+c) + c^2(a+b) + 2abc \\ -(a-b)(b-c)(c-a) &= a^2(b-c) + b^2(c-a) + c^2(a-b) \end{aligned} \tag{4.1.14}$$

4.1.4 Inequalities

While the same number can be added or subtracted from both sides of an inequality, multiplication of both sides of an equation by a negative quantity, or more generally applying a monotonically decreasing function to both sides of an inequality, reverses the sign of the inequality.

Example

$$-(x-4) \geq 1 \tag{4.1.15}$$

which is satisfied for $x < -5$, can be rewritten as

$$(x-4) \leq -1 \tag{4.1.16}$$

Algebraic expressions such as $(x-2)\,y > 0$, which implies $y > 0$ if $x > -2$ but $y < 0$ if $x < -2$, are often negative only for certain variable values, complicating the analysis of algebraic inequalities. Accordingly, answers should be checked by sketching the functions entering into such inequalities.

Example

Since for $x > 2$ and $x < 2$ the function $(x-2)^2$ increases and decreases monotonically, respectively, for $x < 2$, the direction of the inequality obtained by taking the square root of both sides of

$$(x-2)^2 > 4 \tag{4.1.17}$$

changes. Accordingly, the solution of Equation (4.1.17) is $x > 4$ or $x < 0$.

4.1.5 Sum Formulas

Algebraic Series: The sum of N consecutive integers equals N times the average value of the integers, which can be rearranged into pairs of equal value as indicated below:

$$\underbrace{m+(m+1)+\ldots+(m+N-1)}_{N\text{ integers}}=(m+(m+N-1))+(m+1+(m+N-2))+\ldots$$

$$=N\left(\frac{2m+N-1}{2}\right) \qquad (4.1.18)$$

which specializes for $m=1$ to

$$1+2+3+\ldots+N=\frac{N(N+1)}{2} \qquad (4.1.19)$$

Sum of Squares: The sum of the first N squares is computed from the formula

$$1+4+9+\ldots+N^2=\frac{N(N+1)(2N+1)}{6} \qquad (4.1.20)$$

Equation (4.1.20) can be derived by representing the sum of the first N squares by $S(N)$ so that $S(N)-S(N-1)=N^2$ and $S(0)=0$ while additionally $S(N)<N\times N^2=N^3$. Hence, writing $S(N)=aN^3+bN^2+cN+d$ with $a<1$ and $d=0$ from $S(0)=0$,

$$S(N)-S(N-1)=a\left(3N^2-3N+1\right)+b(2N-1)+c=N^2 \qquad (4.1.21)$$

Equating the coefficients of N^2 and setting the coefficients $-3a+2b$ of N and $a-b+c$ of the constant term to zero yields $a=1/3,\ b=1/2,\ c=1/6$, from which Equation (4.1.20) follows.

Geometric Series: The sum of the first N powers of a variable is given by

$$1+a+a^2+a^3+\ldots+a^n=\frac{1-a^{n+1}}{1-a} \qquad (4.1.22)$$

as verified directly through long division

$$\begin{array}{r|llllll}
 & 1 & +a & +a^2 & \ldots & +a^n & \\ \hline
1-a & 1 & & & & & -a^{n+1} \\
 & \underline{1} & \underline{-a} & & & & \\
 & & a & & & & \\
 & & \underline{a} & \underline{-a^2} & & & \\
 & & & & \ddots & & \\
 & & & & & \underline{a^n} & \underline{-a^{n+1}} \\
 & & & & & & 0
\end{array} \qquad (4.1.23)$$

For $a < 1$, this yields the infinite series

$$1 + a + a^2 + a^3 + \ldots = \frac{1}{1-a} \tag{4.1.24}$$

4.1.6 Binomial Theorem

The nth power of a sum of two variables can be written according to the *binomial theorem* as

$$(a+b)^n = {}_nC_n a^n b^0 + {}_nC_{n-1} a^{n-1} b^1 + \ldots + {}_nC_0 a^0 b^n \tag{4.1.25}$$

The *binomial coefficients* are given in terms of the factorial function $n! \equiv n(n-1)(n-2)\ldots 1$ by

$${}_nC_k = \frac{n!}{k!(n-k)!} = \frac{n(n-1)\ldots(n-k+1)}{k!} \tag{4.1.26}$$

so that

$$\begin{aligned} {}_nC_0 &= {}_nC_n = 1 \\ {}_nC_1 &= {}_nC_{n-1} = \frac{n!}{1!(n-1)!} = \frac{n}{1!} \\ {}_nC_2 &= {}_nC_{n-2} = \frac{n!}{2!(n-2)!} = \frac{n(n-1)}{2!} \end{aligned} \tag{4.1.27}$$

For noninteger powers, $n = \alpha$, the series does not terminate and is therefore termed *transcendental*. For $a > b$, the binomial theorem then yields with $\delta = b/a$

$$(a+b)^\alpha = a^\alpha(1+\delta)^\alpha = a^\alpha\left(1 + \alpha\delta + \frac{\alpha(\alpha-1)}{2!}\delta^2 + \frac{\alpha(\alpha-1)(\alpha-2)}{3!}\delta^3 + \cdots\right) \tag{4.1.28}$$

Accordingly, for $\delta \ll 1$,

$$(1+\delta)^\alpha \approx 1 + \alpha\delta + \ldots \tag{4.1.29}$$

while

$$\frac{1}{(1+\delta)^\alpha} = (1+\delta)^{-\alpha} \approx 1 - \alpha\delta + \ldots \tag{4.1.30}$$

4.2 GEOMETRY

Several fundamental results in geometry recur often in physics calculations and should be memorized. Theorems that are trivially derived through vector analysis are discussed in later chapters.

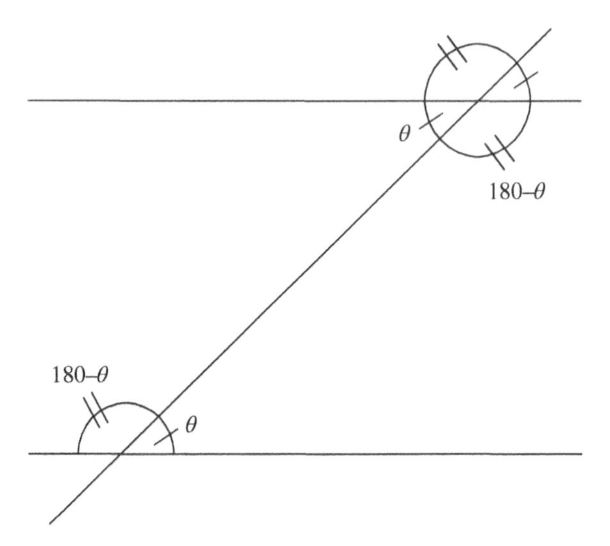

FIGURE 4.1 Corresponding angles.

4.2.1 Angles

The angle between two intersecting rays can be obtained from a circle with vertex at the point of intersection by dividing the (arc)length, s, of the part of the circle included by the rays by its radius, r,

$$\theta = \frac{s}{r} \tag{4.2.1}$$

θ is here expressed in *radians*. Since a full circle thus corresponds to 2π radians,

$$\theta_{\text{degrees}} = \frac{180}{\pi}\theta_{\text{radians}} \tag{4.2.2}$$

If two lines intersect, the angles on either side of one of the two intersecting lines must sum to 180°. For a line intersecting two parallel lines, corresponding angles (angles in Fig. 4.1 distinguished by an identical number of markers) are equal.

4.2.2 Triangles

The equality of corresponding angles together implies from Figure 4.2 that the interior angles of a triangle sum to 180°. A trivial modification of Figure 4.2 further demonstrates that if a side of a triangle is extended beyond a vertex, the *exterior angle* between this line and the adjacent side equals the sum of the opposing two angles.

Two *congruent* (identical) triangles either have the lengths of all sides equal (SSS) or have two sides and their included angle (SAS) or one side and the two adjacent

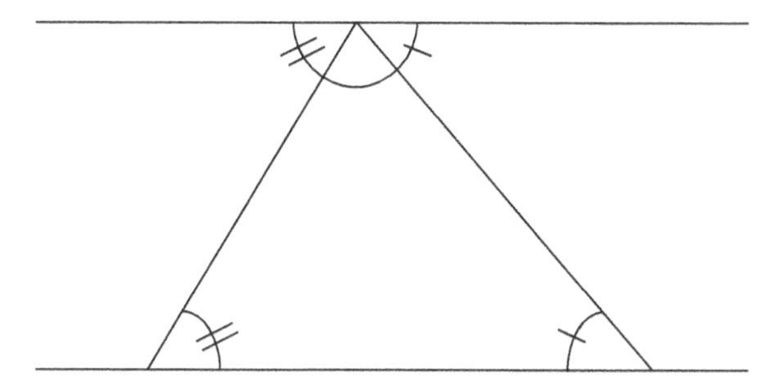

FIGURE 4.2 Proof that the angles of a triangle sum to 180°.

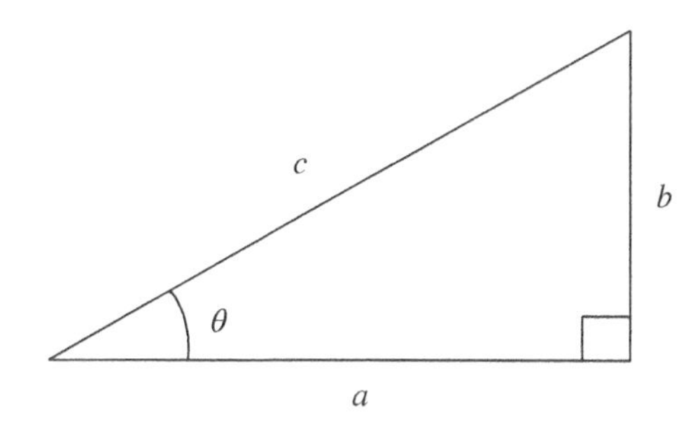

FIGURE 4.3 Right angle triangle.

angles (ASA) identical, as is evident from a drawing. If all three *angles* are identical in the two triangles (AAA), the *ratios* of the lengths of corresponding sides in the two *similar* triangles are instead identical.

The *bisectors* of the angles in any triangle intersect at the *incenter*, around which a circle can be drawn that contacts the three sides of the triangle. Lines extending from each vertex to the midpoint of the opposite side instead intersect at the *centroid*, which is twice as close to the midpoint of each side as it is to the opposite vertex. Finally, lines perpendicular to and passing through the midpoints of each of the triangle's sides intersect at the *circumcenter* situated the same distance from the three vertices.

4.2.3 Right Triangles

Triangles with three equal sides (and therefore angles) are termed equilateral, and those with two equal sides (angles), isosceles; if all angles are less than 90°, they are termed acute; if one angle is greater than 90°, obtuse; and if one angle equals 90°, right. For right triangles, Pythagoras's theorem, which states that the square of the length of the longest side of the triangle equals the sum of the squares of the lengths of the two smaller sides, i.e., $c^2 = a^2 + b^2$, in Figure 4.3 holds.

In Figure 4.3, the cosine, $\cos\theta$, is defined as the ratio of the adjacent leg to the hypotenuse, a/c, while the sine, $\sin\theta$, equals the ratio of the opposite leg to the hypotenuse,

b/c, and the tangent, $\tan\theta$, is formed from the ratio of these quantities or b/a. The reciprocals (the reciprocal of any quantity a is defined as $1/a$) of sin, cos, and tan are denoted, somewhat counterintuitively, csc (cosecant), sec (secant), and cot (cotangent), e.g., $\cot y = 1/\tan y$, $\sec y = 1/\cos y$, and $\csc y = 1/\sin y$.

The formula $c^2 = a^2 + b^2$ possesses integer solutions for certain values of a, b, and c. The smallest of these appear frequently in multiple choice problems and should be memorized, namely,

$$(3,4,5)\ \ (5,12,13)\ \ (8,15,17)\ \ (7,24,25) \tag{4.2.3}$$

The acute angles of a (3,4,5) triangle are approximately 38.7° and 53.1°. Other frequently occurring triangles are the isosceles 45°–45°–90° triangle with lengths $a=\xi, b=\xi, c=\xi\sqrt{2}$ in Figure 4.3 yielding

$$\sin\theta = \sin 45^\circ = \cos 45^\circ = \frac{1}{\sqrt{2}} = \frac{\sqrt{2}}{2}, \quad \tan 45^\circ = 1 \tag{4.2.4}$$

and the 30°–60°–90° triangle with lengths $a=\xi, b=\sqrt{3}\xi, c=2\xi$ so that

$$\sin\theta = \sin 30^\circ = \frac{1}{2}, \quad \cos 30^\circ = \frac{\sqrt{3}}{2}, \quad \tan 30^\circ = \frac{1}{\sqrt{3}} = \frac{\sqrt{3}}{3} \tag{4.2.5}$$

4.2.4 Polygons

An n-sided polygon can be subdivided into n triangles sharing a common vertex within the polygon, each of which has a single side coincident with a facet of the polygon. Since the angles of each triangle sum to 180°, while combining the angles at the shared vertex gives 360°, summing all the internal angles of the polygon yields $180^\circ n - 360^\circ = 180^\circ(n-2)$.

4.2.5 Circles

Considering finally the angles created when lines intersect a circle, the angle between a tangent to a circle and a chord (a line that intersects the circle at two points) passing through the point of tangency equals half the *central angle* between the rays drawn from the center of the circle to the two points of intersection of the chord. If two lines pass through the two points of intersection of a chord to any other point on the circumference of the circle, the *inscribed angle* between these two lines equals half the central angle formed by the chord. Finally, the angle between two chords that intersect inside or outside a circle is half the sum or half the difference of the angles formed by the two arcs that these lines intercept, respectively.

4.3 EXPONENTIAL, LOGARITHMIC FUNCTIONS, AND TRIGONOMETRY

The study of right angle triangles is termed trigonometry. As the trigonometric functions are directly related to complex exponential functions, lengthy geometric analysis can be replaced by compact algebraic derivations.

4.3.1 Exponential Functions

The exponential function is defined through a limiting procedure. Placing \$100 in a bank at 10% interest compounded annually yields \$110 after 1 year, while interest compounded biannually results in \$105 after half a year, and $\$105(1+0.1/2) = \$110.25 > \$110$ after 1 year. For interest compounded n times a year, the amount received after 1 year equals the following nth order polynomial in $0.1/n$:

$$\$100\left(1+\frac{0.1}{n}\right)^n \tag{4.3.1}$$

Hence, if the interest is compounded *continuously* (an infinite number of times per year), the amount is

$$\$100e^{0.1} \equiv \lim_{n\to\infty} \$100\left(1+\frac{0.1}{n}\right)^n \tag{4.3.2}$$

From the binomial theorem,

$$\begin{aligned} e^{\alpha} &= \lim_{n\to\infty}\left(1+\frac{\alpha}{n}\right)^n \\ &= \lim_{n\to\infty}\left(1+\frac{n}{1!}\frac{\alpha}{n}+\frac{n(n-1)}{2!}\left(\frac{\alpha}{n}\right)^2+\cdots\right) \\ &\approx \lim_{n\to\infty}\left(1+\frac{n}{1!}\frac{\alpha}{n}+\frac{n^2}{2!}\left(\frac{\alpha}{n}\right)^2+\cdots\right) \\ &= 1+\frac{\alpha}{1!}+\frac{\alpha^2}{2!}+\cdots \end{aligned} \tag{4.3.3}$$

Functions such as the exponential that are represented by infinite polynomial series are termed *transcendental*.

4.3.2 Inverse Functions and Logarithms

An *inverse function*, $f^{-1}(x)$, is defined such that if $y = f(x)$, $x = f^{-1}(y)$; e.g., the inverse of the square function is the square root function. Graphically, the inverse function is constructed by reflecting $y = f(x)$ through the line $x = y$ since exchanging x and y in $y = f(x)$ yields $x = f(y)$, which is equivalent to $y = f^{-1}(x)$.

The *natural logarithm* (often simply termed logarithm) provides the inverse of the exponential function such that if $\exp(a) = b$, $a = \ln(\exp(a)) = \ln(b)$. The properties of the logarithms follow from the *laws of exponents*:

$$x^a x^b = \underbrace{(x \cdot x \cdot \ldots \cdot x)}_{a \text{ times}} \underbrace{(x \cdot x \cdot \ldots \cdot x)}_{b \text{ times}} = x^{a+b}$$
$$(x^a)^b = x^{ab} \tag{4.3.4}$$

Consequently, if $\exp(\ln x) = x$ and $\exp(\ln y) = y$, $xy = \exp(\ln x)\exp(\ln y) = \exp(\ln x + \ln y)$. Taking the natural logarithm of both sides,

$$\ln(xy) = \ln x + \ln y \tag{4.3.5}$$

which further implies that

$$\ln x^n = n \ln x \tag{4.3.6}$$

The *logarithm to the base b* is defined such that $x = \log_b y$ if $b^x = y$. If the base b is omitted, its implicit value is 10. A logarithm to a base a can be expressed in terms of a logarithm to a different base b by noting that if $b^x = y$, then $x = \log_b y$ while additionally $\log_a(b^x) = x \log_a b = \log_a y$, and hence,

$$\log_b y = \frac{\log_a y}{\log_a b} \tag{4.3.7}$$

For any b, $\log_b(1) = 0$ while $\log_b(x)$ approaches ∞ more slowly than any positive power of x (i.e., $\lim_{x \to \infty} [\log_b(x)/x^n] = 0$ for $n > 0$) as $x \to \infty$ and more slowly than any power of x^{-n} as $x \to 0$.

4.3.3 Hyperbolic Functions

A general function, $f(x)$, can be written as the sum of an *even* function obeying $g(x) = g(-x)$ and an *odd* function with $h(x) = -h(-x)$ according to

$$f(x) = \underbrace{\frac{f(x) + f(-x)}{2}}_{\text{even}} + \underbrace{\frac{f(x) - f(-x)}{2}}_{\text{odd}} \tag{4.3.8}$$

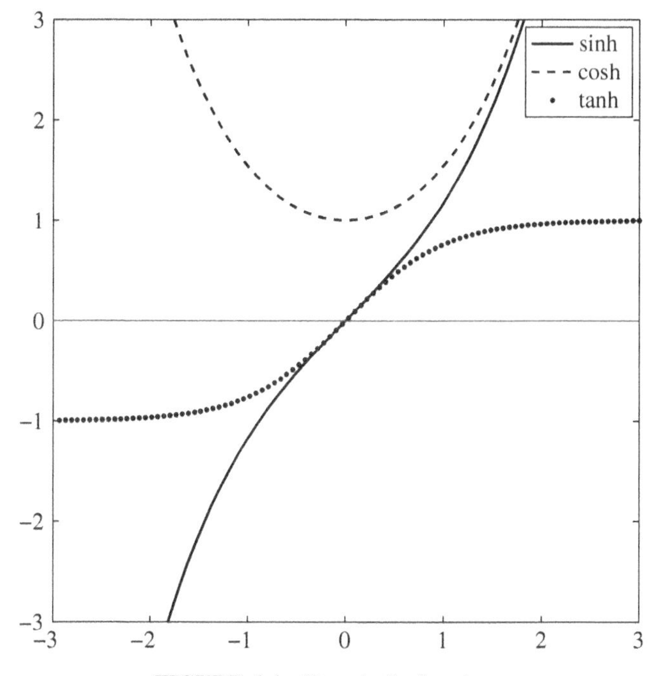

FIGURE 4.4 Hyperbolic functions.

For the exponential function, the even and odd functions are termed the hyperbolic cosine and sine, cosh(x) and sinh(x), respectively. That is,

$$e^x = \underbrace{\frac{e^x + e^{-x}}{2}}_{\cosh x} + \underbrace{\frac{e^x - e^{-x}}{2}}_{\sinh x} = \underbrace{\left(1 + \frac{x^2}{2!} + \frac{x^4}{4!} + \cdots\right)}_{\cosh x} + \underbrace{\left(\frac{x}{1!} + \frac{x^3}{3!} + \cdots\right)}_{\sinh x} \tag{4.3.9}$$

from which

$$\begin{aligned} e^{-x} &= \cosh(-x) + \sinh(-x) = \cosh x - \sinh x \\ 1 = e^x e^{-x} &= (\cosh x + \sinh x)(\cosh x - \sinh x) = \cosh^2 x - \sinh^2 x \end{aligned} \tag{4.3.10}$$

The hyperbolic sine, cosine, and tangent $\tanh(x) = \sinh x / \cosh x$ are graphed in Figure 4.4. The reciprocals of these functions are, respectively, termed cosech (or csch), sech, and cotanh (or coth)

4.3.4 Complex Numbers and Harmonic Functions

If the two solutions of $x^2 + 1 = 0$ are represented by $\pm i$, where i with $i^2 = -1$ is the "imaginary" unit, the *fundamental theorem of algebra* states that a polynomial (with possibly complex coefficients) of nth degree possesses n real or *complex* roots, where m identical roots are counted as m separate values. A *complex number* is represented as $z = a + ib$ with a conjugate and modulus defined by $z^* = a - ib$ and $|z| \equiv \sqrt{zz^*} = \sqrt{a^2 + b^2}$. Further, z can be described by a point in the *complex plane* in which b is plotted on the vertical axis and a on the horizontal axis. The distance from the origin to z equals the modulus, $|z|$. Summing two complex numbers $a_1 + ib_1$ and $a_2 + ib_2$ yields $a_1 + a_2 + i(b_1 + b_2)$, which corresponds graphically to the concatenation of the two vectors or arrows from the origin to the locations of these numbers. Important properties of complex numbers are, where the last line, termed *rationalization*, is obtained by multiplying both numerator and denominator by $a - ib$,

$$\begin{gathered} z_1 = a + ib = z_2 = c + id \text{ implies } a = c, b = d \\ (a + ib)(c + id) = ac - bd + i(ad + bc) \\ \frac{1}{a + ib} = \frac{a - ib}{a^2 + b^2} \end{gathered} \tag{4.3.11}$$

Replacing x in exp (x) by iy leads to the even and odd *harmonic* functions

$$e^{iy} = \cosh iy + \sinh iy = \underbrace{\frac{e^{iy} + e^{-iy}}{2}}_{\cos y} + i\underbrace{\frac{e^{iy} - e^{-iy}}{2i}}_{\sin y} = \underbrace{\left(1 - \frac{y^2}{2!} + \cdots\right)}_{\cos y} + i\underbrace{\left(y - \frac{y^3}{3!} + \cdots\right)}_{\sin y} \tag{4.3.12}$$

which implies

$$\begin{gathered} \cosh(ix) = \cos x \\ \sinh(ix) = i \sin x \end{gathered} \tag{4.3.13}$$

Thus, i acts analogously to the minus sign in $\sinh(-y) = -\sinh y$, $\cosh(-y) = \cosh y$. Figure 4.5 for the sine, cosine, and tangent should be memorized.

As in Equation (4.3.10),

$$\begin{gathered} e^{-ix} = \cos(-x) + i\sin(-x) = \cos x - i\sin x \\ 1 = e^{ix}e^{-ix} = (\cos x + i\sin x)(\cos x - i\sin x) = \cos^2 x + \sin^2 x \end{gathered} \tag{4.3.14}$$

which can also be derived directly from Equations (4.3.10) and (4.3.13):

$$\cos^2 y + \sin^2 y = \cosh^2(iy) + (-i\sinh(iy))^2 = \cosh^2(iy) - \sinh^2(iy) = 1 \tag{4.3.15}$$

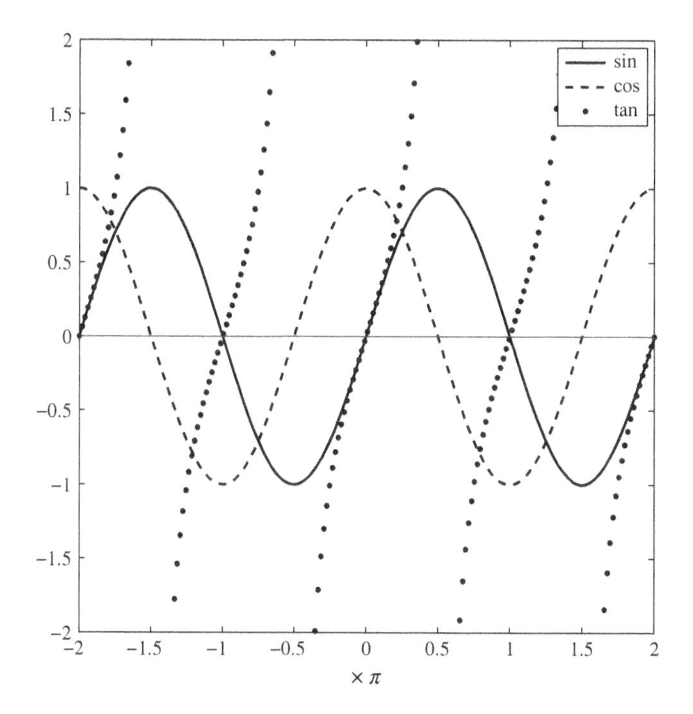

FIGURE 4.5 Harmonic functions.

4.3.5 Inverse Harmonic and Hyperbolic Functions

The inverse trigonometric functions, denoted, e.g., for $y = \sin x$ by $x = \sin^{-1}(y)$ or equivalently $x = \arcsin(y)$, are *multivalued* since all harmonic functions are periodic in x. A single-valued inverse function is constructed by restricting x to its *principal part*, which is the region closest to the origin ($-\pi < y < \pi$ for $y = \arcsin(x)$ and $y = \arccos(x)$, and $-\pi/2 < y < \pi/2$ for $y = \arctan(x)$ and $y = \text{arccot}(x)$). An expression such as $\sin(\tan^{-1}(b/a))$ can be evaluated graphically since the argument of the sine function is the angle, θ whose tangent is b/a. Hence, $\sin\theta = b/\sqrt{a^2 + b^2}$ from Figure 4.6. Again, from this figure, a complex number $z = a + ib$ equals $\sqrt{a^2 + b^2}$ times a unit magnitude complex number $\cos\theta + i\sin\theta = \exp(i\theta)$ (as $|\exp(i\theta)| = (\exp(-i\theta)\exp(i\theta))^{1/2} = 1$) with $\tan\theta = b/a$ or $\theta = \arctan(b/a)$.

The inverse hyperbolic and harmonic functions can be expressed in terms of logarithms. For example, $y = \cosh^{-1}(x)$ implies $\exp(y) + \exp(-y) = 2x$. Writing $\exp(y) = k$, $\exp(-y) = 1/k$ yields $k^2 - 2kx + 1 = 0$ from which $y = \ln(k) = \ln\left(1 + \sqrt{x^2 - 1}\right)$ after applying the quadratic formula (for $\cos^{-1}(x)$, y is replaced by iy). Similarly, $\sinh^{-1}(x) = \ln\left(1 + \sqrt{x^2 + 1}\right)$, while $\tanh^{-1}(x) = (1/2)\log\left((1 + x)/(1 - x)\right)$.

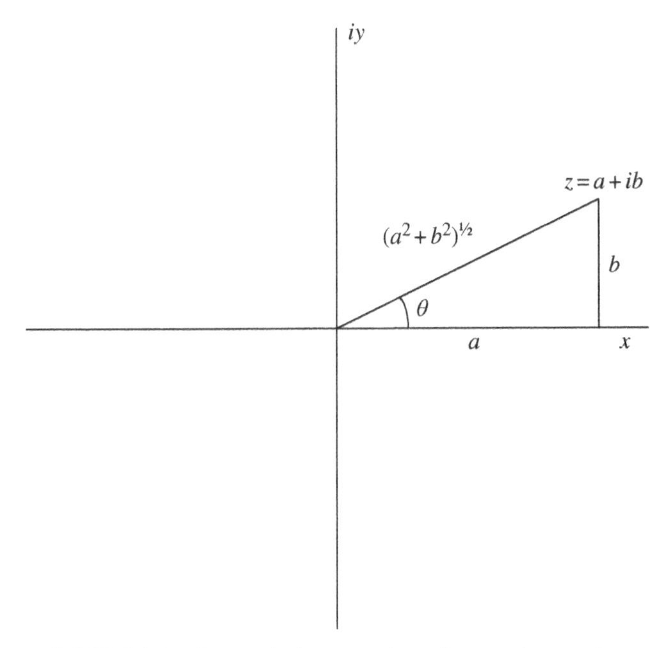

FIGURE 4.6 Geometric interpretation of a complex number.

4.3.6 Trigonometric Identities

The trigonometric (and hyperbolic) functions are elementary combinations of only two exponential functions and are therefore related by numerous identities. The following should be memorized together with their derivations.

- Since $90-\theta$ is the opposite angle to θ in a right angle triangle, from the definition of sine and cosine functions, $\sin(90-\theta)=\cos\theta$. Such identities can also be obtained from graphs of the trigonometric functions. Viewing the sine function graph in the $-\theta$ direction starting from $\theta=90°$ yields the curve for the cosine function. Alternatively, from Equation (4.3.19) below, $\sin(90-\theta)=\sin 90\cos\theta-\cos 90\sin\theta$ with $\sin 90=1,\ \cos 90=0$.
- Dividing both sides of $\sin^2 a+\cos^2 a=1$ by $\cos^2 a$ or $\sin^2 a$ yields, respectively,

$$\begin{aligned}\tan^2 a+1&=\sec^2 a\\ 1+\cot^2 a&=\csc^2 a\end{aligned} \tag{4.3.16}$$

- The "sum formulas" are derived as follows:

$$\begin{aligned} e^{i(a+b)}&=e^{ia}e^{ib}\\ \cos(a+b)+i\sin(a+b)&=(\cos a+i\sin a)(\cos b+i\sin b)\\ &=\cos a\cos b-\sin a\sin b+i(\sin a\cos b+\cos a\sin b)\end{aligned} \tag{4.3.17}$$

Equating real and imaginary parts yields

$$\begin{aligned}\sin(a+b) &= \sin a\cos b + \cos a\sin b\\ \cos(a+b) &= \cos a\cos b - \sin a\sin b\end{aligned} \tag{4.3.18}$$

As well, if $-b$ is substituted for b

$$\begin{aligned}\sin(a-b) &= \sin a\cos b - \cos a\sin b\\ \cos(a-b) &= \cos a\cos b + \sin a\sin b\end{aligned} \tag{4.3.19}$$

These can be remembered from the mnemonic "the signs go with the sines," i.e., a + sign appears in the formula for $\sin(a+b)$, while a − sign is present in $\cos(a+b)$.

- Some important consequences of the sum and difference formulas are

$$\sin a\cos b = \frac{1}{2}(\sin(a+b) + \sin(a-b)) \tag{4.3.20}$$

$$\cos a\cos b = \frac{1}{2}(\cos(a+b) + \cos(a-b)) \tag{4.3.21}$$

$$\sin 2a = \sin(a+a) = 2\sin a\cos a \tag{4.3.22}$$

$$\cos(2a) = \cos(a+a) = \cos^2 a - \sin^2 a = 2\cos^2 a - 1 = 1 - 2\sin^2 a \tag{4.3.23}$$

$$\begin{aligned}\tan(a+b) &= \frac{\sin(a+b)}{\cos(a+b)}\\ &= \frac{(\sin a\cos b + \cos a\sin b)/(\cos a\cos b)}{(\cos a\cos b - \sin a\sin b)/(\cos a\cos b)}\\ &= \frac{\tan a + \tan b}{1 - \tan a\tan b}\end{aligned} \tag{4.3.24}$$

Substituting $a = b/2$ in Equation (4.3.23),

$$\cos b = 1 - 2\sin^2\left(\frac{b}{2}\right) = 2\cos^2\left(\frac{b}{2}\right) - 1 \tag{4.3.25}$$

which leads to the half-angle formulas

$$\begin{aligned}\sin\left(\frac{b}{2}\right) &= \pm\sqrt{\frac{1-\cos b}{2}}\\ \cos\left(\frac{b}{2}\right) &= \pm\sqrt{\frac{1+\cos b}{2}}\end{aligned} \tag{4.3.26}$$

Here the minus sign before $\cos b$ is associated with $\sin(b/2)$ (the signs (sides) switch at half time).

4.4 ANALYTIC GEOMETRY

Geometrical constructs can be represented by algebraic expressions and subsequently manipulated algebraically. Linear and quadratic equations then represent lines or planes and *conic sections*, respectively. The latter possess numerous but infrequently employed properties omitted in the discussion below.

4.4.1 Lines and Planes

A line in the (x, y) plane is represented by a linear equation

$$ax + by = c \tag{4.4.1}$$

The point of intersection of two nonparallel lines corresponds to the (x, y) value that solves the system of the two linear equations describing the lines. A line intercepts the x-axis at the x-intercept c/a obtained by setting $y = 0$ in Equation (4.4.1), while inserting $x = 0$ yields the y-intercept c/b. The *slope*, m, of the line is defined as the ratio of the difference in y values to the difference in x values for any two points, (x_1, y_1) and (x_2, y_2), along the line

$$m = \frac{\Delta y}{\Delta x} = \frac{y_2 - y_1}{x_2 - x_1} = \frac{\frac{1}{b}(\not{c} - ax_2) - \frac{1}{b}(\not{c} - ax_1)}{x_2 - x_1} = -\frac{a}{b} \tag{4.4.2}$$

A line in two dimensions can be specified by its slope, m, and y-intercept, y_0, as

$$y = mx + y_0 \tag{4.4.3}$$

or from the slope and a single point, (x_1, y_1), on the line according to

$$m = \frac{y - y_1}{x - x_1} \tag{4.4.4}$$

or, finally, by any two distinct points from

$$y - y_1 = m(x - x_1) = \left(\frac{y_2 - y_1}{x_2 - x_1}\right)(x - x_1) \tag{4.4.5}$$

To determine the equation of a line that passes through a point (x_1, y_1) *perpendicular* to a second line with slope m, note that a 90° rotation transforms x and y into y and $-x$, respectively. The slope is thus altered from $\Delta y/\Delta x = m$ to $-\Delta x/\Delta y = -1/m$, yielding from Equation (4.4.4)

$$y - y_1 = -\frac{1}{m}(x - x_1) \tag{4.4.6}$$

Lines and planes in three and higher dimensions are most conveniently analyzed with vectors as described in the following chapter. Often, a line is specified *parametrically* by the values of a point (x_0, y_0, z_0) and the *direction cosines*, which are the cosines of the angles, γ_i, that the line makes with each of the coordinate axes, as

$$\begin{aligned} x(\lambda) &= x_0 + \lambda \cos \gamma_x \\ y(\lambda) &= y_0 + \lambda \cos \gamma_y \\ z(\lambda) &= z_0 + \lambda \cos \gamma_z \end{aligned} \tag{4.4.7}$$

where the parameter λ ranges over all real values. A plane is represented as

$$ax + by + cz = d \tag{4.4.8}$$

Two nonparallel planes intersect along a line, while three nonparallel planes that do not share a line of intersection intersect at a point given by the solution of the corresponding system of three equations.

4.4.2 Conic Sections

Quadratic algebraic equations describe the curves obtained from the intersection of a cone with a plane. All points a distance R from (x_1, y_1) form a *circle*

$$(x - x_1)^2 + (y - y_1)^2 = R^2 \tag{4.4.9}$$

A common error is to omit the minus sign in this and similar contexts.

For a *parabola*, the difference between the distance from a point, termed the *focus*, to any point,(x, y), on the parabola and the perpendicular distance from the latter point to a line termed the *principal axis* remains constant. To illustrate, a focus at $(0, a)$ and a principle axis described by $y = -a$ result in the equation

$$\begin{aligned} \sqrt{x^2 + (y-a)^2} - (y+a) &= 0 \\ x^2 + (y-a)^2 &= (y+a)^2 \\ x^2 &= 4ya \\ y &= \frac{x^2}{4a} \end{aligned} \tag{4.4.10}$$

A parabola that opens upward (or downward for negative values of c) possesses the form

$$(y-y_1)=c(x-x_1)^2 \tag{4.4.11}$$

while a parabola that opens to the right or left has the form

$$(x-x_1)=c(y-y_1)^2 \tag{4.4.12}$$

An *ellipse* is characterized by two *focal points or foci*, such that the sum of the distances from any point of the ellipse to both foci is a constant. For example,

$$\sqrt{(x-a)^2+y^2}+\sqrt{(x+a)^2+y^2}=2R \tag{4.4.13}$$

which reduces to the equation of a circle when the distance, $2a$, between the two foci approaches zero. After squaring twice, the above equation can be rewritten as

$$\frac{x^2}{R^2}+\frac{y^2}{(R^2-a^2)}=1 \tag{4.4.14}$$

An ellipse centered at the point (x_1, y_1) can be described by an equation of the form (possibly after rotating the coordinate axes to coincide with those of the ellipse)

$$\frac{(x-x_1)^2}{\alpha^2}+\frac{(y-y_1)^2}{\beta^2}=1 \tag{4.4.15}$$

The quantities 2α and 2β are the diameters of the ellipse measured along the x- and y-directions, respectively. The axes corresponding to the longer and smaller of these are termed the *major and minor axis*, respectively.

Finally, the *hyperbola*

$$\frac{(x-x_1)^2}{\alpha^2}-\frac{(y-y_1)^2}{\beta^2}=1 \tag{4.4.16}$$

is formed from the loci (collection) of points with identical *differences* of their distances from two foci. As the above equation can be solved for any value of x, the hyperbola opens to the right and left, while if the signs of the two terms on the left-hand side of Equation (4.4.16) are reversed, the hyperbola instead opens upward and downward. Rotating by 90° around (x_1, y_1) with the methods of the following section yields for $\alpha=\beta$ a curve that opens in the first and third quadrants, namely,

$$(x-x_1)(y-y_1)=c^2 \tag{4.4.17}$$

4.4.3 Areas, Volumes, and Solid Angles

While the evaluation of areas and volumes generally requires integration, essential formulas include:

- Square and parallelogram: Area = base × height
- Triangle: Area = ½ base × height
- Trapezoid (of any shape, including parallelograms): Area = ½ (lower base + upper base) height
- Circle: Area = $\pi \times (\text{radius})^2$; circumference = $2\pi \times$ radius = $\pi \times$ diameter
- Cube, parallelepiped, and cylinder: Volume = area of base × height
- Pyramid and cone: Volume = (1/3) × area of base × height
- Sphere: Volume = $(4\pi/3) \times (\text{radius})^3$; surface area = $4\pi \times (\text{radius})^2$

Failure to distinguish between radius and diameter frequently leads to errors.

Just as an angle in radians in two dimensions is defined by the formula $\theta = s/R$, where s is the arclength of the segment of a circle of radius, R, centered at the angle's vertex subtended by θ, a three-dimensional *solid angle* $\Omega = A/R$ in *steradians* corresponds to the area, A, of the surface of a sphere of radius, R, centered at the vertex that the solid angle subtends. The entire sphere is therefore subtended by a solid angle of 4π.

5

VECTORS AND MATRICES

Linear equation systems are conveniently described with rectangular arrays of numeric values termed matrices. As algebraic operations on matrices act on groups of elements, their properties are restricted compared to manipulations of single values.

5.1 MATRICES AND MATRIX PRODUCTS

Denoting the element in the ith row and jth column of two matrices **A** and **B** by a_{ij} and b_{ij}, the corresponding element of $\mathbf{C} = \mathbf{A} \pm \mathbf{B}$ equals $a_{ij} \pm b_{ij}$. The (inner) product of two matrices, **AB**, is defined only if the column dimension of **A** equals the row dimension of **B**. Multiplying a $N \times P$ matrix **A** with N rows and P columns with a $P \times M$ matrix **B** results in a $N \times M$ matrix **C** with

$$c_{ij} = \sum_{l=1}^{P} a_{il} b_{lj} \equiv a_{il} b_{lj} \tag{5.1.1}$$

where the *Einstein notation* in which repeated indices are implicitly summed over has been introduced. That is, to obtain c_{ij}, each element in the ith row of the matrix **A** is multiplied by the corresponding element in the jth column of **B**, and these products are then summed.

Matrix multiplication is associative, i.e.,

$$\mathbf{A}(\mathbf{BC}) = (\mathbf{AB})\mathbf{C} \tag{5.1.2}$$

Fundamental Math and Physics for Scientists and Engineers, First Edition.
David Yevick and Hannah Yevick.

but not commutative, i.e., $\mathbf{AB} \neq \mathbf{BA}$ (for square matrices $\mathbf{A}$ and $\mathbf{B}$). The difference between these two quantities, termed the *commutator*, quantifies the degree of noncommutativity and is denoted by

$$[\mathbf{A},\mathbf{B}] = -[\mathbf{B},\mathbf{A}] \equiv \mathbf{AB}-\mathbf{BA} \tag{5.1.3}$$

Two identities satisfied by the commutator are

$$[\mathbf{AB},\mathbf{C}] = \mathbf{ABC}-\mathbf{CAB} = \mathbf{A}[\mathbf{B},\mathbf{C}] + [\mathbf{A},\mathbf{C}]\mathbf{B} \tag{5.1.4}$$

and the *Jacobi identity*

$$[[\mathbf{A},\mathbf{B}],\mathbf{C}] + [[\mathbf{C},\mathbf{A}],\mathbf{B}] + [[\mathbf{B},\mathbf{C}],\mathbf{A}] = 0 \tag{5.1.5}$$

which is proven by summing all 12 terms (the last two sets of four terms can be obtained from the first set through a cyclic permutation of the three matrices according to $\mathbf{A} \rightarrow \mathbf{B} \rightarrow \mathbf{C} \rightarrow \mathbf{A}$).

The $N \times N$ identity matrix $\mathbf{I}$ with elements δ_{ij}, where the *Kronecker delta function* is defined by

$$\delta_{ij} = \begin{cases} 1 & i=j \\ 0 & i \neq j \end{cases} \tag{5.1.6}$$

possesses the property

$$\mathbf{AI} = \mathbf{IA} \tag{5.1.7}$$

for any $N \times N$ matrix $\mathbf{A}$. The inverse of $\mathbf{A}$, $\mathbf{A}^{-1}$, when it exists, is the $N \times N$ matrix defined by

$$\mathbf{AA}^{-1} = \mathbf{A}^{-1}\mathbf{A} = \mathbf{I} \tag{5.1.8}$$

Here,

$$(\mathbf{AB})^{-1} = \mathbf{B}^{-1}\mathbf{A}^{-1} \tag{5.1.9}$$

as is verified by right or left multiplying by $\mathbf{AB}$. Methods for inverting matrices are presented in subsequent sections.

The *transpose*, $\mathbf{A}^T$, of $\mathbf{A}$ reflects $\mathbf{A}$ about its main diagonal yielding elements $a^T{}_{ij} = a_{ji}$. Since $\left[(\mathbf{AB})^T\right]_{ij} = a_{jk}b_{ki} = a^T_{kj}b^T_{ik} = b^T_{ik}a^T_{kj}$,

$$(\mathbf{AB})^T = \mathbf{B}^T\mathbf{A}^T \tag{5.1.10}$$

A *symmetric* matrix satisfies $\mathbf{A} = \mathbf{A}^T$ while for a *skew-symmetric* matrix, $\mathbf{A} = -\mathbf{A}^T$. A matrix $\mathbf{A}$ is decomposed into a sum of symmetric and skew-symmetric parts by

$$\mathbf{A} = \underbrace{\frac{\mathbf{A}+\mathbf{A}^T}{2}}_{\text{symmetric}} + \underbrace{\frac{\mathbf{A}-\mathbf{A}^T}{2}}_{\text{skew symmetric}} \tag{5.1.11}$$

The *Hermitian conjugate* (also called the Hermitian transpose, conjugate transpose, or adjoint), $\mathbf{A}^\dagger$, of $\mathbf{A}$ is generated by a complex conjugating each element of the transpose matrix; i.e., $a_{ij}^\dagger = a_{ji}^*$. As a result of the transposition,

$$(\mathbf{AB})^\dagger = \mathbf{B}^\dagger \mathbf{A}^\dagger \tag{5.1.12}$$

A *Hermitian* matrix satisfies $\mathbf{A} = \mathbf{A}^\dagger$, while for an *anti-Hermitian matrix*, $\mathbf{A} = -\mathbf{A}^\dagger$. A matrix is separated into Hermitian and anti-Hermitian parts according to

$$\mathbf{A} = \underbrace{\frac{\mathbf{A}+\mathbf{A}^\dagger}{2}}_{\text{Hermitian}} + \underbrace{\frac{\mathbf{A}-\mathbf{A}^\dagger}{2}}_{\text{anti-Hermitian}} \tag{5.1.13}$$

5.2 EQUATION SYSTEMS

A general *linear* system of equations can be expressed as a matrix equation of the form $\mathbf{Ax} = \mathbf{b}$ with

$$\begin{pmatrix} a_{11} & \cdots & a_{1n} \\ \vdots & \ddots & \vdots \\ a_{m1} & \cdots & a_{mn} \end{pmatrix} \begin{pmatrix} x_1 \\ \vdots \\ x_n \end{pmatrix} = \begin{pmatrix} b_1 \\ \vdots \\ b_n \end{pmatrix} \tag{5.2.1}$$

The solution $\mathbf{x}$ in Equation (5.2.1) can be obtained by repeating the *elementary operations* described in association with Equation (4.1.4) on the *augmented* coefficient matrix

$$\left(\begin{array}{ccc|c} a_{11} & \cdots & a_{1n} & b_{1n} \\ \vdots & \ddots & \vdots & \vdots \\ a_{m1} & \cdots & a_{mn} & b_{mn} \end{array}\right) \tag{5.2.2}$$

until the matrix $\mathbf{A}$ to the left of the vertical line is transformed into the identity matrix. That is, if Equation (4.1.4) is written as

$$\left(\begin{array}{cc|c} 2 & 3 & 7 \\ 1 & 1 & 3 \end{array}\right) \tag{5.2.3}$$

multiplying the lower equation again by two yields

$$\left(\begin{array}{cc|c} 2 & 3 & 7 \\ 2 & 2 & 6 \end{array}\right) \tag{5.2.4}$$

After further elementary operations, $x_1 = 2, x_2 = 1$ is obtained from the *reduced* augmented matrix

$$\left(\begin{array}{cc|c} 1 & 0 & 2 \\ 0 & 1 & 1 \end{array}\right) \tag{5.2.5}$$

The equation system Equation (5.2.1) can possess zero solutions as in $3x_1 = 0$, $x_1 = 4$, a unique solution, or an infinite number of solutions as in $x_1 + x_2 = 6$, $2x_1 + 2x_2 = 12$. The *row space*, described by all linear combinations of the rows of $\mathbf{A}$, remains invariant with respect to the elementary operations. Hence, the row space for the matrix on the left of the vertical line in Equation (5.2.3) or equivalently Equation (5.2.5) comprises the entire two-dimensional plane, while for $\mathbf{Ax} = \mathbf{b}$ with

$$\mathbf{A} = \begin{pmatrix} 1 & 1 \\ 2 & 2 \end{pmatrix} \quad \mathbf{b} = \begin{pmatrix} 6 \\ 12 \end{pmatrix} \tag{5.2.6}$$

the row space consists of all vectors $(c, c)^T = c_1(1, 1)^T + c_2(2, 2)^T$ that represents a line through the origin. The dimension of the row space is termed the *rank* of the matrix and equals the number of nonzero rows to the left of the vertical line in the reduced augmented matrix. The *kernel* or *null space* is formed from the set of all vectors $\mathbf{x}$ satisfying $\mathbf{Ax} = 0$, which consists of the point $x_1 = x_2 = 0$ of dimension zero in Equation (5.2.3) and the one-dimensional line $x_1 + x_2 = 0$, perpendicular to the line defining the row space, in Equation (5.2.6). In general, the dimension of the null space, or *nullity*, equals the number of columns of $\mathbf{A}$ minus its rank. This is termed the *span* or space of solutions to the equation system as any solution in the null space can be added to a solution of the original linear equation system to yield a new solution. As an example, the difference between the two solutions $(6, 0)^T$ and $(3, 3)^T$ of $\mathbf{Ax} = \mathbf{b}$ with $\mathbf{A}$ and $\mathbf{b}$ given by Equation (5.2.6) yields the vector $(3, -3)^T$. This point satisfies $x_1 + x_2 = 0$ and therefore belongs to the null space of $\mathbf{A}$.

5.3 TRACES AND DETERMINANTS

The *trace* (also written as Sp( )) of a matrix is defined as the sum of its diagonal elements:

$$\mathrm{Tr}(\mathbf{A}) = \sum_i a_{ii} = a_{ii} \tag{5.3.1}$$

The trace possesses the following properties:

$$\begin{aligned} \mathrm{Tr}(\mathbf{A} + \mathbf{B}) &= \mathrm{Tr}(\mathbf{B} + \mathbf{A}) \\ \mathrm{Tr}(\mathbf{A}^T) &= \mathrm{Tr}(\mathbf{A}) \\ \mathrm{Tr}(c\mathbf{A}) &= c\mathrm{Tr}(\mathbf{A}) \\ \mathrm{Tr}(\mathbf{AB}) &= \mathrm{Tr}(\mathbf{BA}) \end{aligned} \tag{5.3.2}$$

the last of which implies that

$$\mathrm{Tr}(\mathbf{ABC}) = \mathrm{Tr}(\mathbf{CAB}) = \mathrm{Tr}(\mathbf{BCA}) \tag{5.3.3}$$

but not $\mathrm{Tr}(\mathbf{ABC}) = \mathrm{Tr}(\mathbf{BAC})$ since $\mathbf{B}$ and $\mathbf{A}$ do not commute.

The *determinant* det(**A**) or |**A**| of a necessarily square matrix is given in Einstein notation by

$$\det(\mathbf{A}) = \varepsilon_{ijkl\ldots} a_{1i} a_{2j} a_{3k} a_{4l} \ldots \tag{5.3.4}$$

where the completely asymmetric *Levi-Civita* symbol $\varepsilon_{ijkl\ldots}$ satisfies

$$\begin{aligned} \varepsilon_{1234\ldots} &= 1 \\ \varepsilon_{ijkl\ldots} &= -\varepsilon_{jikl\ldots} \\ \varepsilon_{iikl\ldots} &= 0 \end{aligned} \tag{5.3.5}$$

The last two relationships apply for any pair of indices. The determinant of a $N \times N$ matrix contains $N!$ terms complicating its evaluation for $N > 3$. *Only for* $N = 2, 3$ can the determinants be simply obtained by subtracting the products of the terms along the left diagonals from those along the right diagonals, where columns are duplicated outside the matrix as required. That is,

$$\begin{vmatrix} a_{11} & a_{12} \\ a_{21} & a_{22} \end{vmatrix} = a_{11}a_{22} - a_{12}a_{21}$$

$$\begin{vmatrix} a_{11} & a_{12} & a_{13} \\ a_{21} & a_{22} & a_{23} \\ a_{31} & a_{32} & a_{33} \end{vmatrix} \begin{matrix} a_{11} & a_{12} \\ a_{21} & a_{22} \\ a_{31} & a_{32} \end{matrix} = \begin{matrix} a_{11}a_{22}a_{33} + a_{12}a_{23}a_{31} + a_{13}a_{21}a_{32} \\ -a_{13}a_{22}a_{31} - a_{11}a_{23}a_{32} - a_{12}a_{21}a_{33} \end{matrix} \tag{5.3.6}$$

From Equation (5.3.4),

$$\det\left(\mathbf{A}^T\right) = \det(\mathbf{A}) \tag{5.3.7}$$

Although $\det(\mathbf{A} + \mathbf{B}) \neq \det(\mathbf{A}) + \det(\mathbf{B})$,

$$\det(\mathbf{AB}) = \det(\mathbf{A})\det(\mathbf{B}) \tag{5.3.8}$$

Equation (5.3.8) follows from the invariance of the determinant under each of the elementary operations employed in, e.g., Equation (5.2.3) to transform the left-hand matrix into a unit matrix. That is, exchanging two rows or two columns of a matrix reverses the sign of the determinant, as evident either from Equation (5.3.4) or by noting that, e.g., the rows of a 2×2 matrix can be interchanged by left multiplying by

$$\mathbf{T}_1 = \begin{pmatrix} 0 & 1 \\ 1 & 0 \end{pmatrix} \tag{5.3.9}$$

with determinant -1, consistent with Equation (5.3.8). Similarly, multiplying a row by a constant, k, increases the determinant by this factor as is evident for the second row by left multiplying with

$$\mathbf{T}_2 = \begin{pmatrix} 1 & 0 \\ 0 & k \end{pmatrix} \tag{5.3.10}$$

for which $\det(\mathbf{T}_2) = k$. Finally, left multiplying by

$$\mathbf{T}_3 = \begin{pmatrix} 1 & k \\ 0 & 1 \end{pmatrix} \tag{5.3.11}$$

with $\det(\mathbf{T}_3) = 1$ corresponds to the elementary operation that adds a multiple k of the second row to the first row. From the discussion surrounding Equation (4.1.4), any invertible matrix $\mathbf{A}$ can be transformed into a unit matrix according to

$$\mathbf{I} = \mathbf{T}_\zeta \ldots \mathbf{T}_\beta \mathbf{T}_\alpha \mathbf{A} \tag{5.3.12}$$

where $\alpha, \beta, \ldots, \zeta$ equal 1, 2, or 3 and $\det(\mathbf{T}_\zeta \ldots \mathbf{T}_\beta\ \mathbf{T}_\alpha) = 1/\det(\mathbf{A})$. Since $\det \mathbf{I} = 1$,

$$\begin{aligned}\det(\mathbf{AB}) &= \det\left(\mathbf{T}_\alpha^{-1}\mathbf{T}_\beta^{-1}\ldots\mathbf{T}_\zeta^{-1}\mathbf{IB}\right) \\ &= \det\left(\mathbf{T}_\alpha^{-1}\right)\det\left(\mathbf{T}_\beta^{-1}\right)\ldots\det\left(\mathbf{T}_\zeta^{-1}\right)\det(\mathbf{IB}) \\ &= \det(\mathbf{A})\det(\mathbf{B})\end{aligned} \tag{5.3.13}$$

Additionally, $\mathbf{AA}^{-1} = \mathbf{I}$ implies

$$\det \mathbf{A}^{-1} = \frac{1}{\det \mathbf{A}} \tag{5.3.14}$$

and hence, $\mathbf{A}^{-1}$ exists only if $\det(\mathbf{A}) \neq 0$.

Elementary operations can be applied to transform the determinant of a matrix $\mathbf{A}$ into an equivalent $|\mathbf{B}|$ for which most elements $b_{ij} = 0$. This enables an efficient expansion

$$\det(\mathbf{B}) = \underbrace{\sum_j b_{ij}C_{ij}}_{\text{for any row } i} = \underbrace{\sum_i b_{ij}C_{ij}}_{\text{for any column } j} \tag{5.3.15}$$

in terms of *cofactors* C_{ij}, defined as $(-1)^{i+j}$ times the determinant formed by eliminating row i and column j from $|\mathbf{B}|$. *Cramer's rule* additionally expresses the solution of the linear system, Equation (5.2.1), as

$$x_1 = \frac{\begin{vmatrix} b_1 & a_{12} & \cdots & a_{n1} \\ \vdots & \vdots & \ddots & \vdots \\ b_n & a_{n2} & \cdots & a_{nn} \end{vmatrix}}{\det(\mathbf{A})},\ x_2 = \frac{\begin{vmatrix} a_{11} & b_1 & \cdots & a_{n1} \\ \vdots & \vdots & \ddots & \vdots \\ a_{n1} & b_n & \cdots & a_{nn} \end{vmatrix}}{\det(\mathbf{A})},\ x_n = \frac{\begin{vmatrix} a_{11} & \cdots & a_{1n-1} & b_1 \\ \vdots & \ddots & \vdots & \vdots \\ a_{n1} & \cdots & a_{nn-1} & b_n \end{vmatrix}}{\det(\mathbf{A})} \tag{5.3.16}$$

The inverse of a matrix **A** is given in terms of the *transpose* of the matrix of its cofactors by

$$\left(\mathbf{A}^{-1}\right)_{ij} = \frac{1}{\det(\mathbf{A})} C_{ij}^T = \frac{1}{\det(\mathbf{A})} C_{ji} \tag{5.3.17}$$

The origin of this formula can be understood by noting that Cramer's rule with, e.g., $\mathbf{b} = (1, 0, \ldots, 0)^T$ yields the first column in the matrix $\mathbf{A}^{-1}$.

Example

$$\begin{pmatrix} 1 & 2 \\ 3 & 4 \end{pmatrix}^{-1} = \frac{1}{-2}\begin{pmatrix} 4 & -2 \\ -3 & 1 \end{pmatrix} = \begin{pmatrix} -2 & 1 \\ 1.5 & -0.5 \end{pmatrix} \tag{5.3.18}$$

5.4 VECTORS AND INNER PRODUCTS

A vector possesses both a *scalar* (numeric) *magnitude* that is independent of the coordinate system and a *direction* that is described differently in different coordinate systems. For example, the *displacement* vector can be represented by an arrow directed from the initial to the final spatial position where the magnitude of the displacement, the scalar *distance*, equals the length of the arrow. Similarly, the *velocity* vector specifies the rate of change of the displacement per unit time, while the *speed* refers to its scalar magnitude. A vector $\vec{A}$ in three dimensions is generally specified by a three-component row or column array with magnitude (length)

$$A \equiv \left|\vec{A}\right| = \sqrt{a_x^2 + a_y^2 + a_z^2} \tag{5.4.1}$$

The *unit vector* in the direction $\vec{A}$ is then

$$\hat{a} = \hat{e}_a \equiv \frac{\vec{A}}{A} = \frac{1}{\sqrt{a_x^2 + a_y^2 + a_z^2}}(a_x, a_y, a_z) \tag{5.4.2}$$

The sum and difference of two vectors $\vec{A} = (a_x, a_y, a_z)$ and $\vec{B} = (b_x, b_y, b_z)$ are given by

$$\vec{A} \pm \vec{B} = (a_x \pm b_x, a_y \pm b_y, a_z \pm b_z) \tag{5.4.3}$$

Graphically, $\vec{A} + \vec{B}$ corresponds to displacing $\vec{A}$ or $\vec{B}$ so that its tail is located at the head of the other vector; the sum then extends from the tail of the first vector to the head of the second as in Figure 5.1.

Since the negative of a vector is formed by reversing its direction, $\vec{A} - \vec{B}$ is represented by the sum of $\vec{A}$ and $\vec{B}$ after the direction of the arrow of $\vec{B}$ is reversed. Two vectors of

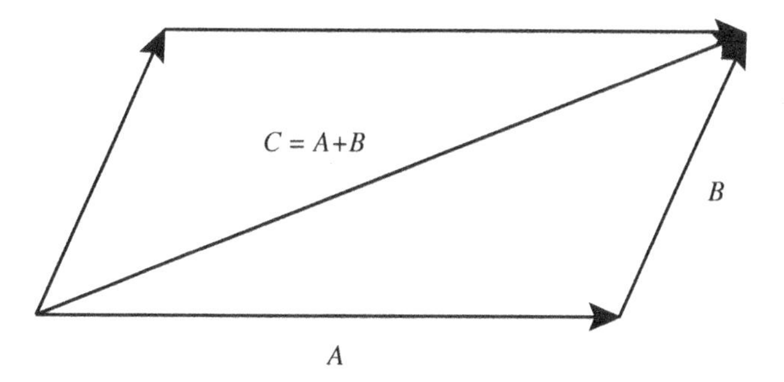

FIGURE 5.1 Addition of two vectors.

the same dimension can be multiplied in three ways to yield either a scalar, a (pseudo) vector, or a matrix (tensor). These are termed the *dot* or *scalar inner* product, the cross product, and the *tensor*, *outer*, or *Helmholtz* product, respectively. The dot product multiplies corresponding components of the two vectors so that in three dimensions

$$\vec{A}\cdot\vec{B} \equiv (a_x, a_y, a_z)\begin{pmatrix} b_x \\ b_y \\ b_z \end{pmatrix} = a_x b_x + a_y b_y + a_z b_z \equiv a_i b_i \tag{5.4.4}$$

If the x-axis of a rectangular coordinate system coincides with the direction of $\vec{A}$, Equation (5.4.4) reduces to $a_x b_x = \left|\vec{A}\right| b_x$, the magnitude of $\vec{A}$ multiplied by the projection of the vector $\vec{B}$ onto $\vec{A}$. Since the dot product represents a scalar that is not altered by rotations or translations of the coordinate axes, denoting the angle between the vectors $\vec{A}$ and $\vec{B}$ by θ_{AB},

$$\vec{A}\cdot\vec{B} = |A||B|\cos\theta_{AB} \tag{5.4.5}$$

while the square of the difference of two vectors equals

$$\left(\vec{A}-\vec{B}\right)^2 \equiv \left(\vec{A}-\vec{B}\right)\cdot\left(\vec{A}-\vec{B}\right) = A^2 + B^2 - 2\vec{A}\cdot\vec{B} = A^2 + B^2 - 2AB\cos\theta_{AB} \tag{5.4.6}$$

As $\vec{C} = \vec{A} - \vec{B}$ is the third side of the triangle formed by $\vec{A}$ and $\vec{B}$, Equation (5.4.6) yields the *law of cosines*

$$\cos\theta_{AB} = \frac{A^2 + B^2 - C^2}{2AB} \tag{5.4.7}$$

A plane *perpendicular* to the x-axis and passing through the point $(x, y, z) = (a, 0, 0)$ is described by

$$x = \hat{e}_x \cdot \vec{r} = a \tag{5.4.8}$$

where the *radius vector*

$$\vec{r} \equiv (x,y,z) \tag{5.4.9}$$

extends from the origin of the coordinate system to (x, y, z) and the unit vector $\hat{e}_x = (1,0,0)$ is oriented along the x-direction. The distance between a point $\vec{r}_1 = (x_1, y_1, z_1)$ and the plane of Equation (5.4.8) equals $x_1 - a = \hat{e}_x \cdot \vec{r}_1 - a$. Since both $\hat{e}_x \cdot \vec{r}$ and a are scalars and thus unchanged under rotations,

$$\hat{e}_{\vec{k}} \cdot \vec{r} = g \tag{5.4.10}$$

similarly represents a plane perpendicular to $\hat{e}_{\vec{k}} = (k_x, k_y, k_z)/\sqrt{k_x^2 + k_y^2 + k_z^2} = (\cos\gamma_x, \cos\gamma_y, \cos\gamma_z)$ corresponding to Equation (4.4.8) with a minimum distance g from the origin. The distance from a point $\vec{r}_1 = (x_1, y_1, z_1)$ to the above plane is found from

$$\hat{e}_{\vec{k}} \cdot \vec{r}_1 - g \tag{5.4.11}$$

5.5 CROSS AND OUTER PRODUCTS

The cross product $\vec{A} \times \vec{B}$ of two vectors $\vec{A}$ and $\vec{B}$ can be represented geometrically, after displacing the tails of both vectors to the origin, as in Figure 5.1, as a (pseudo)vector with a magnitude $|\vec{A} \times \vec{B}|$ equal to the area of the parallelogram described by the endpoints of the vectors $\vec{A}$, $\vec{B}$ and a direction $\hat{e}_{\vec{A}\times\vec{B}}$ determined by the *right-hand rule* in which the vector $\vec{A}$ is turned with the fingers of one's right hand into the vector $\vec{B}$; the thumb then points in the direction of the cross product (out of the paper in Figure 5.1). Accordingly,

$$\vec{A} \times \vec{B} = -\vec{B} \times \vec{A} \tag{5.5.1}$$

The right-hand rule assumes a *right-handed* coordinate system, for which the cross product of $\hat{x}$ and $\hat{y}$ is oriented in the $+\hat{z}$ direction. *This convention must therefore be strictly observed when drawing axes*. Algebraically, the cross product coincides with the determinant

$$\vec{A} \times \vec{B} = \begin{vmatrix} \hat{e}_x & \hat{e}_y & \hat{e}_z \\ a_x & a_y & a_z \\ b_x & b_y & b_z \end{vmatrix} \tag{5.5.2}$$

or, in terms of the three-component Levi-Civita tensor,

$$\left(\vec{A} \times \vec{B}\right)_i = \varepsilon_{ijk} a_j b_k \tag{5.5.3}$$

The *law of sines* can be derived from the area of a triangle written as

$$\text{area} = \frac{1}{2}\left|\vec{A} \times \vec{B}\right| = \frac{1}{2}\left|\vec{B} \times \vec{C}\right| = \frac{1}{2}\left|\vec{C} \times \vec{A}\right| \tag{5.5.4}$$

in which $\vec{A}$, $\vec{B}$, $\vec{C} = \vec{A} + \vec{B}$ are vectors that lie along the three sides of the triangle. Labeling, e.g., the angle between $\vec{A}$ and $\vec{B}$ and therefore opposite to $\vec{C}$ by χ and the angles opposite to $\vec{A}$ and $\vec{B}$ by α and β respectively, Equation (5.5.4) yields $AB \sin\chi = AC \sin\beta = BC \sin\alpha$ or, after division by ABC,

$$\frac{\sin\alpha}{A} = \frac{\sin\beta}{B} = \frac{\sin\chi}{C} \tag{5.5.5}$$

Finally, the *outer product* of $\vec{A}$ and $\vec{B}$ yields the matrix $\left(\vec{A}\otimes\vec{B}\right)_{ij} = a_i b_j$, or $A_i B_j$ in Einstein notation. The cross product can be identified with the antisymmetric component of the outer product in three dimensions since

$$\begin{aligned}\vec{A}\otimes\vec{B} &= \begin{pmatrix} a_1 \\ a_2 \\ a_3 \end{pmatrix}\begin{pmatrix} b_1 & b_2 & b_3 \end{pmatrix} = \begin{pmatrix} a_1b_1 & a_1b_2 & a_1b_3 \\ a_2b_1 & a_2b_2 & a_2b_3 \\ a_3b_1 & a_3b_2 & a_3b_3 \end{pmatrix} \\ &= \frac{1}{2}\left(\vec{A}\otimes\vec{B} + \vec{B}\otimes\vec{A}\right) + \frac{1}{2}\left(\vec{A}\otimes\vec{B} - \vec{B}\otimes\vec{A}\right)\end{aligned} \tag{5.5.6}$$

for which the second antisymmetric part yields

$$\begin{aligned}\frac{1}{2}\left(\vec{A}\otimes\vec{B} - \vec{B}\otimes\vec{A}\right) &= \frac{1}{2}\begin{pmatrix} 0 & a_1b_2-a_2b_1 & a_1b_3-a_3b_1 \\ -(a_1b_2-a_2b_1) & 0 & a_2b_3-a_3b_2 \\ -(a_1b_3-a_3b_1) & -(a_2b_3-a_3b_2) & 0 \end{pmatrix} \\ &= \frac{1}{2}\begin{pmatrix} 0 & \left(\vec{A}\times\vec{B}\right)_3 & -\left(\vec{A}\times\vec{B}\right)_2 \\ -\left(\vec{A}\times\vec{B}\right)_3 & 0 & \left(\vec{A}\times\vec{B}\right)_1 \\ \left(\vec{A}\times\vec{B}\right)_2 & -\left(\vec{A}\times\vec{B}\right)_1 & 0 \end{pmatrix}\end{aligned} \tag{5.5.7}$$

5.6 VECTOR IDENTITIES

Numerous identities resulting from the properties of the Levi-Civita symbol, Equation (5.3.5), relate dot and cross products. From the invariance of ε_{ijk} with respect to cyclic permutations of i, j, k, $\varepsilon_{ijk}A_iB_jC_k = \varepsilon_{kij}C_kA_iB_j = \varepsilon_{jki}B_jC_kA_i$ or, in vector notation,

$$\vec{A}\cdot\left(\vec{B}\times\vec{C}\right) = \vec{C}\cdot\left(\vec{A}\times\vec{B}\right) = \vec{B}\cdot\left(\vec{C}\times\vec{A}\right) \tag{5.6.1}$$

Equation (5.6.1) also follows from

$$\vec{A}\cdot\left(\vec{B}\times\vec{C}\right)=\begin{vmatrix} a_x & a_y & a_z \\ b_x & b_y & b_z \\ c_x & c_y & c_z \end{vmatrix} \tag{5.6.2}$$

since interchanging an even number of rows or columns does not affect the determinant. Further,

$$\varepsilon_{ijk}\varepsilon_{ilm}=\delta_{jl}\delta_{km}-\delta_{jm}\delta_{kl} \tag{5.6.3}$$

as interchanging either j and k or l and m changes the sign of the product as a result of the antisymmetry of ε_{ijk}. Consequently, from Equation (5.6.3)

$$\left[\vec{A}\times\left(\vec{B}\times\vec{C}\right)\right]_i=\varepsilon_{ijk}\varepsilon_{klm}A_jB_lC_m=\varepsilon_{kij}\varepsilon_{klm}A_jB_lC_m=\left[\vec{B}\left(\vec{A}\cdot\vec{C}\right)-\vec{C}\left(\vec{A}\cdot\vec{B}\right)\right]_i \tag{5.6.4}$$

sometimes termed the "BAC – CAB" rule. Equation (5.6.3) further yields the identity

$$\left(\vec{A}\times\vec{B}\right)\cdot\left(\vec{C}\times\vec{D}\right)=\left(\vec{A}\cdot\vec{C}\right)\left(\vec{B}\cdot\vec{D}\right)-\left(\vec{A}\cdot\vec{D}\right)\left(\vec{B}\cdot\vec{C}\right) \tag{5.6.5}$$

5.7 ROTATIONS AND ORTHOGONAL MATRICES

Vector rotation can be described either by rotating the vector in a positive (counterclockwise) angle θ with respect to a fixed coordinate system or by rotating the coordinate basis vectors by $-\theta$ while the vector remains stationary. In two dimensions, rotating a vector $\vec{A}$ that initially describes an angle of θ' with respect to the x-axis yields

$$\begin{aligned} A_{x'}&=A\cos(\theta+\theta')=A(\cos\theta\cos\theta'-\sin\theta\sin\theta')=A_x\cos\theta-A_y\sin\theta \\ A_{y'}&=A\sin(\theta+\theta')=A(\sin\theta\cos\theta'+\cos\theta\sin\theta')=A_x\sin\theta+A_y\cos\theta \end{aligned} \tag{5.7.1}$$

or in terms of the *rotation matrix* $\mathbf{R}(\theta)$,

$$\begin{pmatrix} A_{x'} \\ A_{y'} \end{pmatrix}=\begin{pmatrix} \cos\theta & -\sin\theta \\ \sin\theta & \cos\theta \end{pmatrix}\begin{pmatrix} A_x \\ A_y \end{pmatrix}=\mathbf{R}(\theta)\begin{pmatrix} A_x \\ A_y \end{pmatrix} \tag{5.7.2}$$

If the coordinate system is instead rotated by $+\theta$, θ is replaced by $-\theta$ in Equation (5.7.2). The dot product $(\vec{B}\,|\,\vec{A})$, abbreviated as $(B|A)$ below, of any two vectors is preserved under rotation since

$$(A_{x'}\ A_{y'})\begin{pmatrix} B_{x'} \\ B_{y'} \end{pmatrix}=\left[\mathbf{R}(\theta)\begin{pmatrix} A_x \\ A_y \end{pmatrix}\right]^T\mathbf{R}(\theta)\begin{pmatrix} B_x \\ B_y \end{pmatrix}=(A_x\ A_y)\mathbf{R}^T(\theta)\mathbf{R}(\theta)\begin{pmatrix} B_x \\ B_y \end{pmatrix} \tag{5.7.3}$$

while

$$\mathbf{R}^T(\theta)\mathbf{R}(\theta) = \mathbf{R}(-\theta)\mathbf{R}(\theta) = \begin{pmatrix} \cos\theta & \sin\theta \\ -\sin\theta & \cos\theta \end{pmatrix} \begin{pmatrix} \cos\theta & -\sin\theta \\ \sin\theta & \cos\theta \end{pmatrix} = \mathbf{I} \tag{5.7.4}$$

In general, transformation matrices that preserve the dot product satisfy the orthogonality property

$$\mathbf{R}^T = \mathbf{R}^{-1} \tag{5.7.5}$$

The determinant of an orthogonal matrix equals ± 1 since

$$\det\mathbf{R} = \det\mathbf{R}^T = \det\mathbf{R}^{-1} = \frac{1}{\det\mathbf{R}} \tag{5.7.6}$$

Unit determinant (special orthogonal) matrices correspond to rotations.

Since length is a real scalar quantity, for a complex vector $\vec{x}$, the squared length is identified with

$$\langle x|x\rangle \equiv \vec{x}^{\dagger} \cdot \vec{x} = (x_1^* x_2^* \cdots x_n^*) \begin{pmatrix} x_1 \\ x_2 \\ \vdots \\ x_n \end{pmatrix} = |x_1|^2 + |x_2|^2 + \cdots + |x_n|^2 \tag{5.7.7}$$

where the notation $|x\rangle$ and $\langle x|$ is employed to distinguish intrinsically complex quantities $\vec{x}$ and $\vec{x}^{\dagger}$ from real vectors $|x)$ and $(x|$. For two vectors, since $(\mathbf{AB})^{\dagger} = \mathbf{B}^{\dagger}\mathbf{A}^{\dagger}$, under a transformation $x' = \mathbf{U}x$

$$\langle x'|y'\rangle = \langle \mathbf{U}x|\mathbf{U}y\rangle = \langle x|\mathbf{U}^{\dagger}\mathbf{U}|y\rangle \tag{5.7.8}$$

Therefore, $\langle x'|y'\rangle = \langle x|y\rangle$ if the *unitary condition* $\mathbf{U}^{\dagger}\mathbf{U} = \mathbf{I}$ or

$$\mathbf{U}^{\dagger} = \mathbf{U}^{-1} \tag{5.7.9}$$

is fulfilled. For real vectors, Equation (5.7.7) coincides with Equation (5.7.5) so that all orthogonal matrices are unitary. The *magnitude* $|\det\mathbf{U}| = 1$ since $\det\mathbf{U}\det\mathbf{U}^{\dagger} = 1$ and $\det\mathbf{A} = \det\mathbf{A}^T$ imply

$$\det\mathbf{U} = \frac{1}{\det\mathbf{U}^{\dagger}} = \frac{1}{(\det\mathbf{U}^T)^*} = \frac{1}{(\det\mathbf{U})^*} \tag{5.7.10}$$

5.8 GROUPS AND MATRIX GENERATORS

A *group* comprises a set of elements together with an associative, but not necessarily commutative, product operator $A \odot (B \odot C) = (A \odot B) \odot C$ such that the product of two group elements is a group element, an identity element exists, and every member of the group possesses an inverse. For a *continuous group*, every group element $\mathbf{O}(\alpha)$ can be written as an infinite product of elements that approximate the identity; i.e., $\mathbf{O}(\alpha)=\lim_{N\to\infty}(\mathbf{O}(\alpha/N))^N$ with $\mathbf{O}(0)=\mathbf{I}$. Such groups arise when solving differential equations through repeated infinitesimal steps.

To illustrate, as $N\to\infty$, to order α/N (i.e., neglecting powers of α/N greater than the first), for the skew-symmetric ($\mathbf{A}=-\mathbf{A}^T$) *group generator* $\mathbf{A}$ below, the group element

$$\mathbf{O}\left(\frac{\alpha}{N}\right)=\mathbf{I}+\frac{\alpha}{N}\mathbf{A}=\mathbf{I}+\frac{\alpha}{N}\begin{pmatrix}0 & -1\\ 1 & 0\end{pmatrix} \tag{5.8.1}$$

is orthogonal, e.g., neglecting terms of higher order than α/N,

$$\mathbf{O}^T\left(\frac{\alpha}{N}\right)\mathbf{O}\left(\frac{\alpha}{N}\right)=\begin{pmatrix}1 & \frac{\alpha}{N}\\ -\frac{\alpha}{N} & 1\end{pmatrix}\begin{pmatrix}1 & -\frac{\alpha}{N}\\ \frac{\alpha}{N} & 1\end{pmatrix}=\begin{pmatrix}1+\frac{\alpha^2}{N^2} & 0\\ 0 & 1+\frac{\alpha^2}{N^2}\end{pmatrix}\approx\mathbf{I} \tag{5.8.2}$$

while $\det(\mathbf{O}(\alpha/N))\approx 1$ to this order. A continuous group of orthogonal transformations is constructed through the infinite product for any skew-symmetric $\mathbf{A}$,

$$\mathbf{O}(\alpha)=\lim_{N\to\infty}\left(\mathbf{I}+\frac{\alpha}{N}\mathbf{A}\right)^N\to\mathbf{I}+\frac{\alpha}{1!}\mathbf{A}+\frac{\alpha^2}{2!}\mathbf{A}^2+\cdots=e^{\alpha\mathbf{A}} \tag{5.8.3}$$

with inverse

$$\begin{aligned}\mathbf{O}^{-1}(\alpha)&=\left(\mathbf{I}+\frac{\alpha}{1!}\mathbf{A}+\frac{\alpha^2}{2!}\mathbf{A}^2+\cdots\right)^T\\ &=\mathbf{I}+\frac{\alpha}{1!}\mathbf{A}^T+\frac{\alpha^2}{2!}(\mathbf{A}^2)^T+\cdots\\ &=\mathbf{I}-\frac{\alpha}{1!}\mathbf{A}+\frac{\alpha^2}{2!}\mathbf{A}^2+\cdots\\ &=e^{-\mathbf{A}}=\mathbf{O}(-\alpha)\end{aligned} \tag{5.8.4}$$

In particular, for the two-dimensional matrix of Equation (5.8.1) for which $\mathbf{A}^2=-\mathbf{I}$:

$$\begin{aligned}e^{\alpha\begin{pmatrix}0 & -1\\ 1 & 0\end{pmatrix}}&=\begin{pmatrix}1 & 0\\ 0 & 1\end{pmatrix}+\frac{\alpha}{1!}\begin{pmatrix}0 & -1\\ 1 & 0\end{pmatrix}-\frac{\alpha^2}{2!}\begin{pmatrix}1 & 0\\ 0 & 1\end{pmatrix}-\frac{\alpha^3}{3!}\begin{pmatrix}0 & -1\\ 1 & 0\end{pmatrix}+\cdots\\ &=\left(1-\frac{\alpha^2}{2!}+\cdots\right)\begin{pmatrix}1 & 0\\ 0 & 1\end{pmatrix}+\left(\alpha-\frac{\alpha^3}{3!}+\cdots\right)\begin{pmatrix}0 & -1\\ 1 & 0\end{pmatrix}\\ &=\begin{pmatrix}\cos\alpha & -\sin\alpha\\ \sin\alpha & \cos\alpha\end{pmatrix}=\mathbf{R}(\alpha)\end{aligned} \tag{5.8.5}$$

For complex vectors, unitary transformations are constructed from *skew-Hermitian* generators **S** with $\mathbf{S}^\dagger = -\mathbf{S}$ (or equivalently from i times a Hermitian operator $\mathbf{S} = i\mathbf{H}$) as

$$\mathbf{U}(\alpha) = \lim_{N\to\infty}\left(\mathbf{I} + \frac{\alpha}{N}\mathbf{S}\right)^N \to \mathbf{I} + \frac{\alpha}{1!}\mathbf{S} + \frac{\alpha^2}{2!}\mathbf{S}^2 + \cdots = e^{\alpha\mathbf{S}} = e^{i\alpha\mathbf{H}} \tag{5.8.6}$$

with the inverse transformation, in analogy to Equation (5.8.4),

$$\begin{aligned}\mathbf{U}^{-1}(\alpha) &= \left(\mathbf{I} + \frac{\alpha}{1!}\mathbf{S} + \frac{\alpha^2}{2!}\mathbf{S}^2 + \cdots\right)^\dagger \\ &= \mathbf{I} + \frac{\alpha}{1!}\mathbf{S}^\dagger + \frac{\alpha^2}{2!}(\mathbf{S}^2)^\dagger + \cdots \\ &= \mathbf{I} - \frac{\alpha}{1!}\mathbf{S} + \frac{\alpha^2}{2!}\mathbf{S}^2 + \cdots \\ &= e^{-\alpha\mathbf{S}} = \mathbf{U}(-\alpha)\end{aligned} \tag{5.8.7}$$

5.9 EIGENVALUES AND EIGENVECTORS

An *eigenvector* of a matrix, **M**, is preserved up to a scaling factor, λ_i, termed the corresponding *eigenvalue*, upon multiplication by **M**. That is, for an eigenvector $|\phi_i\rangle$ expressed as a column vector:

$$\mathbf{M}|\phi_i\rangle = \lambda_i|\phi_i\rangle \tag{5.9.1}$$

The method for finding eigenvalues and eigenvectors can be illustrated by inserting

$$\mathbf{M} = \begin{pmatrix} 0 & 1 \\ -1 & 0 \end{pmatrix} \tag{5.9.2}$$

into Equation (5.9.1)

$$\mathbf{M}|\phi\rangle = \begin{pmatrix} 0 & 1 \\ -1 & 0 \end{pmatrix}\begin{pmatrix} x_1 \\ x_2 \end{pmatrix} = \lambda|\phi\rangle = \lambda\begin{pmatrix} x_1 \\ x_2 \end{pmatrix} = \lambda\begin{pmatrix} 1 & 0 \\ 0 & 1 \end{pmatrix}\begin{pmatrix} x_1 \\ x_2 \end{pmatrix} \tag{5.9.3}$$

or equivalently

$$(\mathbf{M} - \lambda\mathbf{I})\begin{pmatrix} x_1 \\ x_2 \end{pmatrix} = \begin{pmatrix} -\lambda & 1 \\ -1 & -\lambda \end{pmatrix}\begin{pmatrix} x_1 \\ x_2 \end{pmatrix} = \begin{pmatrix} 0 \\ 0 \end{pmatrix} \tag{5.9.4}$$

If $(\mathbf{M} - \lambda\mathbf{I})^{-1}$ exists, Equation (5.9.4) implies $x_1 = x_2 = 0$ as evident, e.g., from Equation (5.3.17). However, when

$$\det\begin{pmatrix} -\lambda & 1 \\ -1 & -\lambda \end{pmatrix} = 0 \tag{5.9.5}$$

$(\mathbf{M}-\lambda\mathbf{I})^{-1}$ is not invertible as the first and second rows of $\mathbf{M}-\lambda\mathbf{I}$ are proportional. The rank of the matrix is then less than its dimension, leading to the nontrivial solution $\lambda x_1 = x_2$ or, equivalently, $x_1 = -\lambda x_2$. Equation (5.9.5) accordingly requires that λ satisfies the *characteristic equation* $\lambda^2+1=0$ yielding the two eigenvalues $\lambda_1 = i$, $\lambda_2 = -i$. The corresponding *eigenvectors* are determined directly from $\lambda x_1 = x_2$ or by substituting each eigenvalue into Equation (5.9.3) and solving for x_1 in terms of x_2. For $\lambda_1 = i$,

$$\begin{pmatrix} 0 & 1 \\ -1 & 0 \end{pmatrix}\begin{pmatrix} x_1 \\ x_2 \end{pmatrix} = i\begin{pmatrix} x_1 \\ x_2 \end{pmatrix} \tag{5.9.6}$$

and both rows of the equation system as required yield $x_2 = ix_1$. Setting initially either $x_1 = 1$ or $x_2 = 1$ and subsequently normalizing the resulting eigenvector to unit amplitude result in

$$|\phi_1\rangle = \frac{1}{\sqrt{2}}\begin{pmatrix} 1 \\ i \end{pmatrix} \tag{5.9.7}$$

The eigenvalues of Hermitian and anti-Hermitian matrices are purely real and imaginary, respectively. Further, as a unitary matrix can be written $\mathbf{U} = e^{i\alpha\mathbf{H}}$ with $\mathbf{H}$ Hermitian, each eigenvector $|\phi_i\rangle$ of $\mathbf{H}$ with eigenvalue λ_i is simultaneously an eigenvector of $\mathbf{U}$ with eigenvalue $e^{i\alpha\lambda_i}$; hence, the eigenvalues of a unitary matrix are positioned on the unit circle in the complex plane. Further, the eigenvalues of an even-order $(2N\times 2N)$ skew-symmetric matrix occur in pairs $\pm\lambda_i$, while an odd-order skew-symmetric matrix additionally possesses a zero eigenvalue. A skew-Hermitian matrix similarly possesses pairs of imaginary eigenvalues $\pm i|\lambda_i|$ or zero.

The eigenvectors of Hermitian matrices with different eigenvalues are *orthogonal* and can be *orthonormalized* such that $\langle\phi_i|\phi_j\rangle = \delta_{ij}$. That is, for a normalized eigenvector $|\phi_i\rangle$ with $\mathbf{H}|\phi_i\rangle = \lambda_i|\phi_i\rangle$,

$$\langle\phi_j|\mathbf{H}|\phi_i\rangle = \left(x_{j1}^*, x_{j2}^*, \ldots, x_{jn}^*\right)\mathbf{H}\begin{pmatrix} x_{i1} \\ x_{i2} \\ \vdots \\ x_{in} \end{pmatrix} = \lambda_i\langle\phi_j|\phi_i\rangle \tag{5.9.8}$$

At the same time, however, since $\mathbf{A}^\dagger\mathbf{H}^\dagger = (\mathbf{HA})^\dagger$,

$$\langle\phi_j|\mathbf{H}|\phi_i\rangle = \langle\phi_j|\mathbf{H}^\dagger|\phi_i\rangle = \left(\mathbf{H}\begin{pmatrix} x_{j1} \\ x_{j2} \\ \vdots \\ x_{jn} \end{pmatrix}\right)^\dagger\begin{pmatrix} x_{i1} \\ x_{i2} \\ \vdots \\ x_{in} \end{pmatrix} = \lambda_j^*\langle\phi_j|\phi_i\rangle \tag{5.9.9}$$

Hence, either $\lambda_i = \lambda_j^*$ and λ_i is real or $\langle\phi_i|\phi_j\rangle = 0$.

The *Gram–Schmidt* procedure can be employed to transform N *degenerate* eigenvectors $|\xi_i\rangle$ that possess the same eigenvalue, λ_m, into a set of N orthonormal eigenfunctions $|\phi_i\rangle$ according to

$$
\begin{aligned}
|\phi_1\rangle &= \frac{|\xi_1\rangle}{\langle\xi_1|\xi_1\rangle^{\frac{1}{2}}} \\
|\phi_2'\rangle &= |\xi_2\rangle - \langle\phi_1|\xi_2\rangle|\phi_1\rangle \\
|\phi_2\rangle &= \frac{|\phi_2'\rangle}{\langle\phi_2'|\phi_2'\rangle^{\frac{1}{2}}} \\
|\phi_3'\rangle &= \xi_3 - \langle\phi_2|\xi_3\rangle|\phi_2\rangle - \langle\phi_1|\xi_3\rangle|\phi_1\rangle \\
&\vdots
\end{aligned}
\tag{5.9.10}
$$

The eigenvectors of a Hermitian operator form a *complete set*; i.e., if $\mathbf{H}$ is an $N \times N$ matrix, *any* n-component complex-valued vector can be expressed as a *linear superposition* of the eigenvectors

$$
|V\rangle = \sum_{i=1}^{n} c_n|\phi_n\rangle \tag{5.9.11}
$$

where the real or complex coefficients c_n are obtained by multiplying both sides of the above equation with the vector $\langle\phi_p|$ and employing the orthonormality of the $|\phi_n\rangle$:

$$
c_p = \langle\phi_p|V\rangle \tag{5.9.12}
$$

Combining the two above equations yields for any vector $|V\rangle$

$$
|V\rangle = \sum_{i=1}^{n} |\phi_i\rangle\langle\phi_i|V\rangle \tag{5.9.13}
$$

implying that the identity matrix can be represented by

$$
\mathbf{I} = \sum_{i=1}^{n} |\phi_i\rangle\langle\phi_i| = \begin{pmatrix} x_{11}x_{11}^* & x_{11}x_{12}^* & x_{11}x_{13}^* & \cdots \\ x_{12}x_{11}^* & x_{12}x_{12}^* & x_{12}x_{13}^* & \cdots \\ x_{13}x_{11}^* & x_{13}x_{12}^* & x_{13}x_{13}^* & \cdots \\ \vdots & \vdots & \vdots & \ddots \end{pmatrix} + \begin{pmatrix} x_{21}x_{21}^* & x_{21}x_{22}^* & x_{21}x_{23}^* & \cdots \\ x_{22}x_{21}^* & x_{22}x_{22}^* & x_{22}x_{23}^* & \cdots \\ x_{23}x_{21}^* & x_{23}x_{22}^* & x_{23}x_{23}^* & \cdots \\ \vdots & \vdots & \vdots & \ddots \end{pmatrix} + \ldots \tag{5.9.14}
$$

as is evident by considering the effect of multiplying on the left by $\langle\phi_j|$ and on the right by $|\phi_k\rangle$. Multiplying by the above representation of $\mathbf{I}$ accordingly transforms a vector into its eigenvector expansion. This implies the *Cayley–Hamilton theorem*, which states that a matrix satisfies its characteristic equation, $\lambda^n + a_{n-1}\lambda^{n-1} + \ldots + a_0 = 0$, for $\mathbf{H}$ since for any n-dimensional vector $|V\rangle$,

$$
\begin{aligned}
\left(\mathbf{H}^n + a_{n-1}\mathbf{H}^{n-1} + \ldots + a_0\right)|V\rangle &= \left(\mathbf{H}^n + a_{n-1}\mathbf{H}^{n-1} + \ldots + a_0\right)\sum_{i=1}^{n} c_i|\phi_i\rangle\langle\phi_i|V\rangle \\
&= \sum_{i=1}^{n} c_i\left(\lambda_i^n + a_{n-1}\lambda_i^{n-1} + \ldots + a_0\right)|\phi_i\rangle\langle\phi_i|V\rangle \\
&= 0
\end{aligned}
\tag{5.9.15}
$$

5.10 SIMILARITY TRANSFORMATIONS

Two matrices **A** and **B** are *similar* if an invertible matrix **W** exists such that

$$\mathbf{A} = \mathbf{W}^{-1}\mathbf{B}\mathbf{W} \tag{5.10.1}$$

Similar matrices share the same characteristic polynomial and therefore possess identical eigenvalues as

$$
\begin{aligned}
\det(\mathbf{A} - \lambda\mathbf{I}) &= \det\left(\mathbf{W}^{-1}\mathbf{B}\mathbf{W} - \lambda\mathbf{W}^{-1}\mathbf{W}\right) \\
&= \det\left(\mathbf{W}^{-1}\right)\det(\mathbf{B} - \lambda\mathbf{I})\det(\mathbf{W}) \\
&= \det(\mathbf{B} - \lambda\mathbf{I})\frac{1}{\cancel{\det(\mathbf{W})}}\cancel{\det(\mathbf{W})}
\end{aligned}
\tag{5.10.2}
$$

However, two matrices with the same eigenvalues are not necessarily similar.

A Hermitian matrix **H** is similar to a diagonal matrix **D** with

$$\mathbf{H} = \mathbf{W}^{-1}\mathbf{D}\mathbf{W} \tag{5.10.3}$$

and $D_{ij} = 0$ for $i \neq j$. The matrix **W** can be formed from the eigenvectors of **H** according to

$$\mathbf{H} = \sum_{i,j=1}^{n} |\phi_j\rangle\langle\phi_j|\mathbf{H}|\phi_i\rangle\langle\phi_i| \tag{5.10.4}$$

Since $\langle\phi_j|\mathbf{H}|\phi_i\rangle = \lambda_i\delta_{ij}$, the above equation can be written as

$$\mathbf{H} = \underbrace{\begin{pmatrix} \phi_1 & \phi_2 & & \phi_n \end{pmatrix}}_{W^{-1}} \underbrace{\begin{pmatrix} \lambda_1 & & \\ & \ddots & \\ & & \lambda_n \end{pmatrix}}_{D} \underbrace{\begin{pmatrix} \phi_1^* \\ \vdots \\ \phi_n^* \end{pmatrix}}_{W} \tag{5.10.5}$$

To verify the above expression explicitly, note that the product $\mathbf{H}|\phi_i\rangle$ equals $\lambda_i|\phi_i\rangle$. Hence, **H** produces the correct result when it multiplies any linear combination of eigenfunctions. Further, from the orthonormality of the $|\phi_k\rangle$, $\mathbf{W}^\dagger = \mathbf{W}^{-1}$ so that $\mathbf{W}$ is unitary and

$$\mathbf{H}^{-1} = \underbrace{\begin{pmatrix} \phi_1 \,\Big|\, \phi_2 \,\Big|\, \cdots \,\Big|\, \phi_n \end{pmatrix}}_{W^{-1}} \underbrace{\begin{pmatrix} 1/\lambda_1 & & \\ & \ddots & \\ & & 1/\lambda_n \end{pmatrix}}_{D^{-1}} \underbrace{\begin{pmatrix} \phi_1^* \\ \hline \vdots \\ \hline \phi_n^* \end{pmatrix}}_{W} \tag{5.10.6}$$

The *condition number* of $\mathbf{H}$ approximates the ratio of the largest to the smallest eigenvalue. This provides an estimate of the proximity of $\mathbf{H}$ to a singular matrix and hence the error encountered when inverting the matrix numerically.

6

CALCULUS OF A SINGLE VARIABLE

Calculus extends algebra through the introduction of limits. Inverse problems such as integration reverse the limiting operations and generally require specialized solution techniques.

6.1 DERIVATIVES

While a line cannot be specified by a single point, a unique tangent line does exist at each point on a (differentiable) curve. This apparent contradiction is resolved by defining the tangent as the limit of a secant to the curve as the distance between the points at which the secant is evaluated approaches zero. The derivative operator, where an operator maps one or more functions to a second function in the same manner that a function transforms one or more input values into an output value, applied to a function, $f(x)$, yields the slope of the tangent to the function at each point, x, and is accordingly conventionally identified with the $\Delta x \to 0$ limit of the secant slope

$$\frac{df}{dx} = \lim_{\Delta x \to 0} \frac{f(x+\Delta x) - f(x)}{\Delta x} \tag{6.1.1}$$

Fundamental Math and Physics for Scientists and Engineers, First Edition.
David Yevick and Hannah Yevick.

In numerical computation, the above *forward difference* definition is generally replaced by the more accurate *central difference formula*

$$\frac{df}{dx} = \lim_{\Delta x \to 0} \frac{f(x+\Delta x/2) - f(x-\Delta x/2)}{\Delta x} \tag{6.1.2}$$

As an example, from the binomial theorem,

$$\begin{aligned}
\frac{dx^n}{dx} &= \lim_{\Delta x \to 0} \frac{(x+\Delta x)^n - x^n}{\Delta x} \\
&= \lim_{\Delta x \to 0} \frac{x^n + nx^{n-1}\Delta x + (n(n-1)/2!)x^{n-2}\Delta x^2 + \ldots - x^n}{\Delta x} \\
&= \lim_{\Delta x \to 0} \left(nx^{n-1} + (n(n-1)/2!)x^{n-2}\Delta x + \ldots\right) \\
&= nx^{n-1}
\end{aligned} \tag{6.1.3}$$

The following formulas follow directly from power series expansions:

$$\begin{aligned}
\frac{d\sin x}{dx} &= \cos x \\
\frac{d\cos x}{dx} &= -\sin x \\
\frac{de^x}{dx} &= e^x
\end{aligned} \tag{6.1.4}$$

The derivative is a linear operator since $(f+g)' = f' + g'$, where the prime denotes the derivative with respect to the function argument.

The *product rule* is derived from Equation (6.1.1) according to

$$\begin{aligned}
(fg)' &= \lim_{\Delta x \to 0} \frac{f(x+\Delta x)g(x+\Delta x) - f(x)g(x)}{\Delta x} \\
&= \lim_{\Delta x \to 0} \frac{(f(x) + \Delta x f'(x))(g(x) + \Delta x g'(x)) - f(x)g(x)}{\Delta x} \\
&= \lim_{\Delta x \to 0} \left[\frac{\Delta x(f'(x)g(x) + f(x)g'(x))}{\Delta x} + \Delta x f'(x)g'(x)\right] \\
&= f'(x)g(x) + f(x)g'(x)
\end{aligned} \tag{6.1.5}$$

The *chain rule* is instead applicable to functions of other functions typified by $\sin(x^2)$, which can be represented as $f(g(x))$ with $f(x) = \sin x$ and $g(x) = x^2$. Applying the derivative formula twice,

$$\begin{aligned}\frac{df(g(x))}{dx} &= \lim_{\Delta x\to 0}\frac{f(g(x+\Delta x))-f(g(x))}{\Delta x}\\ &= \lim_{\Delta x\to 0}\frac{f(g(x)+g'(x)\Delta x)-f(g(x))}{\Delta x}\\ &= \lim_{\Delta x\to 0}\frac{f(g(x))+f'(g(x))g'(x)\Delta x-f(g(x))}{\Delta x}\\ &= \frac{df(g(x))}{d(g(x))}\frac{dg(x)}{dx}\end{aligned} \tag{6.1.6}$$

Thus, $d\sin(x^2)/dx = d\sin(x^2)/d(x^2)\times dx^2/dx = 2x\cos(x^2)$, and similarly,

$$\frac{da^x}{dx}=\frac{d\left(e^{\ln a}\right)^x}{dx}=\frac{de^{x\ln a}}{dx}=e^{x\ln a}\ln a=a^x\ln a \tag{6.1.7}$$

The derivative of the quotient of two functions is then given by

$$\begin{aligned}\frac{d}{dx}\left(\frac{f(x)}{g(x)}\right) &= \frac{d}{dx}\left(f(x)(g(x))^{-1}\right)\\ &= f'(x)(g(x))^{-1}+f(x)\left(-(g(x))^{-2}g'(x)\right)\\ &= \frac{f'(x)g(x)-f(x)g'(x)}{g^2(x)}\end{aligned} \tag{6.1.8}$$

as a particular consequence of which

$$\frac{d\tan x}{dx}=\sec^2 x\equiv\frac{1}{\cos^2 x} \tag{6.1.9}$$

Since an inverse function $y=f^{-1}(x)$ is a reflection of $y=f(x)$ through the line $y=x$, its slope equals the reciprocal of $y'(x)$

$$\frac{df^{-1}(x)}{dx}=\frac{dy}{dx}=\frac{1}{\frac{dx}{dy}} \tag{6.1.10}$$

The resulting formula must then be reexpressed in terms of x by inserting $y=f^{-1}(x)$.

Example

$$\frac{d\sin^{-1}x}{dx}=\frac{1}{\frac{d\sin y}{dy}}=\frac{1}{\cos y}=\frac{1}{\cos\left(\sin^{-1}x\right)}=\frac{1}{\sqrt{1-\sin^2\left(\sin^{-1}x\right)}}=\frac{1}{\sqrt{1-x^2}} \quad (6.1.11)$$

In the same manner,

$$\frac{d\ln(x)}{dx}=\frac{d\exp^{-1}(x)}{dx}=\frac{1}{\frac{de^y}{dy}}=\frac{1}{e^y}=\frac{1}{e^{\ln x}}=\frac{1}{x} \quad (6.1.12)$$

For functions expressed in terms of a power of a second function as $f(x)=cg^n(x)$,

$$f'(x)=cng^{n-1}(x)g'(x) \quad (6.1.13)$$

and therefore,

$$\frac{f'(x)}{f(x)}=n\frac{g'(x)}{g(x)} \quad (6.1.14)$$

Example

$k=2\pi/\lambda=2\pi\lambda^{-1}$ implies $dk/k=-d\lambda/\lambda$ or $dk/d\lambda=-k/\lambda=-2\pi/\lambda^2$.

A function is represented *parametrically* by specifying both the independent and dependent variables in terms of a parametric variable t as $(x(t), y(t))$. The derivative of the parametric function with respect to a change in t is obtained from

$$\lim_{\Delta t\to 0}\frac{y(t+\Delta t)-y(t)}{\Delta x(t)}=\lim_{\Delta t\to 0}\frac{\frac{y(t+\Delta t)-y(t)}{\Delta t}}{\frac{x(t+\Delta t)-x(t)}{\Delta t}}=\frac{\frac{dy}{dt}}{\frac{dx}{dt}} \quad (6.1.15)$$

Example

The half circle $y=f(x)=\sqrt{a^2-x^2}$ can be expressed as $(x(t), y(t))=(a\cos t, a\sin t)$ with $\pi>t>0$ since eliminating t yields $y=a\sin(\arccos(x/a))=\sqrt{a^2-x^2}$ (c.f. Eq. 4.3.15). Hence,

$$\frac{dy}{dx}=\frac{\frac{dy}{dt}}{\frac{dx}{dt}}=\frac{a\cos t}{-a\sin t}=-\frac{x}{y} \quad (6.1.16)$$

as can be verified by differentiating $y=\sqrt{a^2-x^2}$ directly.

In the same manner that the derivative yields the *slope* of a curve, the second derivative quantifies the rate in change in slope. If the second derivative curve $f(x)$ is positive, its slope increases with x, and the curve therefore opens upward, corresponding to a positive *curvature*. The second and higher derivative operators can be obtained by iterating, e.g., Equation (6.1.2) as, e.g., applying the central difference formula,

$$\begin{aligned}\frac{d^2f}{dx^2} &= \lim_{\Delta x\to 0}\frac{1}{\Delta x}\left[\left.\frac{df}{dx}\right|_{x+\Delta x/2} - \left.\frac{df}{dx}\right|_{x-\Delta x/2}\right]\\ &= \lim_{\Delta x\to 0}\frac{1}{\Delta x}\left[\left(\frac{f(x+\Delta x)-f(x)}{\Delta x}\right)-\left(\frac{f(x)-f(x-\Delta x)}{\Delta x}\right)\right]\\ &= \lim_{\Delta x\to 0}\frac{f(x+\Delta x)-2f(x)+f(x-\Delta x)}{(\Delta x)^2}\end{aligned} \tag{6.1.17}$$

Thus, if $d^2f/dx^2 > 0$, the average of the function evaluated at two points close to but on either side of a given point exceeds the value of function at the point, indicating an upward curvature.

The *critical points* of a one-dimensional function are the points at which its first derivative vanishes, and therefore, the tangent to its curve parallels the x-axis. If the second derivative is *positive* at a critical point, the function describes a parabola that opens upward near the point that therefore constitutes a *minimum*, while a *negative* second derivative similarly implies that the point constitutes a *maximum*. If the second derivative passes through *zero* at the critical point, the function opens upward on one side and downward on the opposing side of the point, which is then termed a *saddle point*.

A quotient of two functions, $f(x)$ and $g(x)$, for which $f(a) = g(a) = 0$ is evaluated at $x = a$ from

$$\lim_{\Delta x\to 0}\frac{f(a+\Delta x)}{g(a+\Delta x)} = \lim_{\Delta x\to 0}\frac{f(a)+\left.\frac{df(x)}{dx}\right|_{x=a}\Delta x}{g(a)+\left.\frac{dg(x)}{dx}\right|_{x=a}\Delta x} = \frac{\left.\frac{df(x)}{dx}\right|_{x=a}}{\left.\frac{dg(x)}{dx}\right|_{x=a}} \tag{6.1.18}$$

termed *L'Hôpital's rule*. If additionally $f'(a) = g'(a) = 0$, the above procedure is reapplied to the quotient of the two derivatives and, if necessary, further iterated until a well-defined expression emerges.

6.2 INTEGRALS

While differentiation is a local operator involving the values of a function in the immediate vicinity of a given point, its *inverse* operator, integration, requires function values over an extended, global interval. A definite integral transforms its input function termed the *integrand* into a second function corresponding to the area under the integrand between a reference point and the argument value. In one discrete realization, the *rectangular rule*, the integration interval $[a, b]$ is subdivided into N infinitesimal equal length segments, and the areas under the function within each segment are approximated and summed according to

$$I_a(f(x)) = \int_a^x f(x')dx' = \Delta x\left[\lim_{N\to\infty}\sum_{m=0}^{N-1} f(a+m\Delta x)\right] \quad (6.2.1)$$

in which $\Delta x = (x - a)/N$. In numerical implementations, the limit is evaluated as a finite sum, while $f(a + m\Delta x)$ is typically replaced by, e.g., $f(a + (m + 0.5)\Delta x)$, which is termed the *midpoint rule*, to enhance computational accuracy.

The inverse relationship between the integral and derivative operators follows from, by canceling adjacent terms in the sum with the exception of the unpaired first and last contributions,

$$\begin{aligned} I_a\left(\frac{df(x)}{dx}\right) = \int_a^x \frac{df(x')}{dx'}dx' &= \Delta x\left[\lim_{N\to\infty}\sum_{m=0}^{N-1}\left(\frac{f(a+(m+1)\Delta x)-f(a+m\Delta x)}{\Delta x}\right)\right] \\ &= f(a+N\Delta x)-f(a) \\ &= f(x)-f(a) \end{aligned} \quad (6.2.2)$$

which is often termed the *fundamental theorem of calculus*. Similarly,

$$\frac{d}{dx}I_a(f(x)) = \frac{d}{dx}\int_a^x f(x')dx' = \Delta x\left[\lim_{N\to\infty}\left(\frac{\sum_{m=0}^{N} f(a+m\Delta x) - \sum_{m=0}^{N-1} f(a+m\Delta x)}{\Delta x}\right)\right] = f(x) \quad (6.2.3)$$

Thus, an integral can often be performed by inferring which function the integrand is a derivative of.

Example

$d(\log x)/dx = 1/x$ implies $\int_a^x (1/x')dx' = \log(x) - \log(a) = \log(x/a)$.

While an integral between specified upper and lower limits is termed a *definite integral*, an *indefinite integral* is evaluated without limits and yields the general expression whose derivative corresponds to the integrand.

Example

$\int x dx = x^2/2 + \text{const}$, in which the (often suppressed) additive constant is typically determined by specifying the value of the integral for a certain x.

Integrating both sides of the derivative product rule, Equation (6.1.5), over the interval $[a, b]$ yields

$$\begin{aligned}\int_a^b fg' dx &= \int_a^b \frac{d(fg)}{dx} dx - \int_a^b f' g dx \\ &= fg\big|_a^b - \int_a^b f' g dx\end{aligned} \tag{6.2.4}$$

where $f\big|_a^b$ represents $f(a) - f(b)$. This procedure, termed *integration by parts*, simplifies integration of a product of two functions if the effect of differentiating f more than compensates any increase in complexity resulting from integrating g or if, after some number of applications, an expression proportional to the original integral is recovered.

Example

The factorial function is represented as an integral by the *gamma function* that can be extended to noninteger and complex arguments:

$$\begin{aligned}\Gamma(n+1) &\equiv \int_0^\infty \underbrace{e^{-x}}_{g'(x)} \underbrace{x^n}_{f(x)} dx \\ &= \underbrace{-e^{-x}}_{g(x)} \underbrace{x^n}_{f(x)} \Big|_0^\infty - \int_0^\infty \underbrace{-e^{-x}}_{g(x)} \underbrace{nx^{n-1}}_{f'(x)} dx\end{aligned} \tag{6.2.5}$$

as e^{-x} at $x = \infty$ and x^n at $x = 0$ are zero. Hence, $\Gamma(n+1) = n\Gamma(n)$ while $\Gamma(1) = -e^x\big|_0^\infty = 1$, implying that $\Gamma(n+1) = n!$ for integer n. As another illustration, the logarithm can be integrated by

$$\int_a^b \underbrace{\ln x}_{f(x)} \underbrace{1}_{g'(x)} dx = \underbrace{x}_{g(x)} \underbrace{\ln x}_{f(x)} \Big|_a^b - \int_a^b \underbrace{x}_{g(x)} \underbrace{\frac{1}{x}}_{f'(x)} dx = (x \ln x - x)\Big|_a^b \tag{6.2.6}$$

Two partial integrations generate an expression containing the original integral in

$$I=\int_0^a \underbrace{e^{-x}}_{f(x)}\underbrace{\cos x}_{g'(x)}\,dx=\underbrace{e^{-x}\sin x|_0^a}_{e^{-a}\sin a}-\int_0^a \underbrace{-e^{-x}}_{f'(x)=-\tilde{f}(x)}\underbrace{\sin x}_{g(x)=\tilde{g}'(x)}\,dx$$
$$=e^{-a}\sin a+\underbrace{e^{-x}}_{\tilde{f}(x)}\underbrace{(-\cos x)}_{\tilde{g}(x)}\Big|_0^a-I$$
$$I=\frac{1}{2}(e^{-a}(\sin a-\cos a)+1) \tag{6.2.7}$$

which can also be obtained by substituting $\cos x=(\exp(ix)+\exp(-ix))/2$ and integrating over the resulting sum of two exponential functions. Similarly, while integrals over powers of trigonometric functions are normally performed by repeatedly applying double-angle formulas such as $\cos(2x)=2\cos^2 x-1=1-2\sin^2 x$ as in

$$\begin{aligned}I_4&\equiv\int\sin^4 x dx\\&=\frac{1}{4}\int(1-\cos(2x))^2 dx\\&=\frac{1}{4}\int\left(\cos^2(2x)-2\cos 2x+1\right)dx\\&=\frac{1}{8}\int(\cos(4x)-4\cos 2x+3)dx\\&=\frac{1}{32}\sin(4x)-\frac{1}{4}\sin(2x)+\frac{3}{8}x\\&=\frac{1}{8}\left(-2(\sin x)^3\cos x-3\sin x\cos x+3x\right)\end{aligned} \tag{6.2.8}$$

they can also be derived through recursion relations obtained through partial integration such as

$$\begin{aligned}I_n&=\int\underbrace{(\sin^{n-1}x)}_{f(x)}\underbrace{(\sin x)}_{g'(x)}dx\\&=-\sin^{n-1}x\cos x+(n-1)\int\sin^{n-2}x\cos^2 x dx\\&=-\sin^{n-1}x\cos x+(n-1)\int\sin^{n-2}x(1-\sin^2 x)dx\\&=-\sin^{n-1}x\cos x+(n-1)(I_{n-2}-I_n)\\&\Rightarrow nI_n=-\sin^{n-1}x\cos x+(n-1)I_{n-2}\end{aligned} \tag{6.2.9}$$

For $n = 4$, since $I_0 = x$, Equation (6.2.8) is reproduced in the form

$$4I_4 = -\sin^3 x\cos x + 3\left(\frac{1}{2}(-\sin x\cos x + x)\right) \tag{6.2.10}$$

In this context, note that the average of $\sin^2 ax$ or $\cos^2 ax$ over any interval $[x_0, x_0 + \pi/a]$ equals 1/2. While this follows directly from $\cos^2 ax + \sin^2 ax = 1$ together with the observation that the average of $\cos^2 ax$ and $\sin^2 ax$ over a length π/a must be identical, direct integration gives

$$\frac{a}{\pi}\int_{x_0}^{x_0+\frac{\pi}{a}} \cos^2 ax dx = \frac{a}{\pi}\int_{x_0}^{x_0+\frac{\pi}{a}} \frac{\cos(2ax)+1}{2}dx = \frac{a}{\pi}\left(\frac{\sin 2ax}{4}+\frac{1}{2}\right)\Bigg|_{x_o}^{x_o+\frac{\pi}{a}} = \frac{1}{2} \tag{6.2.11}$$

The integral counterpart of the derivative chain rule, Equation (6.1.6), is termed *integration by substitution*. Changing the integration variable from x to $y = g(x)$ yields (observe especially the transformed limits)

$$\int_a^b f(x)dx = \int_{g(a)}^{g(b)} f(g^{-1}(y))\frac{dx}{dy}dy \tag{6.2.12}$$

If $f(g^{-1}(y))dx/dy = f(g^{-1}(y))dg^{-1}(y)/dy$ represents a total differential (the derivative of a function), the integral can be performed.

Example

Substituting $y = x^2$ and therefore $dy = 2x\,dx$ into (or using $xdx/dy = y^{½}d(y^{½})/dy = 1/2$)

$$\int_a^b \cos(x^2)xdx \tag{6.2.13}$$

yields $x\cos x^2(dx/dy) = (\cos y)/2$, which is the total differential of $(\sin y)/2$. Hence,

$$\int_a^b \cos(x^2)xdx = \frac{1}{2}\int_{a^2}^{b^2}\cos y dy = \frac{1}{2}(\sin b^2 - \sin a^2) \tag{6.2.14}$$

$$\int_a^b \cos(x^2)x\,dx = \int_a^b \cos x^2 \frac{1}{2} d(x^2) = \frac{1}{2}\sin x^2\Big|_a^b \tag{6.2.15}$$

To integrate the function $y = \sqrt{a^2 - x^2}$ over the integral $x = [-a, a]$ yielding the area under a half circle, the parameterization $x = a\cos\theta$, $y = a\sin\theta$ can be introduced so that

$$\begin{aligned}\int_{-a}^{a} y\,dx &= \int_{\pi}^{0} y\frac{dx}{d\theta}d\theta \\ &= \int_{\pi}^{0} a\sin\theta(-a\sin\theta)d\theta \\ &= -a^2\int_{\pi}^{0} \sin^2\theta d\theta \\ &= -a^2\int_{\pi}^{0}\frac{1-\cos 2\theta}{2}d\theta = \frac{\pi}{2}a^2\end{aligned} \tag{6.2.16}$$

Integrals of inverse functions are similarly solved by substitution.

Example

Substituting $x = \sin y$ into

$$\int_0^A \frac{1}{\sqrt{1-x^2}}dx \tag{6.2.17}$$

yields $y = \sin^{-1}x$ and $dx = \cos y\,dy$, and hence,

$$\int_{\sin^{-1}0}^{\sin^{-1}A} \frac{\cos y}{\sqrt{1-\sin^2 y}}dy = \int_0^{\sin^{-1}A} dy = \sin^{-1}A \tag{6.2.18}$$

Finally, integrals of a ratio of polynomials $N(x)/D(x)$ where the order of the polynomial $D(x)$ exceeds that of $N(x)$ (otherwise, the two polynomials are divided and the method applied to the remainder term) can be evaluated through *partial fraction decomposition*. Here, $D(x)$ is first factored into a product of lower-order polynomials $D_1(x)D_2(x)\ldots D_n(x)$. The integrand is then written as a sum of terms $N_i(x)/D_i(x)$ where the order of each $N_i(x)$ is less than that of the corresponding $D_i(x)$ and its coefficients are determined by equating the sum to $N(x)/D(x)$.

Example

After factoring the denominator of the integral

$$\int_a^b \frac{3x+2}{x^2+3x+2}dx = \int_a^b \frac{3x+2}{(x+2)(x+1)}dx \tag{6.2.19}$$

the integrand can be written as a sum of lower-order polynomials with unknown coefficients

$$\frac{3x+2}{(x+2)(x+1)} = \frac{A}{x+2} + \frac{B}{x+1} = \frac{A(x+1)+B(x+2)}{(x+2)(x+1)} \tag{6.2.20}$$

which leads to the conditions

$$\begin{aligned} A+B&=3 \\ A+2B&=2 \end{aligned} \tag{6.2.21}$$

or $A=4,\ B=-1$. Therefore,

$$\int_a^b \frac{3x+2}{x^2+3x+2}dx = 4\ln\left(\frac{b+2}{a+2}\right) - \ln\left(\frac{b+1}{a+1}\right) \tag{6.2.22}$$

6.3 SERIES

From the definition of the derivative, $f(x+\Delta x) = (1+\Delta x(d/dx))f(x)$, two values of function $f(x)$ a finite distance $a = N\Delta x$ apart are related in the $N \to \infty$, $\Delta x \to 0$ limit by

$$f(x+a) = \lim_{N\to\infty}\left(1+\frac{a}{N}\frac{d}{dx}\right)^N f(x) = e^{a\frac{d}{dx}}f(x) \tag{6.3.1}$$

Expanding the exponential as in Equation (4.3.3) generates the *Taylor series*

$$f(x+a) = f(x) + \frac{a}{1!}\frac{df(x)}{dx} + \frac{a^2}{2!}\frac{d^2f(x)}{dx^2} + \cdots \tag{6.3.2}$$

Such a power series expansion is valid within a finite or infinite *radius of convergence* about the central point x.

Example

Since x^n increases more slowly with n than $n!$ for large n, the Taylor series of $\exp(x)$ and its even and odd parts, $\cosh(x)$, $\sinh(x)$, $\cos(x)$, and $\sin(x)$, possess infinite convergence radii. The convergence radius of the geometric series obtained by expanding $f(x) = (1 - x)^{-1}$ about $x = 0$ instead equals unity as successive Taylor series terms diverge for real $x > 1$.

The convergence properties of a series $S = \Sigma_{n=1}^{\infty} s_n$ can often be established as follows, where an *absolutely convergent* series converges even if each element is replaced by its absolute value.

- If a second series $\Sigma_{n=1}^{\infty} a_n$ is absolutely convergent and either $|s_n| < c|a_n|$ or $|s_n/s_{n-1}| < c|a_n/a_{n-1}|$ for all $n > n_0$ with c a constant, then S is also absolutely convergent.
- S converges absolutely if $|s_n/s_{n-1}| < 1$ for $n > n_0$ or if $\lim_{n\to\infty} |s_n|^{1/n} < 1$.
- In an alternating series for which successive terms have different signs, if $|s_n|$ decreases monotonically to zero for large n, the series converges.
- If a positive, monotonically decreasing function $f(x)$ can be found such that $f(k) = s_k$ for integer k, then S is convergent if and only if $\int_1^{\infty} f(x)dx < \infty$.
- If the s_n do not approach zero as $n \to \infty$, the series diverges.

7

CALCULUS OF SEVERAL VARIABLES

In multiple dimensions, a function can be differentiated or integrated along different spatial directions. As well, either the function or the derivative or integral operator can correspond to a directed quantity. The multidimensional calculus of scalar functions is examined in this chapter, while vector functions and operators are discussed in subsequent chapters.

7.1 PARTIAL DERIVATIVES

The slope, or *partial derivative*, of a function of several variables, $f(x_1, x_2, \ldots, x_n)$, along the ith coordinate direction equals

$$\frac{\partial f(\vec{x})}{\partial x_i} = \lim_{\Delta x_i \to 0} \frac{f(x_1, x_2, \ldots, x_i + \Delta x_i, \ldots, x_n) - f(x_1, x_2, \ldots, x_i, \ldots, x_n)}{\Delta x_i} \tag{7.1.1}$$

Second-order partial derivatives comprise the second-order partial derivatives $\partial^2 f/\partial x_i^2$ and the second-order mixed derivatives $\partial^2 f/\partial x_i \partial x_j$. The mixed partial derivatives are independent of the order in which the individual derivatives are evaluated, e.g., in two dimensions,

Fundamental Math and Physics for Scientists and Engineers, First Edition.
David Yevick and Hannah Yevick.

$$\begin{aligned}\frac{\partial^2 f(x,y)}{\partial y \partial x} &= \lim_{\Delta y \to 0} \frac{1}{\Delta y}\left(\frac{\partial f(x,y+\Delta y)}{\partial x} - \frac{\partial f(x,y)}{\partial x}\right) \\ &= \lim_{\Delta x, \Delta y \to 0} \frac{1}{\Delta x \Delta y}([f(x+\Delta x, y+\Delta y) - f(x, y+\Delta y)] - [f(x+\Delta x, y) - f(x,y)]) \\ &= \lim_{\Delta x, \Delta y \to 0} \frac{1}{\Delta x \Delta y}([f(x+\Delta x, y+\Delta y) - f(x+\Delta x, y)] - [f(x, y+\Delta y) - f(x,y)]) \\ &= \lim_{\Delta x \to 0} \frac{1}{\Delta x}\left(\frac{\partial f(x+\Delta x,y)}{\partial y} - \frac{\partial f(x,y)}{\partial y}\right) \\ &= \frac{\partial^2 f(x,y)}{\partial x \partial y}\end{aligned} \tag{7.1.2}$$

If a function depends on two or more quantities that are functions of the coordinates, its partial derivative is evaluated with the multidimensional generalization of the chain rule

$$\begin{aligned}&\lim_{\Delta x \to 0} \frac{f(g(x+\Delta x,y), h(x+\Delta x,y)) - f(g(x,y),h(x,y))}{\Delta x} \\ &= \lim_{\Delta x \to 0} \frac{f\left(g(x,y) + \Delta x \frac{\partial g}{\partial x}, h(x,y) + \Delta x \frac{\partial h}{\partial x}\right) - f(g(x,y),h(x,y))}{\Delta x} \\ &\approx \frac{\partial g}{\partial x}\frac{\partial f(g,h)}{\partial g} + \frac{\partial h}{\partial x}\frac{\partial f(g,h)}{\partial h}\end{aligned} \tag{7.1.3}$$

The derivative $\partial y/\partial x|_z$ for (x, y) restricted to a curve in the x–y plane defined by the constraint $z(x, y) = c$ yields the slope of $y(x)$ along the curve, while $\partial x/\partial y|_z$ is the reciprocal of the slope with

$$\left.\frac{\partial y}{\partial x}\right|_{z(x,y)=c} = \frac{1}{\left.\frac{\partial x}{\partial y}\right|_{z(x,y)=c}} \tag{7.1.4}$$

Example

For $z(x, y) = x^2 + y = 3$ at $(1, 2)$, $\partial y/\partial x|_z = -2x|_{x=1} = -2$, while $\partial x/\partial y|_z = -1/(2\sqrt{3-y})|_{y=2} = -1/2$

If x and y are restricted to a curve $z(x, y) = c$, changes in these variables leave $z(x, y)$ unaffected. Accordingly, if $z(x, y)$ is approximated by a planar surface in the vicinity of (x, y),

$$
\begin{aligned}
z(x+\Delta x, y+\Delta y) &= \cancel{z(x,y)} + \left(z(\cancel{x+\Delta x},y) - \cancel{z(x,y)}\right) + \left(z(x+\Delta x, y+\Delta y) - z(\cancel{x+\Delta x},y)\right) \\
&\approx z(x,y) + \left.\frac{\partial z(x,y)}{\partial x}\right|_y \Delta x + \left.\frac{\partial z(x+\Delta x,y)}{\partial y}\right|_{x+\Delta x} \Delta y + \cdots \\
&\approx z(x,y) + \left.\frac{\partial z(x,y)}{\partial y}\right|_x \Delta y + \left.\frac{\partial z(x,y)}{\partial x}\right|_y \Delta x + \frac{\partial^2 z(x,y)}{\partial x \partial y} \Delta x \Delta y + \cdots \\
&\approx z(x,y) + \left.\frac{\partial z(x,y)}{\partial y}\right|_x \Delta y + \left.\frac{\partial z(x,y)}{\partial x}\right|_y \Delta x + \cdots
\end{aligned}
\tag{7.1.5}
$$

where in the last line only first-order terms are retained. Since Δx and Δy are related by the requirement that $z(x+\Delta x, y+\Delta y) = z(x, y) = c$,

$$
z(x+\Delta x, y+\Delta y) - z(x,y) = \Delta z(x,y) = 0 = \left.\frac{\partial z}{\partial x}\right|_y \Delta x + \left.\frac{\partial z}{\partial y}\right|_x \Delta y \tag{7.1.6}
$$

Thus (note carefully the minus sign),

$$
\left.\frac{\partial x}{\partial y}\right|_z = -\frac{\left.\frac{\partial z}{\partial y}\right|_x}{\left.\frac{\partial z}{\partial x}\right|_y} \tag{7.1.7}
$$

Example

In the previous problem, $dz/dy|_x = 1$, $dz/dx|_y = 2x = 2$, yielding again $\partial x/\partial y|_z = -1/2$.

Finally, if a quantity $w(x, y)$ varies subject to the constraint $z(x, y) = c$, Δx and Δy, while still related by Equation (7.1.6), both result from a change in w. Accordingly,

$$
\Delta z = 0 = \left[\left.\frac{\partial z}{\partial x}\right|_y \left.\frac{\partial x}{\partial w}\right|_z + \left.\frac{\partial z}{\partial y}\right|_x \left.\frac{\partial y}{\partial w}\right|_z \right] \Delta w \tag{7.1.8}
$$

which implies, where Equation (7.1.7) is applied in the last step,

$$
\frac{\left.\frac{\partial y}{\partial w}\right|_z}{\left.\frac{\partial x}{\partial w}\right|_z} = -\frac{\left.\frac{\partial z}{\partial x}\right|_y}{\left.\frac{\partial z}{\partial y}\right|_x} = \left.\frac{\partial y}{\partial x}\right|_z \tag{7.1.9}
$$

Example

If $w = x + y$ while, as in the previous two examples, $z = y + x^2 = 3$, at $(1, 2)$,

$$\left.\frac{\partial y}{\partial w}\right|_z = \frac{1}{\left.\frac{\partial w}{\partial y}\right|_z} = \frac{1}{\left.\frac{\partial x}{\partial y}\right|_z + \left.\frac{\partial y}{\partial y}\right|_z} = \frac{1}{-\frac{1}{2}+1} = 2$$

$$\left.\frac{\partial x}{\partial w}\right|_z = \frac{1}{\left.\frac{\partial w}{\partial x}\right|_z} = \frac{1}{\left.\frac{\partial x}{\partial x}\right|_z + \left.\frac{\partial y}{\partial x}\right|_z} = \frac{1}{-2+1} = -1 \tag{7.1.10}$$

The ratio of these two quantities again equals $\partial y/\partial x|_z = -2$.

Finally, if a function $g(x, y)$ is subject to the constraint $z(x, y) = c$, since the constraint can be solved implicitly for either $x(y)$ or $y(x)$,

$$\left.\frac{\partial g}{\partial x}\right|_z = \left.\frac{\partial g(x, y(x))}{\partial x}\right|_z = \left.\frac{\partial g}{\partial x}\right|_y + \left.\frac{\partial g}{\partial y}\right|_x \left.\frac{\partial y}{\partial x}\right|_z$$

$$\left.\frac{\partial g}{\partial y}\right|_z = \left.\frac{\partial g(x(y), y)}{\partial y}\right|_z = \left.\frac{\partial g}{\partial x}\right|_y \left.\frac{\partial x}{\partial y}\right|_z + \left.\frac{\partial g}{\partial y}\right|_x \tag{7.1.11}$$

Often, in evaluating partial derivatives, incorrect variables are held constant.

Example

In transforming derivatives from Cartesian to polar coordinates for which $x = r \cos\theta$, $y = r\sin\theta$, e.g., θ is unchanged by a displacement in the radial direction. Therefore,

$$\frac{\partial}{\partial r} = \left.\frac{\partial x}{\partial r}\right|_\theta \frac{\partial}{\partial x} + \left.\frac{\partial y}{\partial r}\right|_\theta \frac{\partial}{\partial y} = \cos\theta\frac{\partial}{\partial x} + \sin\theta\frac{\partial}{\partial y}$$

$$\frac{\partial}{\partial \theta} = \left.\frac{\partial x}{\partial \theta}\right|_r \frac{\partial}{\partial x} + \left.\frac{\partial y}{\partial \theta}\right|_r \frac{\partial}{\partial y} = -r\sin\theta\frac{\partial}{\partial x} + r\cos\theta\frac{\partial}{\partial y} \tag{7.1.12}$$

This can be rewritten as

$$\begin{pmatrix} \partial/\partial r \\ (1/r)(\partial/\partial\theta) \end{pmatrix} = \begin{pmatrix} \cos\theta & \sin\theta \\ -\sin\theta & \cos\theta \end{pmatrix} \begin{pmatrix} \partial/\partial x \\ \partial/\partial y \end{pmatrix} \equiv \mathbf{R} \begin{pmatrix} \partial/\partial x \\ \partial/\partial y \end{pmatrix} \tag{7.1.13}$$

which implies

$$\begin{pmatrix} \partial/\partial x \\ \partial/\partial y \end{pmatrix} = \mathbf{R}^{-1} \begin{pmatrix} \partial/\partial r \\ (1/r)(\partial/\partial\theta) \end{pmatrix} = \begin{pmatrix} \cos\theta & -\sin\theta \\ \sin\theta & \cos\theta \end{pmatrix} \begin{pmatrix} \partial/\partial r \\ (1/r)(\partial/\partial\theta) \end{pmatrix} \tag{7.1.14}$$

Since $r = \sqrt{x^2+y^2}, \theta = \tan^{-1}(y/x)$, with $d/dx(\arctan x) = 1/(1+x^2)$, Equation (7.1.14) also results from

$$\frac{\partial}{\partial x} = \left.\frac{\partial r}{\partial x}\right|_y \frac{\partial}{\partial r} + \left.\frac{\partial \theta}{\partial x}\right|_y \frac{\partial}{\partial \theta} = \frac{x}{\sqrt{x^2+y^2}}\frac{\partial}{\partial r} + \frac{1}{1+\left(\frac{y}{x}\right)^2}\left(-\frac{y}{x^2}\right)\frac{\partial}{\partial \theta} \quad (7.1.15)$$

Here, the error of identifying $\partial r/\partial x|_y$ with $1/(\partial x/\partial r|_\theta) = 1/\cos\theta$ must be avoided. Then,

$$\begin{aligned}\frac{\partial^2}{\partial x^2} &= \left(\cos\theta\frac{\partial}{\partial r} - \frac{\sin\theta}{r}\frac{\partial}{\partial \theta}\right)\left(\cos\theta\frac{\partial}{\partial r} - \frac{\sin\theta}{r}\frac{\partial}{\partial \theta}\right)\\ &= \cos^2\theta\frac{\partial^2}{\partial r^2} - \cos\theta\sin\theta\frac{\partial}{\partial r}\left(\frac{1}{r}\frac{\partial}{\partial \theta}\right) - \frac{\sin\theta}{r}\frac{\partial}{\partial \theta}\left(\cos\theta\frac{\partial}{\partial r}\right)\\ &\quad + \frac{\sin\theta}{r^2}\frac{\partial}{\partial \theta}\left(\sin\theta\frac{\partial}{\partial \theta}\right)\end{aligned} \quad (7.1.16)$$

Combining this with a similar formula for $\partial^2/\partial y^2$ obtained by substituting $\sin\theta \to \cos\theta$ and $\cos\theta \to -\sin\theta$ in the equation above yields the result, which is derived more simply later,

$$\frac{\partial^2}{\partial x^2} + \frac{\partial^2}{\partial y^2} = \frac{1}{r}\frac{\partial}{\partial r}\left(r\frac{\partial}{\partial r}\right) + \frac{1}{r^2}\frac{\partial^2}{\partial \theta^2} \quad (7.1.17)$$

7.2 MULTIDIMENSIONAL TAYLOR SERIES AND EXTREMA

The Taylor series for functions of several variables is obtained from, as in a single dimension,

$$\begin{aligned}f(x+a,y+b) &= \lim_{N\to\infty}\left(1+\frac{a}{N}\frac{\partial}{\partial x}+\frac{b}{N}\frac{\partial}{\partial y}\right)^N f(x,y)\\ &= e^{a\frac{\partial}{\partial x}+b\frac{\partial}{\partial y}}f(x,y)\\ &\approx \left(1+\frac{a}{1!}\frac{\partial}{\partial x}+\frac{b}{1!}\frac{\partial}{\partial y}+\frac{a^2}{2!}\frac{\partial^2}{\partial x^2}+\frac{2ab}{2!}\frac{\partial^2}{\partial x \partial y}+\frac{b^2}{2!}\frac{\partial^2}{\partial y^2}+\cdots\right)f(x,y)\end{aligned}$$

(7.2.1)

The last line in the above equation can be rewritten as, where, e.g., $f_x \equiv \partial f/\partial x$,

$$f(x+a,y+b) = 1 + af_x + bf_y + \frac{f_{xx}}{2}\left(a+b\frac{f_{xy}}{f_{xx}}\right)^2 + \frac{b^2}{2}\left(f_{yy}-\frac{f_{xy}^2}{f_{xx}}\right) + \cdots \quad (7.2.2)$$

Since the sign of the fourth term after the equals sign coincides with that of f_{xx}, a minimum requires

$$\begin{gathered} f_x = f_y = 0 \\ f_{xx} > 0 \\ f_{yy} - \frac{f_{xy}^2}{f_{xx}} > 0 \Rightarrow f_{xx} f_{yy} - f_{xy}^2 > 0 \end{gathered} \tag{7.2.3}$$

For a maximum, the corresponding conditions are

$$\begin{gathered} f_x = f_y = 0 \\ f_{xx} < 0 \\ f_{yy} - \frac{f_{xy}^2}{f_{xx}} < 0 \Rightarrow f_{xx} f_{yy} - f_{xy}^2 > 0 \end{gathered} \tag{7.2.4}$$

7.3 MULTIPLE INTEGRATION

The integral of a function $f(x)$ from a to b can be recast as a sum or *double integral* over all infinitesimal surface elements $\Delta x \Delta y$ (or $dydx$) within the region bounded by $x = a$, $x = b$, $y = 0$, and $y = f(x)$. This sum can be performed either by first summing the surface elements along the y-direction for each Δx within the integration region and then combining these partial sums or by first summing the surface elements along the x for each Δy. The determination of the optimal order of integration is generally facilitated by drawing the region.

Example

The integral of $f(x) = x$ over the interval $[0, a]$ can be written as a *double integral*:

$$\frac{a^2}{2} = \int_0^a x dx = \int_0^a \left(\int_0^x dy \right) dx \equiv \int_0^a dx \int_0^x dy \tag{7.3.1}$$

If however the elements are first summed along x, since x varies between y and a for a given y,

$$\int_0^a dy \int_y^a dx \equiv \int_0^a \left(\int_y^a dx \right) dy = \int_0^a (a - y) dy = a^2 - \frac{a^2}{2} = \frac{a^2}{2} \tag{7.3.2}$$

Similarly, the area between a circle, $x^2 + (y - a)^2 = a^2$, of radius a centered at $(0, a)$ with $a < 0$ and the x-axis can be written either as

$$2 \int_0^a dx \int_0^{a - \sqrt{a^2 - x^2}} dy \tag{7.3.3}$$

or, equivalently, after reversing the order of integration, as evident from Figure 7.1,

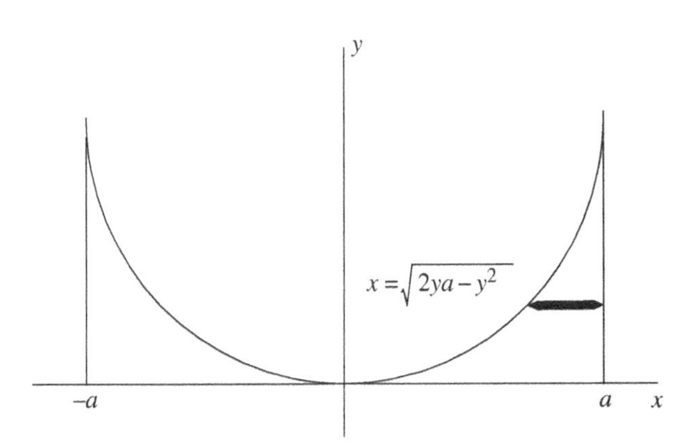

FIGURE 7.1 Integration limits.

$$2\int_0^a dy\int_{(2ya-y^2)^{1/2}}^a dx \tag{7.3.4}$$

The integration order of Equation (7.3.4) simplifies, rather fortuitously, the evaluation of, e.g., the integral of $f(x, y) = xy/(y-a)^2$ over the above region:

$$2\int_0^a dy\frac{y}{(y-a)^2}\underbrace{\left(\int_{(2ya-y^2)^{1/2}}^a xdx\right)}_{\frac{1}{2}x^2\big|_{(2ya-y^2)^{1/2}}^a=\frac{(a-y)^2}{2}}=\int_0^a ydy=\frac{a^2}{2} \tag{7.3.5}$$

If the integration limits in x do not depend on y and vice versa in a double integral over x and y, the integration factors into two separate integrals can be performed in either order.

Example

$$\int_{-a}^a xdx\int_{-b}^b ydy=\int_{-b}^b ydy\int_{-a}^a xdx=b^2a^2 \tag{7.3.6}$$

Accordingly, reversing the procedure,

$$\begin{aligned}\left(\int_{-\infty}^{\infty}e^{-x^2}dx\right)^2&=\int_{-\infty}^{\infty}e^{-x^2}dx\int_{-\infty}^{\infty}e^{-y^2}dy\\&=\int_{-\infty}^{\infty}\int_{-\infty}^{\infty}e^{-(x^2+y^2)}dxdy\\&=\int_0^{\infty}e^{-r^2}rdr\int_0^{2\pi}d\theta\\&=2\pi\int_0^{\infty}\frac{e^{-z}}{2}dz\\&=\pi\end{aligned} \tag{7.3.7}$$

where $z = r^2$. From this result,

$$\int_{-\infty}^{\infty} e^{-ax^2} dx = \frac{1}{\sqrt{a}} \int_{-\infty}^{\infty} e^{-w^2} dw = \sqrt{\frac{\pi}{a}}$$

$$\int_{-\infty}^{\infty} x^2 e^{-ax^2} dx = -\frac{d}{da} \int_{-\infty}^{\infty} e^{-ax^2} dx = -\frac{d}{da}\sqrt{\frac{\pi}{a}} = \frac{\sqrt{\pi}}{2} a^{-3/2} \tag{7.3.8}$$

$$\int_{-\infty}^{\infty} x^4 e^{-ax^2} dx = -\frac{d}{da} \int_{-\infty}^{\infty} x^2 e^{-ax^2} dx = \frac{3}{2} \times \frac{\sqrt{\pi}}{2} a^{-5/2}$$

The latter formulas can also be derived by recasting the integrals as gamma functions after substituting $y = x^2$. Finally, completing the square in the exponent enables the integration of

$$\int_{-\infty}^{\infty} e^{-ax^2} e^{-kx} dx = e^{\frac{k^2}{4a}} \int_{-\infty}^{\infty} e^{\overbrace{-a\left(x+\frac{k}{2a}\right)^2}^{x'}} dx = e^{\frac{k^2}{4a}} \int_{-\infty}^{\infty} e^{-ax'^2} dx' = \sqrt{\frac{\pi}{a}} e^{\frac{k^2}{4a}} \tag{7.3.9}$$

7.4 VOLUMES AND SURFACES OF REVOLUTION

The area of a surface is obtained by double integration. Since the vector $\vec{k} = (a, b, c)$ is perpendicular to each plane $\vec{k} \cdot \vec{r} + d = ax + by + cz + d = 0$ (c.f. Eq. 5.4.10), the plane tangent to the surface $z - f(x, y) = 0$ at (x_0, y_0) is from the Taylor series expansion of $f(x, y)$

$$-(x - x_0)\left.\frac{\partial f}{\partial x}\right|_{x_0, y_0} - (y - y_0)\left.\frac{\partial f}{\partial y}\right|_{x_0, y_0} + z - f(x_0, y_0) = 0 \tag{7.4.1}$$

with the unit normal vector

$$\hat{e}_n = \left(1 + \left(\frac{\partial f}{\partial x}\right)^2 + \left(\frac{\partial f}{\partial y}\right)^2\right)^{-\frac{1}{2}} \left(-\frac{\partial f}{\partial x}, -\frac{\partial f}{\partial y}, 1\right) \tag{7.4.2}$$

The angle between the normal and the z-axis is obtained from the dot product between the above vector and a unit vector in the z-direction:

$$\hat{e}_n \cdot \hat{e}_z = \cos\theta_{\hat{n}, \hat{e}_z} = \left(1 + \left(\frac{\partial f}{\partial x}\right)^2 + \left(\frac{\partial f}{\partial y}\right)^2\right)^{-\frac{1}{2}} \tag{7.4.3}$$

The area of a surface element above the a region $dxdy$ is augmented by a factor of $1/\cos\theta_{n,z}$ relative to the area of its projection on the $x-y$ plane and therefore equals

$$dS = \frac{1}{\cos\theta_{n,z}} dxdy = \sqrt{1 + \left(\frac{\partial f}{\partial x}\right)^2 + \left(\frac{\partial f}{\partial y}\right)^2} dxdy \tag{7.4.4}$$

yielding for the surface integral over a region R in x and y

$$S = \iint_R \sqrt{1 + \left(\frac{\partial f}{\partial x}\right)^2 + \left(\frac{\partial f}{\partial y}\right)^2} dxdy \tag{7.4.5}$$

The volume under the surface is the sum over all infinitesimal volume elements within the region

$$V = \iiint_R dzdxdy = \iint_R zdxdy \tag{7.4.6}$$

The volume of a *solid of revolution* situated between z_1 and z_2 with radius $r = f(z)$ that results when the curve $f(z)$ is rotated around the z-axis is calculated by noting that the volume of the cylindrical shell between z and $z + \Delta z$ equals $\pi r^2 dz$ and therefore

$$V = \pi \int_{z_1}^{z_2} (f(z))^2 dz \tag{7.4.7}$$

Similarly, the surface area of the shell is $2\pi f(z)\sqrt{dz^2 + dr^2}$, yielding for the area of the entire solid

$$S = 2\pi \int_{z_1}^{z_2} f(z)\sqrt{1 + \left(\frac{df(z)}{dz}\right)^2} dz \tag{7.4.8}$$

7.5 CHANGE OF VARIABLES AND JACOBIANS

The manner in which the differential volume element dx^N in N dimensions transforms under variable transformations can be extrapolated from the $N = 2$ case. A transformation from the variables (x, y) to $(f(x, y), g(x, y))$ maps the square region determined by the two sides $[(x_0, y_0), (x_0 + dx_0, y_0)]$ and $[(x_0, y_0), (x_0, y_0 + dy_0)]$ to a parallelogram with corresponding sides $[(f(x_0, y_0), g(x_0, y_0))(f(x_0 + dx, y_0), g(x_0 + dx, y_0))]$ and $[(f(x_0, y_0), g(x_0, y_0))(f(x_0 y_0 + dy), g(x_0 y_0 + dy))]$. The latter two sides are to lowest order given by the vectors in the $x-y$ plane from $(f(x_0, y_0), g(x_0, y_0))$

to $\left(f(x_0,y_0)+\partial f/\partial x|_{x=x_0}dx, g(x_0,y_0)+\partial g/\partial x|_{x=x_0}dx\right)$ and to $\left(f(x_0,y_0)+\partial f/\partial y|_{y=y_0}dy,\right.$ $\left.g(x_0,y_0)+\partial g/\partial y|_{y=y_0}dy\right)$, respectively. The area of the parallelogram equals, according to Equation (5.5.4), the magnitude of the three-dimensional cross product of the two vectors

$$\hat{e}_z dfdg=\begin{vmatrix}\hat{e}_x & \hat{e}_y & \hat{e}_z\\ \partial f/\partial x & \partial f/\partial y & 0\\ \partial g/\partial x & \partial g/\partial y & 0\end{vmatrix}dxdy=\hat{e}_z\left(\frac{\partial f}{\partial x}\frac{\partial g}{\partial y}-\frac{\partial f}{\partial y}\frac{\partial g}{\partial x}\right)dxdy\equiv\hat{e}_z\frac{\partial(f,g)}{\partial(x,y)}dxdy \tag{7.5.1}$$

The last expression introduces the standard notation for the determinant, termed the *Jacobian*. In N dimensions, the analogous $N\times N$ determinant replaces the two-dimensional determinant.

If u and v are functions of x and y, while f and g are functions of u and v, then the Jacobian for f and g in terms of x and y is, by the chain rule, Equation (7.1.3),

$$\begin{pmatrix}\frac{\partial f}{\partial x} & \frac{\partial f}{\partial y}\\ \frac{\partial g}{\partial x} & \frac{\partial g}{\partial y}\end{pmatrix}=\begin{pmatrix}\frac{\partial f}{\partial u}\frac{\partial u}{\partial x}+\frac{\partial f}{\partial v}\frac{\partial v}{\partial x} & \frac{\partial f}{\partial u}\frac{\partial u}{\partial y}+\frac{\partial f}{\partial v}\frac{\partial v}{\partial y}\\ \frac{\partial g}{\partial u}\frac{\partial u}{\partial x}+\frac{\partial g}{\partial v}\frac{\partial v}{\partial x} & \frac{\partial g}{\partial u}\frac{\partial u}{\partial y}+\frac{\partial g}{\partial v}\frac{\partial v}{\partial y}\end{pmatrix}=\begin{pmatrix}\frac{\partial f}{\partial u} & \frac{\partial f}{\partial v}\\ \frac{\partial g}{\partial u} & \frac{\partial g}{\partial v}\end{pmatrix}\begin{pmatrix}\frac{\partial u}{\partial x} & \frac{\partial u}{\partial y}\\ \frac{\partial v}{\partial x} & \frac{\partial v}{\partial y}\end{pmatrix} \tag{7.5.2}$$

Accordingly, from $\det(\mathbf{AB})=\det(\mathbf{A})\det(\mathbf{B})$,

$$\frac{\partial(f,g)}{\partial(x,y)}=\frac{\partial(f,g)}{\partial(u,v)}\frac{\partial(u,v)}{\partial(x,y)} \tag{7.5.3}$$

which further implies

$$\frac{\partial(f,g)}{\partial(u,v)}\frac{\partial(u,v)}{\partial(f,g)}=1 \tag{7.5.4}$$

so that

$$dxdy=\frac{\partial(x,y)}{\partial(f,g)}dfdg=\frac{1}{\frac{\partial(f,g)}{\partial(x,y)}}dfdg \tag{7.5.5}$$

8

CALCULUS OF VECTOR FUNCTIONS

In the same manner as the temperature within a region forms a scalar field, in a vector field such as a velocity or force field, a directed quantity exists throughout a spatial region. Limiting operations on vector fields are described by vector calculus, introduced in this chapter.

8.1 GENERALIZED COORDINATES

In a general N dimensional coordinate system, each point in space is parameterized by a unique set of coordinate values q^j, $j = 1, 2, \ldots, N$ with local coordinate axes directed along unit vectors $\hat{e}_i(q^j)$. An infinitesimal change in a coordinate q^j in an *orthogonal coordinate system* yields a displacement only along the $\hat{e}_j(q^j)$ direction, while in a nonorthogonal coordinate system in which some $\hat{e}_i$ are not perpendicular, the displacement additionally can alter components along additional coordinate directions.

Fundamental Math and Physics for Scientists and Engineers, First Edition.
David Yevick and Hannah Yevick.

Example

In the nonorthogonal two-dimensional coordinate system with basis vectors $\hat{e}_1 = \hat{e}_x$ and $\hat{e}_2 = \left(\hat{e}_x + \hat{e}_y\right)/\sqrt{2}$ oriented at a 45° relative angle, changing $q^1 = x$ by unity yields a displacement of unity along the $\hat{e}_1$ direction and $1/\sqrt{2}$ along the $\hat{e}_2$ direction.

Three important orthogonal coordinate systems are *Cartesian* (x, y, z) coordinates, *cylindrical coordinates*

$$\begin{aligned} x &= r\cos\theta \\ y &= r\sin\theta \\ z &= z \end{aligned} \tag{8.1.1}$$

and spherical polar coordinates

$$\begin{aligned} x &= r\sin\theta\cos\phi \\ y &= r\sin\theta\sin\phi \\ z &= r\cos\theta \end{aligned} \tag{8.1.2}$$

The volume element in these coordinates is obtained from the Jacobian of the transformation from Cartesian coordinates.

Example

For spherical coordinates, the volume element at (r, θ, ϕ) equals $Jdrd\theta d\phi$ with, where the last line corresponds to a cofactor expansion about the last row,

$$J = \begin{vmatrix} \frac{dx}{dr} & \frac{dx}{d\theta} & \frac{dx}{d\phi} \\ \frac{dy}{dr} & \frac{dy}{d\theta} & \frac{dy}{d\phi} \\ \frac{dz}{dr} & \frac{dz}{d\theta} & \frac{dz}{d\phi} \end{vmatrix} = \begin{vmatrix} \sin\theta\cos\phi & r\cos\theta\cos\phi & -r\sin\theta\sin\phi \\ \sin\theta\sin\phi & r\cos\theta\sin\phi & r\sin\theta\cos\phi \\ \cos\theta & -r\sin\theta & 0 \end{vmatrix}$$
$$= \cos\theta\left(r^2\sin\theta\cos\theta\left(\cos^2\phi + \sin^2\phi\right)\right) + r\sin\theta\left(r\sin^2\theta\left(\cos^2\phi + \sin^2\phi\right)\right) = r^2\sin\theta \tag{8.1.3}$$

The above result can also be simply obtained by drawing the volume element. The *spherical angle* in *steradians* is defined by $\Delta s_2 \Delta s_3 / r^2 = \sin\theta\, \Delta\theta\, \Delta\phi$ so that a unit sphere subtends 4π steradians.

Example

Denoting the differential solid angle element in n dimensions by $d\Omega_n$, the volume, V_n, of an n-dimensional hypersphere can be obtained from, where $\Gamma(x)$ represents the gamma function,

$$
\begin{aligned}
(\sqrt{\pi})^n &= \int_{-\infty}^{\infty} \dots \int_{-\infty}^{\infty} e^{-\left(x_1^2 + x_2^3 + \dots x_n^2\right)} dx_1 dx_2 \dots dx_n \\
&= \int r^{n-1} e^{-r^2} d\Omega_n dr \\
&= \int d\Omega_n \int \left(r^2\right)^{\frac{n}{2}-1} e^{-r^2} \frac{dr^2}{2} \\
&= \Omega_n \frac{1}{2} \Gamma\left(\frac{n}{2}\right)
\end{aligned} \tag{8.1.4}
$$

implying that

$$
V_n = \int_0^R \Omega_n r^{n-1} dr = \frac{R^n}{n} \left(\frac{2\pi^{n/2}}{\Gamma(n/2)} \right) \tag{8.1.5}
$$

For integer n, $\Gamma(n/2) = (n/2 - 1)\Gamma$ is applied recursively together with

$$
\Gamma\left(\frac{1}{2}\right) = \int_0^{\infty} e^{-x} \frac{1}{\sqrt{x}} dx = \int_0^{\infty} e^{-x} 2d\sqrt{x} = 2\int_0^{\infty} e^{-y^2} dy = \int_{-\infty}^{\infty} e^{-y^2} dy = \sqrt{\pi} \tag{8.1.6}
$$

Representing the change in *length* associated with a change dq^i in the ith coordinate by ds^i and recalling that for a circle of radius r, the arclength subtended by an central angle $\Delta\theta$ equals $r\Delta\theta$,

$$
\begin{array}{llll}
\text{Cartesian:} & ds^1 = dx & ds^2 = dy & ds^3 = dz \\
\text{Cylindrical:} & ds^1 = dr & ds^2 = r d\phi & ds^3 = dz \\
\text{Spherical:} & ds^1 = dr & ds^2 = r d\theta & ds^3 = r \sin\theta d\phi
\end{array} \tag{8.1.7}
$$

The *scale factors* h_i and associated unit vectors are, respectively, defined by $ds^i = h_i dq^i$ and

$$
\hat{e}_i = \frac{\partial \vec{r}}{\partial s^i} = \frac{1}{h_i} \frac{\partial \vec{r}}{\partial q^i} \tag{8.1.8}
$$

from which unit vectors in one coordinate system can be expressed in other coordinate systems.

Example

Example $\hat{e}_\phi$ in a spherical coordinate system is represented in a rectangular coordinate system as

$$\hat{e}_\phi = \frac{1}{r\sin\theta}\left(\frac{\partial x}{\partial\phi}, \frac{\partial y}{\partial\phi}, \frac{\partial z}{\partial\phi}\right)$$

$$= \frac{1}{r\sin\theta}\left(\frac{\partial(r\sin\theta\cos\phi)}{\partial\phi}, \frac{\partial(r\sin\theta\sin\phi)}{\partial\phi}, \frac{\partial(r\cos\theta)}{\partial\phi}\right) \quad (8.1.9)$$

$$= (-\sin\phi, \cos\phi, 0)$$

In an arbitrary coordinate system, the *metric tensor*, g_{ij}, is defined by writing the squared length element $(ds)^2$ as, where x^i denotes Cartesian coordinates,

$$(ds)^2 = \sum_{i=1}^{n}(dx^i)^2 = \sum_{i,j=1}^{n}\left(\frac{\partial x^i}{\partial q^j}dq^j\right)^2 = \sum_{i,j,l=1}^{n}\frac{\partial x^i}{\partial q^j}\frac{\partial x^i}{\partial q^l}dq^j dq^l \equiv \sum_{j,l=1}^{n} g_{jl}dq^j dq^l \quad (8.1.10)$$

In an orthogonal coordinate system, g_{ij} is diagonal since

$$(ds)^2 = \sum_{i=1}^{n}(h_i dq^i)^2 \equiv \sum_{i,j=1}^{n} g_{ij}dq^i dq^j \quad (8.1.11)$$

Example

For two-dimensional polar coordinates, $x = r\cos\theta$, $y = r\sin\theta$, $ds_1 = dr$, $ds_2 = rd\theta$, while

$$dx = \frac{\partial x}{\partial q^1}dq^1 + \frac{\partial x}{\partial q^2}dq^2 = \cos\theta dr - r\sin\theta d\theta$$

$$dy = \frac{\partial y}{\partial q^1}dq^1 + \frac{\partial y}{\partial q^2}dq^2 = \sin\theta dr + r\cos\theta d\theta \quad (8.1.12)$$

from which $(ds)^2 = (dx)^2 + (dy)^2 = (dr)^2 + r^2(d\theta)^2$ as expected.

While the components of a differential vector transform among coordinate systems according to

$$dx^i = \frac{\partial x^i}{\partial q^j}dq^j \quad (8.1.13)$$

the analogous transformation law for derivatives is

$$\frac{\partial \xi}{\partial x^i} = \frac{\partial q^j}{\partial x^i}\frac{\partial \xi}{\partial q^j} \quad (8.1.14)$$

Quantities such as displacement or velocity that transform as a standard vector according to Equation (8.1.13) are termed *contravariant* and are distinguished when required with upper indices, while quantities that follow the derivative transformation law are labeled *covariant vectors* and possess lower indices.

The inner product or *contraction* of a contravariant and a covariant vector transforms as a scalar

$$a_j b^j = \frac{\partial x^j}{\partial q^m} a'_m \frac{\partial q^p}{\partial x^j} b'^p = \frac{\partial x^j}{\partial q^m}\frac{\partial q^p}{\partial x^j} a'_m b'^p = \delta^m{}_p a'_m b'^p \tag{8.1.15}$$

Hence, multiplication of a contravariant vector by g_{ij} yields a covariant vector since in

$$g_{ij} a^j = \sum_{m,j=1}^{n} \frac{\partial x^m}{\partial q^i}\frac{\partial x^m}{\partial q^j} a^j \tag{8.1.16}$$

the free index i is associated with a covariant derivative. The contravariant metric tensor

$$g^{ij} = \sum_{m=1}^{n} \frac{\partial q^i}{\partial x^m}\frac{\partial q^j}{\partial x^m} \tag{8.1.17}$$

instead transforms a covariant into a contravariant vector. Note that

$$g^{ip} g_{pj} = \sum_{m,l,p=1}^{n} \left(\frac{\partial x^m}{\partial q^i}\frac{\partial x^m}{\partial q^p}\right)\left(\frac{\partial q^p}{\partial x^l}\frac{\partial q^j}{\partial x^l}\right) = \sum_{m,l=1}^{n} \frac{\partial x^m}{\partial q^i}\delta_l^m \frac{\partial q^j}{\partial x^l} = \sum_{m=1}^{n} \frac{\partial x^m}{\partial q^i}\frac{\partial q^j}{\partial x^m} = \frac{\partial q^j}{\partial q^i} = \delta_i^j \tag{8.1.18}$$

so that $g^{ij} = (g_{ij})^{-1}$ while the *mixed* metric tensor equals the Kronecker delta function

$$g_i^j = \frac{\partial q^m}{\partial x^i}\frac{\partial x^j}{\partial q^m} = \delta_i^j \tag{8.1.19}$$

Example

In spherical polar coordinates, the elements of g_{ij} and g^{ij} are given by, in matrix form,

$$g_{ij} = \begin{pmatrix} 1 & 0 & 0 \\ 0 & r^2 & 0 \\ 0 & 0 & r^2\sin^2\theta \end{pmatrix}, \quad g^{ij} = \begin{pmatrix} 1 & 0 & 0 \\ 0 & \frac{1}{r^2} & 0 \\ 0 & 0 & \frac{1}{r^2\sin^2\theta} \end{pmatrix} \tag{8.1.20}$$

8.2 VECTOR DIFFERENTIAL OPERATORS

The basic operators of vector calculus consist of the gradient, which maps a scalar function into a vector field; the divergence, which converts a vector field into a scalar function; the curl, which transforms a vector field into a second (pseudo)vector field; and the Laplacian, which maps a scalar function into a second scalar function.

Gradient: The gradient, $\vec{\nabla} f$, quantifies the magnitude and direction of the slope of a multidimensional function, $f(\vec{r})$, in analogy to a ball placed on a curved surface for which a vector can be associated at each point with the direction opposite that in which the ball rolls with a magnitude given by the slope along this direction. In a rectangular coordinate system, from Equation (7.2.1), after neglecting second-order terms such as $\Delta y \Delta x \partial^2 f / \partial y \partial x$,

$$
\begin{aligned}
f(\vec{r}+\Delta\vec{r}) &= f(x+\Delta x, y+\Delta y, z+\Delta z) \\
&\approx f(x,y,z) + \Delta\vec{r}\cdot\left(\frac{\partial f}{\partial x}, \frac{\partial f}{\partial y}, \frac{\partial f}{\partial z}\right) \\
&\equiv f(\vec{r}) + \Delta\vec{r}\cdot\vec{\nabla} f = f(\vec{r}) + \left|\Delta\vec{r}\right|\left|\vec{\nabla} f\right|\cos\theta_{\Delta\vec{r},\vec{\nabla} f}
\end{aligned}
\tag{8.2.1}
$$

For a fixed $|\Delta\vec{r}|$, the maximum positive change in f occurs when $\theta_{\Delta\vec{r},\vec{\nabla} f} = 0$ so that $\Delta\vec{r}$ and $\vec{\nabla} f$ are parallel. Thus, the *gradient*, $\vec{\nabla} f$, expresses both the magnitude and the direction of the maximum slope of f at each point. The slope of f in the direction of a vector $\vec{n}$ then equals the *directional derivative*:

$$
\frac{df}{dn} \equiv \hat{e}_{\vec{n}}\cdot\vec{\nabla} f
\tag{8.2.2}
$$

In a general orthogonal coordinate system, if $\hat{e}_{\vec{n}}$ is oriented along the direction of the ith unit vector, the gradient must coincide with $\partial f/\partial s_i$, so that

$$
\vec{\nabla} f = \sum_{i=1}^{n} \hat{e}_i \frac{\partial f}{\partial s_i} = \sum_{i=1}^{n} \hat{e}_i \frac{1}{h_i}\frac{\partial f}{\partial q_i}
\tag{8.2.3}
$$

which for spherical coordinates evaluates to

$$
\nabla f = \hat{e}_r \frac{\partial}{\partial r} + \hat{e}_\theta \frac{1}{r}\frac{\partial}{\partial \theta} + \hat{e}_\phi \frac{1}{r\sin\theta}\frac{\partial}{\partial \phi}
\tag{8.2.4}
$$

and for cylindrical coordinates

$$
\nabla f = \hat{e}_r \frac{\partial}{\partial r} + \hat{e}_\phi \frac{1}{r}\frac{\partial}{\partial \phi} + \hat{e}_z \frac{\partial}{\partial z}
\tag{8.2.5}
$$

Divergence: If a vector field is viewed as an abstract flow of the field quantity, a scalar function can be associated with the normalized total flow out of infinitesimal volume elements centered at each point. That is, if $\hat{e}_n$ and $d\vec{S} = \hat{e}_n dS$ denote the *outward normal* to a surface and the infinitesimal *surface element* directed in the direction of $\hat{e}_n$ with a magnitude equal to the element area (e.g., $d\vec{S} = dxdy\hat{e}_z$ for an element in the x–y plane), the *divergence*

$$\vec{\nabla}\cdot\vec{A} \equiv \lim_{\Delta V\to 0}\frac{1}{\Delta V}\oint_S \vec{A}\cdot d\vec{S} = \lim_{\Delta V\to 0}\frac{1}{\Delta V}\oint_S \vec{A}\cdot\hat{e}_n dS \tag{8.2.6}$$

yields the net flow of $\vec{A}$ per unit volume out of the infinitesimal volume element ΔV enclosed by the surface S and therefore quantifies the strength of local sources of $\vec{A}$ inside ΔV. If the boundaries of ΔV coincide with unit vector directions, Equation (8.2.6) becomes

$$\begin{aligned}\vec{\nabla}\cdot\vec{A} &= \lim_{V\to 0}\frac{1}{h_1h_2h_3\Delta q_1\Delta q_2\Delta q_3}\left[\begin{matrix}\Delta q_2\Delta q_3\left\{A_1h_2h_3\Big|_{(q_2+\Delta q_1,q_2,q_3)} - A_1h_2h_3\Big|_{(q_{21},q_2,q_e)}\right\}\\ +\text{cyclic permutations}\end{matrix}\right]\\ &= \frac{1}{h_1h_2h_3}\left[\frac{\partial}{\partial q_1}(A_1h_2h_3) + \text{cyclic permutations}\right]\end{aligned} \tag{8.2.7}$$

In the Cartesian case, $\vec{\nabla}\cdot\vec{A} = \Sigma_{i=1}^{3}\partial A_i/\partial x_i$, while for spherical coordinates,

$$\vec{\nabla}\cdot\vec{A} = \frac{1}{r^2}\frac{\partial}{\partial r}(r^2A_r) + \frac{1}{r\sin\theta}\frac{\partial}{\partial\theta}(\sin\theta A_\theta) + \frac{1}{r\sin\theta}\frac{\partial}{\partial\phi}A_\phi \tag{8.2.8}$$

and for cylindrical coordinates,

$$\vec{\nabla}\cdot\vec{A} = \frac{1}{r}\frac{\partial}{\partial r}(rA_r) + \frac{1}{r}\frac{\partial}{\partial\phi}A_\phi + \frac{\partial}{\partial z}A_z \tag{8.2.9}$$

Laplacian: The *Laplacian* quantifies the curvature of a scalar function in multiple dimensions. In analogy to the second derivative, which is proportional to twice the amount by which the average value of the function evaluated at two points displaced at an infinitesimal amount from a given point exceeds the value of the function at the point itself, in, e.g., three dimensions,

$$\nabla^2 f = \lim_{\Delta x \to 0} \frac{\left\{ \begin{array}{l} f(x+\Delta x, y, z) + f(x, y+\Delta y, z) + f(x, y, z+\Delta z) \\ \quad + f(x-\Delta x, y, z) + f(x, y-\Delta y, z) + f(x, y, z-\Delta z) - 6f(x) \end{array} \right\}}{(\Delta x)^2}$$
$$= \frac{\partial^2 f}{\partial x^2} + \frac{\partial^2 f}{\partial y^2} + \frac{\partial^2 f}{\partial z^2} \qquad (8.2.10)$$

In arbitrary orthogonal coordinates, the Laplacian first transforms f through the gradient operator into a vector field with a flow proportional to its slope and then applies the divergence to generate a scalar field proportional to the outward flow of $\vec{\nabla} f$. Thus if, e.g., $\vec{r}_0$ is a local minimum of f, the gradient $\vec{\nabla} f$ will be directed away from $\vec{r}_0$ at any point in its vicinity and the outward flow will be positive. Combining Equations (8.2.3) and (8.2.7),

$$\nabla^2 f = \vec{\nabla} \cdot \vec{\nabla} f = \frac{1}{h_1 h_2 h_3} \left(\frac{\partial}{\partial q_1} \left(\frac{h_2 h_3}{h_1} \frac{\partial f}{\partial q_1} \right) + \text{cyclic permutations} \right) \qquad (8.2.11)$$

which yields in spherical coordinates

$$\nabla^2 f = \frac{1}{r^2 \sin\theta} \left[\frac{\partial}{\partial r} \left(r^2 \sin\theta \frac{\partial f}{\partial r} \right) + \frac{\partial}{\partial \theta} \left(\sin\theta \frac{\partial f}{\partial \theta} \right) + \frac{\partial}{\partial \phi} \left(\frac{1}{\sin\theta} \frac{\partial f}{\partial \phi} \right) \right]$$
$$= \frac{1}{r^2} \frac{\partial}{\partial r} \left(r^2 \frac{\partial f}{\partial r} \right) + \frac{1}{r^2 \sin\theta} \frac{\partial}{\partial \theta} \left(\sin\theta \frac{\partial f}{\partial \theta} \right) + \frac{1}{r^2 \sin^2\theta} \frac{\partial^2 f}{\partial \phi^2} \qquad (8.2.12)$$

The first term in the above expression is also often written as

$$\frac{1}{r^2} \frac{\partial}{\partial r} \left(r^2 \frac{\partial f}{\partial r} \right) = \frac{1}{r^2} \frac{\partial}{\partial r} \left(r \frac{\partial (rf)}{\partial r} - rf \right)$$
$$= \frac{1}{r^2} \left(\left[r \frac{\partial^2 (rf)}{\partial r^2} + \frac{\partial (rf)}{\partial r} \right] - \frac{\partial (rf)}{\partial r} \right) \qquad (8.2.13)$$
$$= \frac{1}{r} \frac{\partial^2 (rf)}{\partial r^2}$$

In cylindrical coordinates,

$$\nabla^2 f = \frac{1}{r} \frac{\partial}{\partial r} \left(r \frac{\partial f}{\partial r} \right) + \frac{1}{r^2} \frac{\partial^2 f}{\partial \phi^2} + \frac{\partial^2 f}{\partial z^2} \qquad (8.2.14)$$

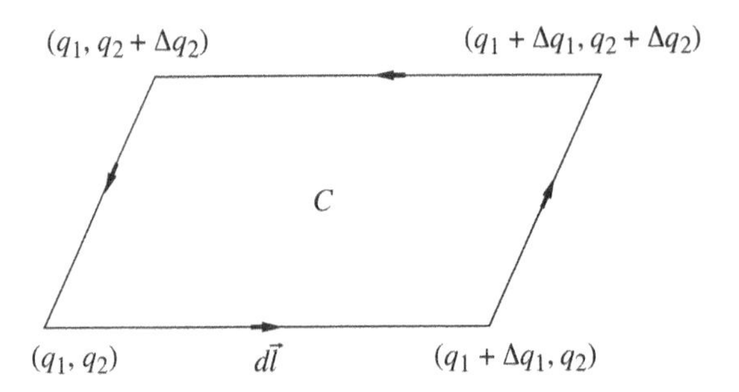

FIGURE 8.1 Computation of curl.

Curl: Finally, the curl evaluates a quantity analogous to the circulation around each point in a fluid. By symmetry, a vector can only be meaningfully assigned to a direction perpendicular to the plane of circulation. As a positive sign is typically associated with the counterclockwise direction, this direction is determined by the *right-hand rule*: wrapping one's right-hand fingers around the circulation in given plane yields the positive direction of the corresponding curl component. A counterclockwise flow in the x–y plane accordingly generates a curl in the $\hat{e}_z$ direction. In an orthogonal coordinate system, the circulation per unit area along a closed curve C equals the integral of the field component along the curve normalized by its area S.

$$\vec{\nabla}\times\vec{A}\equiv\lim_{S\to 0}\frac{1}{S}\oint_{C\subset S}\vec{A}\cdot d\vec{l} \tag{8.2.15}$$

With C taken along the coordinate axes (cf. Fig. 8.1),
the z-component of the curl evaluates to, with $\Delta s_i = h_i\Delta q_i$,

$$\begin{aligned}\left(\vec{\nabla}\times\vec{A}\right)_3 &= \frac{1}{h_1h_2\Delta q_1\Delta q_2}\left\{\begin{matrix} -A_1\Delta s_1\Big|_{\left(q_1+\frac{\Delta q_1}{2},\, q_2+\Delta q_2\right)} + A_1\Delta s_1\Big|_{\left(q_1+\frac{\Delta q_1}{2},\, q_2\right)} \\ -A_2\Delta s_2\Big|_{\left(q_1,\, q_2+\frac{\Delta q_2}{2}\right)} + A_2\Delta s_2\Big|_{\left(q_1+\Delta q_1,\, q_2+\frac{\Delta q_2}{2}\right)}\end{matrix}\right\} \\ &= \frac{1}{h_1h_2\Delta q_1\Delta q_2}\left\{-\frac{\partial}{\partial q_2}(A_1h_1)+\frac{\partial}{\partial q_1}(A_2h_2)\right\}\Delta q_1\Delta q_2 \\ &= \frac{1}{h_1h_2}\left\{\frac{\partial}{\partial q_1}(A_2h_2)-\frac{\partial}{\partial q_2}(A_1h_1)\right\}\end{aligned} \tag{8.2.16}$$

Since the indices 1, 2, and 3 can be cyclically interchanged $(1\to 2\to 3\to 1\to\cdots)$,

$$\vec{\nabla} \times \vec{A} = \frac{1}{h_1 h_2 h_3} \begin{vmatrix} h_1 \hat{e}_1 & h_2 \hat{e}_2 & h_3 \hat{e}_3 \\ \partial/\partial q_1 & \partial/\partial q_2 & \partial/\partial q_3 \\ h_1 A_1 & h_2 A_2 & h_3 A_3 \end{vmatrix} \tag{8.2.17}$$

This yields in spherical coordinates

$$\vec{\nabla} \times \vec{A} = \frac{1}{r\sin\theta}\left(\frac{\partial}{\partial\theta}(\sin\theta A_\phi) - \frac{\partial A_\theta}{\partial\phi}\right)\hat{e}_r + \frac{1}{r}\left(\frac{1}{\sin\theta}\frac{\partial A_r}{\partial\phi} - \frac{\partial r A_\phi}{\partial r}\right)\hat{e}_\theta + \frac{1}{r}\left(\frac{\partial r A_\theta}{\partial r} - \frac{\partial A_r}{\partial\theta}\right)\hat{e}_\phi \tag{8.2.18}$$

and in cylindrical coordinates

$$\vec{\nabla} \times \vec{A} = \left(\frac{1}{r}\frac{\partial A_z}{\partial\phi} - \frac{\partial A_\phi}{\partial z}\right)\hat{e}_r + \left(\frac{\partial A_r}{\partial z} - \frac{\partial A_z}{\partial r}\right)\hat{e}_\phi + \left(\frac{\partial r A_\phi}{\partial r} - \frac{1}{r}\frac{\partial A_r}{\partial\phi}\right)\hat{e}_z \tag{8.2.19}$$

8.3 VECTOR DIFFERENTIAL IDENTITIES

Numerous relationships exist between vector differential operators. Complex expressions, especially those involving tensors, are sometimes most reliably derived by writing out the individual terms in Cartesian coordinates. Alternatively, Einstein notation can be employed together with the properties of the Levi-Civita tensor ε_{ijk}. The most compact procedure, however, expresses each derivative operator as the sum of several individual operators that act on only a single term to its right and assigns a corresponding subscript to the individual operators. The expressions are then simplified while regarding each operator as an algebraic rather than a differential quantity until each derivative acts exclusively on the term indicated by its subscript, e.g.,

$$\vec{\nabla} \times \left(f\vec{A}\right) = \vec{\nabla}_{\vec{A}} \times \left(f\vec{A}\right) + \vec{\nabla}_f \times \left(f\vec{A}\right) = f\vec{\nabla}_{\vec{A}} \times \vec{A} - \vec{A} \times \vec{\nabla}_f f = f\vec{\nabla} \times \vec{A} - \vec{A} \times \vec{\nabla} f \tag{8.3.1}$$

A few common identities are, where all quantities f, g, A, B are functions of position,

$$\vec{\nabla} \times \left(\vec{\nabla} \times \vec{A}\right) = \vec{\nabla}\left(\vec{\nabla} \cdot \vec{A}\right) - \nabla^2 \vec{A} \tag{8.3.2}$$

$$\vec{\nabla} \times \left(\vec{A} \times \vec{B}\right) = \vec{A}\left(\vec{\nabla} \cdot \vec{B}\right) - \left(\vec{A} \cdot \vec{\nabla}\right)B - \vec{B}\left(\vec{\nabla} \cdot \vec{A}\right) + \left(B \cdot \vec{\nabla}\right)\vec{A} \tag{8.3.3}$$

$$\vec{\nabla}(fg) = f\vec{\nabla} g + g\vec{\nabla} f \tag{8.3.4}$$

8.4 GAUSS'S AND STOKES' LAWS AND GREEN'S IDENTITIES

Since the divergence and curl are associated with the flux out of an infinitesimal volume and the circulation around an infinitesimal closed curve, respectively, integrating these quantities over finite regions yields the total flux out of or the net circulation along the region boundary. That is, the integral of $\vec{\nabla} \cdot \vec{A}$ over a finite connected volume V can be expressed as a sum of integrals of the form of Equation (8.2.6) over all enclosed infinitesimal subvolumes. However, in this sum, $\vec{A} \cdot d\vec{S}$ cancels along the internal boundary between each pair of adjacent subvolumes dV_1 and dV_2 as $d\vec{S}$ around dV_1 is oppositely oriented to $d\vec{S}$ of dV_2. The resulting integral is therefore only nonzero along the boundary surface S of V (denoted $S \subset V$), resulting in *Gauss's law*

$$\int_V \vec{\nabla} \cdot \vec{A} dV = \oint_{S \subset V} A \cdot d\vec{S} \tag{8.4.1}$$

Similarly, the integral of a curl over a finite surface can be expressed as a sum over contributions from infinitesimal adjacent surface elements. If the curl evaluated over each element is recast as a line integral along its boundary according to Equation (8.2.15), the contributions of $\vec{A} \cdot d\vec{l}$ to the overall sum from each internal line segment shared by two surface elements $d\vec{S}_1$ and $d\vec{S}_2$ cancel since the $d\vec{l}$ of $d\vec{S}_1$ is directed in the opposite direction to the coincident $d\vec{l}$ of $d\vec{S}_2$. Thus, only the line integral over the boundary remains, leading to *Stokes' law*

$$\int_S \left(\vec{\nabla} \times \vec{A}\right) \cdot d\vec{S} = \oint_{C \subset S} \vec{A} \cdot d\vec{l} \tag{8.4.2}$$

where C is the curve bounding the surface S. The line integral is again performed according to the right-hand rule in that wrapping one's right-hand fingers around C in the direction associated with positive $d\vec{l}$ positions the thumb along the direction corresponding to positive surface elements $d\vec{S}$.

Various integral relations follow from Stokes' and Gauss's laws. For example, if $\vec{A}$ is the product of a scalar field f multiplied by a constant vector $\vec{C}$, from Equation (8.4.1),

$$\int_V \vec{\nabla} \cdot \left(\vec{C}f\right) dV = \vec{C} \cdot \int_V \vec{\nabla} f \, dV = \vec{C} \cdot \oint_S f d\vec{S} \tag{8.4.3}$$

Since $\vec{C}$ is arbitrary,

$$\int_V \vec{\nabla} f dV = \oint_S f d\vec{S} \tag{8.4.4}$$

Green's second identity follows from Gauss's law according to, where $d\vec{S}\cdot\vec{\nabla}$ is often written as $dS\partial/\partial n$, the directional derivative in the direction of the normal to $d\vec{S}$,

$$\begin{aligned}&\oint_S \left(\phi_1 \vec{\nabla}\phi_2 - \phi_2 \vec{\nabla}\phi_1\right)\cdot d\vec{S}\\ &= \int_V \vec{\nabla}\cdot\left(\phi_1 \vec{\nabla}\phi_2 - \phi_2 \vec{\nabla}\phi_1\right) dV\\ &= \int_V \left(\vec{\nabla}\phi_1\cdot\vec{\nabla}\phi_2 + \phi_1\nabla^2\phi_2 - \vec{\nabla}\phi_2\cdot\vec{\nabla}\phi_1 - \phi_2\nabla^2\phi_1\right) dV\\ &= \int_V \left(\phi_1\nabla^2\phi_2 - \phi_2\nabla^2\phi_1\right) dV\end{aligned} \tag{8.4.5}$$

8.5 LAGRANGE MULTIPLIERS

A function $f(x_1, x_2, \ldots, x_n)$ of several variables is *constrained* when subject to auxiliary conditions $g_1(x_1, x_2, \ldots, x_n) = c_1$, $g_2(x_1, x_2, \ldots, x_n) = c_2, \ldots,$. At each stationary point (maxima, minima, and saddle points) of f, the directional derivative, $\hat{e}_{\vec{n}}\cdot\vec{\nabla}f$, must vanish for any $\hat{e}_{\vec{n}}$ along the surface $g_1(x_1, x_2, \ldots, x_n) = c_1$ (otherwise, the function would increase along the surface in the direction of $\hat{e}_{\vec{n}}$). Further, $\hat{e}_{\vec{n}}\cdot\vec{\nabla}g_1 = \vec{\nabla}c_1 = 0$ along this surface, and therefore, for some constant *Lagrange multiplier* λ, $\vec{\nabla}f = -\lambda\vec{\nabla}g_1 = c_1$. Similarly, $\vec{\nabla}(f + \lambda g_1)$ must be proportional to $\vec{\nabla}g_2 = c_2, \ldots$. Hence, the extrema can be found by equating the gradient of $f + \lambda_1 g_1 + \lambda_2 g_2 + \ldots$ to zero and subsequently imposing the initial constraints $g_i = c_i$.

Example

The extremum of $f(x, y) = x^2 + y^2$ subject to $g(x, y) = x + y - 1 = 0$ is obtained from

$$\delta f = \frac{\partial f}{\partial x}\delta x + \frac{\partial f}{\partial y}\delta y = 2x\delta x + 2y\delta y = 0 \tag{8.5.1}$$

However, x and y must further satisfy $g(x, y) = x + y = 1$ so that

$$\frac{\partial g}{\partial x}\delta x + \frac{\partial g}{\partial y}\delta y = \delta x + \delta y = 0 \tag{8.5.2}$$

The expressions in Equations (8.5.1) and (8.5.2) both equal zero and are thus proportional. Hence, for some λ,

$$2x\delta x + 2y\delta y = -\lambda(\delta x + \delta y) \tag{8.5.3}$$

This clearly generates the same equations as $\vec{\nabla}(f+\lambda g)=0$, namely,

$$\begin{aligned} 2x &= -\lambda \\ 2y &= -\lambda \end{aligned} \tag{8.5.4}$$

Eliminating λ from the above equations yields $x=y$, after which the critical point $x=y=1/2$ follows from the constraint $x+y=1$. Note that λ is never actually computed. Standard one-dimensional methods can as well be applied if $y=1-x$ is first substituted into $f(x, y)$.

To find the stationary points of $f(x, y)=x^2+y^2-2xy$ on the line $g(x, y)=x^2+2y-4x-1=0$, the two partial derivatives with respect to x and y of $f(x, y)+\lambda g(x, y)$ are set to 0:

$$\begin{aligned} 2x-2y+\lambda(2x-4) &= 0 \\ 2y-2x+2\lambda &= 0 \end{aligned} \tag{8.5.5}$$

Summing yields $x=1$ for $\lambda \neq 0$. The constraint $g(x, y)=x^2+2y-4x-1=0$ then implies $y=2$.

9

PROBABILITY THEORY AND STATISTICS

In physical systems, inherent uncertainties and fluctuations often produce differing, unpredictable outcomes when a measurement is repeated. The likelihood of an outcome must then be described probabilistically.

9.1 RANDOM VARIABLES, PROBABILITY DENSITY, AND DISTRIBUTIONS

A *random variable* maps the potential outcomes of a system onto differing numerical values. For example, in a coin toss, a random variable X could map heads and tails onto the values 1 and 0, respectively, which is expressed as $X(H) = 1, X(T) = 0$. Similarly, X could represent the square of the number obtained when rolling a dice. Accompanying the random variable is a *probability distribution*, $p(x)$ (often denoted P_X if X only adopts discrete values), defined on the set of possible values of the random variable such that for N repeated measurements,

$$p(X=x) \equiv p(x) = \lim_{N\to\infty}\left(\frac{1}{N}\times \text{number of occurences of } X=x\right) \tag{9.1.1}$$

Hence, $p(x) \geq 0$ and $\Sigma_x p(x) = 1$. For a continuous distribution, such as the distribution of points randomly selected from a line segment, $p(x)dx$ represents the probability of

Fundamental Math and Physics for Scientists and Engineers, First Edition.
David Yevick and Hannah Yevick.

observing a value of X between x and $x+dx$, while the normalization condition adopts the form $\int_{-\infty}^{\infty} p(x)dx = 1$.

9.2 MEAN, VARIANCE, AND STANDARD DEVIATION

The *mean value*, $\bar{f}$, of a function, $f(x)$, of a random variable is computed from

$$\bar{f} = \sum_x p(x)f(x) = \int_{-\infty}^{\infty} p(x)f(x) \tag{9.2.1}$$

The mean value, $\bar{x}$, of the probability distribution, $p(x)$, is computed by setting $f(x) = x$ in the above equation. The *variance* of the distribution

$$\sigma^2 \equiv \overline{(x-\bar{x})^2} = \overline{x^2 - 2x\bar{x} + \bar{x}^2} = \overline{x^2} - \bar{x}^2 \tag{9.2.2}$$

corresponds to the square of its average departure from the mean. The square root of the variance

$$\sigma = \sqrt{\overline{x^2} - \bar{x}^2} \tag{9.2.3}$$

is termed the *standard deviation*. The relative extent of fluctuations about the mean is quantified by $\sigma/\bar{x}$ as, e.g., doubling the values in the distribution doubles both the standard deviation and the mean.

9.3 VARIABLE TRANSFORMATIONS

Given the probability distribution, $p_X(x)$, of a random variable, X, the probability density, $p_y(y)$, of a second random variable $Y = f(X)$ is obtained from

$$p_x(x)dx = p_x(x)\frac{dx}{dy}dy = p_y(y(x))dy \tag{9.3.1}$$

Example

If X and Y describe the amount of collected rain in centiliters and milliliters, respectively, $y = 10x$. If the probability of rainfall between 1.0 cl and $(1.0 + dx)$ cl equals $p_x(x = 1)dx$, the probability of rainfall between 10.0 ml and $(10.0 + dy)$ ml is 10 times smaller in agreement with $p_y(y = 10) = p_x(x = 1)dx/dy = p(x)/10$.

9.4 MOMENTS AND MOMENT-GENERATING FUNCTION

The nth order moment, $\overline{x^n}$, of a probability distribution, $p(x)$, is defined as

$$\overline{x^n} = \int_{-\infty}^{\infty} x^n p(x) dx \tag{9.4.1}$$

A probability distribution function is specified uniquely from all its moments, which can be obtained from the *moment-generating function*

$$M(t) = \int_{-\infty}^{\infty} e^{tx} p(x) dx \tag{9.4.2}$$

according to (since below only the nth term in the Taylor series expansion yields a constant after differentiation that does not vanish as $t \rightarrow 0$)

$$\overline{x^n} = \left.\frac{d^n M(t)}{dt^n}\right|_{t=0} = \lim_{t\to 0} \int_{-\infty}^{\infty} p(x) \frac{d^n}{dt^n}(e^{tx}) dx = \lim_{t\to 0} \int_{-\infty}^{\infty} p(x) \frac{d^n}{dt^n}\left(1 + tx + \frac{t^2 x^2}{2!} + \cdots\right) dx \tag{9.4.3}$$

9.5 MULTIVARIATE PROBABILITY DISTRIBUTIONS, COVARIANCE, AND CORRELATION

A *multivariate probability distribution* is specified by several random variables, such as the probability of simultaneously observing both rain and a temperature above 20°. If the variables are *statistically independent*, implying that the probability of rain is unaffected by temperature,

$$p(x, y) \equiv p(x \text{ and } y) = p_x(x) p_y(y) \tag{9.5.1}$$

Moments then factor into products of the moments of the individual variables, e.g., $\overline{xy} = \bar{x}\bar{y}$. On the other hand, if the probability of rain varies with temperature, Equation (9.5.1) is invalidated and the degree of dependence between the two variables is quantified by the *covariance*

$$\text{cov}(x, y) = \overline{xy} - \bar{x}\bar{y} \tag{9.5.2}$$

9.6 GAUSSIAN, BINOMIAL, AND POISSON DISTRIBUTIONS

Gaussian Distribution: The outcomes of different physical systems often exhibit similar statistical properties described by certain frequently occurring probability distributions. The *central limit theorem* states that the distribution of the *mean values* of repeated sets of statistically independent trials approaches a *Gaussian* or *normal* distribution

$$p(x) = \frac{1}{\sigma\sqrt{2\pi}} e^{-\frac{(x-\bar{x})^2}{2\sigma^2}} \tag{9.6.1}$$

as the number of measurements in each set becomes large. The moment-generating function associated with the Gaussian distribution is determined by completing the square below and subsequently shifting the limits in the integral as in Equations (7.3.8) and (7.3.9):

$$M(t) = \frac{1}{\sigma\sqrt{2\pi}} \int_{-\infty}^{\infty} e^{tx - \frac{(x-\bar{x})^2}{2\sigma^2}} dx = \frac{e^{t\bar{x} + \frac{t^2\sigma^2}{2}}}{\sigma\sqrt{2\pi}} \int_{-\infty}^{\infty} e^{-\frac{(x-\bar{x}-t\sigma^2)^2}{2\sigma^2}} dx = e^{t\bar{x} + \frac{t^2\sigma^2}{2}} \tag{9.6.2}$$

Accordingly,

$$\bar{x} = \left.\frac{dM(t)}{dt}\right|_{t=0} = \left.(\bar{x} + \sigma^2 t) e^{t\bar{x} + \frac{t^2\sigma^2}{2}}\right|_{t=0} = \bar{x} \tag{9.6.3}$$

and, consistent with the definition $\sigma^2 = \overline{x^2} - \bar{x}^2$,

$$\overline{x^2} = \left.\frac{d^2M(t)}{dt^2}\right|_{t=0} = \left.\left(\sigma^2 + (\bar{x} + \sigma^2 t)^2\right) e^{t\bar{x} + \frac{t^2\sigma^2}{2}}\right|_{t=0} = \sigma^2 + \bar{x}^2 \tag{9.6.4}$$

Binomial Distribution: The *binomial distribution* expresses the probability of observing, e.g., m heads after n tosses of a coin for which heads and tails appear with probabilities p and $1 - p$, respectively. Considering events with m heads and $(n - m)$ tails, for $m = 0$, only one sequence exists; for $m = 1$, n separate events yield a single head; etc., where the probability of these occurrences equals $p^m(1 - p)^{n-m}$. In general, n different objects can be arranged in $n!$ ways or *permutations* since n objects can be placed first, any of the remaining $n - 1$ objects can be chosen for the second position, and so on. If, however, the collection consists of several groups of identical objects with $m_1, m_2, m_3, \ldots$ objects in each group only,

$$\frac{n!}{m_1!m_2!\ldots m_n!} \equiv \binom{n}{m_1\ m_2 \ldots\ m_n} \tag{9.6.5}$$

arrangements or *combinations* of the original $n!$ permutations are distinct where $0! = 1$. The denominator of this *multinomial coefficient* indicates that for each distinct arrangement of, e.g., two identical black and two white objects such as (BBWW), $2! \times 2!$ additional permutations would exist if all objects were distinguishable, namely, ($B_1B_2W_1W_2$), ($B_2B_1W_1W_2$), ($B_1B_2W_2W_1$), and ($B_2B_1W_2W_1$), as both white and black objects could then be permuted in 2! additional ways.

For two distinct object types such as heads and tails, Equation (9.6.5) is instead termed the binomial coefficient and is variously written

$$ {}_nC_m = \binom{n}{m} = \frac{n!}{m!(n-m)!} = \frac{\overbrace{n(n-1)(n-2)\ldots}^{m \text{ terms}}}{m!} \tag{9.6.6} $$

while the number of permutations of m distinct objects out of set of n distinct objects is denoted

$$ {}_nP_m = n(n-1)\ldots(n-m+1) \tag{9.6.7} $$

The probability of obtaining m heads and $n-m$ tails is therefore the probability of a single ordered sequence multiplied by the number of sequences with the specified number of heads and tails

$$ p(m) = {}_nC_m\, p^m (1-p)^{n-m} \tag{9.6.8} $$

termed the *binomial distribution*. As required, since $(p+q)^n = \Sigma_{m=0}^{n}\, {}_nC_m\, p^m q^{n-m}$ with $q = 1-p$,

$$ \sum_{m=0}^{n} p(m) = (p+(1-p))^n = 1 \tag{9.6.9} $$

Furthermore,

$$ \overline{m} = \sum_{m=0}^{n} mp(m) = \sum_{m=0}^{n} \frac{n!}{m!(n-m)!} mp^m q^{n-m} = p\frac{\partial}{\partial p}(p+q)^n = np(p+q)^{n-1}\Big|_{q=1-p} = np $$

$$ \begin{aligned} \overline{m^2} &= \sum_{m=0}^{n} \frac{n!}{m!(n-m)!} m^2 p^m q^{n-m} = p\frac{\partial}{\partial p}\left(p\frac{\partial}{\partial p}(p+q)^n\right) = p\frac{\partial}{\partial p}\left(np(p+q)^{n-1}\right) \\ &= p\left[n(p+q)^{n-1} + n(n-1)p(p+q)^{n-2}\right]_{q=1-p} \\ &= np + (n^2-n)p^2 \end{aligned} \tag{9.6.10} $$

and the variance

$$ \sigma^2 = \overline{x^2} - \overline{x}^2 = np + (n^2-n)p^2 - (np)^2 = n(p-p^2) \tag{9.6.11} $$

The associated moment-generating function is given by

$$ M(t) = \overline{e^{mt}} = \sum_{m=0}^{n} \frac{n!}{m!(n-m)!} e^{mt} p^m (1-p)^{n-m} = (e^t p + (1-p))^n \tag{9.6.12} $$

from which, e.g.,

$$\overline{m} = \frac{\partial}{\partial t}(e^t p + (1-p))^n \bigg|_{t=0} = n(e^t p + (1-p))^{n-1} e^t p \bigg|_{t=0} = np \tag{9.6.13}$$

Poisson Distribution: Consider the limit in which the probability of a binary event approaches zero but the possible number of such events becomes infinite. For example, if λ decay events on average occur over a time interval T in a certain system, if T is divided into $N = T/\Delta t$ subintervals, the probability of a decay in each subinterval is λ/N so that the probability of m decays in the time T equals

$$p(m) = {}_N C_m \left(\frac{\lambda}{N}\right)^m \left(1 - \frac{\lambda}{N}\right)^{N-m} = \frac{\overbrace{N(N-1)(N-2)\ldots}^{m \text{ terms}}}{m!} \left(\frac{\lambda}{N}\right)^m \left(1 - \frac{\lambda}{N}\right)^{N-m} \tag{9.6.14}$$

The $\Delta t \to 0$, $N \to \infty$ limit of the above equation generates the *Poisson distribution*

$$p(m) \approx \lim_{N \to \infty} \frac{N^m}{m!} \left(\frac{\lambda}{N}\right)^m \left(1 - \frac{\lambda}{N}\right)^{N-m} \approx e^{-\lambda} \frac{\lambda^m}{m!} \tag{9.6.15}$$

From the Taylor series expansion of the exponential function, $\sum_{m=0}^{\infty} p(m) = 1$ while

$$\begin{aligned}
\overline{m} &= e^{-\lambda} \sum_{m=0}^{\infty} \frac{m\lambda^m}{m!} = e^{-\lambda} \sum_{m=1}^{\infty} \frac{\lambda^m}{(m-1)!} = \lambda e^{-\lambda} \sum_{m=0}^{\infty} \frac{\lambda^m}{m!} = \lambda \\
\overline{m^2} &= e^{-\lambda} \sum_{m=1}^{\infty} \frac{m\lambda^m}{(m-1)!} = e^{-\lambda} \sum_{m=1}^{\infty} \frac{[(m-1)+1]\lambda^m}{(m-1)!} \\
&= e^{-\lambda} \sum_{m=2}^{\infty} \frac{\lambda^m}{(m-2)!} + e^{-\lambda} \sum_{m=1}^{\infty} \frac{\lambda^m}{(m-1)!} = e^{-\lambda}(\lambda^2 e^{\lambda}) + e^{-\lambda}(\lambda e^{\lambda}) = \lambda^2 + \lambda
\end{aligned} \tag{9.6.16}$$

Hence, the variance, $\sigma^2 = \overline{m^2} - (\overline{m})^2$, of the Poisson distribution equals its mean, λ. These results can again be obtained from the moment-generating function

$$\overline{e^{tm}} = e^{-\lambda} \sum_{m=0}^{\infty} \frac{e^{tm} \lambda^m}{m!} = e^{-\lambda} \sum_{m=0}^{\infty} \frac{(e^t \lambda)^m}{m!} = e^{\lambda(e^t - 1)} \tag{9.6.17}$$

since, e.g.,

$$\overline{m} = \frac{\partial}{\partial t} e^{\lambda(e^t - 1)} \bigg|_{t=0} = e^{\lambda(e^t - 1)} \lambda e^t \bigg|_{t=0} = \lambda \tag{9.6.18}$$

and

$$\overline{m^2} = \frac{\partial^2}{\partial t^2} e^{\lambda(e^t - 1)} \bigg|_{t=0} = e^{\lambda(e^t - 1)} \left[\lambda^2 e^{2t} + \lambda e^t \right] \bigg|_{t=0} = \lambda^2 + \lambda \tag{9.6.19}$$

9.7 LEAST SQUARES REGRESSION

In contrast with the direct problem of determining the probability distribution implied by a physical model, *statistics* addresses the inverse problem of determining the probability that a particular model reproduces a set of measurement data and then finding an exact or optimal approximate physical model for the data. In a simple one-dimensional implementation, given a set of m measured data points (x_i, y_i), $i = 1, 2, \ldots, m$, *least squares regression* estimates the optimal coefficients a_l for a model consisting of a *linear* combination of n functions:

$$\tilde{y}(x) = \sum_{l=1}^{n} a_l f_l(x) \tag{9.7.1}$$

With the assumption of identical measurement accuracy at each data point, minimizing the least mean squared error

$$F(a_1, a_2, \ldots, a_n) = \sum_{i=1}^{m} (\tilde{y}(x_i) - y_i)^2 \tag{9.7.2}$$

of the predicted value $\tilde{y}(x_i)$ from the data with respect to each of the n model parameters a_i requires

$$\frac{\partial F}{\partial a_j} = 0 \tag{9.7.3}$$

for $j = 1, 2, \ldots, n$. This generates a system of n equations:

$$2 \sum_{i=1}^{m} (\tilde{y}(x_i) - y_i) f_j(x_i) = 0 \tag{9.7.4}$$

Substituting Equation (9.7.1) for $\tilde{y}(x_i)$,

$$\sum_{i=1}^{m} \left(\sum_{l=1}^{n} a_l f_l(x_i) - y_i \right) f_j(x_i) = 0 \tag{9.7.5}$$

results in a matrix equation $\mathbf{Ma} = \mathbf{b}$ for the a_l with

$$\sum_{l=1}^{n}\underbrace{\left(\sum_{i=1}^{m} f_j(x_i)f_l(x_i)\right)}_{M_{jl}} a_l = \underbrace{\sum_{i=1}^{m} y_i f_j(x_i)}_{b_j} \tag{9.7.6}$$

9.8 ERROR PROPAGATION

Every measurement is subject to error. *Systematic errors* are associated with inherent, reproducible features of the measurement device and can therefore be corrected if their origin is understood. *Random* errors, however, result from intrinsic measurement inaccuracy and can therefore only be mitigated through multiple observations followed by statistical data processing. A repeated measurement that exhibits a large systematic error but small random error possesses a high degree of *precision* but a low *accuracy* as the results are closely grouped around an imprecise value (additionally, if the mean value of the measurements yields a highly accurate result while the standard deviation around the mean is large, the measurements are said to possess a good *trueness*, even though the precision is poor). Errors are typically specified as $\mu \pm \sigma$, where μ and σ represent the mean value and the standard deviation of the measured quantity.

Suppose a result is described by a function $f(\vec{x})$ of a collection of variables $\vec{x}$ that are measured in a series of identical experiments. Denoting an average over all measurements by an overbar, if the observed mean and standard deviation of the jth variable are $\bar{x}_j$ and σ_j, the variance of $f(\vec{x})$ is given by

$$\overline{\sum_j \left(f(x_j) - f(\bar{x}_j)\right)^2} \approx \overline{\sum_j \left(\cancel{f(\bar{x}_j)} + \frac{\partial f}{\partial x_j}(x_j - \bar{x}_j) - \cancel{f(\bar{x}_j)}\right)^2} = \sum_j \left(\frac{\partial f}{\partial x_j}\right)^2 \sigma_j^2 \tag{9.8.1}$$

Examples

If $f(\vec{x})$ equals the sum of the N variables, the root-mean-squared (r.m.s.) error of f is $\sigma_f = \left(\sum_j \sigma_j^2\right)^{1/2}$. Alternatively, for

$$f(\vec{x}) = a\prod_j x_j^{p_j} \tag{9.8.2}$$

the "logarithmic derivative" yields

$$\frac{1}{f}\frac{\partial f}{\partial x_j} = \frac{p_j}{x_j} \tag{9.8.3}$$

from the *relative error* in f, which follows from Equation (9.8.1)

$$\frac{\sigma_f}{f} = \left[\sum_j \left(\frac{p_j \sigma_j}{x_j}\right)^2\right]^{\frac{1}{2}} \tag{9.8.4}$$

9.9 NUMERICAL MODELS

Stochastic (probabilistic) processes are simulated by random number generators. In Octave, `rand` returns a random number between 0 and 1 (the function `randn` similarly returns a random number selected from a Gaussian distribution with $\bar{x}=0$ and $\sigma=1$). A single coin toss in which 1 is assigned to heads and 0 to tails is therefore simulated with `outcome = floor ( 2 * rand );`. The following program performs `numberOfRealizations` experiments in each of which a coin is tossed `numberOfThrows` times and compares the resulting *histogram* (discrete distribution) of the number of heads to the binomial coefficient:

```
numberOfThrows = 30;
numberOfRealizations = 5000;
histogram = zeros( 1, numberOfThrows + 1 );
for loopOuter = 1 : numberOfRealizations
      outcome = 0;
      for loopInner = 1 : numberOfThrows
            outcome = outcome + floor ( 2 * rand );
      end
      histogram( outcome + 1 ) = histogram( outcome + 1 ) + 1;
end

binomialCoefficients = binopdf( [ 0 : numberOfThrows ],...
  numberOfThrows, 0.5 );
histogram = histogram / max( sum( histogram ), 1 );

xAxis = 0 : numberOfThrows;
plot( xAxis, histogram, xAxis, binomialCoefficients );
drawnow
```

10

COMPLEX ANALYSIS

Limiting operations can be applied to complex-valued functions with complex arguments as well as to real functions of a real argument. Performing calculations on functions extended to the complex domain and subsequently specializing to the real limit often results in simplifications relative to the identical calculations on the real functions.

10.1 FUNCTIONS OF A COMPLEX VARIABLE

A function $f(x)$ of a single real variable is transformed into a function of a complex variable by replacing and n an arbitary integer x by $z = x + iy = \sqrt{x^2 + y^2}\exp(i(\theta + 2\pi n)) \equiv r\exp(i(\theta + 2\pi n))$ with $\theta = \arctan(y/x)$ in accordance with Figure 4.6. The function is then generally complex and can be multiple valued.

Fundamental Math and Physics for Scientists and Engineers, First Edition.
David Yevick and Hannah Yevick.

Example

While $f(z) = z^2 = (x + iy)(x + iy) = x^2 + y^2 + 2ixy$ is single valued, $f(x) = \sqrt{x}$ is double valued as

$$f(x) = z^{1/2} = (x + iy)^{1/2} = \left(re^{i(\theta + 2\pi n)}\right)^{1/2} = r^{1/2}e^{i\left(\frac{\theta}{2} + \pi n\right)} = r^{1/2}e^{i\frac{\theta}{2}}(-1)^n \quad (10.1.1)$$

Similarly, the logarithm function possesses infinitely many complex values for each real value:

$$\ln(z) = \ln\left(re^{i(\theta + 2n\pi)}\right) = \ln r + i(\theta + 2n\pi) \quad (10.1.2)$$

The single-valued section of $f(z)$ typically specified by $-\pi < \arg(f(z)) < \pi$ is termed the *principal sheet or branch*, with the values acquired by $f(z)$ on the principal branch termed the *principal values*. For the square root and logarithm functions, the principal branch corresponds to $-2\pi < \arg(z) < 2\pi$ and $-\pi < \arg(z) < \pi$, respectively.

10.2 DERIVATIVES, ANALYTICITY, AND THE CAUCHY–RIEMANN RELATIONS

The derivative of a function of complex arguments can be defined identically to the real derivative

$$\frac{df(z)}{dz} \equiv \lim_{\Delta z \to 0} \frac{f(z + \Delta z) - f(z)}{\Delta z} \quad (10.2.1)$$

However, Equation (10.2.1) must then not depend on the manner in which Δx and Δy approach zero in $\Delta z = \Delta x + i\Delta y$. This *analytic* condition holds if $f(z)$ is nonsingular as $\Delta z \to 0$ and depends only on z, so that $\partial f/\partial z^* = 0$. As x and iy then appear symmetrically in the function argument,

$$\frac{\partial f(x + iy)}{\partial x} = \frac{\partial f(x + iy)}{\partial(iy)} \quad (10.2.2)$$

Writing $f(z) = u(z) + iv(z)$, where $u(z)$ and $v(z)$ are real functions,

$$\frac{\partial u(x + iy)}{\partial x} + i\frac{\partial v(x + iy)}{\partial x} = -i\frac{\partial u(x + iy)}{\partial y} + \frac{\partial v(x + iy)}{\partial y} \quad (10.2.3)$$

Equating real and imaginary parts of the above expression yields the *Cauchy–Riemann relations*

$$\frac{\partial u(x+iy)}{\partial x} = \frac{\partial v(x+iy)}{\partial y}$$
$$\frac{\partial u(x+iy)}{\partial y} = -\frac{\partial v(x+iy)}{\partial x} \tag{10.2.4}$$

Example

The analyticity of $u + iv = x^2 + y^2 + 2ixy$, which originates from $f(z) = z^2$, can be established by verifying that the real and imaginary parts of the function u and v satisfy Equation (10.2.4). Nonanalytic functions such as $|z| = \sqrt{zz^*}$ or $z + z^*$ contain both z and z^* as arguments and violate Equation (10.2.4).

Since the derivative of a complex function is defined identically to the real derivative, the chain and product rules apply. Further, the Cauchy–Riemann relations together with $\partial^2/\partial x\partial y = \partial^2/\partial y\partial x$ yield

$$\frac{\partial}{\partial x}\left(\frac{\partial u}{\partial x}\right) = \frac{\partial}{\partial x}\left(\frac{\partial v}{\partial y}\right) = \frac{\partial}{\partial y}\left(\frac{\partial v}{\partial x}\right) = -\frac{\partial}{\partial y}\left(\frac{\partial u}{\partial y}\right) \tag{10.2.5}$$

Hence, both u and v and hence $f(z)$ satisfy the two-dimensional Laplace equation

$$\left(\frac{\partial^2}{\partial x^2} + \frac{\partial^2}{\partial y^2}\right)\begin{pmatrix} u \\ v \end{pmatrix} = 0 \tag{10.2.6}$$

10.3 CONFORMAL MAPPING

An analytic function $f(z)$ "conformally" maps two lines in the complex plane intersecting at a given angle into two curves that intersect at the same angle. If the two lines are expressed parametrically as

$$z_1(t) = u_1(t) + iv_1(t)$$
$$z_2(t) = u_2(t) + iv_2(t) \tag{10.3.1}$$

their angles of inclination in the complex plane are determined by (since $\arg z = \arctan(v/u)$)

$$\tan^{-1}\left(\frac{dv}{du}\right) = \tan^{-1}\left(\frac{dv/dt}{du/dt}\right) = \arg\left(\frac{du}{dt} + i\frac{dv}{dt}\right) = \arg\left(\frac{dz}{dt}\right) \tag{10.3.2}$$

At the point of intersection t_0, the angle between these lines is therefore arg $(dz_1(t_0)/dt)$ – arg $(dz_2(t_0)/dt)$, while the angle between the corresponding curves $f(z_1(t))$ and $f(z_2(t))$ is

$$\left[\arg\left(\frac{df(z_1(t))}{dt}\right) - \arg\left(\frac{df(z_2(t))}{dt}\right)\right]_{t_0}$$

$$= \left[\arg\left(\frac{df(z_1(t))}{dz_1}\frac{dz_1}{dt}\right) - \arg\left(\frac{df(z_2(t))}{dz_2}\frac{dz_2}{dt}\right)\right]_{t_0} \tag{10.3.3}$$

Since, however, $df/dz|_{t_0}$ is the same for both paths while for any two complex quantities z_1 and z_2

$$\arg(z_1 z_2) = \arg\left(r_1 e^{i\theta_1} r_2 e^{i\theta_2}\right) = \theta_1 + \theta_2 = \arg(z_1) + \arg(z_2) \tag{10.3.4}$$

the angle between the lines is preserved.

10.4 CAUCHY'S THEOREM AND TAYLOR AND LAURENT SERIES

Since an analytic function $f(z)$ depends exclusively on $z = x + iy$, the variations of its real and imaginary parts over a path, $z(t)$, in the complex plane are coupled. If $z(t)$ only passes through points where $F(z)$ is analytic, the integral of the derivative of $F(z)$ is given by the difference of $F(z)$ at its two endpoints since

$$\int_{t_1}^{t_2} \frac{dF(z(t))}{dt} dt = \int_{t_1}^{t_2} \frac{dF(z(t))}{dz}\frac{dz}{dt} dt = \int_{z(t_1)}^{z(t_2)} \frac{dF(z)}{dz} dz = \int_{F(z(t_1))}^{F(z(t_2))} dF = F(z(t_2)) - F(z(t_1)) \tag{10.4.1}$$

Further, if a function $f(z) = u(z) + iv(z)$ is analytic in a connected domain D (a domain without any points or lines where $f(z)$ is singular or otherwise nonanalytic) and C is any contour wholly within D,

$$\oint_C f(z)dz = 0 \tag{10.4.2}$$

since applying Stokes' theorem in a space defined by the two-dimensional (x, y) complex plane together with a third z-dimension perpendicular to this plane so that $d\vec{S}$ below is parallel to $\hat{e}_z$,

$$\oint_C f(z)dz = \oint_C (u+iv)(dx+idy) = \oint_C (udx - vdy) + i(vdx + udy)$$
$$= \oint_C (u,-v,0)\cdot d\vec{l} + i\oint_C (v,u,0)\cdot d\vec{l}$$
$$= \iint_S \left(\vec{\nabla}\times(u,-v,0)\right)\cdot d\vec{S} + i\iint_S \left(\vec{\nabla}\times(v,u,0)\right)\cdot d\vec{S} \tag{10.4.3}$$
$$= -\iint_S \left\{\frac{\partial u}{\partial y} + \frac{\partial v}{\partial x}\right\}dxdy + i\iint_S \left\{\frac{\partial u}{\partial x} - \frac{\partial v}{\partial y}\right\}dxdy = 0$$

where the Cauchy–Riemann relations are applied in the last step.

If $f(z) = 1/z$, termed a *first-order pole*, that is nonanalytic at $z = 0$ is integrated counterclockwise around the origin over a circle of radius, a, so that $z(\theta) = ae^{i\theta}$, $dz(\theta) = iae^{i\theta}d\theta$,

$$\oint_C \frac{1}{z}dz = \oint_C \frac{1}{z(\theta)}\frac{dz(\theta)}{d\theta}d\theta = \int_0^{2\pi} \frac{i\cancel{ae^{i\theta}}}{\cancel{ae^{i\theta}}}d\theta = 2\pi i \tag{10.4.4}$$

Note that an integration contour from a lower limit a to an upper limit b is denoted by an arrow from a to b:

$$\underset{a\longrightarrow b}{\int_a^b f(x)dx} \tag{10.4.5}$$

Thus, when integrating in the counterclockwise direction over C in Equation (10.4.4), the lower limit corresponds to $\theta = 0$, while the upper limit is $\theta = 2\pi$. Additionally, integrals around all singularities of the form $1/z^n$ vanish for $n \neq 1$ (Eq. 6.1.18 with $n \to 1$ can be employed in the last manipulation below):

$$\oint_C \frac{1}{z^n}dz = \oint_C \frac{1}{z^n(\theta)}\frac{dz(\theta)}{d\theta}d\theta = \int_0^{2\pi} \frac{iae^{i\theta}}{a^n e^{in\theta}}d\theta$$
$$= a^{1-n}\int_0^{2\pi} ie^{i(1-n)\theta}d\theta = \left.\frac{a^{1-n}e^{i(1-n)\theta}}{1-n}\right|_0^{2\pi} = 2\pi i\delta_{n1} \tag{10.4.6}$$

Cauchy's theorem accordingly states that if $f(z)$ is analytic inside a contour C,

$$\oint_C \frac{f(z)}{z-z_0}dz = \begin{cases} 0 & \text{if } C \text{ does not enclose } z_0 \\ 2\pi i f(z_0) & \text{if } C \text{ encloses } z_0 \end{cases} \tag{10.4.7}$$

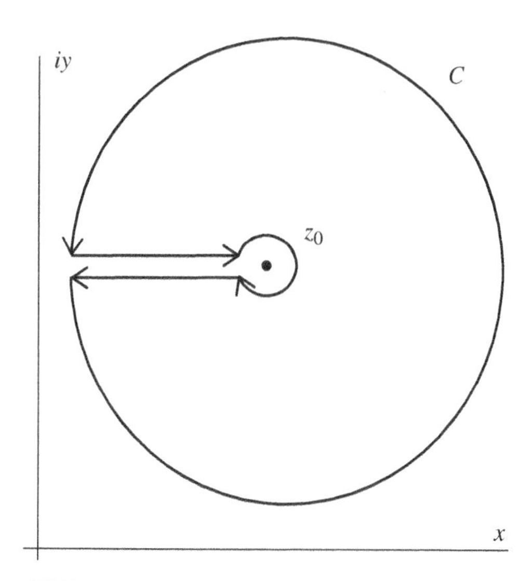

FIGURE 10.1 Contour for Cauchy's theorem.

That is, if C does not enclose z_0, the integral is zero from Equation (10.4.2). However, if C encloses z_0, the integral is evaluated by considering a modified contour including two infinitesimally displaced, oppositely directed line segments extending from C to an inner circle infinitesimally close to and enclosing z_0 as shown in Figure 10.1

The integrals over the two straight line segments are oppositely directed and cancel. Since the modified contour avoids enclosing the singularity at z_0, the counterclockwise integral over its outer loop and the clockwise integral over the infinitesimal circular contour—which is the negative of the counterclockwise integral—surrounding this singular point sum to zero. Approximating the analytic function $f(z)$ in the integral around the infinitesimal circle by its value at z_0, Equation (10.4.7) follows from Equation (10.4.4).

Differentiating both sides of Equation (10.4.7) n times with respect to z_0 yields, if C encloses z_0,

$$f^n(z_0) \equiv \frac{\partial^n f(z_0)}{\partial z_0^n} = \frac{\partial^n}{\partial z_0^n} \frac{1}{2\pi i} \oint_C \frac{f(z)}{z - z_0} dz = \frac{n!}{2\pi i} \oint_C \frac{f(z)}{(z - z_0)^{n+1}} dz \tag{10.4.8}$$

Since z_0 can be anywhere inside C, a function that is analytic within C is infinitely differentiable in this region as its derivatives can be evaluated through the above formula.

If $f(z)$ is instead analytic in an *annular* (ring-shaped) region about z_0, the integral

$$\oint_C \frac{f(z)}{z - (z_0 + h)} dz \tag{10.4.9}$$

is considered in place of Equation (10.4.7), where $z_0 + h$ lies within the annulus and the contour C contains two infinitesimally displaced oppositely directed line segments

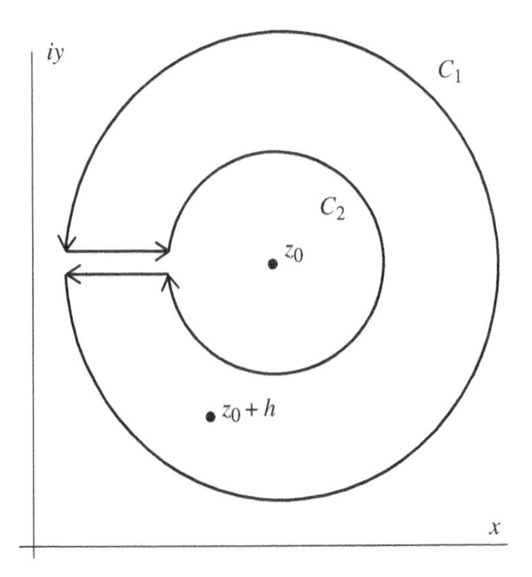

FIGURE 10.2 Laurent series contour.

that connect the contour C_1 along the outer boundary of the region to a contour C_2 along the inner region boundary as displayed in Figure 10.2. For a function $f(z)$ analytic within C, the integral over C yields $2\pi i f(z_0 + h)$ from Cauchy's theorem. However, this equals the sum of the counterclockwise integral over the outer contour and the clockwise integral over the inner contour as the contributions from the lines connecting the singularity with the outer contour are again oppositely directed and therefore cancel. Since replacing the clockwise integral over the inner contour changes by a counterclockwise integral reverses the sign of the term,

$$f(z_0+h)=\frac{1}{2\pi i}\oint_{C_1}\frac{f(z)}{\underbrace{z-z_0-h}_{|z-z_0|>h}}dz-\frac{1}{2\pi i}\oint_{C_2}\frac{f(z)}{\underbrace{z-z_0-h}_{|z-z_0|<h}}dz \tag{10.4.10}$$

Writing the first and second denominator as $(z-z_0)^{-1}(1-h/(z-z_0))^{-1}$ and $(-h)^{-1}(1-(z-z_0)/h)^{-1}$, respectively, and subsequently expanding in convergent power series yield

$$f(z_0+h)=\frac{1}{2\pi i}\left(\sum_{n=0}^{\infty}h^n\oint_{C_1}\frac{f(z')(z'-z_0)^{n+1}}{(z-z_0)^{n+1}}dz+\sum_{n=1}^{\infty}\frac{1}{h^n}\oint_{C_2}(z-z_0)^{n-1}f(z)dz\right) \tag{10.4.11}$$

This *Laurent series* expansion possesses the general form after substituting $z_0 + h = z$

$$f(z)=\sum_{n=-\infty}^{\infty}\left[\frac{1}{2\pi i}\oint_{C_1}\frac{f(z')}{(z'-z_0')^{n+1}}dz\right](z-z_0)^n=\sum_{n=-\infty}^{\infty}a_n(z-z_0)^n \tag{10.4.12}$$

If all $a_n = 0$ for $n < p$ with $p < 0$, $f(z)$ possesses a *pth-order pole* at z_0. If $f(z)$ is nonsingular at z_0, the annular region can be assigned a zero inner radius and $a_n = 0$ for $n < 0$, yielding a *Taylor series* (note that the coefficient of h^n is $f^n(z_0)$ from Eq. (10.4.8)).

Example

A Laurent series with a first-order pole at $z = 0$ that is analytic in the infinite ring $|z| > 0$ is

$$\frac{\sin z}{z} = \frac{1}{z} - \frac{z}{3!} + \frac{z^3}{5!} + \cdots \tag{10.4.13}$$

That a function is uniquely defined by its Laurent series leads to the technique of *analytic continuation*. If a function f_1 that is analytic in a region R and a second function f_2 that is analytic in a region S that partially overlaps R agree in the overlapping region $S \cap R$, the functions are identical since all terms in their series expansions in $S \cap R$ are equal.

10.5 RESIDUE THEOREM

Integrals in the complex domain are often simply evaluated by integrating over a closed contour and applying Cauchy's theorem. Accordingly, integrals over an open contour C_{open} are often transformed into integrals over closed contours by closing the contour over a path on which the integral either vanishes or is simply related to its value over C_{open}. If the integration region contains *branch cuts*, associated with edges of principal sheets, additional issues arise as the contour must be deformed to avoid these edges.

An integral of a function $f(z)$ over a closed contour C enclosing N singularities of $f(z)$ can be recast as the sum of a *singularity-free* integral formed by joining C through infinitesimally displaced, oppositely directed line segments to N *clockwise* integrals around the singularities as in Figure 10.3. Therefore, the *difference* of the integral over C and the sum of the *counterclockwise* integrals over z_i equal zero. Since $f(z)$ is analytic within C except at z_i, it possesses a Laurent series expansion near each z_i. Only the first-order pole term of the form $b_{-1}/(z - z_i)$ yields a nonzero contribution to the integral of each series around the encompassing infinitesimal circle according to Equation (10.4.6). The integral over C thus equals $2\pi i$ *times the sum of the coefficients of all the first-order poles within C.*

From the preceding paragraph, if the function possesses an mth-order pole at $z = z_0$, in the vicinity of this point,

$$f(z) = b_{-m}(z - z_0)^{-m} + b_{-m+1}(z - z_0)^{-m+1} + \ldots + b_{-1}(z - z_0)^{-1} + b_0 + \ldots \tag{10.5.1}$$

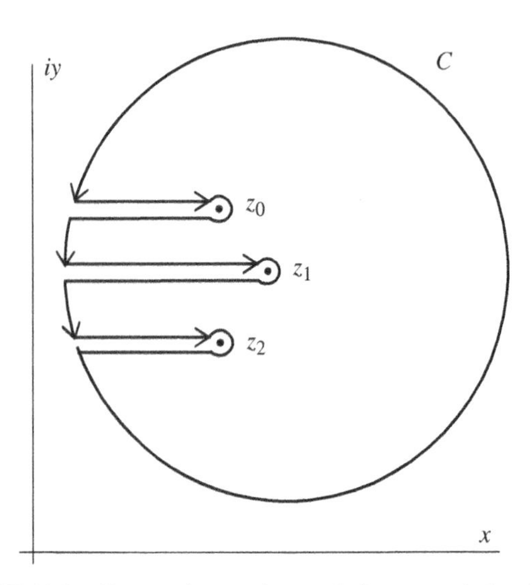

FIGURE 10.3 Contour integration enclosing several singularities.

so that

$$(z-z_0)^m f(z) = b_{-1} + b_{-m+1}(z-z_0) + \ldots + b_{-1}(z-z_0)^{m-1} + b_0(z-z_0)^m + \ldots \quad (10.5.2)$$

only the coefficient or *residue* of the first-order pole term given by

$$b_{-1} = \frac{1}{(m-1)!} \lim_{z \to z_0} \frac{d^{m-1}}{dz^{m-1}} \left((z-z_0)^m f(z) \right) \quad (10.5.3)$$

contributes to the integral around C.

Examples

1. The residue of $(z+1)^2/(z-1)^2$ at $z = 1$ is from Equation (10.5.3):

$$\lim_{z \to 1} \frac{d}{dz} \left((z-1)^2 \left(\frac{z+1}{z-1} \right)^2 \right) = \lim_{z \to 1} 2(z+1) = 4 \quad (10.5.4)$$

as can be obtained directly by expanding about the point $z = 1$ after writing $z = 1 + \varepsilon$

$$\left(\frac{z+1}{z-1} \right)^2 = \left(\frac{1+\varepsilon+1}{\varepsilon} \right)^2 = \frac{4+4\varepsilon+\varepsilon^2}{\varepsilon^2} = \frac{4}{\varepsilon^2} + \frac{4}{\varepsilon} + 1 \quad (10.5.5)$$

The residue, which is the coefficient of $1/(z-1) = 1/\varepsilon$, again evaluates to 4.

2. An integral over an open contour that can be solved with the method of residues is

$$I_k(a) = \int_{-\infty}^{\infty} \frac{e^{ikx}}{x - ia} dx \qquad (10.5.6)$$

for $k > 0$, which possesses a single first-order pole in the integrand at $x = ia$. The contour from $-\infty$ to ∞ can be completed by a half circle in the *positive* half plane that does not affect the integral since

$$\lim_{R\to\infty} \int_0^{\pi} \frac{e^{ikR(\cos\theta + i\sin\theta)}}{R} R d\theta < \lim_{R\to\infty} \int_0^{\pi} e^{-kR\sin\theta} d\theta = 0 \qquad (10.5.7)$$

Over the closed contour, for $a > 0$, $I_k(a)$ equals $2\pi i$ times the residue at ia. On the other hand, if $a < 0$, no singularity exists within the contour and $I_k = 0$. For $a = 0$, the singularity falls on the integration contour. Heuristically, since the contour encloses only the upper half of the singularity, this *Cauchy principal value* or *principal part* integration over the real axis, defined as the integral excluding the region in the immediate vicinity of the singularity, yields half the contribution $(1/2)(2\pi i)$ from a fully enclosed singularity. That is, after replacing $\frac{1}{z}$ by an equivalent real limit, the integral can be written as, where the contour must be completed in the upper half plane and hence only the second term in the integral with a pole in this region contributes,

$$P\int_{-\infty}^{\infty} \frac{e^{ikz}}{z} dz = \lim_{\varepsilon\to 0} \frac{1}{2} \int_{-\infty}^{\infty} e^{ikz} \left(\frac{1}{z + i\varepsilon} + \frac{1}{z - i\varepsilon} \right) dz = \pi i \qquad (10.5.8)$$

Combining the previous results, $I_k(a) = 2\pi i e^{-ka}\theta(a)$, where $\theta(a)$ denotes the *Heaviside step function* that equals 1, 0, and ½ for $a > 0$, $a < 0$, and $a = 0$, respectively. A third analysis of principal value integration is presented in Section 10.6.

3. Symmetries of the integrand can sometimes be employed to complete an open contour as in

$$I = \int_{-\infty}^{\infty} \frac{e^{cx}}{1 + e^{x}} dx \qquad (10.5.9)$$

with $0 < c < 1$. The contour can be closed by the infinite rectangle of Figure 10.4, where the upper segment is the line $\text{Im}(z) = 2\pi i$ and the right and left sides of the graph represent $x = \pm\infty$, since

a. The integrals along the two vertical sections at $x = \pm\infty$ vanish since $e^{(c-1)z} \to 0$ at $z = \infty + iy$ and $e^{cz} \to 0$ at $z = -\infty + iy$.

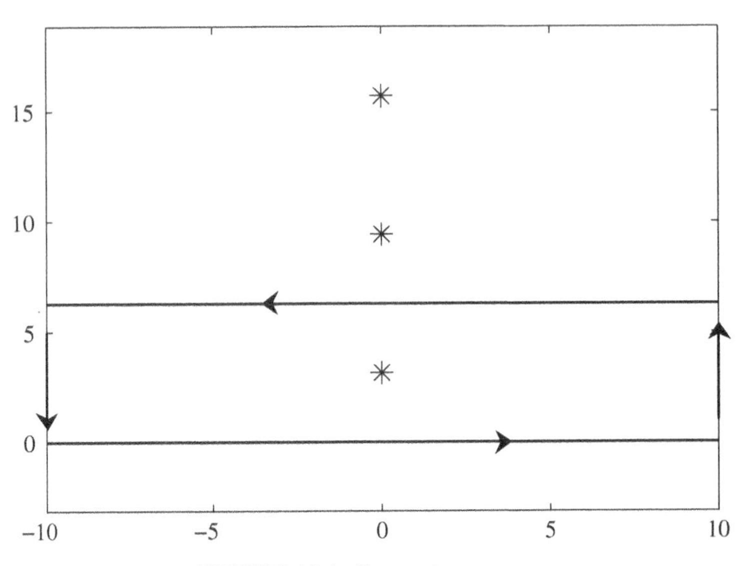

FIGURE 10.4 Integration contour.

b. Along the upper line of the contour, $e^{c(z+2\pi i)} = e^{2\pi ci}e^z$ and $1 + e^{z+2\pi i} = 1 + e^z$ so that (as a result of the reversed direction of integration) $I \to -e^{2\pi ci}I$.

c. C contains a single pole at $z = i\pi$ for which $e^z = -1$. Near the pole,

$$\frac{e^{\pi ci}}{1+e^{i\pi+(z-i\pi)}} = \frac{e^{\pi ci}}{1+(-1)\left(1+(z-i\pi)+(z-i\pi)^2/2!+\dots\right)} \approx -\frac{e^{\pi ci}}{z-i\pi} \tag{10.5.10}$$

Hence, the residue is $-e^{\pi ci}$ so that

$$I - e^{2\pi ci}I = -2\pi i e^{\pi ci} \tag{10.5.11}$$

and finally,

$$I = \frac{-2\pi i e^{\pi ci}}{1-e^{2\pi ci}} = \frac{-2\pi i}{e^{-\pi ci}-e^{\pi ci}} = \frac{\pi}{\sin \pi c} \tag{10.5.12}$$

If $f(z)$ contains p poles of order n and z zeros of order m inside a contour C,

$$\frac{1}{2\pi i}\oint_C \frac{f'(z)}{f(z)}dz = zm - pn \tag{10.5.13}$$

In particular, at a zero of order m, $f(z) = b_m(z-z_0)^m + b_{m+1}(z-z_0)^{m+1} + \dots$ so that as $z - z_0 \to 0$, the contribution from such a singularity is

$$\frac{1}{2\pi i}\oint_C \frac{f'(z)}{f(z)}dz = \frac{1}{2\pi i}\lim_{z-z_0\to 0}\oint_C \frac{mb_m(z-z_0)^{m-1}+(m+1)b_{m+1}(z-z_0)^m+\dots}{b_m(z-z_0)^m+b_{m+1}(z-z_0)^{m+1}+\dots}dz$$

$$= \frac{1}{2\pi i}\oint_C \frac{m}{z-z_0}dz = m \tag{10.5.14}$$

At a pole of order n, $f(z) = b_{-n}(z-z_0)^{-n} + b_{-n+1}(z-z_0)^{-n+1} + \dots$, replacing m by $-n$ above.

10.6 DISPERSION RELATIONS

If $f(z) = u(z) + iv(z)$ is analytic and vanishes faster than $1/|z|$ as $|z| \to \infty$ in the upper half plane,

$$\frac{1}{2\pi}\oint_C \frac{f(z')}{z'-x}dz' = 0 \tag{10.6.1}$$

where C is taken along the real axis but avoids the pole at $z' = x$ by, e.g., traversing a half circle of infinitesimal radius above the singularity in the clockwise direction as illustrated in Figure 10.5. The contribution to the integral over this section is $-\pi i f(x)$, namely, half the result obtained by fully encircling the singularity with an additional negative sign associated with the contour direction.

As the integral vanishes over the infinite semicircle, its principal part, P, along the real axis that arises from the contribution away from the pole, Equation (10.6.1) becomes

$$P\int_{-\infty}^{\infty} \frac{u(x')+iv(x')}{x'-x}dx' = \pi(iu(x)-v(x)) \tag{10.6.2}$$

Equating real and imaginary terms leads to the *dispersion relations*

$$u(x) = \frac{1}{\pi}P\int_{-\infty}^{\infty} \frac{v(x')}{x'-x}dx'$$
$$v(x) = -\frac{1}{\pi}P\int_{-\infty}^{\infty} \frac{u(x')}{x'-x}dx' \tag{10.6.3}$$

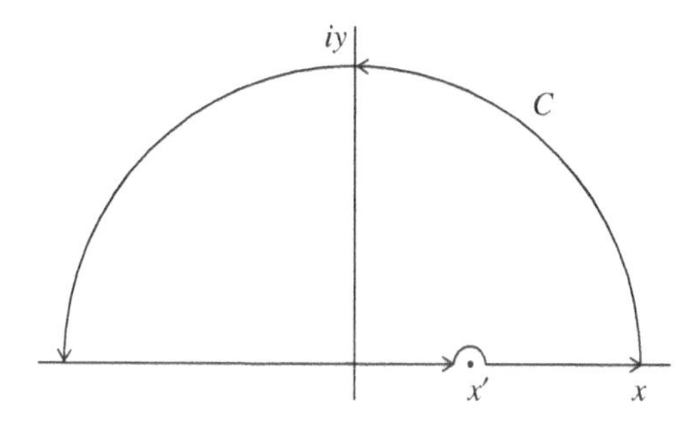

FIGURE 10.5 Contour for principal part integration.

10.7 METHOD OF STEEPEST DECENT

Contour integrals over exponential functions can often be approximated by integrals in the complex plane over Gaussian functions. The contour is then deformed to pass through saddle points along directions of constant phase.

Example

For the gamma function of Equation (6.2.5)

$$\Gamma(n+1) = n! = \int_0^\infty e^{-x} x^n dx = \int_0^\infty e^{-x+n\ln x} dx \tag{10.7.1}$$

the maximum value of the integrand occurs at

$$\frac{d}{dx}(-x+n\ln x) = -1 + \frac{n}{x} = 0 \tag{10.7.2}$$

Expanding the exponent in a power series about the maximum $x = n$ yields up to second order

$$\int_0^\infty e^{-x+n\ln x} dx \approx \int_0^\infty e^{-n+n\ln n+\frac{(x-n)^2}{2!}\left(-\frac{n}{x^2}\right)\big|_{x=n}} dx = e^{-n+n\ln n} \int_0^\infty e^{-\frac{1}{n}\frac{(x-n)^2}{2!}} dx \tag{10.7.3}$$

Integrating along the real axis insures that the phase is constant over the contour and that the integral adopts a Gaussian form. That is, the complex function $e^{-c(x-n)^2} = e^{-cz'^2}$ with $z' = z - n$ possesses a maximum along real z' but a minimum along imaginary z' about $z' = 0$. Thus, the integrand describes a saddle point in the complex plane; such that for the real axis contour the result originates from a small region of zero phase around the extremum. Since the integrand is small at the $x = 0$ lower limit for large n, this limit can further be replaced by $-\infty$. Substituting $x' = (x-n)/\sqrt{2n}$ for which $dx' = dx/\sqrt{2n}$ yields *Stirling's approximation*

$$n! = \sqrt{2\pi n} e^{-n+n\ln n} = \sqrt{2\pi n} n^n e^{-n} \tag{10.7.4}$$

which is generally cited in its lowest order form $\ln n! \approx n \ln n - n$.

The method of steepest decent can be applied to integrals over functions for which the path of maximum negative curvature is oriented in any complex direction since a closed contour can be deformed within any singularity-free region without altering the integral. In practical calculations, integrals of the form

$$I(t) = \int_C R(z) e^{tI(z)} dz \tag{10.7.5}$$

are performed by applying the formula

$$I(t) = \sum_{m=1}^{M} I(t) = \sqrt{2\pi} \sum_{m=1}^{M} \left| t \frac{\partial^2 I(z_m)}{\partial z^2} \right|^{-\frac{1}{2}} R(z_m) e^{tI(z_m) + \theta_m} \tag{10.7.6}$$

which results from Gaussian integration over a line $z = z_m + te^{i\theta_m}$ along the direction of maximum negative curvature through each saddle point z_m $(m = 1, 2, \ldots, M)$. Equation (10.7.6) reproduces Equation (10.7.5) by setting $t = n$, $R = 1$, $I = -x/n + \ln x$, $\theta = 0$ and $z_1 = n$.

11

DIFFERENTIAL EQUATIONS

As noted earlier, differentiation is a direct operation that yields $f(x)$ from $Dy(x) = f(x)$, with $D = d/dx$, while integration solves the inverse problem of determining $y(x)$ given $f(x)$. This chapter considers inverse problems for more involved differential operators.

11.1 LINEARITY, SUPERPOSITION, AND INITIAL AND BOUNDARY VALUES

A differential equation of *order m* and *degree n* contains derivatives of up to order m as well as powers of derivatives such as $(Dy)^2$ up to the nth degree. The coefficients of the derivative terms are in general functions of the independent and dependent variables x and y. A *linear* differential equation $Ly(x) = f(x)$ is of first degree with coefficients that are independent of y so that L obeys

$$\begin{aligned} L(ay) &= aL(y) \\ L(y_1 + y_2) &= L(y_1) + L(y_2) \end{aligned} \tag{11.1.1}$$

The first property implies that the form of a solution is independent of its amplitude, while the second implies that any sum or *superposition* of individual solutions is

Fundamental Math and Physics for Scientists and Engineers, First Edition.
David Yevick and Hannah Yevick.

a solution. The form of an *inhomogeneous* linear kth-order differential equation is given by

$$\sum_{n=0}^{k} c_n(x) D^n y = f(x) \tag{11.1.2}$$

while $f(x) = 0$ for a *homogeneous* equation.

The solution of an inhomogeneous differential equation is a superposition of a particular solution of Equation (11.1.2), with a solution of the corresponding homogeneous equation. The homogeneous solution, which does not affect the right-hand side of the above equation, is then uniquely determined by the boundary conditions. That is, an n:th-order differential equation provides a single relationship among $y(x)$ and its first n derivatives at each point x leaving n degrees of freedom undetermined. A unique solution is therefore obtained by imposing n conditions at either the initial, final, or boundary values of x.

Example

The solution of $Dy = \cos x$ is the sum of the unique inhomogeneous solution, $y_{\text{particular}} = \sin x$, and the nonunique homogeneous solution of $Dy = 0$, $y_{\text{general}} = c$. A unique solution is obtained by specifying the value of $y(x)$ at some initial or final point $y(a) = y_0$.

11.2 NUMERICAL SOLUTIONS

A first-order differential equation of the form

$$Dy = y(y - 2.5) \tag{11.2.1}$$

relates the slope of $y(x)$ to a function of its amplitude. The following Octave function file, `derpol.m`, displays the *slope field*, which displays the slope, $Dy(x)$, as a vector $(dx, dy)/\sqrt{(dx)^2 + (dy)^2}$ at each point on a grid of (x, y) values:

```
function [xSlopeProjection, ySlopeProjection] = derpol(x, y)
  mySlope = y .* ( y - 2.5 );
  xSlopeProjection = 1 ./ sqrt( 1 + mySlope.^2 );
  ySlopeProjection = mySlope ./ sqrt( 1 + mySlope.^2 );
endfunction
```

The calling program is placed in a second file, `vecfield.m`,

```
[ xf, yf ] = meshgrid( -5 : 0.5 : 5, -5 : 0.5 : 5 );
[ xProjection, yProjection ] = derpol( xf, yf );
```

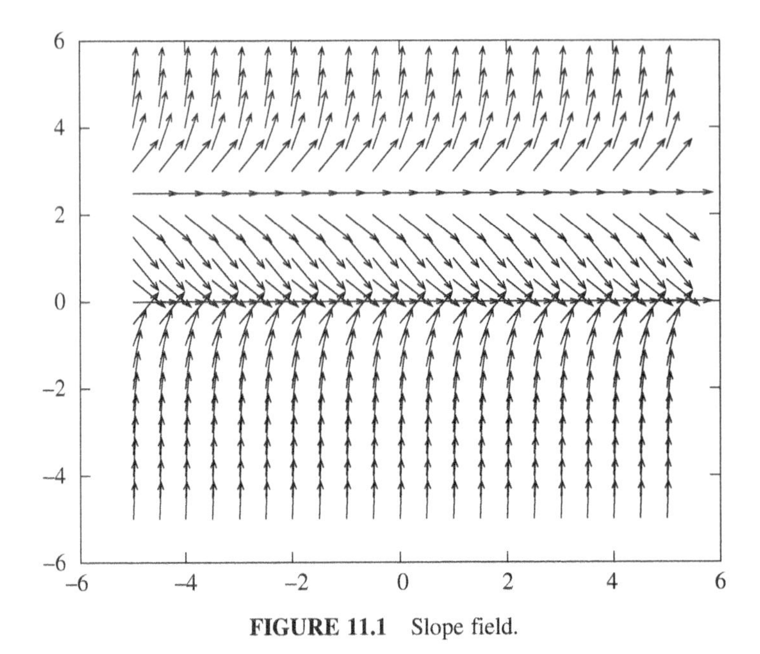

FIGURE 11.1 Slope field.

```
quiver( xf, yf, xProjection, yProjection );
print( "quiver", "-demf" );
```

and is run after navigating to the directory containing both files by typing `vecfield` at the Octave command prompt. The result (Fig. 11.1) clarifies the dependence of the solution on the initial conditions; for an initial value of y between $-\infty$ and 2.5, the solution evolves along the arrows to the *stable fixed point* $y = 0$ from which any small deviation of the solution diminishes with increasing x. The value $y = 2.5$ represents an *unstable fixed point* for which an infinitesimal perturbation instead grows with x.

Equation (11.2.1) can be solved numerically with *Euler's method* by applying the discrete forward difference representation of the derivative (Equation (6.1.1)). This corresponds to numerical integration and requires a single initial condition

$$\begin{aligned} D^{+}y(x) &\equiv \frac{y(x+\Delta x)-y(x)}{\Delta x} = y(x)(y(x)-2.5) \\ y(x+\Delta x) &= y(x)(1+(y(x)-2.5)\Delta x) \end{aligned} \tag{11.2.2}$$

as is implemented numerically by

```
y(1) = 1;
numberOfSteps = 100;
deltaX = 0.1;
xAxis = deltaX * 1 : numberOfSteps;
for loop = 2 : numberOfSteps
  y(loop) = y(loop - 1) * ( 1 + ( y(loop - 1) - 2.5 ) * deltaX );
end
plot( xAxis, y )
```

Second- or higher-order differential equations can be written as a system of first-order equations. For

$$D^2y = -cDy - ky \tag{11.2.3}$$

introducing the "velocity" $v = Dy$ yields

$$\begin{aligned} Dy &= v \\ Dv &= -cv - ky \end{aligned} \tag{11.2.4}$$

or equivalently, in matrix form,

$$D\begin{pmatrix} y \\ v \end{pmatrix} = \begin{pmatrix} 0 & 1 \\ -k & -c \end{pmatrix}\begin{pmatrix} y \\ v \end{pmatrix} \tag{11.2.5}$$

Subsequently, from the forward finite difference representation of the derivative operator,

$$\begin{aligned} y(x+\Delta x) &= y(x) + v\Delta x \\ v(x+\Delta x) &= v(x) + (-cv(x) - ky(x))\Delta x \end{aligned} \tag{11.2.6}$$

which is implemented after specifying initial conditions for y and v as follows:

```
y(1) = 1;
v(1) = 0;
c = 0.1;
k = 0.5;
numberOfSteps = 100;
deltaX = 0.1;
for loop = 2 : numberOfSteps
  y(loop) = y(loop - 1) + v(loop - 1) * deltaX;
  v(loop) = v(loop - 1) + ( -c * v(loop - 1) - k * y(loop - 1) )
            * deltaX;...
end
plot( y, v )
figure( 2 )
plot( y )
```

The first figure corresponds to the *phase space* plot of the velocity as a function of position, while the second displays the solution y. Since this numerical solution procedure generates unphysical divergences in energy-conserving systems, each calculation should be verified by examining the dependence of the result on the magnitude of Δx.

11.3 FIRST-ORDER DIFFERENTIAL EQUATIONS

Although even if $f(x, y)$ is a function only of x in the general first-order differential equation

$$Dy = f(x, y) \tag{11.3.1}$$

the integral expression for $y(x)$ can be analytically intractable, for certain $f(x, y)$, straightforward solution methods exist. A function $f(x, y)$ of the form $X(x)Y(y)$ yields a *separable* equation that can in many cases be solved by integrating according to

$$\int_{y(x_0)}^{y(x)} \frac{dy}{Y(y)} = \int_{x_0}^{x} X(x)dx \tag{11.3.2}$$

Example

For $f(x, y) = yx^2$,

$$\begin{aligned} \int \frac{1}{y} dx &= \int x^2 dy \\ \log y &= \frac{x^3}{3} + c \\ y &= e^c e^{x^3/3} = c' e^{x^3/3} \end{aligned} \tag{11.3.3}$$

The constant c' is determined from the initial conditions.

When $f(x, y)$ is not a product $X(x)Y(y)$, Equation (11.3.1) may still be separable after appropriate transformations. Unfortunately, these effectively constitute more involved versions of the variable substitutions employed in integration and are therefore often not apparent. If, however, $f(x, y)$ can be expressed as a ratio, $-m(x, y)/n(x, y)$, such that Equation (11.3.1) becomes

$$m(x, y)dx + n(x, y)dy = 0 \tag{11.3.4}$$

where additionally $m(x, y)$ and $n(x, y)$ are both *homogeneous of degree p*, e.g., $n(\alpha x, \alpha y) = \alpha^p n(x, y)$, then

$$m(x, y)dx + n(x, y)dy = x^n \left(m\left(\frac{x}{x}, \frac{y}{x}\right)dx + n\left(\frac{x}{x}, \frac{y}{x}\right)dy \right) = 0 \tag{11.3.5}$$

Since m and n on the right-hand side of Equation (11.3.5) then only depend on x through the ratio $w = y/x$ for which $dy = wdx + xdw$, the separable equation

$$m(w)dx + n(w)(wdx + xdw) = (m(w) + n(w)w)dx + n(w)xdw = 0 \tag{11.3.6}$$

results.

Alternatively, if a function $k(x, y)$ can be identified in Equation (11.3.4) such that

$$\begin{aligned}\frac{\partial k}{\partial x} &= m(x,y)\\ \frac{\partial k}{\partial y} &= n(x,y)\end{aligned} \tag{11.3.7}$$

then

$$\begin{aligned}0 &= m(x,y)dy + n(x,y)dx\\ &= \frac{\partial k(x,y)}{\partial y}dy + \frac{\partial k(x,y)}{\partial x}dx\\ &= dk(x,y)\end{aligned} \tag{11.3.8}$$

which yields

$$k(x,y) = c \tag{11.3.9}$$

From Equation (11.3.7), such differential equations, which are termed *exact*, must fulfill the condition

$$\frac{\partial m(x,y)}{\partial y} = \frac{\partial^2 k(x,y)}{\partial y \partial x} = \frac{\partial^2 k(x,y)}{\partial x \partial y} = \frac{\partial n(x,y)}{\partial x} \tag{11.3.10}$$

Nonexact equations can sometimes be recast into a separable form through a transformation of variables or multiplied with an *integrating factor* $F(x, y)$ such that

$$F(x,y)(Dy - f(x,y)) = d(\mu(x,y)) = 0 \tag{11.3.11}$$

Example

For $f(x, y) = yx^2$ solved in Equation (11.3.3), multiplication by $e^{x^3/3}/y^2$ yields

$$\frac{e^{\frac{x^3}{3}}}{y^2}\left(\frac{dy}{dx} - x^2 y\right) = -\left[e^{\frac{x^3}{3}}\frac{d}{dx}\left(\frac{1}{y}\right) + \frac{x^2}{y}e^{\frac{x^3}{3}}\right] = -\frac{d}{dx}\left(\frac{e^{\frac{x^3}{3}}}{y}\right) = 0 \tag{11.3.12}$$

which again yields $e^{x^3/3} = cy$.

An integration factor for first-order differential equations of the form

$$Dy + a(x)y = b(x) \tag{11.3.13}$$

is given by

$$e^{\int_l^x a(x')dx'} \tag{11.3.14}$$

which yields

$$D\left(ye^{\int_l^x a(x')dx'}\right) = b(x)e^{\int_l^x a(x')dx'} \tag{11.3.15}$$

Consequently, for the boundary condition $y(l) = 0$,

$$\begin{aligned} ye^{\int_l^x a(x')dx'} &= \int_l^x b(x'')e^{\int_l^{x''} a(x')dx'} dx'' + c \\ y &= e^{-\int_l^x a(x')dx'} \int_l^x b(x'')e^{\int_l^{x''} a(x')dx'} dx'' \end{aligned} \tag{11.3.16}$$

11.4 WRONSKIAN

A set of n solutions, $\{\phi_i\}$, to the nth-order linear differential equation

$$D^n y + f_1(x)D^{n-1}y + \ldots f_{n-1}(x)Dy + f_n(x)y = 0 \tag{11.4.1}$$

are termed *linearly dependent* on an interval if a set of constants exist such that

$$\Phi = \sum_{i=1}^{n} c_i \phi_i = 0 \tag{11.4.2}$$

throughout the interval so that one or more of the solutions can be expressed as a linear combination of the remaining functions; otherwise, the solutions are termed *linearly independent*. Linear independence is insured if the *Wronskian*

$$W(\phi_1, \phi_2, \ldots, \phi_n) = \begin{vmatrix} \phi_1 & \phi_2 & \cdots & \phi_n \\ \phi_1' & \phi_2' & \cdots & \phi_n' \\ \vdots & \vdots & \ddots & \vdots \\ \phi_1^{(n-1)} & \phi_2^{(n-1)} & \cdots & \phi_n^{(n-1)} \end{vmatrix} \tag{11.4.3}$$

where $\phi^{(p)}$ denotes the pth derivative of ϕ, differs from zero throughout the interval. That is, if, a set of constants exist for which Equation (11.4.2) is satisfied for at least one point, in the interval, repeatedly repeatedly differentiating both sides of this equation indicates that in this case not only Φ but also its first $n - 1$ derivatives are zero at these points. This matches the unique zero solution to the differential equation; hence, $\Phi = 0$ and the functions are linearly dependent with zero Wronskian throughout the interval as e.g. the first column of W can be replaced by a linear combination of the other columns.

11.5 FACTORIZATION

If all $f_j(x)$ in Equation (11.4.1) are *constants*, the equation can be factorized as

$$(D-b_n)(D-b_{n-1})\dots(D-b_1)y(x)=0 \tag{11.5.1}$$

If all b_i differ, since the operators $(D-b_i)$ commute, the general solution is a superposition of the solutions of $Dy_i = b_i y_i$ with $i = 1, 2, \dots, n$, i.e.,

$$y=\sum_{i=1}^{n} c_i e^{b_i x} \tag{11.5.2}$$

However, when m of b_i are identical, the relevant equation for these factors becomes

$$(D-b_l)^m y(x)=0 \tag{11.5.3}$$

Writing Equation (11.5.3) as $(D-b_l)[(D-b_l)^{m-1}y(x)]=0$ yields initially

$$(D-b_l)^{m-1}y(x)=\tilde{c}_{m-1}e^{b_l x} \tag{11.5.4}$$

Multiplying on both sides by the integrating factor $e^{-b_l x}$,

$$\begin{aligned} e^{-b_l x}(D-b_l)(D-b_l)^{m-2}y(x) &= \tilde{c}_{m-1} \\ D\left[e^{-b_l x}(D-b_l)^{m-2}\right]y(x) &= \tilde{c}_{m-1} \end{aligned} \tag{11.5.5}$$

After both sides are integrated

$$\begin{aligned} e^{-b_l x}(D-b_l)^{m-2}y(x) &= \tilde{c}_{m-1}x+\tilde{c}_{m-2} \\ (D-b_l)^{m-2}y(x) &= (\tilde{c}_{m-1}x+\tilde{c}_{m-2})e^{b_l x} \end{aligned} \tag{11.5.6}$$

and the procedure repeated an additional $m-2$ times the final solution is obtained in the form

$$y=\left(c_{m-1}x^{m-1}+c_{m-2}x^{m-2}+\dots+c_0\right)e^{b_l x} \tag{11.5.7}$$

11.6 METHOD OF UNDETERMINED COEFFICIENTS

Considering next *particular* solutions of *inhomogeneous* linear differential equations, if the inhomogeneous term is composed of functions whose form remains invariant upon differentiation, the solution can be sought as a linear superposition of such functions with undetermined coefficients. The coefficients are determined by substituting the proposed solution into the equation.

Example

To solve

$$(D^2+4)y=4x^2-2 \tag{11.6.1}$$

since derivatives of polynomials remain polynomials, a possible trial solution is

$$ax^2+bx+c \tag{11.6.2}$$

Inserting this into Equation (11.6.1) yields

$$2a+4ax^2+4bx+4c=4x^2-2 \tag{11.6.3}$$

Equating corresponding coefficients leads to $a=1,\ b=0,\ c=-1$.

The method similarly applies to harmonic inhomogeneous terms as the form of such functions is unchanged upon differentiation. That is, for an equation of the form $p(D)y=e^x+e^{2x}$, where p represents a polynomial, $A\exp(x)+B\exp(2x)$ is a possible trial function since no new terms are obtained by differentiating the exponential functions. If, however, one of these exponential functions corresponds to an l-fold root of $p(D)$, the trial function adopts the form of x^l times the exponential together with all terms (with undetermined coefficients) obtained by differentiating this expression. Similarly, for an equation $p(D)y=\cos kx$, a possible solution is given by $A\cos kx+B\sin kx$, unless again $p(D)$ possesses multiple roots (alternatively, $2\cos(kx)=\exp(ikx)+\exp(-ikx)$ can be employed together with $A'\exp(ikx)+B'\exp(-ikx)$).

11.7 VARIATION OF PARAMETERS

A trial expression for the particular solution of an inhomogeneous linear differential equation of order n with constant coefficients, e.g., Equation (11.1.2), with all $c_n(x)$ constant functions and zero replaced by $g(x)$ after the equals sign, can be expressed as a sum

$$\Phi=\sum_{i=1}^{n}\phi_i(x)t_i(x) \tag{11.7.1}$$

of products of each of the n linearly independent solutions, $\phi_i(x),\ i=1,2,\ldots,n$, to the homogeneous equation with unknown functions, $t_i(x)$. The $t_i(x)$ can then be constructed such that the first $n-1$ derivatives acting on Φ lead to these derivatives acting only on the $\phi_i(x)$, while the nth derivative generates the inhomogeneous term $g(x)$. That is, the first derivative of Φ gives

$$\Phi'=\sum_{i=1}^{n}\left(\phi_i'(x)t_i(x)+\phi_i(x)t_i'(x)\right) \tag{11.7.2}$$

Imposing the condition

$$\sum_{i=1}^{n} \phi_i(x) t_i'(x) = 0 \tag{11.7.3}$$

leads to

$$\Phi' = \sum_{i=1}^{n} \phi_i'(x) t_i(x) \tag{11.7.4}$$

Repeating this procedure results in, where $\phi_i^{(m)} \equiv d^m \phi_i / dx^m$,

$$\sum_{i=1}^{n} \phi_i^{(m)}(x) t_i'(x) = 0, \quad m = 0, 1, \ldots, n-2 \tag{11.7.5}$$

while the inhomogeneous term is incorporated at the last step by setting

$$\sum_{i=1}^{n} \phi_i^{(n-1)}(x) t_i'(x) = g(x) \tag{11.7.6}$$

Example

The homogeneous solution to the equation

$$y'' + y = \cos x \tag{11.7.7}$$

is given by $y = c_1 \cos x + c_2 \sin x$. Accordingly, taking $y = t_1(x)\cos x + t_2(x)\sin x$ as the particular solution and imposing the constraints, obtained with $n = 2$ in the above equations,

$$\begin{aligned} t_1' \cos x + t_2' \sin x &= 0 \\ -t_1' \sin x + t_2' \cos x &= \cos x \end{aligned} \tag{11.7.8}$$

yields $t_1' = -\sin x \cos x = -\sin 2x/2$ and $t_2' = \cos^2 x = (\cos 2x + 1)/2$ or after integration $t_1 = \cos 2x/4$ and $t_2 = (\sin 2x + 2x)/4$. The particular solution therefore equals

$$\begin{aligned} \Phi &= \sum_{i=1}^{2} \phi_i t_i \\ &= \frac{1}{4}\cos x \cos 2x + \frac{1}{4}\sin x \sin 2x + \frac{1}{2} x \sin x \\ &= \frac{1}{4}\cos x\left(1 - 2\sin^2 x\right) + \frac{1}{4}\sin x(2 \sin x \cos x) + \frac{1}{2} x \sin x \\ &= \frac{1}{4}\cos x + \frac{1}{2} x \sin x \end{aligned} \tag{11.7.9}$$

11.8 REDUCTION OF ORDER

If a homogeneous solution of an nth-order linear differential equation is known, the equation can be replaced by a new equation of order $n-1$. To illustrate, if $\phi(x)$ is a homogeneous solution of the general linear inhomogeneous second-order equation

$$\left(D^2+p(x)D+q(x)\right)y(x)=r(x) \tag{11.8.1}$$

and $y(x)$ is expressed as $w(x)\phi(x)$, after substitution

$$\phi w''+(2\phi'+p(x)\phi)w'+(\phi''+p(x)\phi'+q(x)\phi)w=r(x) \tag{11.8.2}$$

Since, however, $\phi(x)$ is a homogeneous solution, the third term on the left-hand side of the above equation vanishes. Substituting $u(x)$ for $w'(x)$ then yields the first-order differential equation

$$\phi u'+(2\phi'+p(x)\phi)u(x)=r(x) \tag{11.8.3}$$

If $u(x)$ can be determined, $w(z)$ is obtained by solving the first-order equation $w'(z)=u(z)$.

11.9 SERIES SOLUTION AND METHOD OF FROBENIUS

Differential equations can often be solved by expanding both the solution and the functions that appear in the equation as power series in the independent variable, x. In the *method of Frobenius*, if $P(x)$ and $Q(x)$ in

$$\left(D^2+P(x)D+Q(x)\right)y(x)=0 \tag{11.9.1}$$

possess convergent power series expansions at the origin, which is then termed a *regular* point, expanding $P(x)$, $Q(x)$, and $y(x)$ around $x=0$ and setting the coefficients of equal powers of x to zero yield a *recursion relation* among the Taylor series coefficients of $y(x)$.

Example

Inserting

$$y=\sum_{m=0}^{\infty}c_m x^m \tag{11.9.2}$$

into $(D-1)y=0$ gives

$$\sum_{m=0}^{\infty}(c_{m+1}(m+1)-c_m)x^m=0 \tag{11.9.3}$$

Since x^m are linearly independent, the coefficient of each x^j must be zero separately; therefore,

$$c_{m+1}=\frac{c_m}{m+1} \tag{11.9.4}$$

which reproduces the power series expansion of $c_0 \exp(x)$.

At a *regular singular point*, $P(x)$ and $Q(x)$ are singular at the origin, but $xP(x)$ and $x^2Q(x)$ are nonsingular so that

$$P(x)=\sum_{m=-1}^{\infty}P_m x^m \quad Q(x)=\sum_{m=-2}^{\infty}Q_m x^m \tag{11.9.5}$$

At singular points, y and its derivatives are restricted to a set (possibly empty) of special values in order for a simple Taylor series solution to exist.

Example

The equation $D^2y=2y/x^2$ possesses the solutions $y=x^2$ and $y=1/x$, where however only the first of these is valid if the finite curvature condition that $y(x)\to 0$ for $x\to 0$ least as rapidly as x^2 is imposed. In this more general case, the trial solution is therefore taken as

$$y=x^k\sum_{m=0}^{\infty}a_m x^m \tag{11.9.6}$$

where k can be integer, real, or complex. Additionally, if one of the solutions $\tilde{y}$ to Equation (11.8.1) contains a logarithmic singularity, it is obtained from the regular solution y by replacing the above expression by

$$\tilde{y}=\tilde{a}_{-1}y\ln x+x^k\sum_{m=0}^{\infty}\tilde{a}_m x^m \tag{11.9.7}$$

11.10 SYSTEMS OF EQUATIONS, EIGENVALUES, AND EIGENVECTORS

An nth-order inhomogeneous linear differential equation

$$\left(a_nD^n+a_{n-1}D^{n-1}+\ldots+a_0\right)y=f(x) \tag{11.10.1}$$

with constant coefficients can be written as a system of n first-order equations (cf. Eq. (11.2.4))

$$\begin{aligned} w_1(x) &= \frac{dy}{dx} \\ w_2(x) &= \frac{dw_1}{dx} \\ w_3(x) &= \frac{dw_2}{dx} \\ &\vdots \\ \frac{dw_{n-1}}{dx} &= -a_{n-1}w_{n-1} - a_{n-2}w_{n-2} - \ldots - a_0 y + f(x) \end{aligned} \tag{11.10.2}$$

or equivalently as the matrix equation

$$\frac{d}{dx}\begin{pmatrix} y \\ w_1 \\ w_2 \\ \vdots \\ w_{n-1} \end{pmatrix} = \begin{pmatrix} 0 & 1 & 0 & \cdots & 0 \\ 0 & 0 & 1 & \cdots & 0 \\ 0 & 0 & 0 & \cdots & 0 \\ \vdots & \vdots & \vdots & \ddots & 1 \\ -a_0 & -a_1 & -a_2 & \cdots & -a_{n-1} \end{pmatrix}\begin{pmatrix} y \\ w_1 \\ w_2 \\ \vdots \\ w_{n-1} \end{pmatrix} + \begin{pmatrix} 0 \\ 0 \\ 0 \\ \vdots \\ f(x) \end{pmatrix} \tag{11.10.3}$$

with the initial condition

$$\vec{w}(x_0) \equiv \begin{pmatrix} y(x_0) \\ w_1(x_0) \\ w_2(x_0) \\ \vdots \\ w_{n-1}(x_0) \end{pmatrix} \tag{11.10.4}$$

or symbolically

$$\frac{d\,\vec{w}(x)}{dx} = \mathbf{M}\,\vec{w}(x) + \vec{f}(x) \tag{11.10.5}$$

The homogeneous equation is solved by

$$\vec{w}(x) = e^{\mathbf{M}(x-x_0)}\,\vec{w}(x_0) \tag{11.10.6}$$

while after multiplying by the integration factor $e^{-\mathbf{M}x}$, Equation (11.10.5) becomes

$$\frac{d}{dx}\left(e^{-\mathbf{M}x}\,\vec{w}(x)\right) = e^{-\mathbf{M}x}\,\vec{f}(x) \tag{11.10.7}$$

Consequently, the general solution is given by

$$\vec{w}(x) = e^{\mathbf{M}(x-x_0)}\vec{w}(x_0) + e^{\mathbf{M}x}\int_{x_o}^{x} e^{-\mathbf{M}x}\vec{f}(x')dx' \tag{11.10.8}$$

Example

For $(D-1)\ y = e^x$ with initial conditions $y(x_0) = 1$, $dy(x_0)/dx = 0$

$$\mathbf{M} = \begin{pmatrix} 0 & 1 \\ 1 & 0 \end{pmatrix} \quad \vec{f}(x) = \begin{pmatrix} 0 \\ e^x \end{pmatrix} \quad \vec{w}(x_0) = \begin{pmatrix} 1 \\ 0 \end{pmatrix} \tag{11.10.9}$$

so that

$$e^{\pm\mathbf{M}x} = \begin{pmatrix} 1 & 0 \\ 0 & 1 \end{pmatrix} \pm x\begin{pmatrix} 0 & 1 \\ 1 & 0 \end{pmatrix} + \frac{x^2}{2!}\begin{pmatrix} 1 & 0 \\ 0 & 1 \end{pmatrix} \pm \ldots = \begin{pmatrix} \cosh x & \pm\sinh x \\ \pm\sinh x & \cosh x \end{pmatrix} \tag{11.10.10}$$

The solution is then simply obtained from the following intermediate results:

$$e^{\mathbf{M}(x-x_0)}\vec{w}(x_0) = \begin{pmatrix} \cosh(x-x_0) \\ \sinh(x-x_0) \end{pmatrix}$$

$$\int_{x_o}^{x} e^{-\mathbf{M}x'}\vec{f}(x')dx' = \begin{pmatrix} -\int_{x_o}^{x}\left(e^{x'}\sinh x'\right)dx' \\ \int_{x_o}^{x}\left(e^{x'}\cosh x'\right)dx' \end{pmatrix} = \frac{1}{2}\begin{pmatrix} -\int_{x_o}^{x}\left(e^{2x'}-1\right)dx' \\ \int_{x_o}^{x}\left(e^{2x'}+1\right)dx' \end{pmatrix} \tag{11.10.11}$$

12

TRANSFORM THEORY

When linear differential equations of second and higher order are supplemented by appropriate physical boundary conditions, only certain "eigenfunctions" either (i) satisfy both the differential equation and the boundary conditions or (ii) do not change in form along directions in which both the differential equation and the boundary conditions remain invariant. Any function satisfying the boundary conditions can then be expressed as a linear superposition of the eigenfunctions. Expansions in the harmonic eigenfunctions of the wave equation in uniform regions over finite and infinite intervals are termed Fourier series and Fourier transforms, respectively.

12.1 EIGENFUNCTIONS AND EIGENVECTORS

To illustrate the homogeneous linear differential equation

$$\left(D^2 + k^2\right)y(x) = 0 \tag{12.1.1}$$

is solved by any linear combination of real or complex exponential functions as these possess a curvature that varies as the negative of their amplitude. However, additionally imposing boundary conditions on a finite interval such as the *Dirichlet* conditions

Fundamental Math and Physics for Scientists and Engineers, First Edition.
David Yevick and Hannah Yevick.

$$y(x=0)=y(x=L)=0 \tag{12.1.2}$$

limits the solution space to the *eigenvector* solutions

$$y_m = c_m \sin k_m x \tag{12.1.3}$$

with a discrete *spectrum of eigenvalues*

$$k_m = \frac{\pi m}{L} \tag{12.1.4}$$

Since $c_1 \sin(-k_m x) = -c_1 \sin k_m x = c_1' \sin k_m x$, the m are conventionally restricted to positive values. As Equation (12.1.4) encompasses all solutions, the *completeness property* states that any continuous function that satisfies the given boundary conditions at both 0 and L can be written as a *Fourier series*:

$$\phi = \sum_{m=1}^{\infty} a_m \sin k_m x \tag{12.1.5}$$

As the length, L, of the interval approaches infinity, the spacing between the k_m vanishes, the eigenvalue spectrum becomes continuous, and the Fourier series is replaced by a Fourier transform.

12.2 STURM–LIOUVILLE THEORY

In general, linear second-order differential equations can be transformed into the form of the *Sturm–Liouville* equation by incorporation of a suitable integrating factor:

$$Ly + \lambda w(x)y \equiv \frac{d}{dx}\left(f(x)\frac{dy}{dx}\right) - g(x)y + \lambda w(x)y = 0 \tag{12.2.1}$$

Here, $w(x) > 0$ is termed the *weight factor*. For (i) "mixed" boundary conditions,

$$\begin{aligned} c_1 y + d_1 \frac{dy}{dx} &= 0 \quad x=a \\ c_2 y + d_2 \frac{dy}{dx} &= 0 \quad x=b \end{aligned} \tag{12.2.2}$$

(termed Dirichlet if $d_{1,2} = 0$ or *Neumann* boundary conditions for $c_{1,2} = 0$), (ii) periodic *periodic* boundary conditions, (assuming $f(a) = f(b)$)

$$\begin{aligned} y(a) &= y(b) \\ \frac{dy(a)}{dx} &= \frac{dy(b)}{dx} \end{aligned} \tag{12.2.3}$$

or (iii) if $f(a) = f(b) = 0$, only certain eigenvalues, λ, and eigenfunctions, y_m, are permitted that satisfy the eigenvector equation

$$Ly_m = \frac{d}{dx}\left(f(x)\frac{dy_m}{dx}\right) - g(x)y_m = -\lambda_m w(x)y_m \tag{12.2.4}$$

The eigenfunctions form a complete orthogonal set *relative to the weights* $w(x)$; i.e.,

$$\int_a^b w(x)y_i^*(x)y_j(x)dx = \delta_{ij} \tag{12.2.5}$$

Example

The eigenfunctions of *Legendre equation*

$$\frac{\partial}{\partial\mu}\left((1-\mu^2)\frac{\partial P(\mu)}{\partial\mu}\right) + l(l+1)P(\mu) = \frac{\partial}{\partial\theta}\left(\sin\theta\frac{\partial P(\theta)}{\partial\theta}\right) + l(l+1)\sin\theta P(\theta) = 0 \tag{12.2.6}$$

are orthogonal over $[0, \pi]$ relative to the weight $w(\theta) = \sin\theta$, which satisfies (iii) above. Indeed,

$$\int_0^\pi P_i(\theta)P_j(\theta)\sin\theta d\theta = \int_{-1}^1 P_i(\mu)P_j(\mu)\underbrace{d\mu}_{d(\cos\theta)} = \delta_{ij} \tag{12.2.7}$$

If any of the conditions (i) to (iii) are satisfied, the boundary terms below vanish, implying that L is *self-adjoint*:

$$\int_a^b y_n^* Ly_m dx = \underbrace{\int_a^b y_n^*\frac{d}{dx}\left(f(x)\frac{dy_m}{dx}\right)dx}_{\underbrace{-\int_a^b \frac{dy_n^*}{dx}f(x)\frac{dy_m}{dx}dx}_{\int_a^b \frac{d}{dx}\left(f(x)\frac{dy_n^*}{dx}\right)y_m dx - \cancel{y_m f(x)\frac{dy_n^*}{dx}\Big|_a^b}} + \cancel{y_n^* f(x)\frac{dy_m}{dx}\Big|_a^b}} - \int_a^b g(x)y_n^* y_m dx = \int_a^b (Ly_n^*)y_m dx \tag{12.2.8}$$

That $\lambda_n = \lambda_n^*$, i.e., all eigenvalues are real, thus follows from setting $n = m$ in

$$\int_a^b y_n^* Ly_m dx = -\lambda_m\int_a^b w(x)y_n^* y_m dx = \int_a^b (Ly_n^*)y_m dx = -\lambda_n^*\int_a^b w(x)y_n^* y_m dx \tag{12.2.9}$$

Equation (12.2.9) then implies that either $\lambda_n = \lambda_m^*$ or $\int_a^b w(x) y_n^* y_m dx = 0$ so that y_n and y_m are orthogonal with respect to the weight factor $w(x)$ for differing eigenvalues.

As well, the y_n form a complete set of eigenfunctions, meaning that any function satisfying the boundary conditions can be written as a linear superposition

$$f(x) = \sum_{m=0}^{\infty} c_m y_m(x) \tag{12.2.10}$$

of eigenfunctions with

$$c_m = \frac{\int_a^b f(x') y_m^*(x') w(x') dx'}{\int_a^b |y_m(x')|^2 w(x') dx'} \tag{12.2.11}$$

Orthogonality and completeness are related since if Equation (12.2.1) is restricted to a discrete space of N points, any function can be represented as a superposition of N linearly independent eigenfunctions.

12.3 FOURIER SERIES

A *Fourier series* representation of $f(x)$ results from Equations (12.2.10) and (12.2.11) if periodic boundary conditions are imposed on Equation (12.1.1) over an interval $[a, a + L]$. In terms of real even cosine and odd sine functions,

$$f(x) = \chi_0 + \sum_{n=1}^{\infty} \sigma_n \sin\left(\frac{2n\pi x}{L}\right) + \chi_n \cos\left(\frac{2n\pi x}{L}\right) \tag{12.3.1}$$

or equivalently, expressed as complex exponential functions,

$$\begin{aligned} f(x) &= \sum_{n=0}^{\infty}\left(\sigma_n\left(\frac{e^{i\frac{2n\pi x}{L}} - e^{-i\frac{2n\pi x}{L}}}{2i}\right) + \chi_n\left(\frac{e^{i\frac{2n\pi x}{L}} + e^{-i\frac{2n\pi x}{L}}}{2}\right)\right) \\ &= \sum_{n=0}^{\infty}\left(\underbrace{\left(\frac{\chi_n - i\sigma_n}{2}\right)}_{\varepsilon_m} e^{i\frac{2n\pi x}{L}} + \underbrace{\left(\frac{\chi_n + i\sigma_n}{2}\right)}_{\varepsilon_{-m}} e^{-i\frac{2n\pi x}{L}}\right) \\ &= \sum_{m=-\infty}^{\infty} \varepsilon_m e^{i\frac{2m\pi x}{L}} \end{aligned} \tag{12.3.2}$$

Since

$$\int_a^{a+L} \sin\frac{2\pi nx}{L}\sin\frac{2\pi mx}{L} = \frac{L}{2}(\delta_{nm}-\delta_{n0}\delta_{m0})$$
$$\int_a^{a+L} \cos\frac{2\pi nx}{L}\cos\frac{2\pi mx}{L} = \frac{L}{2}(\delta_{nm}+\delta_{n0}\delta_{m0}) \tag{12.3.3}$$
$$\int_a^{a+L} e^{i\frac{2\pi nx}{L}}e^{-i\frac{2\pi mx}{L}} = L\delta_{nm}$$

the *Fourier coefficients* are given by, where the constant σ_0 above is arbitrary as $\sin(0)=0$,

$$\chi_0 = \frac{1}{L}\int_a^{a+L} f(x)dx$$
$$\chi_n = \frac{2}{L}\int_a^{a+L} f(x)\cos\frac{2\pi nx}{L}dx$$
$$\sigma_n = \frac{2}{L}\int_a^{a+L} f(x)\sin\frac{2\pi nx}{L}dx \tag{12.3.4}$$
$$\varepsilon_n = \frac{1}{L}\int_a^{a+L} f(x)e^{-i\frac{2\pi nx}{L}}dx$$

If $f(x)$ is complex, some or all of the χ_m, σ_m are complex. For real $f(x)$, χ_m and σ_m are real while $\varepsilon_{-m}=\varepsilon_m^*$ (insuring that $\varepsilon_{-m}\exp(-2\pi imx/L)+\varepsilon_m\exp(2\pi imx/L)$ is real). Further, $\sigma_m=0$ ($\chi_m=0$) for all m if $f(x)$ is an even (odd) function and the lower interval limit $a=-L/2$.

Parseval's theorem for exponential series states

$$\int_a^{a+L} g^*(x)f(x)dx = \sum_{n,m=-\infty}^{\infty}\int_a^{a+L}\left(\varepsilon_n^{(g)}\right)^*\varepsilon_m^{(f)}e^{i\frac{2\pi(m-n)x}{L}}dx$$
$$= L\sum_{n,m=-\infty}^{\infty}\left(\varepsilon_n^{(g)}\right)^*\varepsilon_m^{(f)}\delta_{mn} \tag{12.3.5}$$
$$= L\sum_{n=-\infty}^{\infty}\left(\varepsilon_n^{(g)}\right)^*\varepsilon_n^{(f)}$$

Setting $g=f$ indicates that $|\varepsilon_n|^2$ can be considered as the "power" of the nth Fourier component.

Since each term in the Fourier series does not change if x is replaced by $x+L$, the Fourier series *periodically extends* $f(x)$ with period L beyond the interval $\mathrm{I}=[a, a+L]$. Thus, the accuracy of the Fourier series is greatest if $f(x)$ obeys periodic boundary conditions at the endpoints of I or if it is defined on an *infinite interval* but is *periodic* with period L. Further, since the series employs functions that are infinitely differentiable, the fewest terms are required for a given level of precision in regions where $f(x)$ possesses many continuous derivatives. Close to a discontinuity in $f(x)$ or its derivatives (or a similar discontinuity between $f(a)$ and $f(a+L)$ since the representation is periodic), the "Gibbs phenomenon," in which a Fourier series with a finite number of terms successively overestimates and underestimates $f(x)$ as x approaches the position of the discontinuity, limits the accuracy of the Fourier series. This behavior is analyzed in the following example by considering the *sum rule* that can always be obtained by evaluating the Fourier series for a function $f(x)$ at any point.

Example

As the function

$$f(x) = \mathrm{sgn}(x) = \begin{cases} -1 & x<0 \\ 1 & x>0 \end{cases} \tag{12.3.6}$$

is odd, its Fourier series on $[-\pi, \pi]$ only contains sine terms with

$$\sigma_m = \left(\frac{2}{2\pi}\right) 2\int_0^{\pi} \sin(mx)dx = -\frac{2}{m\pi}\cos(mx)\Big|_0^{\pi} = -\frac{2}{m\pi}((-1)^m - 1) = \frac{4}{m\pi}, \; m \text{ odd} \tag{12.3.7}$$

Equations (12.3.1), (12.3.6), and (12.3.7) yield the sum rule for any $0 < x < \pi$

$$\frac{\pi}{4} = \sum_{n=1,3,\ldots}^{\infty} \frac{1}{n} \sin mx \tag{12.3.8}$$

for any $0 < x < L/2$. For small x (or for x near $L/2$), a large number of terms in the above sum are initially positive, summing to a value much larger than $\pi/4$ before mx exceeds π, changing the sign of the subsequent terms. For $x = \pi/2$, halfway between the discontinuities of the periodic extension of $f(x)$ at the origin and the boundaries, the terms in the sum evaluate to $(-1)^n/n$ so that successive terms instead always alternate signs, yielding the fastest rate of convergence.

12.4 FOURIER TRANSFORMS

A Fourier series in $L \to \infty$, $n \to \infty$ limit is termed a *Fourier transform*. With $k = 2\pi n/L$ held finite, the exponential Fourier series coefficient of Equation (12.3.4) becomes, where F represents the Fourier transform operation,

$$\sqrt{2\pi} f(k) \equiv \sqrt{2\pi} F(f(x)) \equiv L c_n = \int_{-L/2}^{L/2} f(x) e^{-i\frac{2\pi n x}{L}} dx \to \int_{-\infty}^{\infty} f(x) e^{-ikx} dx \tag{12.4.1}$$

where the factor of $\sqrt{2\pi}$ leads to a symmetric Fourier transform pair (an alternate convention omits the $\sqrt{2\pi}$ and instead replaces the factor $1/\sqrt{2\pi}$ in the inverse Fourier transform, Eq. (12.4.2), by $1/2\pi$). With $dn = L dk/2\pi$, Equation (12.4.1) similarly generates the inverse Fourier transform

$$f(x) = F^{-1}(f(k)) \equiv \sum_{n=-\infty}^{\infty} \varepsilon_n e^{i\frac{2\pi n x}{L}} = \int_{-\infty}^{\infty} \varepsilon_n e^{i\frac{2\pi n x}{L}} dn = \frac{L}{2\pi}\int_{-\infty}^{\infty} \varepsilon_n e^{i\frac{2\pi n x}{L}} dk \to \frac{1}{\sqrt{2\pi}}\int_{-\infty}^{\infty} f(k) e^{ikx} dk \quad (12.4.2)$$

Fourier transforming the derivative of a localized function with $f(x) \to 0$ for $x \to \pm\infty$

$$F\left(\frac{df(x)}{dx}\right) = \frac{1}{\sqrt{2\pi}}\int_{-\infty}^{\infty} \underbrace{e^{-ikx}}_{g(x)} \underbrace{\frac{df(x)}{dx}}_{h'(x)} dx = \frac{1}{\sqrt{2\pi}}\left[-\int_{-\infty}^{\infty} \underbrace{-ike^{ikx}}_{g'(x)} \underbrace{f(x)}_{h(x)} dx + \underbrace{e^{-ikx}}_{g(x)} \underbrace{f(x)}_{h(x)}\Bigg|_{-\infty}^{\infty}\right] = ikf(k) \quad (12.4.3)$$

By extension, the Fourier transform of $d^n f/dx^n$ equals $(ik)^n f(k)$.

12.5 DELTA FUNCTIONS

Combining the direct and inverse Fourier transforms yields

$$f(x) = \frac{1}{\sqrt{2\pi}}\int_{-\infty}^{\infty} f(k)e^{ikx}dk = \frac{1}{2\pi}\int_{-\infty}^{\infty}\int_{-\infty}^{\infty} f(x')e^{ik(x-x')}dx'dk \quad (12.5.1)$$

If the order of integration is reversed,

$$f(x) = \int_{-\infty}^{\infty} f(x')\left(\frac{1}{2\pi}\int_{-\infty}^{\infty} e^{ik(x-x')}dk\right)dx' = \int_{-\infty}^{\infty} f(x')\delta(x-x')dx' \quad (12.5.2)$$

where the *delta function* (distribution)

$$\delta(x-x') \equiv \frac{1}{2\pi}\int_{-\infty}^{\infty} e^{ik(x-x')}dk \quad (12.5.3)$$

vanishes except at $x = x'$, where it acquires an infinite value such that its integral over all x' is unity. The normalization $1/2\pi$ in Equation (12.5.3) (and in the definition of the Fourier transform) can be verified by

$$\frac{1}{2\pi}\int_{-\infty}^{\infty} e^{ikx}dk = \frac{1}{2\pi}\lim_{\kappa\to\infty}\int_{-\kappa}^{\kappa} e^{ikx}dk = \lim_{\kappa\to\infty}\frac{\sin\kappa x}{\kappa\pi x} \tag{12.5.4}$$

which, after setting $x' = \kappa x$ and therefore $dx' = \kappa dx$, yields, upon integration,

$$\int_{-\infty}^{\infty}\frac{\sin\kappa x}{\kappa\pi x}dx = \frac{1}{\pi}\int_{-\infty}^{\infty}\frac{\sin x'}{x'}dx' = \frac{1}{\pi i}\mathrm{Im}\int_{-\infty}^{\infty}\frac{e^{ix'}}{x'}dx' = \frac{1}{\pi i}\pi i e^{ix}\Big|_{x=0} = 1 \tag{12.5.5}$$

where the contour of the integral of e^{ix}/x bisects the pole and therefore only acquires half the value of the residue at the origin. Unless the derivatives of the delta function explicitly enter into a calculation, $\delta(x-x')$ can be defined as the limit of any sequence of functions that integrate to unity and approach zero for $x \neq x'$ such as a rectangle of width Δx and height $1/\Delta x$ as $\Delta x \to 0$. In multiple dimensions,

$$\delta^n(\vec{r}-\vec{r}') = \frac{1}{(2\pi)^n}\int_{-\infty}^{\infty} e^{i\vec{k}\cdot(\vec{r}-\vec{r}')}d^nk = \delta(x-x')\delta(y-y')\ldots \tag{12.5.6}$$

Some properties of the one-dimensional delta function are first, with $k' = ak$

$$\delta(a(x-x')) = \frac{1}{2\pi}\int_{-\infty}^{\infty} e^{ika(x-x')}dk = \begin{cases} \dfrac{1}{2\pi}\displaystyle\int_{-\infty}^{\infty} e^{ik'(x-x')}\frac{dk'}{a} = \frac{1}{a}\delta(x-x') & a>0 \\ \dfrac{1}{2\pi}\displaystyle\int_{\infty}^{-\infty} e^{ik'(x-x')}\frac{dk'}{a} = -\frac{1}{a}\delta(x-x') & a<0 \end{cases} \tag{12.5.7}$$

That is, since the area under the delta function viewed as a rectangle equals unity, if the units of x are changed from, e.g., meters to centimeters, as the width of the rectangle is a hundred times larger in terms of the new units, its height must be reduced by a factor of a hundred to generate an area of one in the new system. Alternatively, the delta function possesses units of $1/L$ from $\int\delta(x)dx = 1$, which again implies Equation (12.5.7).

The previous result together with a Taylor expansion of $f(x)$ about its zeros, x_i, yields

$$\delta(f(x)) = \sum_{i=\text{zeros of } f(x)}\frac{1}{2\pi}\int_{-\infty}^{\infty} e^{ik\frac{df(x_i)}{dx}(x-x_i)}dk = \sum_{i=\text{zeros of } f(x)}\frac{1}{\left|\frac{df(x_i)}{dx}\right|}\delta(x-x_i) \tag{12.5.8}$$

where the sum, as indicated, is taken over the zeros (assumed to be of the form $a(x-x_i)$) of $f(x)$. Derivatives of the delta function possess the property

$$\int_{-\infty}^{\infty} \underbrace{f(x)}_{g(x)} \underbrace{\delta'(x-a)}_{h'(x)} dx = \underbrace{f(x)}_{g(x)} \underbrace{\delta(x-a)}_{h(x)} \Big|_{-\infty}^{\infty} - \int_{-\infty}^{\infty} \underbrace{f'(x)}_{g'(x)} \underbrace{\delta(x-a)}_{h(x)} dx = -f'(a) \quad (12.5.9)$$

and by extension

$$\int_{-\infty}^{\infty} f(x) \frac{d^n \delta(x-a)}{dx^n} dx = (-1)^n \frac{d^n f}{dx^n}\Big|_{x=a} \quad (12.5.10)$$

The delta function can be expressed in terms of the eigenfunctions of any Sturm–Liouville equation according to Equations (12.2.10) and (12.2.11), which combined yield

$$\begin{aligned} f(x) &= \sum_{i=0}^{\infty} \frac{\int_a^b f(x') y_i^*(x') y_i(x) w(x') dx'}{\int_a^b |y_i(x')|^2 w(x') dx'} \\ &= \int_a^b f(x') \sum_{i=0}^{\infty} \left(\frac{y_i^*(x') y_i(x) w(x')}{\int_a^b |y_i(x')|^2 w(x') dx} \right) dx' \end{aligned} \quad (12.5.11)$$

and hence

$$\delta(x-x') = \sum_{i=0}^{\infty} \left(\frac{y_i^*(x') y_i(x) w(x')}{\int_a^b |y_i(x')|^2 w(x') dx'} \right) \quad (12.5.12)$$

The inverse Fourier transform of a product of two Fourier transforms yields a *convolution* in space. Conversely, the Fourier transform of a convolution yields a product of transformed functions:

$$\begin{aligned} \int_{-\infty}^{\infty} f(x') g(x-x') dx' &= \frac{1}{2\pi} \int_{-\infty}^{\infty} \int_{-\infty}^{\infty} e^{ik'x'} f(k') \int_{-\infty}^{\infty} e^{ik''(x-x')} g(k'') dx' dk' dk'' \\ &= \int_{-\infty}^{\infty} \int_{-\infty}^{\infty} f(k') \left(\frac{1}{2\pi} \int_{-\infty}^{\infty} e^{i(k'-k'')x'} dx' \right) e^{ik''x} g(k'') dk' dk'' \\ &= \int_{-\infty}^{\infty} f(k') \int_{-\infty}^{\infty} e^{ik''x} \delta(k'-k'') g(k'') dk' dk'' \\ &= \int_{-\infty}^{\infty} e^{ik'x} f(k') g(k') dk' \end{aligned} \quad (12.5.13)$$

Parseval's relation is derived similarly:

$$\int_{-\infty}^{\infty} f(x)h^*(x)dx = \int_{-\infty}^{\infty} \frac{1}{\sqrt{2\pi}} \int_{-\infty}^{\infty} e^{ik'x} f(k')dk' \frac{1}{\sqrt{2\pi}} \int_{-\infty}^{\infty} e^{-ik''x} h^*(k'')dk''dx$$
$$= \int_{-\infty}^{\infty} f(k') \int_{-\infty}^{\infty} \delta(k'-k'')h^*(k'')dk'dk''$$
$$= \int_{-\infty}^{\infty} f(k')h^*(k')dk' \tag{12.5.14}$$

12.6 GREEN'S FUNCTIONS

The solution to a linear differential equation with specified boundary conditions and an inhomogeneous delta function term can be employed to find the solution of the inhomogeneous differential equation

$$Ly(\vec{r}) = f(\vec{r}) \tag{12.6.1}$$

subject to the same boundary conditions. In particular, by linearity, $y(\vec{r})$ must be a sum of contributions from the individual points in the source distribution $f(\vec{r}')d^3\vec{r}'$ according to

$$y(\vec{r}) = \int_{-\infty}^{\infty} f(\vec{r}')G(\vec{r},\vec{r}')dr' \tag{12.6.2}$$

However, since L acts only on the variable $\vec{r}$,

$$Ly(\vec{r}) = \int_{-\infty}^{\infty} f(\vec{r}')LG(\vec{r},\vec{r}')d^3r' = f(\vec{r}) = \int_{-\infty}^{\infty} = f(\vec{r}')\delta(\vec{r}-\vec{r}')d^3r' \tag{12.6.3}$$

Accordingly, since $f(\vec{r})$ is arbitrary, the *Green's function* $G(x, x')$ must obey the boundary conditions and the defining equation

$$LG(\vec{r},\vec{r}') = \delta(\vec{r}-\vec{r}') \tag{12.6.4}$$

Examples

1. The Green's function for the first-order differential equation describing linear-free particle motion in the presence of friction

$$m\frac{dv}{dt} + \alpha v = F(t) \tag{12.6.5}$$

solves

$$m\frac{dG(t,t')}{dt}+\alpha G(t,t')=\delta(t-t') \quad (12.6.6)$$

The delta function term here corresponds to a unit impulse, yielding a momentum change $I=\Delta p=1$ at $t=t'$. Causality therefore yields the single required boundary condition $G(t,t')=0$ for $t<t'$ as a disturbance at t' only affects $y(t)$ at later times t. The solution to Equation (12.6.6) for $t>t'$ where $\delta(t-t')=0$ is the homogeneous expression

$$G(t,t')=ce^{-\frac{\alpha}{m}t} \quad (12.6.7)$$

To determine c for $t>t'$, Equation (12.6.6) can be integrated from a time, $t=t'-\delta$, just before t' to $t=t'+\delta$. Since G is finite, the integral over the second term on the left-hand side approaches zero as $\delta\to 0$ leaving

$$\int_{t'-\delta}^{t'+\delta} m\frac{dG(t,t')}{dt}dt=mce^{-\frac{\alpha}{m}t'}=\int_{t'-\delta}^{t'+\delta}\delta(t-t')dt=1 \quad (12.6.8)$$

so that $c=\exp(\alpha t'/m)/m$, consistent with the interpretation of the inhomogeneous term as a unit impulse. Accordingly, where $\theta(t)=1$ for $t>0$ and zero for $t<0$,

$$G(t,t')=\frac{1}{m}e^{-\frac{\alpha}{m}(t-t')}\theta(t-t') \quad (12.6.9)$$

Since time is homogeneous, G depends only on $t-t'$ and not on t and t' separately.

2. The Green's function of a free particle of mass m in one dimension again describes a unit impulse

$$m\frac{d^2G}{dt^2}=\delta(t-t') \quad (12.6.10)$$

applied at time t'. Except at $t=t'$, the Green's function must satisfy the homogeneous equation and can again only depend on $t-t'$, implying $G=c_1(t-t')+c_2$. Again, integrating Equation (12.6.10) as in the previous problem and employing $G(t-t'<0)=c_2=0$ so that $dG/dt|_{t=t'-\varepsilon}=0$,

$$\int_{t'_-}^{t'_+}\frac{d^2G}{dt^2}dt=\left.\frac{dG}{dt}\right|_{t=t'+\varepsilon}-\left.\frac{dG}{dt}\right|_{t=t'-\varepsilon}=\frac{1}{m}\int_{t'_-}^{t'_+}\delta(t-t')dx=\frac{1}{m} \quad (12.6.11)$$

as the momentum and velocity changes at t' are 1 and $1/m$, respectively. The physical *retarded or casual Green's function* therefore reflects the acquisition of unit momentum at time t':

$$G_{\text{retarded}}(t-t')=\frac{1}{m}(t-t')\theta(t-t') \quad (12.6.12)$$

the second derivative of which diverges at t' and is zero elsewhere. The normally unphysical *advanced Green's function* for which the impulse instead halts a moving particle at $t = t'$ instead satisfies the boundary condition $G(t - t' > 0) = 0$:

$$G_{\text{advanced}}(t, t') = -\frac{1}{m}(t - t')\theta(t' - t) \qquad (12.6.13)$$

Another procedure for constructing $G(t - t')$ Fourier transforms both sides of Equation (12.6.10):

$$\frac{1}{\sqrt{2\pi}}\int_{-\infty}^{\infty} \underbrace{\left(\frac{d^2G}{dt^2}\right)}_{g'(t)} \underbrace{e^{-i\omega t}}_{h(t)} dt = \frac{1}{\sqrt{2\pi}}\left(i\omega \int_{-\infty}^{\infty} \underbrace{\frac{dG}{dt}}_{\tilde{g}'(t)} \underbrace{e^{-i\omega t}}_{\tilde{h}(t)} dt + \underbrace{\frac{dG}{dt}}_{g(t)} \underbrace{e^{-i\omega t}}_{h(t)} \Bigg|_{-\infty}^{\infty} \right)$$

$$= \frac{1}{\sqrt{2\pi}}\left(-\omega^2 \int_{-\infty}^{\infty} e^{-i\omega t} \underbrace{G}_{\tilde{g}(t)} dt + i\omega G e^{-i\omega t} \Bigg|_{-\infty}^{\infty} \right)$$

$$(12.6.14)$$

where an artificial infinitesimal negative imaginary component $- i\varepsilon$ is added to ω to eliminate the upper limits of the surface terms while $G(-\infty) = 0$ for casual boundary conditions eliminates the lower limits. Setting Equation (12.6.14) equal to $e^{-i\omega t'}/\sqrt{2\pi}$, the Fourier transform of $\delta(t - t')$, yields

$$G(\omega) = -\frac{e^{i\omega t'}}{\sqrt{2\pi}\omega^2} \qquad (12.6.15)$$

For $t - t' > 0$, inverse Fourier transforming, where $- i\varepsilon$ is again added to ω and the contour is closed in the upper half plane over which $\exp(i\alpha) \to 0$ with $\alpha = (t - t')\omega$ as $|\omega| \to \infty$, yields

$$\begin{aligned} G_{\text{retarded}}(t - t') &= -\frac{1}{2\pi}\int_{-\infty}^{\infty} e^{i\omega t} \frac{e^{-i\omega t'}}{(\omega - i\varepsilon)^2} d\omega \\ &= -\frac{(t - t')}{2\pi}\int_{-\infty}^{\infty} \frac{e^{i\alpha}}{(\alpha - i\varepsilon)^2} d\alpha \\ &= -\frac{(t - t')}{2\pi}\frac{2\pi i}{1!}\lim_{\alpha \to i\varepsilon \approx 0} \frac{de^{i\alpha}}{d\alpha} \\ &= (t - t') \end{aligned} \qquad (12.6.16)$$

For $(t - t') < 0$, the contour is closed in the lower half plane excluding the pole at $\omega = i\varepsilon$, yielding $G_{\text{retarded}}(t - t') = 0$. Evaluating the contour integral with $\varepsilon < 0$ yields instead $G_{\text{advanced}}(t - t')$.

The Green's function for the Sturm–Liouville equation is obtained by expanding $G(x,x') = \sum_{i=1}^{\infty} c_i(x')y_i(x)$ in the normalized eigenfunctions of the homogeneous equation

$$\left(\frac{d}{dx}\left(f(x)\frac{d}{dx}\right) - g(x) + \lambda w(x)\right)\sum_{i=1}^{\infty} c_i(x')y_i(x) = \delta(x-x')$$
$$\sum_{i=1}^{\infty}(-\lambda_i + \lambda)c_i(x')w(x)\,y_i(x) = \delta(x-x') \tag{12.6.17}$$

Multiplying both sides of the lower equation by $y_j^*(x)$, integrating over all space and employing the orthogonality of the y_i, results in $c_j(x') = y_j^*(x')/(\lambda - \lambda_i)$ and therefore

$$G(x,x') = \sum_{i=1}^{\infty} \frac{y_i(x)y_i^*(x')}{\lambda - \lambda_i} \tag{12.6.18}$$

Example

The Green's function for the one-dimensional Helmholtz equation on [0, 1]

$$\frac{d^2y(x)}{dx^2} + k^2y(x) = 0 \tag{12.6.19}$$

with boundary conditions $y(0) = y(1) = 0$ leading to normalized eigenfunctions and eigenvalues $y_n(x) = \sqrt{2}\sin(k_n x)$ with $k_n = n\pi$ is given by

$$G(x,x') = 2\sum_{n=1}^{\infty} \frac{\sin(n\pi x')\sin(n\pi x)}{k^2 - n^2\pi^2} \tag{12.6.20}$$

For $k = n\pi$ for any n, the Green's function is infinite since a solution $y(x)$ does not exist to

$$\frac{d^2y}{dx^2} + n^2\pi^2 y = \sin n\pi x \tag{12.6.21}$$

Equation (12.6.20) coincides with the Green's function of Equation (12.6.10). which, to be continuous at $x = x'$ and satisfy the given boundary conditions must be of the form

$$G(x,x') = \begin{cases} ax(1-x') & x < x' \\ a(1-x)x' & x > x' \end{cases} \tag{12.6.22}$$

where the derivative possesses a unit discontinuity so that at $x = x'$

$$-ax - a(1-x') = 1 \tag{12.6.23}$$

and therefore $a = -1$. Hence, Equation (12.6.22) can be written as, with $x_< =$ $\min(x,x')$ and $x_> = \max(x,x')$,

$$G(x,x') = x(x'-1)\theta(x'-x) + x'(x-1)\theta(x-x') = x_<(x_>-1) \tag{12.6.24}$$

The above form is consistent with the symmetry of the domain under the transformation $x, x' \leftrightarrow 1-x, 1-x'$. While $G(x,x') = G(x',x)$, G does not depend only on $x-x'$ since the boundary conditions are not translationally invariant. Multiplication of both sides of the expansion

$$G(x,x') = \sum_{m=0}^{\infty} \sigma_m \sin(m\pi x') \tag{12.6.25}$$

by $\sin(p\pi x')$ and integrating over x' from 0 to 1 as follows reproduces Equation (12.6.20) with $k=0$ as

$$\begin{aligned}
c_p &= 2\left[(x-1)\int_0^x x' \sin p\pi x' dx' + x\int_x^1 (x'-1)\sin p\pi x' dx'\right] \\
&= 2\left[x\int_0^1 x' \sin p\pi x' dx' - \int_0^x x' \sin p\pi x' dx' - x\int_x^1 \sin p\pi x' dx'\right] \\
&= 2\left[-x\frac{\cos p\pi}{p\pi} - \left\{-x\frac{1}{p\pi}\cos p\pi x + \frac{\sin p\pi x}{p^2\pi^2}\right\} - x\left\{-\frac{\cos p\pi}{p\pi} + \frac{\cos p\pi x}{p\pi}\right\}\right] \\
&= -\frac{2\sin p\pi x}{p^2\pi^2}
\end{aligned} \tag{12.6.26}$$

12.7 LAPLACE TRANSFORMS

The *Laplace transform*

$$\mathcal{L}_f(s) \equiv \int_o^{\infty} f(x)e^{-sx}dx \tag{12.7.1}$$

replaces the complex exponential function in the Fourier transform by a real exponential. While Equation (12.7.1) is explicitly real for real $f(x)$, this complicates the evaluation of the inverse transform, which is therefore often obtained by referring to precomputed tables.

If $|f(x)e^{-s_0 x}|$ remains finite as $x \to \infty$, the function $f(x)$ is termed *exponential of order* s_0 and the following relationships hold for $\text{Re}(s) > s_0$:

$$\mathcal{L}_{e^{-ax}f}(s) = \int_0^{\infty} e^{-ax}f(x)e^{-sx}dx = \mathcal{L}_f(s+a) \tag{12.7.2}$$

$$\mathcal{L}_{\frac{df}{dx}}(s) = \int_o^{\infty} \underbrace{\frac{df}{dx}}_{g'(x)} \underbrace{e^{-sx}}_{h(x)} dx = \underbrace{f(x)}_{g(x)} \underbrace{e^{-sx}}_{h(x)} \Big|_0^{\infty} - \int_o^{\infty} \underbrace{f(x)}_{g(x)} \underbrace{(-s)e^{-sx}}_{h'(x)} dx = -f(0) + s\mathcal{L}_f(s) \tag{12.7.3}$$

This relation can be iterated to obtain the Laplace transforms of higher-order derivatives, e.g.,

$$\begin{aligned}\mathcal{L}_{\frac{d^2f}{dx^2}}(s) &= -\frac{df}{dx}(0) + s\mathcal{L}_{\frac{df}{dx}}(s) \\ &= -\frac{df}{dx}(0) + s\left(-f(0) + s\mathcal{L}_f(s)\right)\end{aligned} \tag{12.7.4}$$

and with $d^0f/dx^0 \equiv f$,

$$\mathcal{L}_{\frac{d^nf}{dx^n}}(s) = s^n\mathcal{L}_f(s) - \sum_{j=1}^{n} s^{j-1}\frac{d^{n-j}f}{d^{n-j}x}(0) \tag{12.7.5}$$

Examples

$$\mathcal{L}_1(s) = \int_0^\infty e^{-sx}dx = \frac{1}{s} \tag{12.7.6}$$

$$\mathcal{L}_{e^{ax}}(s) = \int_0^\infty e^{-sx}e^{ax}dx = \frac{1}{a-s}e^{(a-s)x}\Big|_0^\infty = \frac{1}{s-a} \tag{12.7.7}$$

for $\mathrm{Re}(s) > a$. From this expression,

$$\mathcal{L}_{\cos kx}(s) = \frac{1}{2}\left(\frac{1}{s-ik} + \frac{1}{s+ik}\right) = \frac{s}{s^2+k^2} \tag{12.7.8}$$

Inverse Laplace transforms are often evaluated by transforming a quotient of algebraic functions into a sum of partial fractions whose inverse Laplace transform can be obtained from tables.

Example

Laplace transforming the differential equation

$$\left(D^2+4\right)y(x) = 0, \quad y(0) = 2, \quad y'(0) = 2 \tag{12.7.9}$$

according to Equation (12.7.5) replaces differential by algebraic expressions so that

$$\begin{aligned}\left(s^2\mathcal{L}_y(s) - y'(0) - sy(0)\right) + 4\mathcal{L}_y(s) &= 0 \\ \left(s^2\mathcal{L}_y(s) - 2 - 2s\right) + 4\mathcal{L}_y(s) &= 0\end{aligned} \tag{12.7.10}$$

or

$$\mathcal{L}_y(s) = \frac{2s}{s^2+4} + \frac{2}{s^2+4} \qquad (12.7.11)$$

Equation (12.7.8) and the corresponding result for the sine function then yield $y = \sin 2x + 2\cos 2x$.

12.8 Z-TRANSFORMS

Laplace transforming a function defined on a discrete set of points at locations 0, T, $2T$, ... according to

$$f(x) = \sum_{k=0}^{\infty} f_k \delta(x - kT) \qquad (12.8.1)$$

and setting $z = e^{sT}$ yields the *z-transform*

$$\mathcal{Z}_f(z) = \sum_{k=0}^{\infty} f_k z^{-k} \qquad (12.8.2)$$

Power series expansions can be manipulated from tables of forward and reverse z-transforms.

Examples

1. The z-transform of $f_k = 1$ yields the sum, $(1 - 1/z)^{-1}$, of a geometric series.
2. Since replacing $1/z$ by z in Equation (12.8.2) leads to a standard power series, $f_k = 1/k!$ implies $\mathcal{Z}_f(z) = e^{1/z}$.

13

PARTIAL DIFFERENTIAL EQUATIONS AND SPECIAL FUNCTIONS

Differential equations involving two or more variables are termed partial differential equations. While numerical solution methods are required for general source distributions, medium properties, and boundary conditions, analytic results are often accessible for structures with the symmetries of orthogonal coordinate systems. Thus, a homogeneous sourceless medium within a rectangular boundary can be described in rectangular coordinates as a superposition of products of harmonic and hyperbolic functions. The analogous functions for other orthogonal coordinate systems are termed "special functions," the simplest of which are summarized in this chapter.

13.1 SEPARATION OF VARIABLES AND RECTANGULAR COORDINATES

Partial differential equations involving multiple variables are often simply analyzed in a discrete representation.

Fundamental Math and Physics for Scientists and Engineers, First Edition.
David Yevick and Hannah Yevick.

Example

The *Laplace equation* in Cartesian coordinates

$$\nabla^2 V = \frac{\partial^2 V}{\partial x^2} + \frac{\partial^2 V}{\partial y^2} = 0 \tag{13.1.1}$$

for $\Delta x = \Delta y = \Delta \alpha$ takes the form, by extension of Equation (6.1.17),

$$\frac{V(x+\Delta\alpha, y) + V(x-\Delta\alpha, y) + V(x, y+\Delta\alpha) + V(x, y+\Delta\alpha) - 4V(x,y)}{(\Delta\alpha)^2} = 0 \tag{13.1.2}$$

Equation (13.1.2) restricts V at the center of four equidistantly spaced surrounding points to the average of V over these points, thus coupling values of V along both the x - and y - directions. A unique solution results if boundary conditions are supplied on all boundaries of a closed surface by imposing this restriction iteratively as

$$V_n(x,y) = \frac{V_{n-1}(x+\Delta\alpha, y) + V_{n-1}(x-\Delta\alpha, y) + V_{n-1}(x, y+\Delta\alpha) + V_{n-1}(x, y+\Delta\alpha)}{4} \tag{13.1.3}$$

in which $V_n(x, y)$ is interpreted as the solution after n computation steps and the boundary conditions are further imposed at each step. Thus, V within a square of unit length with $V = 1$ along the line $x = 0$ and $V = 0$ on the other boundaries is obtained from the following Octave program, where $V_{n-1}(x_i, y_j)$ must be stored before calculating $V_n(x_i, y_j)$ so that it is not overwritten before, e.g., $V_n(x_{i+1}, y_j)$ is evaluated:

```
clear all
numberOfPoints = 20;
numberOfSteps = 2000;
grid = zeros( numberOfPoints, numberOfPoints );
gridSave = zeros( numberOfPoints, numberOfPoints );
grid(1, :) = 1;
grid(numberOfPoints, :) = 0;
grid(:, numberOfPoints) = 0;
grid(:, 1) = 0;
for stepLoop = 1 : numberOfSteps;
  gridSave = grid;
  for outerLoop = 2 : numberOfPoints - 1
       for innerLoop = 2 : numberOfPoints - 1
         grid(outerLoop, innerLoop) = 0.25 * ( ...
              gridSave(outerLoop, innerLoop + 1) + ...
              gridSave(outerLoop, innerLoop - 1) + ...
              gridSave(outerLoop + 1, innerLoop) + ...
              gridSave(outerLoop - 1, innerLoop) );
       end
  end
end;
mesh(grid)
```

If the symmetries of the partial differential equation, medium, boundaries, and source distributions coincide with those of an orthogonal coordinate system ξ, ψ, ζ, the method of *separation of variables* can often be employed. In three dimensions, six boundary surfaces (planes) are defined by $\xi = c_1, c_2$, $\psi = c_3, c_4$, and $\zeta = c_5, c_6$. Since the Laplacian operator is linear, the principle of superposition holds, and the potential inside a region for which potential functions $V_1^{\text{boundary}}, \ldots, V_6^{\text{boundary}}$ are specified on each of the boundary surfaces is equal to the sum of the solutions of six problems such that in the mth problem the potential at the mth boundary equals V_m^{boundary}, while the potential on the other boundaries is zero. The solution to each of these problems can then be expressed as a product of functions of the individual variables, i.e.,

$$V_m(\xi,\psi,\zeta) = \Xi(\xi)\Psi(\psi)Z(\zeta) \tag{13.1.4}$$

Examples

A problem in which nonzero boundary conditions are specified on all four sides of a homogenous two-dimensional square medium is recast as the sum of four independent problems each of which employs nonzero boundary conditions on a single side of the square and zero boundary conditions on the remaining three sides. Thus, the solution of the two-dimensional Laplace equation on a square region extending from (0, 0) to (L, L) with boundary conditions on the four lines $x = 0$, $x = L$, $y = 0$, $y = L$ given by $V^{\text{boundary}}(0, y) = V^{\text{boundary}}(x, 0) = 0$, $V^{\text{boundary}}(x, L) = \sin \pi x/L$, and $V^{\text{boundary}}(0, y) = \sin 2\pi y/L$ by superposition can be expressed as $V_1 + V_2$, where V_1 and V_2 solve the Laplace equation with boundary conditions $V_1^{\text{boundary}}(0,y) = V_1^{\text{boundary}}(L,y) = V_1^{\text{boundary}}(x,0) = 0$, $V_1^{\text{boundary}}(x,L) = \sin \pi x/L$ and $V_2^{\text{boundary}}(L,y) = V_2^{\text{boundary}}(x,0) = V_2^{\text{boundary}}(x,L) = 0$, $V_2^{\text{boundary}}(0,y) = \sin 2\pi y/L$, respectively. With $V_1 = X_1(x)Y_1(y)$, a single Fourier component $X_1(x) = \sin \pi x/L$ satisfies the boundary condition along the side of the square at $y = L$ and additionally insures that $V_1(x, y)$ vanishes at $x = 0, L$ for all values of y. Inserting V_1 into the Laplace equation indicates that $Y_1(y)$ satisfies

$$\cancel{\sin\left(\frac{\pi x}{L}\right)} \frac{d^2 Y_1(y)}{dy^2} = -Y_1(y)\frac{d^2}{dx^2}\left[\sin\left(\frac{\pi x}{L}\right)\right] = \left(\frac{\pi}{L}\right)^2 \cancel{\sin\left(\frac{\pi x}{L}\right)} Y_1(y) \tag{13.1.5}$$

The boundary condition at $y = 0$ and the condition $Y_1(L) = 1$ are therefore satisfied by the solution $Y_1(y) = \sinh(\pi y/L)/\sinh(\pi)$ of Equation (13.1.5). For V_2, the potential must instead be zero at $y = L$. Since the differential equation remains invariant under the transformation $x \leftrightarrow L - x$, this is accomplished by setting $V_2 = \sin(2\pi y/L)\sinh(2\pi(L - x)/L)/\sinh(2\pi)$.

If arbitrary boundary conditions are imposed on $V(x, y)$ along the sides of a square, these can be expanded in Fourier series, and the superposition principle again applied to express the solution as a sum over the potentials arising from each Fourier component individually. Thus, for $V^{\text{boundary}}(0, y) = V^{\text{boundary}}(\pi, y) = V^{\text{boundary}}(x, 0) = 0$, $V^{\text{boundary}}(x, \pi) = V_0$, the Fourier expansion of $V^{\text{boundary}}(x, \pi)$

is given by (the general case in which the boundaries are at $x' = 0$ and L is obtained by the transformation $x \rightarrow \pi x'/L, y \rightarrow \pi y'/L$)

$$V^{\text{boundary}}(x, \pi) = \sum_{n=1}^{\infty} c_n \sin nx = V_0 \tag{13.1.6}$$

Multiplying both sides of the above equation by $\sin mx$ and integrating over $[0, \pi]$ according to

$$\begin{aligned} &\int_0^{\pi} \sin mx \sin nx dx = \delta_{mn} \frac{\pi}{2} \\ &\int_0^{\pi} \sin mx V(x, L) dx = -\frac{V_0}{m} \cos mx \Big|_0^{\pi} = -\frac{V_0}{m}((-1)^m - 1) \end{aligned} \tag{13.1.7}$$

yields

$$V^{\text{boundary}}(x, \pi) = V_0 \sum_{m \text{ odd}}^{\infty} \frac{4}{m\pi} \sin mx \tag{13.1.8}$$

and consequently, from the superposition principle, following the calculation of Equation (13.1.5), the solution satisfying the three boundary conditions is given by

$$V(x, y) = \sum_{m \text{ odd}} \frac{4V_0}{m\pi \sinh m\pi} \sin mx \sinh my \tag{13.1.9}$$

In the *diffusion equation*, the velocity of decay of a field $y(x, t)$ is proportional to its curvature:

$$\frac{dy(x, t)}{dt} = D \frac{d^2 y(x, t)}{dx^2} \tag{13.1.10}$$

Consequently, if the curvature is negative, $y(x, t)$ decays exponentially to zero with time. The temperature distribution of an inhomogeneously heated solid rod with an initial temperature profile given by $y(x, t = 0) = T(0) \sin \pi x/L$ in °C and its two ends at 0 and L held at $T = 0$ at all t is determined by inserting $y(x, t) = T(t)X(x)$ where $X(x) = \sin \pi x/L$ to satisfy the initial condition into Equation (13.1.10) so that

$$\frac{dT(t)}{dt} \sin \frac{\pi x}{L} = -D \left(\frac{\pi}{L} \right)^2 \sin \frac{\pi x}{L} T(t) \tag{13.1.11}$$

and therefore

$$T(t) = T(0) e^{-D\left(\frac{\pi}{L}\right)^2 t} \tag{13.1.12}$$

An arbitrary initial condition $y(x, 0) = y_0(x)$ can be decomposed through Fourier analysis into a superposition

$$y(x, 0) = \sum_{n=1}^{\infty} c_n \sin \frac{n\pi x}{L} \tag{13.1.13}$$

with

$$c_n = \frac{2}{L}\int_{-L/2}^{L/2} y(x,0)\sin\frac{\pi n x}{L}dx \tag{13.1.14}$$

The problem can then be viewed as a superposition of an infinite number of subproblems such that the initial condition for the nth problem is given by $y_n(x, t=0) = c_n \sin n\pi x/L$. Hence,

$$y(x,t) = \sum_{n=1}^{\infty} c_n e^{-D\left(\frac{n\pi x}{L}\right)^2 t}\sin\frac{n\pi x}{L} \tag{13.1.15}$$

Numerically, the diffusion equation can be solved by applying the forward finite difference approximation to the time derivative and the centered approximation to the second spatial derivative:

$$\frac{T(x_i,t_{i+1})-T(x_i,t_i)}{\Delta t} = D\frac{T(x_{i+1},t_i)-2T(x_i,t_i)+T(x_{i-1},t_i)}{(\Delta x)^2}$$

$$T(x_i,t_{i+1}) = T(x_i,t_i) + \frac{D\Delta t}{(\Delta x)^2}(T(x_{i+1},t_i)-2T(x_i,t_i)+T(x_{i-1},t_i)) \tag{13.1.16}$$

as in the program below. Numerical stability is only ensured for sufficiently small $D\Delta t/(\Delta x)^2$:

```
numberOfPoints = 50;
numberOfTimeSteps = 1000;
deltaTime = 0.02;
windowLength = 10;
diffusionConstant = 0.5;
deltaX = windowLength / ( numberOfPoints - 1 );
equationConstant = diffusionConstant * deltaTime / ...
   ( deltaX * deltaX );
position = ( 0 : numberOfPoints - 1 ) * deltaX;
temperature = sin ( pi * position / windowLength );

for outerLoop = 0 : numberOfTimeSteps
   if (mod(outerLoop, 100) == 0)
      plot(position, temperature);
      drawnow;
   end
   tempSaveNew = 0;
   for loop = 2 : numberOfPoints - 1
      tempSave = temperature(loop);
```

```
        temperature(loop) = temperature(loop) + ...
            equationConstant * ( tempSaveNew - 2 * ...
            temperature(loop) + temperature(loop + 1) );
        tempSaveNew = tempSave;
    end
end
```

In three dimensions for a potential function specified on the surfaces of a rectangular prism, inserting $V(x, y, z) = X(x)Y(y)Z(z)$ into the Laplace equation and dividing the result by $V(x, y, z)$ yields

$$\frac{1}{X(x)}\frac{d^2X(x)}{dz^2} \equiv \Xi(x) = -\frac{1}{Y(y)Z(z)}\left(Y(y)\frac{d^2Z(z)}{dx^2} + Z(z)\frac{d^2Y(y)}{dy^2}\right) \equiv \Sigma(y,z) \quad (13.1.17)$$

Since, e.g., all partial derivatives of $\Xi(x)$ and consequently of $\Sigma(y, z)$ with respect to y and z are then zero, these both must equal an invariant *separation constant*. This separation constant should be set to $-k_x^2$ if the boundary conditions are specified on a $y=$ constant or $z=$ constant face so that a Fourier series in x is required to represent the potential along the boundary and $+k_x^2$ otherwise. Subsequently, in the same manner,

$$\frac{1}{Z(z)}\frac{d^2Z(z)}{dz^2} \mp k_x^2 \equiv \mathrm{N}(z) = \frac{1}{Y(y)}\frac{d^2Y(y)}{dy^2} \equiv \Psi(y) = \pm k_y^2 \quad (13.1.18)$$

For example, if the potential is specified on the $z = 0$ face of a cube and is zero on the remaining faces, *sinusoidal* functions that satisfy

$$\begin{aligned} \frac{d^2X_m(x)}{dx^2} &= -k_{x,m}^2X_m(x) \\ \frac{d^2Y_n(y)}{dx^2} &= -k_{y,n}^2Y_n(y) \end{aligned} \quad (13.1.19)$$

where X_n and Y_n are zero at $x = 0,\ L_x$ and $y = 0,\ L_y$, respectively, are required so that V at $z = 0$ can be expanded as

$$V(x,y,z=0) = \sum_{n,m=1}^{\infty} c_{mn}X_m(x)Y_n(y)Z_{mn}(z=0) \quad (13.1.20)$$

Each function $Z_{mn}(z)$ then corresponds to the *hyperbolic* solution with $Z_{mn}(L_z) = 0$ and $Z_{mn}(0) = 1$ of

$$\frac{d^2 Z_{mn}(z)}{dz^2} = k_{z,mn}^2 = \left(k_{x,m}^2 + k_{y,n}^2\right) Z_{mn}(z) \tag{13.1.21}$$

Separation of variables can also be employed to determine Green's functions.

Example

Following the formalism of Equation (12.6.18), the Green's function of the Poisson equation with zero boundary conditions on the surface of a rectangular region bounded by the planes at $x = 0, \pi$ and $y = 0, \pi$ can be constructed from the normalized eigenfunctions

$$V_{mn} = X_m(x) Y_n(y) = \frac{2}{\pi} \sin mx \sin ny \tag{13.1.22}$$

of the two-dimensional Helmholtz equation

$$\frac{\partial^2 V(x,y)}{\partial x^2} + \frac{\partial^2 V(x,y)}{\partial y^2} + k^2 V(x,y) = 0 \tag{13.1.23}$$

which satisfy

$$\nabla^2 V_{mn} = -\left(m^2 + n^2\right) V_{mn} \equiv -k_{mn}^2 V_{mn} \tag{13.1.24}$$

From Equation (12.6.18), the Green's function is then obtained by setting $k^2 = 0$ in

$$G_{\text{Helmholtz}}\left(\vec{r}, \vec{r}'\right) = \frac{4}{\pi^2} \sum_{n,m=1}^{\infty} \frac{\sin mx \sin ny \sin mx' \sin ny'}{k^2 - k_{mn}^2} \tag{13.1.25}$$

Alternatively, this Green's function can be constructed in a different form by expanding in a complete set of eigenfunctions of the Poisson equation that solve the boundary conditions in the two regions $y < y'$ and $y > y'$ (or equivalently in $x < x'$ and $x > x'$) as

$$G\left(\vec{r}, \vec{r}'\right) \equiv \sum_{n=1}^{\infty} g_n\left(\vec{r}', y\right) \sin nx = \begin{cases} \displaystyle\sum_{n=1}^{\infty} [u_n(x', y') \sinh ny'] \sin nx \sinh n(\pi - y) & y > y' \\ \displaystyle\sum_{n=1}^{\infty} [d_n(x', y') \sinh n(\pi - y')] \sin nx \sinh ny & y < y' \end{cases} \tag{13.1.26}$$

As a result of the symmetric form employed for the y' dependence within the square brackets, continuity of G at $y = y'$ implies

$$u_n(x', y') = d_n(x', y') \tag{13.1.27}$$

Additionally, from the equation for Green's function,

$$\sum_{m=1}^{\infty}\left(-m^2 g_m+\frac{\partial^2 g_m}{\partial y^2}\right)\sin mx=\delta(x-x')\delta(y-y') \tag{13.1.28}$$

After multiplying both sides by $\sin nx$ and integrating over the interval $[0, \pi]$,

$$\frac{\pi}{2}\left(-n^2 g_n+\frac{\partial^2 g_n}{\partial y^2}\right)=\sin nx'\delta(y-y') \tag{13.1.29}$$

indicating that $d_n(x', y') = C_n(y')\sin nx'$. Further, the above equation yields

$$\frac{\pi}{2}\lim_{\Delta y\to 0}\left(\left.\frac{\partial g_n}{\partial y}\right|_{y=y'+\Delta y}-\left.\frac{\partial g_n}{\partial y}\right|_{y=y'-\Delta y}\right)=\sin nx' \tag{13.1.30}$$

which, from Equations (13.1.26) and (13.1.27), reduces to, after canceling $\sin nx'$,

$$\frac{n\pi C_n(y')}{2}(-\sinh ny'\cosh n(\pi-y')-\sinh n(\pi-y')\cosh ny')=1 \tag{13.1.31}$$

Applying $\sinh(a+b)=\sinh a\cosh b+\cosh a\sinh b$ leads to $C_n(y')=-2(n\pi\sinh n\pi)^{-1}$ and consequently, with $y_> = \max(y, y')$ and $y_< = \min(y, y')$,

$$G(\vec{r},\vec{r}')=-\frac{2}{\pi}\sum_{n=1}^{\infty}\frac{1}{n\sinh n\pi}\sin nx\sin nx'\sinh n(\pi-y_>)\sinh ny_< \tag{13.1.32}$$

13.2 LEGENDRE POLYNOMIALS

In spherical coordinates, the Laplace equation adopts the form of Equation (8.2.12). For problems with boundaries that lie along planes with constant spherical coordinate values, separation of variables assumes a solution of the form $V(r, \theta, \phi) = R(r)Y(\theta, \phi)$, which yields

$$\frac{1}{R_l(r)}r\frac{d^2}{dr^2}(rR_l(r))=l(l+1)=-\frac{1}{Y(\theta,\phi)}\left(\frac{1}{\sin\theta}\frac{\partial}{\partial\theta}\left(\sin\theta\frac{\partial Y(\theta,\phi)}{\partial\theta}\right)+\frac{1}{\sin^2\theta}\frac{\partial^2 Y(\theta,\phi)}{\partial\phi^2}\right) \tag{13.2.1}$$

The form of the separation constant $l(l+1)$ is motivated by the general solution, $R_l(r)\approx cr^l+dr^{-(l+1)}$, of the radial equation, which can be written as

$$\frac{d^2}{dr^2}(rR_l(r))=\frac{l(l+1)}{r^2}(rR_l(r)) \tag{13.2.2}$$

The constant l adopts noninteger values if boundary conditions are imposed along cones $\theta=\theta_{1,2}$ or on planes $\phi=\phi_{1,2}$. In azimuthally symmetric problems since $\partial^2 Y(\theta,\phi)/\partial\phi^2=0$,

$$\frac{1}{\sin\theta}\frac{d}{d\theta}\left(\sin\theta\frac{dP_l(\cos\theta)}{d\theta}\right)+l(l+1)P_l(\cos\theta)=0 \tag{13.2.3}$$

Here, $P_l(\cos\theta)$ is termed the lth *Legendre polynomial*. Substituting $x=\cos\theta$ for which $d/d\theta=(dx/d\theta)(d/dx)=-\sin\theta(d/dx)$ yields an alternative equation for $P(x)$

$$\frac{d}{dx}\left(\underbrace{(1-x^2)}_{\sin^2\theta}\frac{dP_l(x)}{dx}\right)+l(l+1)P_l(x)=0 \tag{13.2.4}$$

The most general azimuthally symmetric solution to Equation (13.2.1) on the domain $0\le\theta\le\pi$ is

$$V(r,\theta)=\sum_{l=0}^{\infty}\left(c_l r^l+d_l r^{-(l+1)}\right)P_l(\cos\theta) \tag{13.2.5}$$

The Legendre polynomials can be generated from the electric potential of a point charge at $\vec{r}'$, $V(r)=1/|\vec{r}-\vec{r}'|$, which satisfies the Laplace equation away from $\vec{r}'$. For $\vec{r}'$ along the z-axis, the angle between $\vec{r}$ and $\vec{r}'$ coincides with the polar angle θ, and hence,

$$\phi(r,\theta)=\frac{1}{|\vec{r}-\vec{r}'|}=\frac{1}{(r^2+r'^2-2rr'\cos\theta)^{\frac{1}{2}}}=\frac{1}{(r^2+r'^2-2rr'x)^{\frac{1}{2}}} \tag{13.2.6}$$

Since for a charge on the axis of symmetry the potential is on the axis of symmetry, it must be of the form of Equation (13.2.5). For $\vec{r}$ also along the z-axis, expanding in a Taylor series

$$\frac{1}{|\vec{r}-\vec{r}'|}=\frac{1}{|r-r'|}=\frac{1}{r_>}\left(1-\frac{r_<}{r_>}\right)^{-1}=\sum_{i=0}^{\infty}\frac{r_<^l}{r_>^{l+1}} \tag{13.2.7}$$

With $\theta=0$, $x\equiv\cos\theta=1$, the Legendre polynomials are normalized according to

$$P_l(1)=1 \tag{13.2.8}$$

and Equation (13.2.5) becomes $V(r,0)=\sum_{l=0}^{\infty}\left(c_l r^l+d_l r^{-(l+1)}\right)$. Identifying like powers of r with Equation (13.2.7),

$$\begin{aligned} c_l&=\frac{1}{(r')^{l+1}}=\frac{1}{r_>^{l+1}} & d_l&=0 & r&<r' \\ c_l&=0 & d_l&=(r')^l=r_<^l & r&>r' \end{aligned} \tag{13.2.9}$$

Inserting Equation (13.2.9) into Equation (13.2.5) then results in an expansion of the potential in terms of the Legendre polynomials for arbitrary $\vec{r}$ with $\vec{r}'$ positioned along z

$$\frac{1}{|\vec{r}-\vec{r}'|} = \sum_{l=0}^{\infty} \frac{r_<^l}{r_>^{l+1}} P_l(\cos\theta) \tag{13.2.10}$$

Setting $t \equiv r_</r_>$ yields the *generating function* for the Legendre polynomials $G_P(x, t)$

$$\frac{1}{|\vec{r}-\vec{r}'|} = \frac{1}{\sqrt{r^2 - 2rr'\cos\theta + r'^2}} = \frac{1}{r_>} \frac{1}{\sqrt{1-2xt+t^2}} \equiv \frac{1}{r_>} G_P(x,t) = \frac{1}{r_>} \sum_{l=0}^{\infty} t^l P_l(x) \tag{13.2.11}$$

Expanding $(1-2xt+t^2)^{-1/2}$ in a power series and identifying the coefficient of t^l with $P_l(x)$ (or equivalently employing $P_l(x) = (1/l!)\partial^l G_p/\partial^l t|_{t=0}$) gives

$$\begin{aligned} P_0(x) &= 1 \\ P_1(x) &= x \\ P_2(x) &= \frac{1}{2}(3x^2 - 1) \end{aligned} \tag{13.2.12}$$

These can be obtained directly from *Rodriguez's formula*, whose derivation is algebraically involved

$$P_l(x) = \frac{1}{2^l l!} \frac{d^l}{dx^l} (x^2 - 1)^l \tag{13.2.13}$$

The Legendre polynomial $P_l(x)$ is real with highest power x^l, and subsequent powers $x^{l-2}, x^{l-4}, \ldots$ and is therefore even for even l and odd for odd l.

The $P_l(x)$ can also be obtained with the *method of Frobenius*, which employs the trial solution

$$T(x) = x^\alpha \sum_{n=0}^{\infty} a_n x^n \tag{13.2.14}$$

For single-valued solutions on $[-1, 1]$, l is an integer and $\alpha = 0$, which gives after Equation (13.2.14) is inserted into the Legendre equation

$$\begin{aligned} &\sum_{n=1}^{\infty} a_n \{(1-x^2)n(n-1)x^{n-2} - 2nx^n + l(l+1)x^n\} \\ &= \sum_{n=1}^{\infty} a_n \{n(n-1)x^{n-2} - (n(n+1) - l(l+1))x^n\} = 0 \end{aligned} \tag{13.2.15}$$

Equating identical powers of n in the last line of the above equation generates the recursion relation

$$a_{n+2} = \frac{n(n+1) - l(l+1)}{(n+2)(n+1)} a_n \tag{13.2.16}$$

The series, which is normalized after it is computed, terminates when $n = l$. For even and odd l, the lowest term is a_0 and a_1, respectively. For example, for $l = 2$, the terminating solution is obtained by setting $a_{n=0} = -1/2$ to insure that $P_2(0) = 1$ leading to $a_2 = -3a_0$ and $a_4 = 0$.

Significant properties and derivations of Legendre polynomials are given below. A recursion relation is obtained by differentiating the generating function with respect to x

$$\begin{gathered} \frac{\partial G_p}{\partial x} = \frac{t}{(1-2xt+t^2)^{3/2}} = \frac{t}{(1-2xt+t^2)} \sum_{l=0}^{\infty} P_l(x) t^l = \sum_{l=0}^{\infty} P_l{}'(x) t^l \\ t \sum_{l=0}^{\infty} P_l(x) t^l = (1-2xt+t^2) \sum_{l=0}^{\infty} P_l{}'(x) t^l \end{gathered} \tag{13.2.17}$$

Hence, after equating identical powers of t,

$$P'_{l+1}(x) - 2xP'_l(x) + P'_{l-1}(x) = P_l(x) \tag{13.2.18}$$

Differentiating G_p instead with respect to t,

$$\begin{gathered} \frac{\partial G_p}{\partial t} = \frac{x-t}{(1-2xt+t^2)^{3/2}} = \frac{x-t}{(1-2xt+t^2)} \sum_{l=0}^{\infty} P_l(x) t^l = \sum_{l=0}^{\infty} l P_l(x) t^{l-1} \\ (x-t) \sum_{l=0}^{\infty} P_l(x) t^l = (1-2xt+t^2) \sum_{l=0}^{\infty} l P_l(x) t^{l-1} \end{gathered} \tag{13.2.19}$$

and again equating similar powers of t leads to, where $P_{-1}(x) = 0$, $P_0(x) = 1$,

$$\begin{gathered} (l+1)P_{l+1}(x) - 2xlP_l(x) + (l-1)P_{l-1}(x) = xP_l(x) - P_{l-1}(x) \\ (l+1)P_{l+1}(x) - x(2l+1)P_l(x) + lP_{l-1}(x) = 0 \end{gathered} \tag{13.2.20}$$

Differentiating Equation (13.2.20), solving first for $P'_{l-1}(x)$ and then $P'_{l+1}(x)$, and then substituting each equation in turn into Equation (13.2.18) generate the two relations

$$\begin{gathered} lP_l(x) = xP_l{}'(x) - P'_{l-1}(x) \\ (l+1)P_l(x) = -xP_l{}'(x) + P'_{l+1}(x) \end{gathered} \tag{13.2.21}$$

Adding these yields a result that can be employed in integration, namely,

$$(2l+1)P_l(x) = P'_{l+1}(x) - P'_{l-1}(x) = \frac{d}{dx}(P_{l+1}(x) - P_{l-1}(x)) \tag{13.2.22}$$

Raising and lowering operators that transform P_l into $P_{l\pm 1}$ are obtained by replacing, e.g., $l \to l-1$ in the second equation of Equation (13.2.21) and substituting for $P'_{l-1}(x)$ from the first equation:

$$P_{l-1}(x) = xP_l(x) + \left(\frac{1-x^2}{l}\right)P'_l(x) \tag{13.2.23}$$

Alternatively, setting $l \to l+1$ in the first equation and eliminating $P'_{l+1}(x)$ with the second equation,

$$P_{l+1}(x) = xP_l(x) - \left(\frac{1-x^2}{l+1}\right)P'_l(x) \tag{13.2.24}$$

The orthogonality of the Legendre polynomials is insured by the properties of the Sturm–Liouville equation, while their normalization follows from, recalling that log $(1+\delta) = \delta - \delta^2/2 + \delta^3/3 + \ldots$,

$$\begin{aligned}
\sum_{l,k=0}^{\infty} t^{l+k}\int_{-1}^{1} P_l(x)P_k(x)dx &= \sum_{l=0}^{\infty} t^{2l}\int_{-1}^{1}(P_l(x))^2 dx \\
&= \int_{-1}^{1}\frac{1}{1+t^2-2xt}dx \\
&= -\frac{1}{2t}\log(1+t^2-2xt)\Big|_{x=-1}^{x=1} = -\frac{1}{2t}\log\left(\frac{(1-t)^2}{(1+t)^2}\right) = \frac{1}{t}\log\left(\frac{1+t}{1-t}\right) \\
&= \frac{1}{t}[\log(1+t) - \log(1-t)] = \frac{1}{t}\sum_{l=1}^{\infty}\frac{t^l}{l}\left((-1)^{l+1}+1\right) = 2\sum_{l=0}^{\infty}\frac{t^{2l}}{2l+1}
\end{aligned} \tag{13.2.25}$$

Equating identical powers of t,

$$\int_{-1}^{1}(P_l(x))^2 dx = \frac{2}{2l+1} \tag{13.2.26}$$

Thus, any function on the interval $[-1, 1]$ can be expanded as

$$\begin{aligned}
f(x) &= \sum_{l=0}^{\infty}\varpi_l P_l(x) \\
\varpi_l &= \frac{2l+1}{2}\int_{-1}^{1} f(x')P_l(x')dx'
\end{aligned} \tag{13.2.27}$$

Since $x^j, j = 0, 1, \ldots, m-1$ form a complete basis set for polynomials of order $m-1$ on $[-1, 1]$ and the m Legendre polynomials of order less than m are orthogonal and of order less than x^m, any polynomial of order less than m can be written as a linear superposition of the first m Legendre polynomials.

Example

The potential inside a hollow insulating sphere of radius R with potential V_0 on the upper half and zero on the lower half is derived by observing that since $V(r, x)$ is finite within the sphere, $d_i = 0$ in Equation (13.2.5). Accordingly, on the surface of the sphere,

$$V^{\text{boundary}}(R,x) = \sum_{l=0}^{\infty} c_l R^l P_l(x) = \begin{cases} V_0 & 0<x<1 \\ 0 & -1<x<1 \end{cases} \tag{13.2.28}$$

Equations (13.2.22) and (13.2.27) then yield

$$\begin{aligned} \sum_{l=0}^{\infty} c_l R^l \int_{-1}^{1} P_m(x) P_l(x) dx &= V_0 \int_0^1 P_m(x) dx \\ \sum_{l=0}^{\infty} c_l R^l \left(\frac{2}{2m+1} \delta_{ml} \right) &= \frac{V_0}{2m+1} \int_0^1 \frac{d}{dx} (P_{m+1}(x) - P_{m-1}(x)) dx \end{aligned} \tag{13.2.29}$$

With P(1) = 1, the recursion relation, Equation (13.2.20) with $x = 0$ for the lower limit leads to

$$c_m = \frac{1}{2R^m} V_0 (P_{m-1}(0) - P_{m+1}(0)) = -\frac{1}{2R^m} V_0 \left(1 + \frac{m+1}{m} \right) P_{m+1}(0) \quad m>0 \tag{13.2.30}$$

With these c_m values, noting that $c_0 = V_0/2$, Equation (13.2.5) becomes

$$V(r,x) = \frac{V_0}{2} \left(1 - \sum_{l=1}^{\infty} \left(\frac{2l+1}{l} \right) \left(\frac{r}{R} \right)^l P_{l+1}(0) P_l(x) \right) \tag{13.2.31}$$

13.3 SPHERICAL HARMONICS

From a diagram or from Equation (8.1.8), the unit vectors in spherical and rectangular coordinates are related by

$$\begin{aligned} \hat{e}_\phi &= -\hat{e}_x \sin\phi + \hat{e}_y \cos\phi \\ \hat{e}_\theta &= -\hat{e}_z \sin\theta + \cos\theta (\hat{e}_x \cos\phi + \hat{e}_y \sin\phi) \\ \hat{e}_r &= \hat{e}_z \cos\theta + \sin\theta (\hat{e}_x \cos\phi + \hat{e}_y \sin\phi) \end{aligned} \tag{13.3.1}$$

The angular component of the Laplacian operator equals minus $1/r^2$ times the square of the dimensionless operator, proportional to the quantum mechanical *angular momentum operator*,

$$\vec{L} = -i\vec{r}\times\vec{\nabla} = -i\begin{vmatrix} \hat{e}_r & \hat{e}_\theta & \hat{e}_\phi \\ r & 0 & 0 \\ \frac{\partial}{\partial r} & \frac{1}{r}\frac{\partial}{\partial\theta} & \frac{1}{r\sin\theta}\frac{\partial}{\partial\phi} \end{vmatrix} = \hat{e}_\theta \underbrace{\frac{i}{\sin\theta}\frac{\partial}{\partial\phi}}_{L_\theta} + \hat{e}_\phi \underbrace{\left(-i\frac{\partial}{\partial\theta}\right)}_{L_\phi} \tag{13.3.2}$$

Hence, inserting Equation (13.3.1),

$$\begin{aligned}\vec{L} &= i\left[(\hat{e}_x\sin\phi - \hat{e}_y\cos\phi)\frac{\partial}{\partial\theta} + (\hat{e}_x\cos\phi + \hat{e}_y\sin\phi)\frac{\cos\theta}{\sin\theta}\frac{\partial}{\partial\phi} - \hat{e}_z\frac{\partial}{\partial\phi}\right] \\ &= i\left[\hat{e}_x\left(\sin\phi\frac{\partial}{\partial\theta} + \cos\phi\cot\theta\frac{\partial}{\partial\phi}\right) + \hat{e}_y\left(-\cos\phi\frac{\partial}{\partial\theta} + \sin\phi\cot\theta\frac{\partial}{\partial\phi}\right) - \hat{e}_z\frac{\partial}{\partial\phi}\right] \\ &= \vec{L}_x + \vec{L}_y + \vec{L}_z \end{aligned} \tag{13.3.3}$$

where in Cartesian components

$$L_x = -i\left(y\frac{\partial}{\partial z} - z\frac{\partial}{\partial y}\right),\ L_y = -i\left(z\frac{\partial}{\partial x} - x\frac{\partial}{\partial z}\right),\ L_z = -i\left(x\frac{\partial}{\partial y} - y\frac{\partial}{\partial x}\right) \tag{13.3.4}$$

while calculating $\partial\hat{e}_\phi/\partial\phi$ and the analogous derivatives directly from Equation (13.3.1),

$$\begin{aligned} L^2 = \vec{L}\cdot\vec{L} &= -\left(\hat{e}_\theta\frac{1}{\sin\theta}\frac{\partial}{\partial\phi} - \hat{e}_\phi\frac{\partial}{\partial\theta}\right)\cdot\left(\hat{e}_\theta\frac{1}{\sin\theta}\frac{\partial}{\partial\phi} - \hat{e}_\phi\frac{\partial}{\partial\theta}\right) \\ &= -\left\{\frac{1}{\sin^2\theta}\frac{\partial^2}{\partial\phi^2} + \frac{\partial^2}{\partial\theta^2} + \frac{1}{\sin^2\theta}\underbrace{\hat{e}_\theta\cdot\frac{\partial\hat{e}_\theta}{\partial\phi}}_{0}\frac{\partial}{\partial\phi} - \frac{1}{\sin\theta}\underbrace{\hat{e}_\theta\cdot\frac{\partial\hat{e}_\phi}{\partial\phi}}_{-\cos\theta}\frac{\partial}{\partial\theta}\right. \\ &\qquad \left. - \frac{1}{\sin\theta}\underbrace{\hat{e}_\phi\cdot\frac{\partial\hat{e}_\theta}{\partial\theta}}_{0}\frac{\partial}{\partial\phi} + \underbrace{\hat{e}_\phi\cdot\frac{\partial\hat{e}_\phi}{\partial\theta}}_{0}\frac{\partial}{\partial\theta}\right\} \\ &= -\left(\frac{1}{\sin\theta}\frac{\partial}{\partial\theta}\left(\sin\theta\frac{\partial}{\partial\theta}\right) + \frac{1}{\sin^2\theta}\frac{\partial^2}{\partial\phi^2}\right) \end{aligned} \tag{13.3.5}$$

However, since in $-\left(\vec{r}\times\vec{\nabla}\right)\left(\vec{r}\times\vec{\nabla}\right)$ the first gradient acts on both the second $\vec{r}$ (denoted temporarily below as $\vec{\tilde{r}}$) and the second $\vec{\nabla}$, distinguishing these two operations by subscripts and applying $\left(\vec{A}\times\vec{B}\right)\cdot\left(\vec{C}\times\vec{D}\right)=\left(\vec{A}\cdot\vec{C}\right)\left(\vec{B}\cdot\vec{D}\right)-\left(\vec{A}\cdot\vec{D}\right)\left(\vec{B}\cdot\vec{C}\right)$ yields

$$\begin{aligned}
-L^2 &= \left(\vec{r}\times\vec{\nabla}_{\tilde{r}}\right)\cdot\left(\vec{\tilde{r}}\times\vec{\nabla}\right)+\left(\vec{r}\times\vec{\nabla}_{\tilde{\nabla}}\right)\cdot\left(\vec{r}\times\vec{\nabla}\right)\\
&= r_i\tilde{r}_i(\nabla_{\tilde{r}})_j\nabla_j-\left(\vec{r}\cdot\vec{\nabla}\right)\left(\vec{\tilde{r}}\cdot\vec{\nabla}_{\tilde{r}}\right)+r^2\nabla^2-\left(\vec{r}\cdot\vec{\nabla}_{\tilde{\nabla}}\right)\vec{r}\cdot\vec{\nabla}\\
&= r_i\underbrace{\left(\nabla_j r_i\right)}_{\delta_{ji}}\nabla_j-\underbrace{\left(\vec{\nabla}\cdot\vec{r}\right)}_{3}\left(\vec{r}\cdot\vec{\nabla}\right)+r^2\nabla^2-r^2\frac{\partial}{\partial r^2}\\
&= r^2\nabla^2-r^2\frac{\partial}{\partial r^2}-2r\frac{\partial}{\partial r}\\
&= r^2\nabla^2-\frac{\partial}{\partial r}r^2\frac{\partial}{\partial r}
\end{aligned} \tag{13.3.6}$$

Therefore, where the second terms equal $-l(l+1)/r^2$, if separation of variables can be applied,

$$\nabla^2=\frac{1}{r^2}\frac{\partial}{\partial r}r^2\frac{\partial}{\partial r}-\frac{L^2}{r^2}=\frac{1}{r^2}\frac{\partial}{\partial r}r^2\frac{\partial}{\partial r}+\frac{1}{r^2}\left(\frac{1}{\sin\theta}\frac{\partial}{\partial\theta}\left(\sin\theta\frac{\partial}{\partial\theta}\right)+\frac{1}{\sin^2\theta}\frac{\partial^2}{\partial\phi^2}\right) \tag{13.3.7}$$

For nonaxially symmetric problems, separating variables in the Laplace equation for spherical coordinates $V(r,\theta,\phi)=R(r)Y(\theta,\phi)$ with $R_l(r)\approx cr^l+dr^{-(l+1)}$ as in Equation (13.2.2) leads to

$$\sin\theta\frac{\partial}{\partial\theta}\left(\sin\theta\frac{\partial Y}{\partial\theta}\right)+l(l+1)\sin^2\theta Y=-\frac{\partial^2 Y}{\partial\phi^2} \tag{13.3.8}$$

Further setting $Y_{lm}(\theta,\phi)=P_l^m(\theta)\Phi_m(\phi)$ and introducing the separation constant results in

$$\frac{d^2\Phi_m(\phi)}{d\phi^2}=-m^2\Phi_m \tag{13.3.9}$$

with the general solution

$$\Phi_m(\phi)=e_1e^{im\phi}+e_2e^{-im\theta} \tag{13.3.10}$$

If the problem region extends from $\phi=0$ to $\phi=2\pi$, m must be an integer to ensure that $\Phi_m(\phi)$ is single valued. The *associated Legendre polynomials* $P_l^m(\theta)$ then obey the equation

$$\frac{1}{\sin\theta}\frac{\partial}{\partial\theta}\left(\sin\theta\frac{\partial P_l^m(\theta)}{\partial\theta}\right)+l(l+1)P_l^m-\frac{m^2}{\sin^2\theta}P_l^m=0 \tag{13.3.11}$$

However, the $Y_{lm}(\theta,\phi)$ can again be conveniently determined through *raising* and *lowering (ladder) operators* that relate $Y_{lm\pm1}(\theta,\phi)$ and $Y_{lm}(\theta,\phi)$. From Equations (13.3.8) and (13.3.9), the $Y_{lm}(\theta,\phi)$ are simultaneous eigenfunctions of L^2 and L_z with

$$\begin{aligned}L_zY_{lm}(\theta,\phi)&=mY_{lm}(\theta,\phi)\\ L^2Y_{lm}(\theta,\phi)&=l(l+1)Y_{lm}(\theta,\phi)\end{aligned} \tag{13.3.12}$$

The factor of $e^{\pm i\phi}$ required to compensate for the differing ϕ dependence of Y_{lm} and $Y_{lm\pm1}$ is obtained by combining L_x and L_y according to

$$\begin{aligned}L_\pm\equiv L_x\pm iL_y&=\pm(\cos\phi\pm i\sin\phi)\frac{\partial}{\partial\theta}+i\cot\theta(\cos\phi\pm i\sin\phi)\frac{\partial}{\partial\phi}\\ &=e^{\pm i\phi}\left(\pm\frac{\partial}{\partial\theta}+i\cot\theta\frac{\partial}{\partial\phi}\right)\end{aligned} \tag{13.3.13}$$

In particular, since, e.g.,

$$[L_z,L_x]=-\left(x\frac{\partial}{\partial y}-y\frac{\partial}{\partial x}\right)\left(y\frac{\partial}{\partial z}-z\frac{\partial}{\partial y}\right)+\left(y\frac{\partial}{\partial z}-z\frac{\partial}{\partial y}\right)\left(x\frac{\partial}{\partial y}-y\frac{\partial}{\partial x}\right)=\left(z\frac{\partial}{\partial x}-x\frac{\partial}{\partial z}\right)=iL_y \tag{13.3.14}$$

or, after performing the cyclic substitutions $x\to y\to z\to x$,

$$\begin{aligned}[L_i,L_j]&=i\varepsilon_{ijk}L_k\\ [L_z,L_\pm]&=\pm L_\pm\end{aligned} \tag{13.3.15}$$

which in turn implies, $L_\pm Y_{lm}=c_\pm Y_{lm\pm1}$ for some constants c_+ and c_- from

$$L_z(L_\pm Y_{lm})=(m\pm1)(L_\pm Y_{lm})=(L_\pm L_z\pm L_\pm)Y_{lm} \tag{13.3.16}$$

To determine these normalization constants, note first that

$$\begin{aligned}L^2=L_x^2+L_y^2+L_z^2&=\frac{1}{2}\{(L_x+iL_y)(L_x-iL_y)+(L_x-iL_y)(L_x+iL_y)\}+L_z^2\\ &=\frac{1}{2}\{L_+L_-+L_-L_+\}+L_z^2\end{aligned} \tag{13.3.17}$$

and

$$[L_+, L_-] = [L_x + iL_y, L_x - iL_y] = -i[L_x, L_y] + i[L_y, L_x] = 2L_z \quad (13.3.18)$$

yield

$$L^2 = L_+ L_- + L_z(L_z - 1) = L_- L_+ + L_z(L_z + 1) \quad (13.3.19)$$

Since the Y_{lm} are normalized according to unity according to, where Y_{lm} is denoted by $|l, m\rangle$,

$$\langle l, m | l', m' \rangle \equiv \int Y_{lm}^*(\theta, \phi) Y_{l'm'}(\theta, \phi) d\Omega \equiv \int Y_{lm}^*(\theta, \phi) Y_{l'm'}(\theta, \phi) \sin\theta d\theta d\phi = \delta_{ll'} \delta_{mm'} \quad (13.3.20)$$

Equation (13.3.19) together with the hermiticity of $\vec{L}$ (e.g., $\int f^*(\vec{r}) L_x g(\vec{r}) d^3x = \int [L_x f(\vec{r})]^* g(\vec{r}) d^3x$, after partial integration), which implies $L_\pm^\dagger = L_\mp$, determines the normalization of Y_{lm+1} relative to that of Y_{lm} from

$$\begin{aligned}[l(l+1) - m(m-1)]\langle l, m | l, m \rangle &= c_+^2 \langle l, m | L_- | l, m+1 \rangle \langle l, m+1 | L_+ | l, m \rangle \\ &= c_+^2 |\langle l, m+1 | L_+ | l, m \rangle|^2 \\ &= c_+^2 \langle l, m+1 | l, m+1 \rangle^2\end{aligned} \quad (13.3.21)$$

From this and a similar calculation for c_-, where the upper (+) sign is chosen by convention,

$$\begin{aligned}c_+ &= \pm(l(l+1) - m(m+1))^{1/2} = \pm((l-m)(l+m+1))^{1/2} \\ c_- &= \pm(l(l+1) - m(m-1))^{1/2} = \pm((l+m)(l-m+1))^{1/2}\end{aligned} \quad (13.3.22)$$

The spherical harmonic functions can be obtained from the raising and lowering operators since

$$L_+ Y_{ll} = (l(l+1) - l(l+1))^{1/2} Y_{ll} + 1 = 0 \quad (13.3.23)$$

is equivalent to

$$e^{i\phi}\left(\frac{\partial}{\partial\theta} + i\cot\theta\frac{\partial}{\partial\phi}\right) P_l^l(\theta) e^{il\phi} = 0 \quad (13.3.24)$$

Hence,

$$\left(\frac{\partial}{\partial\theta} - l\cot\theta\right) P_l^l(\theta) = 0 \quad (13.3.25)$$

and therefore,

$$\left(\frac{1}{\cos\theta}\frac{\partial}{\partial\theta} - \frac{l}{\sin\theta}\right) P_l^l(\theta) = \left(\frac{\partial}{\partial(\sin\theta)} - \frac{l}{\sin\theta}\right) P_l^l(\theta) = 0 \quad (13.3.26)$$

with the solution

$$P_l^l(\theta) = A_l \sin^l\theta \tag{13.3.27}$$

The absolute normalization factor, A_l, for Y_{ll} is then, where $n!!$ denotes $n(n-2)$ $(n-4)\ldots$,

$$A_l = \left(2\pi\int_0^\pi \sin^{2l+1}\theta d\theta\right)^{-\frac{1}{2}} = \pm\left(4\pi\frac{2l!!}{(2l+1)!!}\right)^{-1/2} \tag{13.3.28}$$

where the recursion relation

$$\begin{gathered}\int_0^\pi \underbrace{\sin^{2l}\theta}_{g(x)}\underbrace{\sin\theta}_{h'(x)}d\theta = 2l\int_0^\pi \sin^{2l-1}\theta\underbrace{\cos\theta\cos\theta}_{1-\sin^2\theta}d\theta - \sin^{2l}\theta\cos\theta\Big|_0^\pi \\ (2l+1)\int_0^\pi \sin^{2l+1}\theta d\theta = 2l\int_0^\pi \sin^{2l-1}\theta d\theta\end{gathered} \tag{13.3.29}$$

together with $\int_0^\pi \sin\theta d\theta = 2$ is employed.

Example

From

$$Y_{1\pm1}(\theta,\phi) = \mp\sqrt{\frac{3}{8\pi}}\sin\theta e^{\pm i\phi} \tag{13.3.30}$$

where the sign of Y_{11} is set by convention, it follows that

$$\begin{aligned}((l+m)(l-m+1))^{1/2}Y_{10} &= \sqrt{2\cdot1}Y_{10} = L_-Y_{11} \\ &= -\sqrt{\frac{3}{8\pi}}e^{-i\phi}\left(-\frac{\partial}{\partial\theta} + i\cot\theta\frac{\partial}{\partial\phi}\right)\sin\theta e^{i\phi}\end{aligned} \tag{13.3.31}$$

leading to

$$Y_{10} = -\frac{1}{\sqrt{2}}\sqrt{\frac{3}{8\pi}}(-\cos\theta - \cos\theta) = \sqrt{\frac{3}{4\pi}}\cos\theta \tag{13.3.32}$$

Functions of angular coordinates can be expanded in spherical harmonics as

$$f(\theta,\phi) = \sum_{l=0}^{\infty}\sum_{m=-l}^{l} A_{lm}Y_{lm}(\theta,\phi), \quad A_{lm} = \int Y_{lm}^*(\theta',\phi')f(\theta',\phi')d\Omega' \tag{13.3.33}$$

from which the general solution to the Laplace equation in spherical coordinates

$$V(r,\theta,\phi)=\sum_{l=0}^{\infty}\sum_{m=-l}^{l}\left(A_{lm}r^{l}+\frac{B_{lm}}{r^{l+1}}\right)Y_{lm}(\theta,\phi) \tag{13.3.34}$$

can be determined, where $A_{lm}=0$ for $l>0$ for functions V that approach zero at infinity, while $B_{lm}=0$ if V is finite at $r=0$. If V is specified on two concentric shells, twice as many constants are required to match both sets of boundary conditions, and all A_{lm} and B_{lm} can appear.

Some additional useful properties of the spherical harmonics are

$$Y_{l0}(\theta,\phi)=\sqrt{\frac{2l+1}{4\pi}}P_{l}(\cos\theta) \tag{13.3.35}$$

$$Y_{lm}(0,\phi)=Y_{lm}(\pi,\phi)=0 \quad \text{for } m\neq 0 \tag{13.3.36}$$

$$Y_{l,-m}(\theta,\phi)=(-1)^{m}Y_{lm}^{*}(\theta,\phi) \tag{13.3.37}$$

Upon inversion through the coordinate system origin,

$$Y_{lm}(\pi-\theta,\pi+\phi)=(-1)^{l}Y_{lm}(\theta,\phi) \tag{13.3.38}$$

Finally, the *addition theorem* states, where γ is the angle between the directions (θ, ϕ) and (θ', ϕ')

$$P_{l}(\cos\gamma)=\frac{4\pi}{2l+1}\sum_{m=-l}^{l}Y_{lm}^{*}(\theta',\phi')Y_{lm}(\theta,\phi) \tag{13.3.39}$$

which takes the form for $\gamma=0$

$$\sum_{m=-l}^{l}|Y_{lm}(\theta,\phi)|^{2}=\frac{2l+1}{4\pi} \tag{13.3.40}$$

13.4 BESSEL FUNCTIONS

The Laplace equation in cylindrical coordinates

$$\nabla^{2}V=\frac{1}{r}\frac{\partial}{\partial r}\left(r\frac{\partial V}{\partial r}\right)+\frac{1}{r^{2}}\frac{\partial^{2}V}{\partial\phi^{2}}+\frac{\partial^{2}V}{\partial z^{2}}=0 \tag{13.4.1}$$

is separated with $V = R(r)\Phi(\phi)Z(z)$ leading to

$$\frac{1}{rR}\frac{\partial}{\partial r}\left(r\frac{\partial R}{\partial r}\right)+\frac{1}{r^2\Phi}\frac{\partial^2\Phi}{\partial\phi^2}+\frac{1}{Z}\frac{\partial^2 Z}{\partial z^2}=0 \tag{13.4.2}$$

Introducing the separation constants (note that the dimension of k is $[1/D]$), where the sign of k^2 is chosen such that the solution for $k \neq 0$ varies sinusoidally in both ϕ and r,

$$\frac{\partial^2\Phi}{\partial\phi^2}=-m^2\Phi \qquad \frac{\partial^2 Z}{\partial z^2}=k^2 Z \tag{13.4.3}$$

If the problem domain extends over the entire interval $\phi = [0, 2\pi]$, for the solution to be single valued m must be an integer while $f_0 = 0$ below

$$\Phi(\phi)=\begin{cases} e_0+f_0\phi & m=0 \\ e_m\sin m\phi+f_m\cos m\phi & m\neq 0 \end{cases} \tag{13.4.4}$$

The resulting radial function $R_m(kr)$ is then the *ordinary solution* of the *Bessel equation*

$$\frac{1}{r}\frac{\partial}{\partial r}\left(r\frac{\partial R}{\partial r}\right)+\left(k^2-\frac{m^2}{r^2}\right)R=0 \tag{13.4.5}$$

or, equivalently in terms of the dimensionless variable $\rho = kr$,

$$\frac{\partial}{\partial\rho}\left(\rho\frac{\partial R_m(\rho)}{\partial\rho}\right)+\left(\rho-\frac{m^2}{\rho}\right)R_m(\rho)=\rho\frac{\partial^2 R_m(\rho)}{\partial\rho^2}+\frac{\partial R_m(\rho)}{\partial\rho}+\left(\rho-\frac{m^2}{\rho}\right)R_m(\rho)=0 \tag{13.4.6}$$

If $k = 0$ in Equation (13.4.5), $Z(z) = c_0 + d_0 z$, while

$$R_m(r)=\begin{cases} g_0+h_0\ln\rho & m=0 \\ g_m\rho^m+h_m\rho^{-m} & m\neq 0 \end{cases} \tag{13.4.7}$$

yielding solutions of the form

$$V(r,\theta,z)=(c_0+d_0 z)\left\{g_0+h_0\ln\rho+\sum_{m=1}^{\infty}(e_m\sin m\theta+f_m\cos m\theta)(g_m\rho^m+h_m\rho^{-m})\right\} \tag{13.4.8}$$

For $k \neq 0$, as $\rho \to 0$, the term $\rho - m^2/\rho \approx -m^2/\rho$ in the Bessel equation resulting in Equation (13.4.7). Hence, a series expansion of solutions that are finite and vary as ρ^m for $\rho \to 0$ is obtained by setting

$$R_m(\rho)=\rho^m\sum_{n=0}^{\infty}a_n^{(m)}\rho^n \tag{13.4.9}$$

for which

$$\sum_{n=0}^{\infty}(n+m)(n+m-1)a_n^{(m)}\rho^{n+m-1}+\sum_{n=0}^{\infty}(n+m)a_n^{(m)}\rho^{n+m-1}$$
$$+\sum_{n=0}^{\infty}a_n^{(m)}\rho^{n+m+1}-m^2\sum_{n=0}^{\infty}a_n^{(m)}\rho^{n+m-1}=0 \tag{13.4.10}$$

Equating terms with equal powers of ρ generates the recursion relation

$$\left[(n+m)(n+m-1)+(n+m)-m^2\right]a_n^{(m)}=\left[(n+m)^2-m^2\right]a_n^{(m)}=\left(n^2+2nm\right)a_n^{(m)}=-a_{n-2}^{(m)} \tag{13.4.11}$$

Since the coefficient of $a_0^{(m)}$, and hence $a_{-2}^{(m)}$, equals zero, the series terminates. Further,

$$a_{2n}^{(m)}=\frac{-1}{2n(2n+2m)}\frac{-1}{(2n-2)(2n-2+2m)}a_{2n-4}^{(m)}$$
$$=(-1)^n\left(\frac{1}{2^n n!}\right)\left(\frac{1}{2^n(n+m)(n+m-1)\cdots(m+1)}\right)a_0^{(m)} \tag{13.4.12}$$

The convention

$$a_0^{(m)}\equiv\frac{1}{2^m m!} \tag{13.4.13}$$

simplifies the series coefficients yielding the *Bessel functions*

$$J_m(\rho)\equiv R_m(\rho)=\rho^m\sum_{n=0}^{\infty}a_{2n}^{(m)}\rho^{2n}=\sum_{n=0}^{\infty}\frac{(-1)^n}{n!(n+m)!}\left(\frac{\rho}{2}\right)^{m+2n} \tag{13.4.14}$$

which are finite for all ρ. The corresponding solution J_{-m}, equals $(-1)^m$ times $J_m(\rho)$ since Equation (13.4.11) results in $a_{2m}^{(-m)}=\infty$ unless the lowest-order term starts at $n=2m$, reproducing the above series. For noninteger m, however, $(n+m)!$ is replaced by $\Gamma(n+m+1)$, the power series does not terminate at $n=0$, and J_m and J_{-m} are linearly independent. As a result, the second independent solution for integer m is therefore given by the linear combination $c(J_\mu+(-1)^{m+1}J_{-\mu})$ of the two noninteger solutions as $\mu\rightarrow m$. The components of $J_{\pm\mu}$ that are identical with J_m cancel, leaving the orthogonal *Neumann function* $N_m(\rho)$.

A generating function for J_m can be constructed from the solution $\exp(iky-kz)$ to the three-dimensional Laplace equation in which k denotes the separation constant in Equation (13.4.3). As this solution is finite at $\rho = 0$ and oscillates in ρ and ϕ, it can be expressed as

$$e^{-kz+iky} = e^{-kz}e^{i\rho\sin\phi} = e^{-kz}\sum_{m=-\infty}^{\infty} g_m J_m(\rho)e^{im\phi} \tag{13.4.15}$$

From the orthogonality of the complex exponential functions,

$$2\pi g_m J_m(\rho) = \int_0^{2\pi} e^{-im\phi}e^{i\rho\sin\phi}d\phi \tag{13.4.16}$$

Equating the coefficients of the leading ρ^m term as $\rho \to 0$ of the power series expansion of both sides,

$$g_m\left(\frac{\rho}{2}\right)^m \frac{1}{m!} = \frac{1}{2\pi}\frac{(i\rho)^m}{m!}\int_0^{2\pi} e^{-im\phi}(\sin\phi)^m d\phi \tag{13.4.17}$$

Since the $\exp(im\phi)/(2i)^m$ term appearing in the expansion of $\sin^m\phi = (\exp(\phi) + \exp(-\phi))^m$ cancels the $\exp(-im\phi)$ angular variation of the integrand, $g_m = 1$ yielding the generating function

$$e^{i\rho\sin\phi} = \sum_{m=-\infty}^{\infty} J_m(\rho)e^{im\phi} \tag{13.4.18}$$

Equivalently with $t = \exp(i\phi)$ so that $\sin\phi = (t + 1/t)/2i$,

$$e^{\frac{\rho}{2t}(t^2-1)} = \sum_{l=-\infty}^{\infty} J_m(\rho)t^m \tag{13.4.19}$$

From the generating function, *recursion relations* can be derived for the Bessel functions and their derivatives. In particular, taking the derivative of both sides of the above equation with respect to ρ,

$$\frac{1}{2}\left(t - \frac{1}{t}\right)e^{\frac{\rho}{2}\left(t-\frac{1}{t}\right)} = \frac{1}{2}\sum_{m=-\infty}^{\infty} J_m(\rho)\left(t^{m+1} - t^{m-1}\right) = \sum_{m=-\infty}^{\infty}\frac{dJ_m(\rho)}{d\rho}t^m \tag{13.4.20}$$

Equating the coefficients of the t^m terms yields the recursion relation

$$J_{m-1}(\rho) - J_{m+1}(\rho) = 2\frac{dJ_m(\rho)}{d\rho} \tag{13.4.21}$$

If the generating function is instead differentiated with respect to t,

$$\frac{\rho}{2}\left(1 + \frac{1}{t^2}\right)e^{\frac{\rho}{2}\left(t-\frac{1}{t}\right)} = \frac{\rho}{2}\sum_{m=-\infty}^{\infty} J_m(\rho)\left(t^m + t^{m-2}\right) = \sum_{m=-\infty}^{\infty} mJ_m(\rho)t^{m-1} \tag{13.4.22}$$

which gives after identifying the coefficients of t^{m-1},

$$J_{m-1}(\rho)+J_{m+1}(\rho)=\frac{2m}{\rho}J_m(\rho) \tag{13.4.23}$$

Subtracting Equation (13.4.21) from Equation (13.4.23) yields

$$J_{m+1}(\rho)=\frac{m}{\rho}J_m(\rho)-\frac{dJ_m(\rho)}{d\rho} \tag{13.4.24}$$

Multiplying both sides by ρ^{-m} and combining terms on the right-hand side,

$$\frac{J_{m+1}(\rho)}{\rho^m}=-\frac{d}{d\rho}\left(\frac{J_m(\rho)}{\rho^m}\right) \tag{13.4.25}$$

Hence, $J_1(\rho)=-dJ_0(\rho)/d\rho$ for $m=0$, while, after iterating,

$$\frac{J_m}{\rho^m}=-\frac{1}{\rho}\frac{d}{d\rho}\left(\frac{J_{m-1}}{\rho^{m-1}}\right)=\left(\frac{1}{\rho}\frac{\partial}{\partial\rho}\right)^2\left(\frac{J_{m-2}}{\rho^{m-2}}\right)=\ldots=(-1)^m\left(\frac{1}{\rho}\frac{d}{d\rho}\right)^m J_0(\rho) \tag{13.4.26}$$

That J_m and N_m are cylindrical analogues of cos and sin follows from the $\rho\to\infty$ limit of Equation (13.4.5),

$$\frac{\partial^2 R}{\partial\rho^2}+\frac{1}{\rho}\frac{\partial R}{\partial\rho}+\left(1-\frac{m^2}{\rho^2}\right)R\approx\frac{d^2R}{d\rho^2}+R=0 \tag{13.4.27}$$

for which $R\to\chi\cos(\rho+\delta)$. This is improved by setting $R\approx\exp(-i\rho)\rho^\alpha$ in the Bessel equation,

$$-\cancel{\rho^\alpha}-2i\alpha\rho^{\alpha-1}-\frac{i}{\rho}\rho^\alpha+\cancel{\rho^\alpha}+O\left(\rho^{\alpha-2}\right)=0 \tag{13.4.28}$$

so that neglecting nonleading terms proportional to $\rho^{\alpha-2}$ yields $\alpha=-1/2$. Thus, the following cylindrical functions can be constructed and employed similarly to the specified analogue harmonic and exponential functions, where underlined quantities remain finite.

Harmonic function	Corresponding cylindrical function	Behavior as $\rho\to 0$	Behavior as $\rho\to\infty$
$\cos\rho$, $\sin\rho$	Ordinary Bessel function $J_m(\rho)$	$\underline{\frac{1}{\Gamma(m+1)}\left(\frac{\rho}{2}\right)^m}$	$\underline{\sqrt{\frac{2}{\pi\rho}}\cos\left(\rho-\frac{m\pi}{2}-\frac{\pi}{4}\right)}$
$\sin\rho$, $\cos\rho$	Neumann function $N_m(\rho)$	$\begin{cases}\frac{2}{\pi}\left(\ln\frac{\rho}{2}+0.577\right) & m=0\\ -\frac{\Gamma(m)}{\pi}\left(\frac{2}{\rho}\right)^m & m>0\end{cases}$	$\underline{\sqrt{\frac{2}{\pi\rho}}\sin\left(\rho-\frac{m\pi}{2}-\frac{\pi}{4}\right)}$

(*continued*)

Harmonic function	Corresponding cylindrical function	Behavior as $\rho \to 0$	Behavior as $\rho \to \infty$
$e^{i\rho} = \cos\rho + i\sin\rho$	Hankel function of the first kind $H_m^{(1)}(\rho) = J_m(\rho) + iN_m(\rho)$	$iN_m(\rho)$	$\sqrt{\frac{2}{\pi\rho}}e^{i\left(\rho - \frac{m\pi}{2} - \frac{\pi}{4}\right)}$
$e^{-i\rho} = \cos\rho - i\sin\rho$	Hankel function of the second kind $H_m^{(2)}(\rho) = J_m(\rho) - iN_m(\rho)$	$-iN_m(\rho)$	$\sqrt{\frac{2}{\pi\rho}}e^{-i\left(\rho - \frac{m\pi}{2} - \frac{\pi}{4}\right)}$
$\cosh\rho$, $\sinh\rho$	Modified Bessel function of the first kind $I_m(\rho) = i^{-m}J_m(i\rho)$	$\frac{1}{\Gamma(m+1)}\left(\frac{\rho}{2}\right)^m$	$\frac{1}{\sqrt{2\pi\rho}}e^{\rho}$
$e^{-\rho}$	Modified Bessel function of the second kind $K_m(\rho) = \frac{\pi}{2}i^{m+1}H_m^{(1)}(i\rho)$	$\begin{cases} -\ln\frac{\rho}{2} - 0.577 & m=0 \\ \frac{\Gamma(m)}{2}\left(\frac{2}{\rho}\right)^m & m>0 \end{cases}$	$\sqrt{\frac{\pi}{2\rho}}e^{-\rho}$

The recursion relations are slightly modified for the I_m and K_m functions to reflect the differing normalization. Thus, replacing ρ by $i\rho$ in Equation (13.4.21) and applying $J_m(i\rho) = i^m I_m(\rho)$ results in

$$I_{m-1}(\rho) - I_{m+1}(\rho) = \frac{2m}{\rho}I_m(\rho) \tag{13.4.29}$$

Further, all J_m and N_m are orthogonal from the Sturm–Liouville theory, while *for each* m, the $J_m(rx_{mn}/L)$ with x_{mn} the nth zero of $J_m(\rho)$ are orthogonal on $[0, L]$, similarly to $\sin(xp_n/L)$ with $p_n = n\pi$ the nth zero of $\sin x$, according to

$$\int_0^L J_m\left(x_{mn}\frac{r}{L}\right)J_m\left(x_{ml}\frac{r}{L}\right)r\,dr = \delta_{nl}\frac{L^2}{2}\left(J_{m+1}(x_{ml})\right)^2 \tag{13.4.30}$$

If the problem domain includes the origin as a regular (nonsingular) point, only J_m or I_m are present, while if infinity is a regular point, I_m is excluded. When $V^{\text{boundary}}(z, \phi)$ is specified on the cylinders $r = r_<, r_>$ and is zero on the remaining surfaces, the separation constant $k^2 < 0$ and the solution takes the form of a sum of products of functions of the form of $a_n \sin k_n z + b_n \cos k_n z$ with $c_m \exp(im\phi) + b_m \exp(im\phi)$ and $d_{mn}I_m(k_n r) + e_{mn}K_m(k_n r)$, where all $d_{mn} = 0$ and $e_{mn} = 0$ for $r_> = \infty$ and $r_< = 0$, respectively. The potential along the cylindrical boundaries can in this manner be expressed in a two-dimensional Fourier series. When $V^{\text{boundary}}(r, \phi)$ instead differs from zero on one of the two planes $z = a_{1,2}$, the solution inside the region $r_< < r < r_>$ requires $k^2 > 0$ and is composed of a sum over products of $c_m \exp(im\phi) + b_m \exp(im\phi)$ with $d_{mn}J_m(\alpha_{mn}r) + e_{mn}N_m(\alpha_{mn}r)$ and $a \sinh(\alpha_{nm}z) + b\cosh(\alpha_{mn}z)$, where conventionally e_{mn} is set to zero if $r_> = \infty$ or $r_< = 0$. The expansion in J_m and N_m with harmonic functions of ϕ provides an analogy to a two-dimensional Fourier series.

Example

For the potential inside a cylinder of length π over which $V_{\text{boundary}} = 0$ except for $V_{\text{boundary}}(R, \phi, z) = 3\sin(4\phi)\sin(z)$, $I(r)$ replaces $\sinh x$ for the rectangular problem so that

$$V(r,\phi,z) = \sum_{m=0}^{\infty}\sum_{n=0}^{\infty}(\alpha_{mn}\cos nz + \beta_{mn}\sin nz)(A_{mn}\cos m\phi + B_{mn}\sin m\phi)I_m(nr) \tag{13.4.31}$$

Setting $V(R, \phi, z) = V_{\text{boundary}}(R, \phi, z)$ and employing the orthogonality of the harmonic functions,

$$V(r,z,\phi) = \frac{3}{I_4(R)}I_4(r)\sin z \sin(4\phi) \tag{13.4.32}$$

13.5 SPHERICAL BESSEL FUNCTIONS

The Bessel functions of half-integer argument termed the spherical Bessel functions arise, e.g., in spherical solutions of the wave equation

$$\nabla^2 E(r,t) - \frac{1}{v^2}\frac{\partial^2 E(r,t)}{\partial t^2} = 0 \tag{13.5.1}$$

Separating variables yields a linear superposition of solutions $E_{lm,\omega}(\vec{r},t) = R_l(r)Y_{lm}(\phi,\theta)T_\omega(t)$ with $T''(t) = -\omega^2 T(t)$ and $k^2 = \omega^2/v^2$. Accordingly, $T_\omega(t) = \exp(i\omega t)$ and

$$\frac{1}{r}\frac{\partial^2}{\partial r^2}(rR_l(r)) + \left[k^2 - \frac{l(l+1)}{r^2}\right]R_l(r) = 0 \tag{13.5.2}$$

The above equation can be recast as the Bessel equation by substituting $R_l = Z_l/\sqrt{r}$, which ensures a $r^{-1/2}$ behavior of Z_l as $r \to \infty$, consistent with the Bessel function properties

$$\begin{aligned}\frac{1}{\sqrt{r}}\left(\sqrt{r}Z_l'' + 2\left(\frac{1}{2}\right)\frac{Z_l'}{\sqrt{r}} + \left(\frac{1}{2}\right)\left(-\frac{1}{2}\right)\frac{Z_l}{r^{3/2}}\right) + \left[k^2 - \frac{l(l+1)}{r^2}\right]Z_l &= \\ Z_l'' + \frac{1}{r}Z_l' + \left[k^2 - \left(l + \frac{1}{2}\right)^2\frac{1}{r^2}\right]Z_l &= 0\end{aligned} \tag{13.5.3}$$

In terms of the dimensionless variable $\rho = kr$, the normalized *spherical Bessel function* is therefore defined and normalized such that (recall that $J_l(\rho) \to \sqrt{2/\pi\rho}\cos(\rho - l\pi/2 - \pi/4)$ for large ρ)

$$j_l(\rho) = \sqrt{\frac{\pi}{2\rho}}Z_l(\rho) = \frac{1}{\sqrt{\rho}}J_{l+1/2}(\rho) \tag{13.5.4}$$

with similar formulas for the spherical Neumann and Hankel functions. For example, as $r \to \infty$,

$$h_l^{(1)}(\rho) = \sqrt{\frac{\pi}{2\rho}} H_{l+1/2}^{(1)}(\rho) \to (-i)^{l+1} \frac{e^{i\rho}}{\rho} \tag{13.5.5}$$

The recursion relations among the spherical Bessel functions are obtained from those of the standard Bessel functions, e.g.,

$$j_{l+1} + j_{l-1} = \sqrt{\frac{\pi}{2\rho}} \left(J_{l+3/2} + J_{l-1/2} \right) = \frac{2\left(l + \frac{1}{2}\right)}{\rho} \sqrt{\frac{\pi}{2\rho}} J_{l+\frac{1}{2}} = \frac{2l+1}{\rho} j_l \tag{13.5.6}$$

$$j_{l+1} - j_{l-1} = \sqrt{\frac{\pi}{2\rho}} \left(-2J'_{l+1/2} \right) = -\frac{2}{\sqrt{\rho}} \left(\sqrt{\rho} j_l \right)' = -2j'_l - \frac{j_l}{\rho} \tag{13.5.7}$$

Eliminating j_l from the two above equations yields again

$$(l+1) j_{l+1}(\rho) - l j_{l-1}(\rho) = -(2l+1) j'_l \tag{13.5.8}$$

while eliminating j_{l-1} from Equations (13.5.6) and (13.5.8) regenerates the standard Bessel function ladder relation of Equation (13.4.24)

$$j_{l+1} = \frac{l}{\rho} j_l - j_l' \tag{13.5.9}$$

Consequently, as also follows directly from Equation (13.4.25) with $m \to l - 1/2$, Equation (13.5.9) leads to the identical Rodrigues-type formula as for the Bessel functions

$$j_l = -\rho^{l-1} \frac{d}{d\rho} \left(\frac{j_{l-1}}{\rho^{l-1}} \right) = (-1)\rho^l \left(\frac{1}{\rho} \frac{\partial}{\partial \rho} \right) \left(\frac{1}{\rho^{l-1}} j_{l-1} \right) = \ldots = (-\rho)^l \left(\frac{1}{\rho} \frac{d}{d\rho} \right)^l j_0(\rho) \tag{13.5.10}$$

Since for $l = 0$ Equation (13.5.1) can be rewritten as

$$\frac{1}{\rho} \frac{\partial^2}{\partial \rho^2} (\rho R) = -R \tag{13.5.11}$$

for which the general solution is given by $\rho R = A \cos \rho + B \sin \rho$, the normalized solution that remains finite at the origin (by l'Hopital's rule) is

$$j_0(\rho) = \frac{\sin \rho}{\rho} \tag{13.5.12}$$

from which all successive $j_l(\rho)$ follow from Equation (13.5.10). The second set of spherical functions, $n_l(\rho)$, often termed spherical Neumann functions are instead generated by employing $-\cos\rho$ in place of $\sin\rho$ above. Appropriate linear combinations of these then yield the spherical Hankel functions, $h_l^{(1)} = j_l + in_l$ and $h_l^{(2)} = j_l - in_l$. The lowest-order spherical functions are consequently

$$
\begin{aligned}
j_0 &= \frac{\sin\rho}{\rho} & j_1 &= \frac{\sin\rho}{\rho^2} - \frac{\cos\rho}{\rho} \\
n_0 &= -\frac{\cos\rho}{\rho} & n_1 &= -\frac{\cos\rho}{\rho^2} - \frac{\sin\rho}{\rho} \\
\begin{matrix} h_0^{(1)} \\ h_0^{(2)} \end{matrix} &= \frac{e^{\pm i\rho}}{\pm i\rho} & \begin{matrix} h_1^{(1)} \\ h_1^{(2)} \end{matrix} &= -\frac{e^{\pm i\rho}}{\rho}\left(1 \pm \frac{i}{\rho}\right)
\end{aligned}
\tag{13.5.13}
$$

For large ρ, noting that the $1/\rho$ term in j_n is obtained by repeatedly differentiating $\sin\rho$ in j_0,

$$
j_n(\rho) \to \frac{1}{\rho}\sin\left(\rho - \frac{n\pi}{2}\right) \tag{13.5.14}
$$

As $\rho \to 0$, the leading ρ^l term results from applying Equation (13.5.10) to the ρ^{2l} term in the power series expansion of $\sin\rho/\rho$, as can be easily understood by applying, e.g., $[(1/\rho)d/d\rho]^2$ to $(x - x^3/3! + x^5/5! + \ldots)/x$. Hence,

$$
j_l \to (-\rho)^l \left(\frac{1}{\rho}\frac{d}{d\rho}\right)^l (-1)^l \frac{\rho^{2l}}{(2l+1)!} = \rho^l \left(\frac{1}{\rho}\frac{d}{d\rho}\right)^{l-1} \frac{\rho^{2l-2}(2l)}{(2l+1)!} = \ldots = \frac{\rho^l}{(2l+1)!!} \tag{13.5.15}
$$

A plane wave can be expanded as, with $x = \cos\theta$,

$$
e^{ikz} = e^{i\rho\cos\theta} = e^{i\rho x} = \sum_{l=0}^{\infty} c_l j_l(\rho) P_l(x) \tag{13.5.16}
$$

After employing the orthogonality of the P_l, the c_l are determined from the $\rho \to 0$ behavior of

$$
\begin{aligned}
\lim_{\rho\to 0} c_l j_l(\rho) = c_l \frac{\rho^l}{(2l+1)!!} &= \left(\frac{2l+1}{2}\right) \lim_{\rho\to 0} \int_{-1}^{1} e^{i\rho x} P_l(x)\,dx \\
&= \left(\frac{2l+1}{2}\right) \lim_{\rho\to 0} \int_{-1}^{1} \sum_{n=0}^{\infty} i\rho^{n/n!} \left\{ \sum_{j=0}^{n} a_{nj} P_{nj}(x) \right\} P_l(x)\,dx
\end{aligned}
\tag{13.5.17}
$$

where the a_{nj} are obtained by expanding x^n in terms of the Legendre polynomials of order $\le n$. From the orthogonality of the $P_n(x)$, the leading nonzero term in the integral on the right-hand side of Equation (13.5.17) occurs when $n = l$ for which the above equation simplifies to

$$c_l \frac{\rho^l}{(2l+1)!!} = i^l \frac{\rho^l}{l!} \left(\frac{2l+1}{2}\right) \int_{-1}^{1} P_l(x) x^l dx \tag{13.5.18}$$

However, from Equation (13.2.13), the coefficient of x^l in $P_l(x)$, which is obtained by taking d^l/dx^l of the single x^{2l} term arising from $(x^2 - 1)^l$, equals $(1/2^l l\,!)((2l)\,!/l\,!)$. Hence,

$$c_l \frac{(2l)!!\rho^l}{(2l+1)!(l!)} = i^l \frac{\rho^l}{l!} \left(\frac{2l+1}{2}\right) \int_{-1}^{1} P_l(x) \left[\frac{2^l (l!)^2}{(2l)!} P_l(x) + \sum_{n=0}^{l-1} a_n P_l(x)\right] dx \tag{13.5.19}$$

This yields finally $c_l = i^l(2l+1)$ after applying $(2l)\,!! = 2^l l\,!$. Accordingly,

$$e^{ikz} = \sum_{l=0}^{\infty} i^l (2l+1) j_l(\rho) P_l(\cos\theta) \tag{13.5.20}$$

Finally, if $\rho_{l+1/2,n}$ is the nth zero of $J_{l+1/2}(\rho) = 0$, then from Equation (13.4.30),

$$\begin{aligned}\int_0^L j_l\left(\rho_{l+1/2,n}\frac{r}{L}\right) j_l\left(\rho_{l+1/2,m}\frac{r}{L}\right) r^2 dr &= \delta_{nm}\frac{L^3}{2}\left(j_l'\left(\rho_{l+1/2,m}\right)\right)^2 \\ &= \delta_{nm}\frac{L^3}{2}\left(j_{l+1}\left(\rho_{l+1/2,m}\right)\right)^2\end{aligned} \tag{13.5.21}$$

14

INTEGRAL EQUATIONS AND THE CALCULUS OF VARIATIONS

While a differential equation associates a function with its derivatives, an integral equation relates a function to its integrals (integrodifferential equations contain both integrals and derivatives). As integral equations incorporate boundary conditions, they can in certain cases be solved by determining their global extrema through the calculus of variations, which converts the problem into the solution of a local differential equation.

14.1 VOLTERRA AND FREDHOLM EQUATIONS

Fredholm equations contain an integral with constant limits, while the integration limits are variables in *Volterra equations*. The equation is termed an *integral equation of the first (second) kind* if the function to be determined appears only inside and both inside and outside the integral, respectively.

Example

If $f(x)$ is specified, the Fourier transform

$$f(x) = \frac{1}{\sqrt{2\pi}} \int_{-\infty}^{\infty} f(k) e^{-ikx} dk \tag{14.1.1}$$

Fundamental Math and Physics for Scientists and Engineers, First Edition.
David Yevick and Hannah Yevick.

represents a Fredholm equation of the first kind for $f(k)$ where $\exp(ikx)/\sqrt{2\pi}$ is termed the *kernel* of the equation and is generally designated by $K(x, t)$, while

$$\phi(x) = f(x) + \int_a^x K(x,t)\phi(t)dt \tag{14.1.2}$$

where $f(x)$ is known and $\phi(x)$ is the function to be determined is a Volterra equation of the second kind.

Boundary conditions, e.g., $\phi(a) = f(a)$ in Equation (14.1.2), are implicit in the above formulas. In general, differential equations subject to initial conditions transform into Volterra integral equations, while if boundary conditions are instead specified, Fredholm equations result.

Example

A differential equation such as $(D^2 + 1)y = 0$ with boundary conditions $y(0) = y_0, y'(0) = y'_0$ is transformed into a Volterra equation of the second kind for $y(x)$ by incorporating the two boundary conditions through two lower integration limits according to

$$\begin{aligned} y' - y'_0 &= -\int_0^x y dx' \\ y &= y'_0 x + y_0 - \int_0^x \int_0^{x'} y(x'') dx'' dx' \end{aligned} \tag{14.1.3}$$

Interchanging the order of integration then recasts the equation into a Volterra form

$$\begin{aligned} y &= y'_0 x + y_0 - \int_0^x y(x'') \left[\int_{x'}^x dx' \right] dx'' \\ &= y'_0 x + y_0 - \int_0^x y(x'')(x - x'') dx'' \end{aligned} \tag{14.1.4}$$

That this expression satisfies the differential equation and boundary conditions follows from $\frac{d}{dx}\int_{x_0}^x f(x,x')dx' = f(x,x) + \int_{x_0}^x \frac{df(x,x')}{dx} dx'$

Integral equations are often solved through iteration. For example, if the magnitude of the integral is small in Equation (14.1.2) compared to the remaining terms, $f(t)$ can first be substituted for $\phi(t)$. Repeating this procedure for the modified $\phi(t)$ leads to the series

$$\begin{aligned}
\phi^{(0)}(x) &= f(x) \\
\phi^{(1)}(x) &= f(x) + \int_a^x K(x,t)\phi^{(0)}(t)dt = f(x) + \int_a^x K(x,t)f(t)dt \\
\phi^{(2)}(x) &= f(x) + \int_a^x K(x,t)\phi^{(1)}(t)dt \\
&= f(x) + \int_a^x K(x,t)f(t)dt + \int_a^x K(x,t)\int_a^t K(x,t')f(t')dt'dt
\end{aligned} \tag{14.1.5}$$

Integral equations can be solved in special cases such as that of separable kernels such as $K(x,t) = \sum_i X_i(x)T_i(t)$ and can also generally be converted to matrix equations by discretizing the integration variable and solved numerically.

14.2 CALCULUS OF VARIATIONS THE EULER-LAGRANGE EQUATION

Variational calculus enables the determination of the extrema of integral expressions typified by

$$I = \int_{\vec{x}(t_1),P}^{\vec{x}(t_2)} f(t, x_i(t), \dot{x}_i(t))dt \tag{14.2.1}$$

over a path P extending between two points $\vec{x}(t_1)$ and $\vec{x}(t_2)$ at different times for a system described by the variables $x_i(t)$, $i = 1, 2, \ldots, n_d$ and their derivatives $\dot{x}_i(t) \equiv dx_i/dt$. While I is a global quantity that receives contributions from an infinite set of points, any *local* change in the paths defining its *extrema* leaves I unchanged to first order in the same manner that the derivative and hence first-order variation of a function $f(x)$ of a single variable with respect to x equals zero at its extrema. Therefore, if I is discretized at M times, its extrema are invariant with respect to first-order changes in all $2n_dM$ coordinates. The global problem can therefore be transformed into that of identifying the solutions of a local differential equation that pass through. That is, since any first-order displacement, $\delta x_m(t)$, along the extremal paths leaves I invariant,

$$\delta I = \sum_m \int_{\vec{x}(t_1)}^{\vec{x}(t_2)} \left(\frac{\partial L(\dot{x}_k, x_k, t)}{\partial \dot{x}_m} \delta \dot{x}_m(t) + \frac{\partial L(\dot{x}_k, x_k, t)}{\partial x_m} \delta x_m(t) + \frac{\partial L(\dot{x}_k, x_k, t)}{\partial t} \right) dt = 0 \tag{14.2.2}$$

However, $\delta\dot{x}_m = d(\delta x_m)/dt$ is determined by $\delta x_m(t)$ and can accordingly be eliminated by partial integration of the first term in Equation (14.2.2):

$$\sum_m \int_{\vec{x}(t_1)}^{\vec{x}(t_2)} \left(-\frac{d}{dt}\frac{\partial L(\dot{x}_k, x_k, t)}{\partial \dot{x}_m} + \frac{\partial L(\dot{x}_k, x_k, t)}{\partial x_m} \right) \delta x_m(t)dt \\ + \left(\sum_m \frac{\partial L(\dot{x}_k, x_k, t)}{\partial \dot{x}_m} \delta x_m(t) + L(\dot{x}_k, x_k, t) \right) \Bigg|_{\vec{x}(t_1)}^{\vec{x}(t_2)} = 0 \tag{14.2.3}$$

where the variation $\delta\vec{x}(t) = 0$ at the endpoints. Since each $\delta x_m(t)$ is an *arbitrary* function at all other times,

$$-\frac{d}{dt}\frac{\partial L(\dot{x}_k, x_k, t)}{\partial \dot{x}_m} + \frac{\partial L(\dot{x}_k, x_k, t)}{\partial x_m} + L(\dot{x}_k, x_k, t_2) - L(\dot{x}_k, x_k, t_1) = 0 \tag{14.2.4}$$

Thus, the requirement that the path is an extremum of the action transforms the *global* condition on I into a local condition on L.

Example

The integrated distance between two points equals

$$I = \int_{\vec{x}_1}^{\vec{x}_2} dl = \int_{\vec{x}_1}^{\vec{x}_2} \sqrt{dx_1^2 + dx_2^2} = \int_{\vec{x}_1}^{\vec{x}_2} \sqrt{\dot{x}_1^2 + \dot{x}_2^2}\, dt \tag{14.2.5}$$

The shortest distance between the points is therefore given by the straight line solution of

$$\frac{d}{dt}\frac{\partial\sqrt{\dot{x}_1^2 + \dot{x}_2^2}}{\partial \dot{x}_i} = \frac{1}{\sqrt{\dot{x}_1^2 + \dot{x}_2^2}}\frac{\partial^2 \dot{x}_i}{\partial t^2} = 0, \quad i = 1, 2 \tag{14.2.6}$$

15

PARTICLE MECHANICS

Classical mechanics describes the motion of macroscopic bodies moving at small velocities compared to the speed of light. While the classical law of motion is simply formulated, if several particles interact, solving the resulting coupled equation system requires numerical, approximate, or statistical techniques in the absence of a high degree of symmetry.

15.1 NEWTON'S LAWS

The classical mechanics of particles are governed by *Newton's laws of motion*:

- An object observed from an *inertial frame* (a frame that moves at a uniform velocity with respect to a fixed point in space) in the absence of external forces proceeds with constant velocity in a straight line or remains at rest.
- Every particle has a fixed *mass*, which is a measure of a particle's *inertia* or resistance to change in velocity. The mass is given according to Newton's equation by the ratio of the total force on the object, which is the directed, vector sum of the component forces, to the object's acceleration.
- If a rigid body A exerts a force on a nonaccelerating rigid body B in a given direction, body B exerts a force of equal magnitude on body A in the opposing direction. Otherwise, the force pair would accelerate the center of mass of the

Fundamental Math and Physics for Scientists and Engineers, First Edition.
David Yevick and Hannah Yevick.

two bodies or induce rotational motion violating momentum conservation. For *central forces*, these forces lie along the line connecting the two bodies. Thus, while a person standing on a floor initially accelerates by bending the floor downward, this motion ceases when the upward force balances his weight.

Newton's equation expressed in meter-kilogram-second (mks) units relates the force in *newtons* to the *acceleration* in m/s^2. In centimeter-gram-second (cgs) units, force is instead measured in *dynes* with 1 N = 10^5 dyn (1 kg-m/s^2)(10^3 g/kg)(10^2 cm/m). That is,

$$\vec{F}_{\text{total}} = \sum_i \vec{F}_i = m\vec{a} \quad \left[\frac{MD}{T^2}\right] \tag{15.1.1}$$

where *acceleration* is defined as the second time derivative of the displacement (position) vector

$$\vec{a} = \frac{d^2\vec{r}}{dt^2} \quad \left[\frac{D}{T^2}\right] \tag{15.1.2}$$

A more general form of Equation (15.1.1), valid for time-dependent particle masses, can be obtained by introducing the *velocity vector* as the first derivative of the displacement vector with respect to time

$$\vec{v} = \frac{d\vec{r}}{dt} \quad \left[\frac{D}{T}\right] \tag{15.1.3}$$

The unit of velocity is consequently meters per second, clarifying the terminology for the units of acceleration, which, since $\vec{a} = d\vec{v}/dt$, are termed meters per second per second. Multiplying the velocity vector of a particle by its mass yields the *momentum* vector,

$$\vec{p} = m\vec{v} \tag{15.1.4}$$

in terms of which Equation (15.1.1) becomes

$$\vec{F} = \frac{d\vec{p}}{dt} \tag{15.1.5}$$

Equation (15.1.5) is then applicable to systems with time-varying masses.

15.2 FORCES

The most frequently occurring forces are simple functions of position. The *gravitational force* on a particle of mass m at a point $\vec{r}$ induced by a particle of M at $\vec{r}'$ is

$$\vec{F} = -\frac{GMm}{|\vec{r}-\vec{r}'|^2}\hat{e}_{\vec{r}-\vec{r}'} \quad \left[\frac{MD}{T^2}\right] \tag{15.2.1}$$

As $\hat{e}_{\vec{r}-\vec{r}'}$ denotes a unit vector in the direction from $\vec{r}'$ to $\vec{r}$ (from M to m), the force is central. Here,

$$G=6.67\times 10^{-11}\frac{N-m^2}{\text{kg}^2} \quad \left[\frac{D}{MT^2}\right] \tag{15.2.2}$$

is termed the *gravitational constant*. Close to the surface of a planet, the force can be approximated simply by

$$\vec{F} = -mg\hat{e}_z \tag{15.2.3}$$

in which the unit vector $\hat{e}_z$ points vertically upward from the planet surface and the *local gravitational acceleration* is defined as

$$g=\frac{GM_{\text{planet}}}{R^2_{\text{planet}}} \quad \left[\frac{MD}{T^2}\right] \tag{15.2.4}$$

For the earth, $M_{\text{earth}}=6\times 10^{24}$ kg, at a sea level radius, $R_{\text{earth}}=6731$ km, the gravitational acceleration $g=9.8$ m/s^2. The magnitude of the gravitational force on an object is termed its *weight*. From Equation (15.1.1), *the same force is required to accelerate an object independent of its weight*. Thus, although on the moon a person weighs less and can run faster, the destructive forces from a collision with a wall at a given velocity are identical to those on earth. *Mass and weight, which quantify inertia and force, respectively, are often confused since scales, which measure forces, are incorrectly calibrated in mass units such as kilograms*. Thus, 1 kg on a scale in fact indicates a force of 9.8 N, which corresponds to a 1.0 kg mass only at sea level, since g decreases with height.

An object on a inclined plane slanted at an angle θ from the *horizontal* is only accelerated by the gravitational force component $mg\cos(90-\theta)=mg\sin\theta$ parallel to the floor as the perpendicular (normal) component, $mg\cos\theta$, is balanced by the equal and opposite force, $\vec{F}_{\text{normal}}$, of the floor on the body. Similarly, an object on a horizontal floor pulled at an angle θ from the *normal* is only accelerated by the force component $|F|\sin\theta$ along the floor's surface. The frictional force exerted by a surface on an object is proportional to $\vec{F}_{\text{normal}}$. The *coefficient of static friction*, μ_{static}, defined by

$$\left|\vec{F}_{\text{initiate motion}}\right| = \mu_{\text{static}}\left|\vec{F}_{\text{normal}}\right| \tag{15.2.5}$$

establishes the magnitude of the force, $\vec{F}_{\text{initiate motion}}$, required to set a body in motion. Subsequently, in the standard model of friction, the body experiences *dynamic friction* in a direction opposite to its velocity with

$$\vec{F}_{\text{dynamic}} = -\mu_{\text{dynamic}}\left|\vec{F}_{\text{normal}}\right|\hat{e}_{\vec{v}} \tag{15.2.6}$$

Generally, $\mu_{\text{static}} > \mu_{\text{dynamic}}$, since the body and floor surfaces tend to conform or bond chemically at rest. In many physical systems however, especially gases and fluids, the frictional force is instead either proportional to or a nonlinear function of velocity.

Additional force types include *tension*, $\vec{T}$, which is the outward force applied to each end of a linear solid such as a string. The tension is constant along an ideal rod or string such that if a point along the string is held fixed and the string severed, the force required to hold the new ends stationary remain equal to $|\vec{T}|$. Finally, the *linear restoring force* of an ideal spring varies as

$$\vec{F} = -k\left(x - x_{\text{equilibrium}}\right)\hat{e}_x \tag{15.2.7}$$

in which $x_{\text{equilibrium}}$ represents the equilibrium position and k is termed the spring constant.

15.3 NUMERICAL METHODS

Newton's law provides a second-order differential equation that describes the evolution of the displacement $\vec{r}$ of a massive body with time. Since an nth order differential equation can be recast as a coupled system of n first-order equations, introducing the velocity vector yields the system of two vector equations

$$\begin{aligned} \frac{d\vec{r}}{dt} &= \vec{v} \\ \frac{d\vec{v}}{dt} &= \frac{\vec{F}}{m} \end{aligned} \tag{15.3.1}$$

or in matrix notation

$$\frac{d}{dt}\begin{pmatrix} \vec{r} \\ \vec{v} \end{pmatrix} = \begin{pmatrix} 0 & 1 \\ 0 & 0 \end{pmatrix}\begin{pmatrix} \vec{r} \\ \vec{v} \end{pmatrix} + \begin{pmatrix} 0 \\ \vec{F}/m \end{pmatrix} \tag{15.3.2}$$

For finite Δt, the forward finite difference approximation for the derivative operator (Eq. 6.1.1) results in

$$\begin{aligned} \vec{r}(t+\Delta t) &= \vec{r}(t) + \Delta t\,\vec{v}(t) \\ \vec{v}(t+\Delta t) &= \vec{v}(t) + \frac{\Delta t}{m}\vec{F}\left(\vec{r}(t+\Delta t), v(t)\right) \end{aligned} \tag{15.3.3}$$

The initial conditions are provided by specifying $\vec{r}$ and $\vec{v}$ at $t=0$. Employing the updated value of the position in the second equation reduces or removes spurious numerical divergences (see pp. 198–199 of D. Yevick, *A First Course in Computational Physics and Object-Oriented Programming with C++*, Cambridge University Press (2005), for a discussion). The code for one-dimensional motion subject to a linear restoring force with $ma = -kx$ is:

```
dT = input ( 'input dT ' )
springConstant = input ( 'input springConstant' )
mass = input ( 'input mass' )
```

```
Particle.position(1) = 0.;
Particle.velocity(1) = 1.;
for loop = 2 : 2000
  Particle.position(loop) = Particle.positionn(loop - 1) +...
  Particle.velocity(loop - 1) * dT;
    Particle.velocity(loop) = Particle.velocity(loop - 1) +...
       - dT * springConstant * Particle.position(loop) / mass;
end
plot( dT* [ 1 : 2000 ], Particle.position )
```

15.4 WORK AND ENERGY

The *work* performed by a force on an object equals the change in the *total energy of the particle and its environment*, including, e.g., heat and electromagnetic radiation radiated by the particle. *Only the component of the force along the direction of particle motion (e.g.,* $\vec{v}$*) contributes to work and changes the particle energy*, since if $\vec{F}\cdot\vec{v} = m(d\vec{v}/dt)\cdot\vec{v} = m/2(dv^2/dt) = 0$, the particle's velocity and hence energy remain constant. Accordingly, the *work performed by the force on a body*, which is the *negative of the work the body performs against the force*, as it moves along a path C between $\vec{r}_i$ and $\vec{r}_f$ is

$$W = \int_C \vec{F}\cdot d\vec{r} \quad \left[\frac{MD^2}{T^2}\right] \tag{15.4.1}$$

which equals

$$\int_{\vec{r}_i}^{\vec{r}_f} \vec{F}\cdot d\vec{r} = \int_{t_i}^{t_f} m\frac{d\vec{v}}{dt}\cdot\frac{d\vec{r}}{dt}dt = \int_{\vec{v}_i}^{\vec{v}_f} m\frac{d\vec{v}}{dt}\cdot\vec{v}\,dt = \frac{m}{2}\int_{\vec{v}_i}^{\vec{v}_f}\frac{d\vec{v}^{\,2}}{dt}dt = \frac{m}{2}\left(v_f^2 - v_i^2\right) \equiv \Delta K \tag{15.4.2}$$

which is termed the change in *kinetic energy*. Here, $\vec{F}\cdot d\vec{r} = F_x dx + F_y dy + F_z dz$ where the differential elements dx, dy, dz are evaluated along the path.

Work and energy are measured in *Joules* in mks units, and *ergs* in cgs units with 1 J = 10^7 erg. The *power* supplied by $\vec{F}$ equals the rate at which work is performed,

$$P = \frac{dW}{dt} \quad \left[\frac{MD^2}{T^3}\right] \tag{15.4.3}$$

The mks and cgs units of power are, respectively, the *Watt* and *dyne* with 1 W = 10^7 dyn. The energy stored in, e.g., a battery is often specified in units such as watt-hours (Wh).

For *conservative* forces, the work, W, in Equation (15.4.1) does not depend on the path C. Hence, for every closed path C passing through each point,

$$\oint_{C \subset S} \vec{F} \cdot d\vec{r} = \int_{S} \left(\vec{\nabla} \times \vec{F}\right) \cdot d\vec{S} = 0 \tag{15.4.4}$$

As C can be made infinitesimally small at any point in space, Equation (15.4.4) implies that $\vec{\nabla} \times \vec{F} = 0$ locally. Since the integral of the force is a unique function of position, labeling the integral of $\vec{F} \cdot d\vec{r}'$ over a path from some reference point $\vec{r}_0$ to a position $\vec{r}$ by $U(\vec{r}_0) - U(\vec{r})$ leads to, where the paths in both integrals in the first line below coincide up to the point $\vec{r}$,

$$\begin{aligned} -U\left(\vec{r} + \hat{e}_x \Delta x\right) + U\left(\vec{r}\right) &= \int_{\vec{r}_0}^{\vec{r} + \hat{e}_x \Delta x} \vec{F}(r') \cdot dr' - \int_{\vec{r}_0}^{\vec{r}} \vec{F}(r') \cdot dr' \\ &= \int_{\vec{r}}^{\vec{r} + \hat{e}_x \Delta x} \vec{F}(r') \cdot dr' \approx F_x\left(\vec{r}\right) \Delta x \end{aligned} \tag{15.4.5}$$

Thus, a single-valued *potential energy function* $U(\vec{r})$ can be defined as

$$\vec{F} = -\vec{\nabla} U(\vec{r}) \tag{15.4.6}$$

From the definition of the directional derivative, Equation (8.2.2), $\vec{F} \cdot \delta \vec{r}$, where $\delta \vec{r} = (\delta x, \delta y, \delta z)$ is an infinitesimal displacement between $\vec{r}$ and $\vec{r} + \delta \vec{r}$ equals the *potential energy* difference $-\left(U\left(\vec{r} + \delta \vec{r}\right) - U\left(\vec{r}\right)\right) = -\delta U\left(\vec{r}\right)$. The minus sign indicates that the force associated with a potential that increases with x is directed along $-\hat{e}_x$, consistent with a ball rolling backward on an upward-sloping surface.

Examples

The potential function for an object located at small distances from the earth's surface follows from $\vec{F} \cdot d\vec{r} = (-mg\hat{e}_z) \cdot d\vec{r} = -mgdz$ and, thus taking $U = 0$ at $z = 0$, typically the surface,

$$U(z) = -\int_0^z (-mg)dz = mgz \tag{15.4.7}$$

For a spring, $\vec{F} \cdot d\vec{r} = (-kx\hat{e}_x) \cdot d\vec{r} = -kxdx$ where the distance x is measured from the equilibrium (rest) position of the spring, leading to

$$U = \frac{1}{2}kx^2 \tag{15.4.8}$$

If time-independent conservative forces act on a system, the total energy, E, of the system defined as the sum of the potential and kinetic energy remains constant. This is evident from Equation (15.4.1), which can be rewritten for a conservative force as, where from the definition of the directional derivative, dU represents the change in potential energy evaluated along an infinitesimal section of C,

$$\Delta K = W = \int_C \vec{F} \cdot d\vec{r} = -\int_C \vec{\nabla} U \cdot d\vec{r} = -\int_C dU = -\Delta U \tag{15.4.9}$$

which yields, with E a constant total energy,

$$K + U = E \tag{15.4.10}$$

Note that as *only differences in energy are measurable, the zero of E and therefore U can be chosen arbitrarily.*

Example

For a particle, if the coordinate l is oriented along the direction of $\vec{F}$ so that $\vec{F} \cdot d\vec{l} = Fdl = -d\vec{l} \cdot \vec{\nabla} U = -dU$, with $v_l = dl/dt$,

$$F = -\frac{dU}{dl} = m\frac{d^2l}{dt^2} = m\left(\frac{dl}{dt}\frac{d}{dl}\right)\frac{dl}{dt} = mv_l\frac{dv_l}{dl} = \frac{m}{2}\frac{dv_l^2}{dl} \tag{15.4.11}$$

and therefore, since the velocity components perpendicular to $\hat{e}_{\vec{l}}$ are constant,

$$\frac{d}{dl}\left(\frac{1}{2}mv^2 + U\right) = \frac{dE}{dl} = 0 \tag{15.4.12}$$

15.5 LAGRANGE EQUATIONS

While Newton's laws of motion provide a *local* relationship that can be iterated numerically or solved analytically to find the trajectory of a particle, variational calculus recasts the problem into that of determining the extrema of an integral over a *global* trajectory. In particular, if $\vec{r}(t)$ denotes the classical trajectory of a particle between fixed initial and final points $\vec{r}(t_1)$ and $\vec{r}(t_2)$ as a function of time, then for any small deviation $\delta\vec{r}(t)$ of the path from this trajectory that preserves $\vec{r}(t_1)$ and $\vec{r}(t_2)$,

$$\int_{t_1}^{t_2} \left(\vec{F} - \frac{d}{dt}(m\vec{v})\right) \cdot \delta\vec{r}(t)dt = 0 \tag{15.5.1}$$

For a conservative force, $\vec{F} \cdot \delta\vec{r} = -\vec{\nabla} U \cdot \delta\vec{r} = -\delta U$, and hence,

$$\int_{t_1}^{t_2} \left(-\delta U - \frac{d}{dt}(m\vec{v} \cdot \delta\vec{r}(t)) + m\vec{v} \cdot \frac{d}{dt}\delta\vec{r}(t)\right)dt = 0 \tag{15.5.2}$$

Integrating the second, total differential, term yields $m(\vec{v}(t_2) \cdot \delta\vec{r}(t_2) - \vec{v}(t_1) \cdot \delta\vec{r}(t_1))$, which vanishes as $\delta\vec{r}(t) = 0$ at both endpoints. With $d(\delta\vec{r})/dt = \delta\vec{v}$ and $\vec{v} \cdot \delta\vec{v} = \delta v^2/2$ and denoting $\dot{\vec{r}} = d\vec{r}/dt$,

$$\delta \int_{t_1}^{t_2} (T - U(\vec{r}))dt \equiv \delta \int_{t_1}^{t_2} L(\vec{r}, \dot{\vec{r}})dt \equiv \delta S = 0 \tag{15.5.3}$$

where S and $L(\vec{r}\dot{\vec{r}})$ are termed the *action* and *Lagrangian*, respectively. Equation (15.5.3) can be applied to any coordinate system provided that every accessible system configuration corresponds to a unique set of coordinate values. If the number, m, of these *generalized coordinates* is greater than the number of degrees of freedom, n, an additional $m-n$ constraint relations are required to determine the system uniquely. When these relations can be integrated, the system is termed *holonomic*; otherwise, it is *nonholonomic*. Generalized coordinates are often distinguished by the notation q_j and $\dot{q}_j$.

Reversing the above derivation yields a differential equation for the classical paths in terms of L. In particular, with $\delta x_j(t)$ again as an arbitrary first-order path variation, in N dimensions for the slightly more involved case of a time-dependent lagrangian,

$$\delta S = \sum_{j=1}^{N} \int_{t_1}^{t_2} \left(\frac{\partial L(\dot{\vec{x}}, \vec{x}, t)}{\partial \dot{x}_j} \delta \dot{x}_j(t) + \frac{\partial L(\dot{\vec{x}}, \vec{x}, t)}{\partial x_j} \delta x_j(t) + \frac{\partial L(\dot{\vec{x}}, \vec{x}, t)}{\partial t} \right) dt = 0 \tag{15.5.4}$$

With $\delta \dot{x}_j = d(\delta x_j)/dt$, integration by parts to eliminate $\delta \dot{x}_j$ in favor of δx_j yields

$$\sum_{j=1}^{N} \int_{t_1}^{t_2} \left(\left(-\frac{d}{dt} \frac{\partial L(\dot{\vec{x}}, \vec{x}, t)}{\partial \dot{x}_j} + \frac{\partial L(\dot{\vec{x}}, \vec{x}, t)}{\partial x_j} \right) \delta x_j(t) + \frac{\partial L(\dot{\vec{x}}, \vec{x}, t)}{\partial t} \right) dt$$

$$+ \sum_{j=1}^{N} \frac{\partial L(\dot{\vec{x}}, \vec{x}, t)}{\partial \dot{x}_j} \delta x_j(t) \Big|_{t_1}^{t_2} = 0 \tag{15.5.5}$$

Since $\Delta x_j(t)$, except at both endpoints where $\Delta x_j = 0$, constitutes an *arbitrary* infinitesimal function in every coordinate direction, $L(\dot{\vec{x}}, \vec{x}, t)$ satisfies the m *Lagrange equations*

$$\frac{d}{dt} \frac{\partial L(\dot{\vec{x}}, \vec{x}, t)}{\partial \dot{x}_j} = \frac{\partial L(\dot{\vec{x}}, \vec{x}, t)}{\partial x_j} + L(\dot{\vec{x}}, \vec{x}, t_2) - L(\dot{\vec{x}}, \vec{x}, t_1) \tag{15.5.6}$$

Constraints on the system motion can be incorporated into L through Lagrange multipliers as illustrated in the problems below. For nonconservative forces, such as that on a charged particle in a magnetic field, L can still be defined as $T-W$ if a *work function* can be related to the force by

$$F_j = \left(\frac{\partial}{\partial x_j} - \frac{d}{dt} \frac{\partial}{\partial \dot{x}_j} \right) W(\vec{r}, \dot{\vec{r}}, t) \tag{15.5.7}$$

The difficulty of the Lagrangian approach generally resides in formulating the kinetic energy. This can often be facilitated by introducing an orthogonal coordinate system.

Examples

1. The time-independent Lagrangian $L\left(\dot{\vec{x}}, \vec{x}\right) = T\left(\dot{\vec{x}}\right) - U\left(\vec{x}\right) = m\dot{\vec{x}}^2/2 - U\left(\vec{x}\right)$ generates Newton's equation upon insertion into Equation (15.5.6).
2. For the system in Figure 15.1 in which the right pulley can move in a vertical direction while the position of the left pulley is fixed, if T represents the string tension and x_1 and x_2 the vertical positions of the left and right mass, respectively, applying Newton's law of motion directly by setting the sum of the forces on each weight equal to its acceleration yields

$$\begin{aligned} T - m_1 g &= m_1 \ddot{x}_1 \\ 2T - m_2 g &= m_2 \ddot{x}_2 \end{aligned} \tag{15.5.8}$$

with $\ddot{x}_1 = -2\ddot{x}_2$. In the Lagrangian formalism, if the left mass is displaced a distance x_1 upward from its initial position, the cord is lengthened by $x_1/2$ on both sides of the pulley, and the right mass therefore moves a distance $x_2 = -x_1/2$. Employing x_1 as the generalized variable,

$$L = T - U = \frac{1}{2} m_1 \ddot{x}_1^2 + \frac{1}{8} m_2 \ddot{x}_1^2 - \left(m_1 g x_1 - \frac{1}{2} m_2 g x_1 \right) \tag{15.5.9}$$

yielding the Lagrange equation of motion

$$\left(m_1 + \frac{m_2}{4}\right)\ddot{x}_1 = -\left(m_1 - \frac{m_2}{2}\right)g \tag{15.5.10}$$

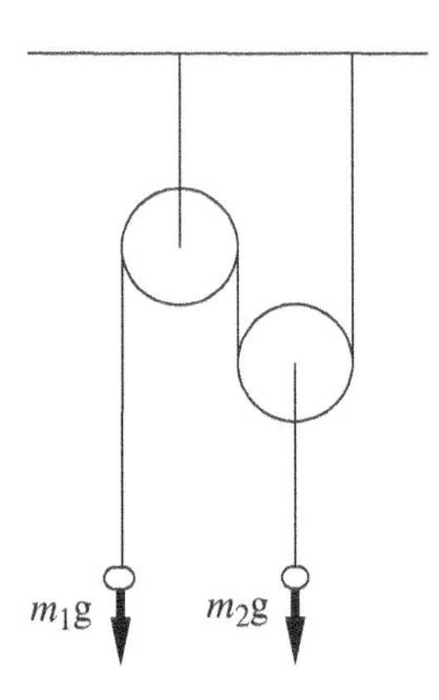

FIGURE 15.1 Two pulley system.

Alternatively, this result can be obtained by employing different variables x_1 and x_2 for the two mass positions and introducing the constraint $x_1 + 2x_2 = 0$ through a Lagrange multiplier according to

$$L = T - U = \frac{1}{2}m_1\ddot{x}_1^2 + \frac{1}{2}m_2\ddot{x}_2^2 - (m_1 g x_1 + m_2 g x_2) + \lambda(x_1 + 2x_2) \qquad (15.5.11)$$

The Lagrange equations then adopt the form

$$\begin{aligned} m_1\ddot{x}_1 + m_1 g &= \lambda \\ m_2\ddot{x}_2 + m_2 g &= 2\lambda \end{aligned} \qquad (15.5.12)$$

The Lagrange multiplier thus corresponds to the force supplied by the spring tension. Eliminating λ,

$$m_1\ddot{x}_1 + m_1 g = \frac{1}{2}(m_2\ddot{x}_2 + m_2 g) \qquad (15.5.13)$$

which, together with the constraint equation, $x_1 + 2x_2 = 0$ and therefore $\ddot{x}_1 = -2\ddot{x}_2$ gives Equation (15.5.10).

3. To demonstrate the advantages of introducing an orthogonal coordinate system, for a pendulum with mass m and length a attached to the end of a similar pendulum, if the pendulums describe angles of θ_1 and θ_2 with respect to the $-\hat{e}_y$ direction, then, e.g., $y_2 = -a(\cos\theta_1 + \cos\theta_2)$ with respect to the rest position and

$$\begin{aligned} L &= \frac{1}{2}ma^2\dot{\theta}_1^2 + mga\cos\theta_1 + mga(\cos\theta_1 + \cos\theta_2) \\ &\quad + \frac{1}{2}ma^2\left(\left[\frac{d}{dt}(\sin\theta_1 + \sin\theta_2)\right]^2 + \left[\frac{d}{dt}(\cos\theta_1 + \cos\theta_2)\right]^2\right) \\ &= ma^2\left(\dot{\theta}_1^2 + \frac{1}{2}\dot{\theta}_2^2 + \dot{\theta}_1\dot{\theta}_2\cos(\theta_1 - \theta_2)\right) + mga(2\cos\theta_1 + \cos\theta_2) \end{aligned}$$

$$(15.5.14)$$

4. Similarly, the path of a ball rolling down a spiral ramp of radius R with $z = c\theta$ can be parameterized in cylindrical coordinates for which $d\hat{e}_r(t)/dt = \dot{\theta}\hat{e}_\theta = (\dot{z}/c)\hat{e}_\theta$ (see Section 15.9), and hence,

$$\begin{aligned} L &= \frac{1}{2}m\dot{\vec{r}}^2 - mgz = \frac{1}{2}m\left(\frac{d}{dt}(z\hat{e}_z + R\hat{e}_r(t))\right)^2 \\ &\quad - mgz = m\left(\frac{1}{2}\left[\dot{z}^2 + \left(\frac{R}{c}\dot{z}\right)^2\right] - gz\right) \end{aligned} \qquad (15.5.15)$$

15.6 THREE-DIMENSIONAL PARTICLE MOTION

Newton's laws in a temporally and spatially invariant gravitational field $\vec{F} = -mg\hat{e}_z$ yield, where $\vec{d}_0$ and $\vec{v}_0$ are the displacement and velocity at $t = t_0$, respectively,

$$\begin{aligned} \vec{v} &= -g(t-t_0)\hat{e}_z + \vec{v}_0 \\ \vec{d} &= -\frac{1}{2}g(t-t_0)^2\hat{e}_z + \vec{v}_0(t-t_0) + \vec{d}_0 \end{aligned} \tag{15.6.1}$$

From the potential energy difference $mg(z - z_0)$ of a particle in moving from an initial height z_0 to z, energy conservation further yields

$$v^2 - v_0^2 = -2g(z - z_0) \tag{15.6.2}$$

Example

If a particle is launched at the origin with a velocity $v_0 = |\vec{v}_0|$ at an angle θ with respect to the horizontal, its subsequent horizontal position is given by $x = v_0 t \cos\theta$ since $F_x = 0$, while its vertical position equals $z = v_0 t \sin\theta - gt^2/2$. Accordingly, the particle hits the ground when $z = 0$ yielding $t = 2v_0 \sin\theta/g$ and $x = 2v_0^2 \sin\theta \cos\theta/g$. This is simulated by the following program where the x - and z -components of the particle position and particle velocity are, respectively, stored in the `particle` structure. The calculation terminates when z becomes negative:

```
dT = input ( 'input dT' )
theta = input ( 'input theta' )
theta = theta * %pi / 180;

% Initial conditions on position, velocity
Particle.position(:,1) = [ 0 ; 0 ];                      % x, z
Particle.velocity(:,1) = [ cos( theta ) ; ...
  sin( theta ) ] * 200;                                   % Vx, Vz

for loop = 2 : 2000
  acceleration = [ 0 ; -10 ];
  Particle.velocity(:,loop)=Particle.velocity(:,loop-1)+...
     acceleration * dT;
  Particle.position(:,loop)=Particle.position(:,loop-1)+...
     Particle.velocity(:,loop-1) * dT;
  if ( Particle.position(2,loop) < 0 ) break; end;
end
plot( Particle.position(1,:), Particle.position(2,:) );
```

15.7 IMPULSE

The momentum of a particle acted on by a time-varying force between times t_0 and $t_0 + \Delta t$ changes by

$$\Delta p = \int_{t_0}^{t_0 + \Delta t} F(t) dt \tag{15.7.1}$$

For small time intervals Δt, this momentum change can be specified as

$$I_{\Delta t} \equiv \Delta p \approx F_{\text{average}} \Delta t \tag{15.7.2}$$

where $I_{\Delta t}$ is termed the *impulse* and F_{average} signifies the average force over Δt.

15.8 OSCILLATORY MOTION

Subsequent to examining free space motion with zero applied force and motion in a constant force field, forces that vary proportionally with distance are now considered. If the force on a particle passing through its equilibrium position $x_{\text{equilibrium}}$ at which $\vec{F} = 0$ is oppositely directed and proportional to the displacement from $x_{\text{equilibrium}}$, *periodic motion* results. Thus, for an ideal spring with an attached mass m,

$$F = m \frac{d^2 \left(x - x_{\text{equilibrium}}\right)}{dt^2} = m \frac{d^2 x}{dt^2} = -k\left(x - x_{\text{equilibrium}}\right) \tag{15.8.1}$$

The coordinate origin is often chosen such that $x_{\text{equilibrium}} = 0$. Accordingly, the curvature of the particle position graphed as a function of time varies as the negative of its displacement from the equilibrium position. The real functions with this property are the cosine and sine functions implying *simple harmonic motion*. Indeed, introducing $\omega_0 = \sqrt{k/m}$ and rewriting Equation (15.8.1) as

$$\left(\frac{d}{dt} - i\omega_0\right)\left(\frac{d}{dt} + i\omega_0\right) x(t) = 0 \tag{15.8.2}$$

yield the general complex solution $a \exp(i\omega_0 t) + b \exp(-i\omega_0 t)$, the terms of which must combine to generate a real and hence physical solution

$$x(t) = A \cos(\omega_0 t) + B \sin(\omega_0 t) = C \cos(\omega_0 t + \delta) \tag{15.8.3}$$

with A, B, C, and δ real. Enhancing the spring stiffness and thus restoring force or decreasing the spring mass and thus inertia increases the acceleration and hence ω_0. To relate C and δ *to* A and B in Equation (15.8.3), with $C' = \sqrt{A^2 + B^2}$ and $\tan \delta = B/A$,

$$\begin{aligned} C'\left(\frac{A}{C'} \cos(\omega_0 t) + \frac{B}{C'} \sin(\omega_0 t)\right) &= C'(\cos\delta \cos(\omega_0 t) + \sin\delta \sin(\omega_0 t)) \\ &= C' \cos(\omega_0 t - \delta) \end{aligned} \tag{15.8.4}$$

Equivalently, the $\cos(\omega_0 t)$ and $\sin(\omega_0 t)$ components of $x(t)$ can be represented by the x and iy components of a complex vector or *phasor*, $c = C\exp(-i\delta)$, that rotates in the positive angular direction in the complex plane with angular velocity ω_0. For Equation (15.8.4),

$$C\cos(\omega_0 t + \delta) = \text{Re}\left(Ce^{-i\delta}e^{i\omega_0 t}\right) = \text{Re}\left(ce^{i\omega_0 t}\right) \tag{15.8.5}$$

While the *oscillation period of a harmonic oscillator is amplitude independent*, both the form of $x(t)$ and the period vary with amplitude in an *anharmonic* oscillator with a *nonlinear* restoring force.

Example

The restoring force in a pendulum consisting of mass m suspended from a massless arm of length R equals the gravitational force component, $-mg\sin\theta\hat{e}_\theta$, perpendicular to the pendulum arm, where θ denotes the angle between the arm and the negative vertical direction. Since at an angle θ the mass has traveled a distance equal to the arclength $s = R\theta$ from its equilibrium position,

$$ma = m\frac{d^2 s}{dt^2} = mR\frac{d^2\theta}{dt^2} = -mg\sin\theta \approx -mg\left(\theta - \frac{\theta^3}{3!} + \ldots\right) \tag{15.8.6}$$

Therefore, only small angular displacements yield simple harmonic motion with $\omega_0 = \sqrt{g/R}$.

A linear oscillator in the presence of a dissipative, velocity-dependent retarding force and an oscillatory driving force is described by

$$m\frac{d^2\tilde{x}(t)}{dt^2} + \alpha\frac{d\tilde{x}(t)}{dt} + k\tilde{x}(t) = F\cos(\omega t) \tag{15.8.7}$$

Introducing complex notation $\tilde{x}(t) = \text{Re}\,x(t)$ and dividing by m, with $f = F/m$, $a = \alpha/m$, $\omega_0^2 = k/m$,

$$\frac{d^2 x}{dt^2} + a\frac{dx}{dt} + \omega_0^2 x = \left(\frac{d}{dt} - i\omega_1\right)\left(\frac{d}{dt} - i\omega_2\right)x(t) = fe^{i\omega t} \tag{15.8.8}$$

with

$$i\omega_{1,2} = \frac{-a \pm \sqrt{a^2 - 4\omega_0^2}}{2} \tag{15.8.9}$$

The solution to Equation (15.8.8)

$$x(t) = \underbrace{c_1 e^{i\omega_1 t} + c_2 e^{i\omega_2 t}}_{x_{\text{homogeneous}}(t)} + \underbrace{\frac{fe^{i\omega t}}{\omega_0^2 - \omega^2 + ia\omega}}_{x_{\text{particular}}(t)} \tag{15.8.10}$$

is a sum of the homogeneous solution found by setting the right-hand side to zero and the particular solution obtained by inserting $x(t) = c_{\text{particular}}e^{i\omega t}$ into Equation (15.8.8) with the right-hand side present and solving for $c_{\text{particular}}$. The constants c_1 and c_2 then follow from the initial conditions. Since for $a \ll \omega_0$, the homogeneous solution is composed of *transient* damped oscillating terms that decay as $\exp(-at/2)$, as $t \to \infty$ only the particular solution remains. This term possesses a characteristic *Lorentzian line shape* as a function of ω with a maximum at the *resonant frequency* near ω_0. The *Q-factor* is defined either as 2π times the ratio of the stored energy at the resonance frequency to the energy dissipated per cycle or alternatively as the frequency at resonance divided by the full width at half maximum (FWHM), i.e., the spacing between the two points with half the maximum value, of a curve of the oscillator energy as a function of frequency. Since the dependence of the energy associated with the particular term in the vicinity of ω_0 is given by $|x(t)|^2$ and is therefore proportional to $\left|\omega_0^2 - \omega^2 + ia\omega\right|^{-2} = 1\Big/\left[\left(\omega_0^2 - \omega^2\right)^2 + a^2\omega^2\right]$, when $a \ll \omega_0$, the approximate energy half width equals $2\Delta\omega$ with $\left(\omega_0^2 - \left(\omega_0^2 + 2\Delta\omega\omega_0\right)\right)^2 = a^2\omega_0^2$. Therefore, $\Delta\omega_{\text{FWHM}} = 2\Delta\omega \approx a$ and $Q = f_0 / \Delta f_{\text{FWHM}} = \omega_0 / \Delta\omega_{\text{FWHM}} = \omega_0 / a$.

The eigenfrequencies and eigenfunctions of N-coupled oscillators are analyzed by inserting $x_k = c_k e^{-i\omega t}$ with $k = 1, 2, \ldots, N$ into the equations of motion. Second-order time derivatives are then replaced by multiplication with $-\omega^2$, transforming the differential equation system into an algebraic system.

Example

Recalling that for a spring $U = kx^2/2$, for two identical harmonic oscillators joined to opposing walls by springs with spring constant κ_{11} and linked with a spring with spring constant κ_{12}, with each x_i measured from its equilibrium position,

$$L = T - U = \frac{m}{2}\left(\dot{x}_1^2 + \dot{x}_2^2\right) - \frac{1}{2}\left(\kappa_{11}\left(x_1^2 + x_2^2\right) + \kappa_{12}(x_1 - x_2)^2\right) \tag{15.8.11}$$

leading to

$$\begin{aligned} m\frac{d^2x_1}{dx^2} + (\kappa_{11} + \kappa_{12})x_1 - \kappa_{12}x_2 = 0 \\ m\frac{d^2x_2}{dx^2} + (\kappa_{11} + \kappa_{12})x_2 - \kappa_{12}x_1 = 0 \end{aligned} \tag{15.8.12}$$

With $x_k = c_k e^{-i\omega t}$, the resulting algebraic system possesses nontrivial solutions for

$$\begin{vmatrix} -m\omega^2 + (\kappa_{11} + \kappa_{12}) & -\kappa_{12} \\ -\kappa_{12} & -m\omega^2 + (\kappa_{11} + \kappa_{12}) \end{vmatrix} = 0 \tag{15.8.13}$$

or $m\omega_{\gtrless}^2 = (\kappa_{11} + \kappa_{12}) \pm \kappa_{12}$ (as can be seen by insertion). The symmetric eigenvector corresponding to the eigenvalue $\omega_<$ is $x_1 = x_2$, so that the restoring force is not influenced by the term $\kappa_{12}(x_2 - x_1)$ and $\omega_<$ therefore does not depend on κ_{12}. For $\omega = \omega_>$, the eigenvector is given by $x_1 = -x_2$ and the oscillation is antisymmetric. The restoring force provided by κ_{12} is then doubled compared to a system with a stationary second mass for which $x_2 = 0$, enhancing the effective stiffness and hence oscillation frequency.

Consider next N masses spaced at distance D are connected by rods that effectively act as springs with force constants κ with the first and last mass similarly attached to immovable walls. Noting that if $x_{k+1} - x_k$ is larger than 0, the kth mass is pulled to the right, while if $x_k - x_{k-1}$ is larger than 0, the mass is instead pulled to the left, for each mass

$$m\frac{d^2x_k}{dt^2} = \kappa(x_{k+1} - x_k) - \kappa(x_k - x_{k-1}), \quad k = 1, 2, \ldots, N \tag{15.8.14}$$

Since the two endpoints are fixed, $x_0 = x_{N+1} = 0$. With $x_k = \tilde{x}_k \cos(\omega t + \delta)$ and $\omega_0^2 = k/m$,

$$-\frac{\omega^2}{\omega_0^2}\tilde{x}_k = \tilde{x}_{k+1} - 2\tilde{x}_k + \tilde{x}_{k-1} \tag{15.8.15}$$

The right-hand side of Equation (15.8.15) possesses the same structure as the discrete second difference operator suggesting oscillatory solutions of the form $\tilde{x}_k = a \sin k\mu + b \cos k\mu$. Additionally, imposing zero boundary conditions at $k = 0, N + 1$ results in

$$x_k^{(l)} = c_l \sin \mu_l k \cos(\omega_l t + \delta_l) \tag{15.8.16}$$

with $\mu_l \equiv l\pi/(N + 1)$ and $l = 1, 2, \ldots, N$. To determine the ω_l, Equation (15.8.16) is inserted into Equation (15.8.15), as $\sin \mu_l (k + 1) + \sin \mu_l (k - 1) = 2 \sin \mu_l k \cos \mu_l$ yields, after canceling $\sin \mu_l k$,

$$-\frac{\omega_l^2}{\omega_0^2} = -2(1 - \cos \mu_l) = -4 \sin^2 \frac{\mu_l}{2} \tag{15.8.17}$$

The above formalism can be applied to a solid object such as a rod of cross-sectional area A if the rod is first considered as a series of N coupled rods of length D and mass m each of which provides a restoring force κ. As *doubling the length, D, of each section halves the restoring force associated with an elongation Δx while doubling A doubles this force*, the intrinsic material (as opposed to both material and geometric) properties are expressed by *Young's modulus* $Y \equiv \kappa D/A$. Defining a position coordinate along the rod z such that $z = z_n = n\Delta z$, in the limit $N \to \infty$, $D \to 0$, $m \to 0$ with the density $\rho = m/DA$ and total equilibrium length of the rod ND held fixed,

Equation (15.8.14) transforms into the *wave equation*, where $v=\sqrt{Y/\rho}$ is the (phase) velocity of wave propagation.

$$\frac{m}{AD}\frac{d^2x_k}{dt^2}=\frac{\kappa D}{A}\left(\frac{x_{k+1}-2x_k+x_{k-1}}{D^2}\right)\Rightarrow\frac{\partial^2 x(z,t)}{\partial t^2}=v^2\frac{\partial^2 x(z,t)}{\partial z^2} \tag{15.8.18}$$

15.9 ROTATIONAL MOTION ABOUT A FIXED AXIS

A body displaced an angle θ on a circle of radius R traverses an arclength $s=R\theta$. Therefore, defining the *angular velocity* and *angular acceleration* as the rate of change of θ in radians/second (rad/s) (recall that the angle θ, which is specified as the ratio of two quantities with dimensions of length is dimensionless) and the corresponding rate of change of the angular velocity in rad/s^2,

$$\begin{aligned}\omega&=\frac{d\theta}{dt}\\ \alpha&=\frac{d\omega}{dt}=\frac{d^2\theta}{dt^2}\end{aligned} \tag{15.9.1}$$

yields $\vec{v}=R\omega\hat{e}_\theta$ and $\vec{a}=R\alpha\hat{e}_\theta$ for the *instantaneous* velocity and acceleration of a particle traveling along a circle of *fixed radius R*. To determine $\vec{v}$ and $\vec{a}$ for general two-dimensional motion with $\vec{r}(t)=r(t)\hat{e}_r(t)$ requires the time derivatives of the polar coordinate basis vectors which are from Figure 15.2:

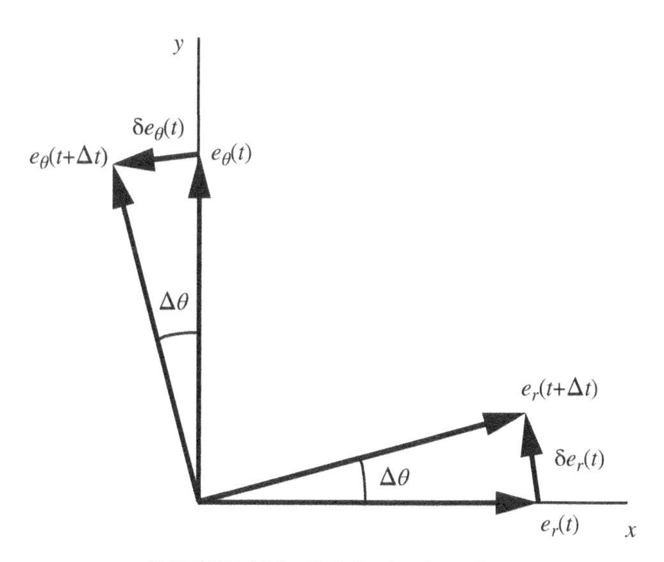

FIGURE 15.2 Infinitesimal rotation.

$$\frac{d}{dt}\hat{e}_r(t) = \frac{\Delta\theta \cdot |\hat{e}_r(t)|}{\Delta t}\hat{e}_\theta(t) = \omega\hat{e}_\theta(t)$$
$$\frac{d}{dt}\hat{e}_\theta(t) = \frac{\Delta\theta \cdot |\hat{e}_\theta(t)|}{\Delta t}(-\hat{e}_r(t)) = -\omega\hat{e}_r(t) \tag{15.9.2}$$

as can be confirmed algebraically from

$$\hat{e}_r(t) = \cos\theta(t)\hat{e}_x + \sin\theta(t)\hat{e}_y$$
$$\hat{e}_\theta(t) = -\sin\theta(t)\hat{e}_x + \cos\theta(t)\hat{e}_y \tag{15.9.3}$$

since, e.g., differentiating $\hat{e}_r$ in Equation (15.9.3) with respect to time, and comparing with the formula for $\hat{e}_\theta$, reproduces the first equation in Equation (15.9.2). Accordingly, the time derivative of the particle position is

$$\frac{d}{dt}(r(t)\hat{e}_r(t)) = \frac{dr(t)}{dt}\hat{e}_r(t) + r(t)\frac{d\hat{e}_r(t)}{dt} = \dot{r}(t)\hat{e}_r(t) + \omega(t)r(t)\hat{e}_\theta(t) \tag{15.9.4}$$

The second derivative yields, similarly,

$$\frac{d^2(r\hat{e}_r)}{dt^2} = \ddot{r}\hat{e}_r + \alpha r\hat{e}_\theta + 2\omega\dot{r}\hat{e}_\theta - \omega^2 r\hat{e}_r \tag{15.9.5}$$

For exclusively radial motion, only the first term in the above equation is present, while for angular motion, two terms contribute, the angular acceleration $\alpha r\hat{e}_\theta$ and the "centrifugal acceleration" in the $-\hat{e}_r$ direction with magnitude $\omega^2 r = v^2/r$. Accordingly, an inward force $F = -mv^2\hat{e}_r/r$ is required to maintain a particle in circular motion at a constant speed. Finally, the *Coriolis acceleration*, $2\omega\dot{r}\hat{e}_\theta$, arises when both the radial and angular velocities differ from zero. Indeed, displacing a particle outward on a radial spoke of a spinning wheel requires an additional force component that both increases the particle's linear velocity to maintain $\omega = v_\theta/r$ and further compensates for the associated directional change in the difference between the velocity vectors at the smaller and larger radii.

Example

Southward winds in the Northern Hemisphere increase their distance from the earth's axis diminishing their angular velocity, while the opposite occurs for northward winds resulting in a clockwise circulation.

Alternatively, from Equation (15.9.4), the Lagrangian expressed in polar coordinates is given by

$$L = \frac{m}{2}\left(\dot{r}^2 + r^2\dot{\theta}^2\right) - V \tag{15.9.6}$$

from which Equation (15.9.5) follows from the Lagrangian equations of motion,

$$\frac{\partial}{\partial t}\left(\frac{\partial L}{\partial \dot{r}}\right) = \frac{\partial}{\partial t}(m\dot{r}) = m\ddot{r} = \frac{\partial L}{\partial r} = -\frac{\partial V}{\partial r} + mr\dot{\theta}^2 = F_r + mr\dot{\theta}^2$$

$$\frac{\partial}{\partial t}\left(\frac{\partial L}{\partial \dot{\theta}}\right) = \frac{\partial}{\partial t}(mr^2\dot{\theta}) = mr^2\ddot{\theta} + 2mr\dot{r}\dot{\theta} = \frac{\partial L}{\partial \theta} = -\frac{\partial V}{\partial \theta} = r\underbrace{\left(-\hat{e}_\theta \frac{1}{r}\frac{\partial V}{\partial \theta}\right)}_{-(\vec{\nabla}V)_\theta} = rF_\theta \tag{15.9.7}$$

15.10 TORQUE AND ANGULAR MOMENTUM

Force and momentum can similarly be divided into radial and angular components. The radial momentum equals $\vec{p}\cdot\hat{e}_r$, while the *angular momentum* is defined as

$$\vec{l} = \vec{r}\times\vec{p} \quad \left[\frac{MD^2}{T}\right] \tag{15.10.1}$$

(note the presence of $\vec{r}$ in place of $\hat{e}_r$). Since $d\vec{r}/dt$ and $\vec{p}$ are parallel,

$$\vec{\tau} \equiv \frac{d\vec{l}}{dt} = \vec{r}\times\frac{d\vec{p}}{dt} = \vec{r}\times\vec{F} \quad \left[\frac{MD^2}{T^2}\right] \tag{15.10.2}$$

Here, $\vec{\tau}$ is termed the *torque* and $|\vec{r}|\sin\theta_{\vec{r},\vec{F}}$ the *lever arm*, where $\theta_{\vec{r},\vec{F}}$ is the angle between the force and the radius vectors. *If the vector sum of all torques on a body is zero, angular forces are absent and the angular momentum is a constant of the motion.*

Example

If a string extending from $\vec{r}=0$ to a mass rotating in a circle of radius r with a speed v_θ is shortened by pulling inward at the origin, the particle velocity increases such that $l = mv_\theta r$ remains constant, since a torque cannot be applied at $\vec{r}=0$.

The form of the equations for *a point mass in circular motion at radius R around a pivot subject to a constant torque* through the pivot in the direction perpendicular to the plane of motion as determined by the right-hand rule coincides with those for linear motion with a constant applied force. That is, from $|\vec{l}| = mRv_\theta = mR^2\omega$,

$$\tau \equiv |\vec{\tau}| = \frac{dl}{dt} = mR^2\frac{d^2\theta}{dt^2} \tag{15.10.3}$$

Accordingly, in analogy to Equation (15.6.1),

$$\begin{aligned}\omega &= \frac{\tau}{mR^2}(t-t_0)+\omega_0 \\ \theta &= \frac{1}{2}\frac{\tau}{mR^2}(t-t_0)^2+\omega_0(t-t_0)+\theta_0\end{aligned} \tag{15.10.4}$$

with $\theta_0 = \theta(t_0)$ and $\omega_0 = \omega(t_0)$. The kinetic energy and power are then given by

$$T = \frac{1}{2}mv_\theta^2 = \frac{1}{2}mR^2\left(\frac{d\theta}{dt}\right)^2 \tag{15.10.5}$$

and

$$P = \frac{dT}{dt} = mR^2\frac{d^2\theta}{dt^2}\frac{d\theta}{dt} = \tau\omega \tag{15.10.6}$$

15.11 MOTION IN ACCELERATING REFERENCE SYSTEMS

Often, as for an observer on the earth's surface, an accelerating coordinate system proves most natural; however, time derivatives of vectors such as velocity and acceleration then differ from those in an inertial system. That is, if $\hat{e}_k^{\text{space}}$ and $\hat{e}_k^{\text{body}}(t)$, are respectively, the basis vectors of a fixed rectangular and an accelerating coordinate system,

$$\vec{v}(t) = \sum_{k=1}^{3} v_k^{\text{space}}(t)\hat{e}_k^{\text{space}} = \sum_{k=1}^{3} v_k^{\text{body}}(t)\hat{e}_k^{\text{body}}(t) \tag{15.11.1}$$

Differentiating with respect to time,

$$\begin{aligned}\frac{d\vec{v}}{dt} &= \sum_{k=1}^{3}\frac{dv_k^{\text{space}}(t)}{dt}\hat{e}_k^{\text{space}} \equiv \left.\frac{d}{dt}\right|_{\text{space}}\vec{v}(t) \\ &= \sum_{k=1}^{3}\left(\frac{dv_k^{\text{body}}(t)}{dt}\hat{e}_k^{\text{body}}(t) + v_k^{\text{body}}(t)\frac{d\hat{e}_k^{\text{body}}(t)}{dt}\right) \\ &\equiv \left.\frac{d}{dt}\right|_{\text{body}}\vec{v}(t) + \sum_{k=1}^{3} v_k^{\text{body}}(t)\frac{d\hat{e}_k^{\text{body}}(t)}{dt}\end{aligned} \tag{15.11.2}$$

where derivatives denoted space and body are evaluated with respect to the inertial and noninertial reference frames as their components are multiplied by the corresponding basis vectors. The direction of the basis vectors is preserved in linearly accelerating but not in rotating frames for which $d\hat{e}_k(t)/dt$ is perpendicular to both $\vec{\omega}$ and $\hat{e}_k(t)$ and further possesses a magnitude $\sin\theta_{\hat{e}_k,\vec{\omega}}d\theta/dt$, proportional to the component of the unit vector perpendicular to the rotation axis. Since $d\theta = \omega dt$,

$$\sum_{k=1}^{3} v_k^{\text{body}} \frac{d\hat{e}_k^{\text{body}}(t)}{dt} = \vec{\omega} \times \sum_{k=1}^{3} v_k^{\text{body}} \hat{e}_k^{\text{body}}(t) \tag{15.11.3}$$

and therefore, when applied to any vector quantity,

$$\left.\frac{d}{dt}\right|_{\text{space}} = \left.\frac{d}{dt}\right|_{\text{body}} + \hat{\omega} \times \tag{15.11.4}$$

Similarly, for time-dependent $\hat{\omega}$, since $d(\vec{\omega} \times \vec{A})/dt = d\vec{\omega}/dt \times \vec{A} + \vec{\omega} \times d\vec{A}/dt$,

$$\left.\frac{d^2}{dt^2}\right|_{\text{space}} = \left(\left.\frac{d}{dt}\right|_{\text{body}} + \hat{\omega}\times\right)\left(\left.\frac{d}{dt}\right|_{\text{body}} + \hat{\omega}\times\right) = \left.\frac{d^2}{dt^2}\right|_{\text{body}} + \frac{d\hat{\omega}}{dt} \times + 2\hat{\omega} \times \left.\frac{d}{dt}\right|_{\text{body}} + \hat{\omega} \times (\hat{\omega} \times \tag{15.11.5}$$

In a rotating and accelerating coordinate system with origin at $\vec{r}_0(t)$ relative to fixed space axes, the position vector of a moving body in the space axes is given by $\vec{r}_{\text{space}} = \vec{r}_0 + \vec{r}_{\text{body}}$. Accordingly,

$$\vec{a}_{\text{space}} - \vec{a}_0 = \vec{a}_{\text{body}} + \frac{d\hat{\omega}}{dt} \times \vec{r}_{\text{body}} + 2\hat{\omega} \times \vec{v}_{\text{body}} + \hat{\omega} \times \left(\hat{\omega} \times \vec{r}_{\text{body}}\right) \tag{15.11.6}$$

The second, third, and fourth terms on the right-hand side of the above expression represent the transverse (angular), Coriolis and centrifugal acceleration, respectively.

15.12 GRAVITATIONAL FORCES AND FIELDS

The *gravitational field* is defined such that the force on an infinitesimally small test mass m is given by $m\vec{g}$. Hence, the field at $\vec{r}$ of a point mass M at $\vec{r}'$ obeys the *inverse square law*

$$\vec{g} = -\frac{GM}{\left(\vec{r} - \vec{r}'\right)^2} \hat{e}_{\vec{r}-\vec{r}'} \tag{15.12.1}$$

Gravitational fields obey the *superposition principle* that field arising from a collection of objects is the vector sum of the fields associated with each individual object.

For any closed surface S enclosing a point mass M, each surface element $d\vec{S}$ is normal to the surface and directed outward from the enclosed volume so that $\hat{e}_{\vec{r}-\vec{r}'} \cdot d\vec{S} = \cos\theta_{\vec{r}-\vec{r}', d\vec{S}} dS$. However, viewed from $\vec{r}$, the solid angle $d\Omega$ is determined by the projection of $d\vec{S}$ onto a plane perpendicular to the radius vector subtended by $d\vec{S}$ according to $d\Omega = \cos\theta_{\vec{r}-\vec{r}', d\vec{S}} dS / \left|\vec{r} - r'\right|^2$ and consequently

$$\int_S \vec{g} \cdot d\vec{S} = -\int_S GM d\Omega = -4\pi GM \tag{15.12.2}$$

If S does not enclose the origin, the enclosed mass and hence the surface integral are zero. Since M can be located anywhere within S, if S contains a distribution of mass elements with mass $dM = \rho(\vec{r})d^3r$ each of which contributes $-4\pi G\rho(\vec{r})dV$ to the integral,

$$\int_S \vec{g} \cdot d\vec{S} = -4\pi G \iiint_V \rho(r')d^3r' = -4\pi GM_{\text{enclosed}} \tag{15.12.3}$$

in which M_{enclosed} represents the total mass enclosed by S. Equation (15.12.3) constitutes a *global* relationship between the enclosed mass and the field integrated over an entire *Gaussian surface* S. However, just as the local behavior of $f(x)$ cannot be inferred from a global property such as the value of its integral over an interval unless the functional dependence of $f(x)$ on x is specified, $\vec{g}$ at a *local* point cannot be found from its surface integral unless the field distribution is either constant or is described by a known analytic function on S.

Example

For a spherical body with uniform density centered at the origin with total mass M and radius R, the gravitational field is constant by symmetry over the spherical surface $|\vec{r}| = r$. For $r > R$,

$$\vec{g} = -\frac{GM_{\text{enclosed}}}{r^2}\hat{e}_r \tag{15.12.4}$$

so that the gravitational force is identical to the force generated by a point mass at its center with the mass, M_{body}, of the entire body. On the other hand, inside the body, $r < R$ and the spherical Gaussian surface S encloses a volume fraction $V_r/V_R = r^3/R^3$ of the total mass of the body. Consequently, the gravitational field varies linearly with r according to

$$\vec{g} = -\frac{G}{r^2}\left(\frac{M_{\text{body}}r^3}{R^3}\right)\hat{e}_{\vec{r}} = -\frac{GM_{\text{body}}r}{R^3}\hat{e}_{\vec{r}} \tag{15.12.5}$$

The *gravitational potential energy*, U, follows from $\vec{F} \cdot d\vec{r} = mg\hat{e}_r \cdot d\vec{r}$. With the convention that $U(\infty) = 0$ and for a point mass M (note the successive minus signs),

$$U(r) - U(\infty) = U(r) = -\int_\infty^r \vec{F} \cdot dr = -\int_\infty^r \left(-\frac{GMm}{r^2}\right)dr = -\frac{GMm}{r} \tag{15.12.6}$$

Similarly, the integral of the gravitational field yields the *gravitational potential*

$$\Upsilon(r) = \frac{U(r)}{m} = -\frac{GM}{r} \tag{15.12.7}$$

The gravitational potential and potential energy resemble an infinitely deep depression sloping toward the origin that provides a radially inward force. The superposition principle obeyed by the gravitational field implies that the gravitational potential generated by several bodies is the sum of the gravitational potentials of each body in isolation; the negative of the gradient of the total potential yields the gravitational field.

15.13 CELESTIAL MECHANICS

Planets orbit an effectively stationary star conserving both angular momentum and energy. The conservation of angular momentum, L, implies *Kepler's second law* that *planets sweep out equal areas in equal times*. In particular, with the distance from the planet to the star and the angular coordinate in the plane of the planetary motion denoted by r and θ, respectively, over a short time Δt, the planet is displaced from $\vec{r}$ to $\vec{r} + \vec{v}\Delta t$. The area of the triangle formed by these two vectors, ΔA, is only a function of L/m according to

$$\Delta A = \frac{1}{2}\left|\vec{r}\times(\vec{r}+\vec{v}\Delta t)\right| = \frac{1}{2}\left|\vec{r}\times\vec{v}\Delta t\right| = \frac{L\Delta t}{2m} \tag{15.13.1}$$

The energy of a planetary orbit equals, from $\vec{L} = \vec{r}\times\vec{p} = mvr\sin\theta\hat{e}_{\vec{r}\times\vec{p}} = mrv_\theta\hat{e}_{\vec{r}\times\vec{p}}$,

$$E = T + U = \frac{m}{2}\left(v_r^2 + v_\theta^2\right) + U = \frac{m}{2}\left(\frac{dr}{dt}\right)^2 + \frac{L^2}{2mr^2} - \frac{GMm}{r} \tag{15.13.2}$$

Depending on E and L, the orbit describes one of the four conic sections. For $E<0$, the path describes an ellipse with inner and outer turning points at the points where $v_r = 0$ given by the roots of the quadratic equation

$$Er^2 + GMmr - \frac{L^2}{2m} = 0 \tag{15.13.3}$$

which describes a circle when the roots coincide, while $E=0$ and $E>0$ are associated with parabolas or hyperbolas, respectively. The equation for conic sections in polar coordinates possesses the form

$$r = \frac{a}{1 + b\cos\theta} \tag{15.13.4}$$

indicating that a conic section can be generated by a harmonic oscillator equation in $1/r$ and hence in $v_\theta = L/mr$. Since $v_\theta \propto r^{-1}$ implies $v_\theta^{-1}dv_\theta/d\theta = -r^{-1}dr/d\theta$,

$$\frac{dv_\theta}{d\theta} = -\frac{v_\theta}{r}\left[\frac{dr}{d\theta}\right] = -\left(\frac{r\omega}{r}\right)\left[\left(\frac{d\theta}{dt}\right)^{-1}\frac{dr}{dt}\right] = -\frac{dr}{dt} \tag{15.13.5}$$

Inserting the above equation and $L = mrv_\theta$ into Equation (15.13.2),

$$\frac{m}{2}\left(\left(\frac{dv_\theta}{d\theta}\right)^2 + v_\theta^2\right) = E + \frac{GMm^2}{L} v_\theta \tag{15.13.6}$$

After d/dv_θ is applied to both sides together with $(dv_\theta/d\theta)(d/dv_\theta)(dv_\theta/d\theta) = d^2v_\theta/d\theta^2$, the desired harmonic oscillator equation with a constant driving term results

$$m\left(\frac{d^2v_\theta}{d\theta^2} + v_\theta\right) = \frac{GMm^2}{L} \tag{15.13.7}$$

The inhomogeneous and homogeneous solutions to Equation (15.13.7) are $v_\theta = GMm/L$ and a sum of sine and cosine functions. For initial conditions such that only the cosine function is present

$$v_\theta = \frac{L}{mr} = \frac{GMm}{L} + c \cdot \cos\theta \tag{15.13.8}$$

a conic section equation is obtained in the form

$$r = \frac{L^2/GMm^2}{1 + e\cos\theta} \tag{15.13.9}$$

with *eccentricity* $e = cL/GMm$. With Equation (15.13.8) inserted into Equation (15.13.6), the constant E at $\theta = 0$ for which $dv_\theta/d\theta = 0$ equals

$$E = \frac{m}{2}\left(-\left(\frac{GMm}{L}\right)^2 + c^2\right) \tag{15.13.10}$$

which implies

$$c = \frac{GMme}{L} = \sqrt{\frac{2E}{m} + \left(\frac{GMm}{L}\right)^2}, \quad e = \sqrt{\frac{2E}{m}\left(\frac{L}{GMm}\right)^2 + 1} \tag{15.13.11}$$

From Equation (15.13.9), a circle of radius L^2/GMm^2 is obtained for $e = 0$, while $0 < e < 1$ yields an ellipse with the sum of inner ($r_<$) and outer ($r_>$) turning points given by

$$r_> + r_< = \frac{\frac{2L^2}{GMm^2}}{1 - e^2} = \frac{\frac{2L^2}{GMm^2}}{-\frac{2EL^2}{m^3G^2M^2}} = -\frac{GMm}{E} \tag{15.13.12}$$

Parabolic motion is obtained for $e = 1$, and therefore, $E = 0$ and $r_> = \infty$, while hyperbolic motion results if $e > 1$ and $E > 0$ as the radius of the motion approaches infinity when $\cos\theta \to -1/e > -1$.

Planetary problems are conveniently solved in *astronomical units*, in which the units of distance, mass, and time are the earth–sun distance, 1.496×10^{11} m; the solar mass, $M_{\text{sun}} = 1.989 \times 10^{30}$ kg; and a day, 86,400 s (a useful fact in this context is that a year is approximately $\pi \times 10^7$ s). The gravitational constant, which in astronomical units is normally expressed in terms of a new parameter $k^2 \equiv GM_{\text{sun}}$, then becomes, where astronomical units are distinguished by primes, e.g., $r = 1.45 \times 10^{11} r'$,

$$G\left[\frac{D^3}{MT^3}\right] = 6.673 \times 10^{-11}\left[\frac{D^3}{MT^3}\right] = k^2\left[\frac{D'^3}{M'T'^3}\right]\frac{\left(1.496 \times 10^{11}\right)^3}{\left(1.989 \times 10^{30}\right)\left(8.64 \times 10^4\right)^2} \tag{15.13.13}$$

Accordingly,

$$k^2 = \frac{\left(6.673 \times 10^{-11}\right)\left(1.989 \times 10^{30}\right)\left(8.64 \times 10^4\right)^2}{\left(1.496 \times 10^{11}\right)^3}\left[\frac{D'^3}{M'T'^3}\right] = (0.0172)^2 \tag{15.13.14}$$

while $m'_{\text{earth}} = 1/333{,}000$ and the force exerted by the sun on the earth equals

$$\vec{F}'_{\text{earth}} = m'_{\text{earth}}\,\vec{a}'_{\text{earth}} = -k^2 m'_{\text{earth}}\hat{e}_{r'} \tag{15.13.15}$$

15.14 DYNAMICS OF SYSTEMS OF PARTICLES

While the motion of a system of several particles cannot in general be calculated analytically, certain combinations of variables are either invariant or obey simple evolution equations. The equation of motion for each particle in a system of N particles is given by

$$\frac{d\vec{p}_j}{dt} = \vec{F}_j^{\,\text{ext}} + \sum_{k=1,k\neq j}^{N} \vec{F}_{kj}^{\,\text{int}}, \quad j = 1, 2, \ldots, N \tag{15.14.1}$$

The first term represents the total external force on the jth particle, while the kth term in the sum corresponds to the internal force on this particle resulting from interaction with the kth particle. Although the force on each particle thus depends on the location of all other particles, Newton's third law states that the force of particle i on particle j is the negative of that of j on i. Hence, after summing over j,

$$\frac{d}{dt}\sum_{j=1}^{N} m_j\vec{r}_j = \sum_{j=1}^{N}\vec{F}_j^{\,\text{ext}} + \sum_{k,j=1,k\neq j}^{N}\vec{F}_{kj}^{\,\text{int}} \tag{15.14.2}$$

the sum over internal forces vanishes and

$$\frac{d}{dt}\sum_{j=1}^{N} m_j \vec{r}_j = m_{\text{tot}} \frac{d}{dt}\left(\frac{1}{m_{\text{tot}}}\sum_{j=1}^{N} m_j \vec{r}_j\right) \equiv m_{\text{tot}} \frac{d}{dt}\vec{r}_{\text{cm}} = \sum_{j=1}^{N} \vec{F}_j^{\text{ext}} \tag{15.14.3}$$

where

$$m_{\text{tot}} = \sum_{j=1}^{N} m_j \tag{15.14.4}$$

denotes the total mass of the system and $\vec{r}_{\text{cm}}$ is termed the *center of mass* position. Equation (15.14.3) also expresses the *momentum conservation law* that *if the sum of the external forces on a system is zero, the total momentum of the system is conserved.*

Example

The exhaust of a rocket traveling in a straight line in a gravity and external force free region travels at a relative velocity $\vec{v}_{\text{exhaust}}$ with respect to the rocket frame. For a rocket velocity $v_{\text{rocket}}(t)\hat{e}_x$, observed from a fixed reference frame, the time rate of change of the total momentum composed of the sum of the rocket momentum $M_{\text{rocket}} v_{\text{rocket}}(t)\hat{e}_x$ and the exhaust momentum $m_{\text{exhaust}}(v_{\text{rocket}}(t) - v_{\text{exhaust}})\hat{e}_x$ equals zero. Thus, after neglecting $m_{\text{exhaust}}\, dv_{\text{rocket}}/dt$,

$$\vec{F}_{\text{external}} = 0 = \frac{d}{dt}(M_{\text{rocket}} v_{\text{rocket}}(t)) + \frac{dm_{\text{exhaust}}}{dt}(v_{\text{rocket}}(t) - v_{\text{exhaust}}) \tag{15.14.5}$$

Applying $dM_{\text{rocket}}/dt = -dm_{\text{exhaust}}/dt$ then yields $M_{\text{rocket}} dv_{\text{rocket}}/dt = -v_{\text{exhaust}}$ dM_{rocket}/dt or

$$\frac{dM_{\text{rocket}}}{M_{\text{rocket}}} = -\frac{dv_{\text{rocket}}}{v_{\text{exhaust}}} \tag{15.14.6}$$

and as v_{exhaust} remains constant, $v_{\text{rocket}}(t) = v_{\text{exhaust}} \log(M_{\text{rocket}}(0)/M_{\text{rocket}}(t)) + v_{\text{rocket}}(0)$.

Similarly, the external torque equals the change of the total angular momentum of the system as

$$\frac{d\vec{L}}{dt} = \sum_{j=1}^{N} \vec{r}_j \times \vec{F}_j = \sum_{j=1}^{N} \vec{r}_j \times \vec{F}_j^{\text{ext}} + \sum_{k,j=1,k\neq j}^{N} \vec{r}_j \times \vec{F}_{jk}^{\text{int}} = \sum_{j=1}^{N} \vec{r}_j \times \vec{F}_j^{\text{ext}} = \vec{\tau}^{\text{ext}} \tag{15.14.7}$$

and hence, *the angular momentum is conserved in a system with zero net external torque*.

The kinetic energy of a system equals the sum of the kinetic energy of the center of mass and that of the particles about the center of mass from

$$\begin{aligned} T &= \frac{1}{2}\sum_{j=1}^{N} m_j\left(\frac{d\vec{r}_{\rm cm}}{dt}+\frac{d(\vec{r}_j-\vec{r}_{\rm cm})}{dt}\right)^2 \\ &= \frac{1}{2}\sum_{j=1}^{N} m_j\left(\left(\frac{d\vec{r}_{\rm cm}}{dt}\right)^2+\underbrace{2\frac{d(\vec{r}_j-\vec{r}_{\rm cm})}{dt}\frac{d\vec{r}_{\rm cm}}{dt}}_{0}+\left(\frac{d(\vec{r}_j-\vec{r}_{\rm cm})}{dt}\right)^2\right) \end{aligned} \qquad (15.14.8)$$

since from the definition of the center of mass, the middle term vanishes. Similarly, the angular momentum, $\vec{L}$, of a system about a point, P, equals that of the center of mass about this point with the vector sum of $\vec{L}$ for each particle about the center of mass. That is, following the above derivation, with P at the origin, yields

$$\begin{aligned} \vec{L} &= \sum_{j=1}^{N}\left\{m_j(\vec{r}_{\rm cm}+(\vec{r}_j-\vec{r}_{\rm cm}))\times\frac{d}{dt}(\vec{r}_{\rm cm}+(\vec{r}_j-\vec{r}_{\rm cm}))\right\} \\ &= \left(\sum_{j=1}^{N} m_j\right)\vec{r}_{\rm cm}\times\frac{d}{dt}\vec{r}_{\rm cm}+\sum_{j=1}^{N} m_j(\vec{r}_j-\vec{r}_{\rm cm})\times\frac{d}{dt}(\vec{r}_j-\vec{r}_{\rm cm}) \end{aligned} \qquad (15.14.9)$$

For a particle to experience zero acceleration and thus maintain a constant velocity, the component forces on the particle must sum vectorially to zero, which is termed *mechanical equilibrium*. If a body is motionless with respect to an inertial reference frame, it is further said to be in *static equilibrium* with respect to this frame. For an extended object, this requires that the sum of the forces and torques in each coordinate direction must both be zero.

Examples

1. For a ladder of mass m and length L resting along a frictionless wall at an angle θ to a horizontal floor, for the net forces to cancel in the horizontal, x, and vertical, y, directions,

$$\begin{aligned} &F_{\text{floor normal},y}-mg=0 \\ &F_{\text{wall normal},x}-F_{\text{floor frictional},x}=0 \end{aligned} \qquad (15.14.10)$$

Computing the torque about the contact point of the ladder with the ground so that $\vec{r}\times\vec{F}_{\rm floor}=0$, since the effective gravitation force on the ladder acts on its center of mass,

$$\tau_z = F_{\text{wall normal},x}L\sin\theta-\frac{mgL}{2}\cos\theta=0 \qquad (15.14.11)$$

2. If a falling ladder of length L is placed on a frictionless floor, no force is exerted on the ladder parallel to the floor, and hence, the center of mass of the ladder falls vertically. Accordingly, if the bottom of the ladder is initially a distance d

from a vertical line drawn from the center of the ladder, when the ladder hits the ground, the bottom will have moved a distance $L/2-d$.

3. The equilibrium equations for a mass held by both a horizontal string, 1, under tension T_1 and a second string, 2, under tension T_2 that describes an angle θ with respect to the surface of a horizontal ceiling to which it is attached are given in the vertical and horizontal directions, respectively, by

$$\begin{aligned} mg - T_2 \sin\theta = 0 \\ T_1 - T_2 \cos\theta = 0 \end{aligned} \tag{15.14.12}$$

4. Finally, if a person with mass m stands a distance d from one support of a bridge of mass M and length L, noting that the weight of the bridge acts at its center of mass, the torque relative to this support equals $g(md + ML/2)$. In equilibrium, a countervailing torque $-LF_{\text{support}}$ must be supplied by the opposite support implying $F_{\text{support}} = g(md/L + M/2)$.

The time average of the kinetic energy

$$\langle T \rangle \equiv \frac{1}{\tau}\int_{t_0}^{t_0+\tau} T dt \tag{15.14.13}$$

over either a single period of a periodic orbit or a time τ over which a spatially limited system effectively samples all physically accessible states can be simply related to the time average, $\langle U \rangle$, of U by expressing $\langle T \rangle$ as the difference of a total differential and a term containing the forces

$$\begin{aligned} 2\langle T \rangle &= \left\langle \sum_{j=1}^{N} \vec{p}_j \cdot \vec{p}_j \right\rangle = \left\langle \frac{d}{dt}\left(\sum_{j=1}^{N} \vec{r}_j \cdot \vec{p}_j \right) \right\rangle - \left\langle \sum_{j=1}^{N} \vec{r}_j \cdot \frac{d\vec{p}_j}{dt} \right\rangle \\ &\equiv \left\langle \frac{d\varpi}{dt} \right\rangle - \left\langle \sum_{j=1}^{N} \vec{r}_j \cdot \vec{F}_j \right\rangle \end{aligned} \tag{15.14.14}$$

The quantity ϖ is termed the *virial*. Since

$$\left\langle \frac{d\varpi}{dt} \right\rangle = \frac{1}{\tau}\int_{t_0}^{t_0+\tau} \frac{d\varpi}{dt} dt = \frac{1}{\tau}(\varpi(\tau + t_0) - \varpi(t_0)) \to 0 \tag{15.14.15}$$

either over a period of the motion or as $\tau \to \infty$ for a bounded system,

$$2\langle T \rangle = -\left\langle \sum_{j=1}^{N} \vec{r}_j \cdot \vec{F}_j \right\rangle = \left\langle \sum_{j=1}^{N} \vec{r}_j \cdot \vec{\nabla} U_j \right\rangle \tag{15.14.16}$$

Example

For a single particle in a potential $U = cr^n$,

$$2\langle T\rangle = \left\langle \vec{r}\cdot\hat{e}_r \frac{dU}{dr}\right\rangle = \left\langle r\left(cnr^{n-1}\right)\right\rangle = n\langle U\rangle \quad (15.14.17)$$

In a gravitational force field, $n = -1$ and therefore $2\langle T\rangle = -\langle U\rangle$, implying $E = \langle U\rangle/2$.

15.15 TWO-PARTICLE COLLISIONS AND SCATTERING

If particles with masses m_j, $j = 1, 2$ collide, in the absence of external forces, the total vector momentum must be unchanged before and after the collision, yielding *three* equations relating the initial and final velocities $\vec{v}_j^{\text{initial}}$ and $\vec{v}_j^{\text{final}}$:

$$m_1\vec{v}_1^{\text{initial}} + m_2\vec{v}_2^{\text{initial}} = m_1\vec{v}_1^{\text{final}} + m_2\vec{v}_2^{\text{final}} \quad (15.15.1)$$

The kinetic energies are then related by

$$\frac{1}{2}m_1\left(\vec{v}_1^{\text{initial}}\right)^2 + \frac{1}{2}m_2\left(\vec{v}_2^{\text{initial}}\right)^2 = \frac{1}{2}m_1\left(\vec{v}_1^{\text{final}}\right)^2 + \frac{1}{2}m_2\left(\vec{v}_2^{\text{final}}\right)^2 + Q \quad (15.15.2)$$

For an *elastic* collision, $Q = 0$, while for *inelastic* collisions, the particles can experience an energy loss, $Q > 0$, as through friction, or an energy gain, $Q < 0$, as in an explosion.

In the *center of mass frame*, the center of mass of the two particles is stationary. Consequently, at all times, $m_1\vec{r}_1 + m_2\vec{r}_2 = 0$ and therefore $m_1\vec{v}_1 + m_2\vec{v}_2 = 0$, while the particles are situated at

$$\begin{aligned} \vec{r}_1^{\text{ cm}} &= \vec{r}_1 - \vec{r}^{\text{ cm}} = \vec{r}_1 - \frac{m_1\vec{r}_1 + m_2\vec{r}_2}{m_1 + m_2} = \frac{m_2}{m_1 + m_2}(\vec{r}_1 - \vec{r}_2) \\ \vec{r}_2^{\text{ cm}} &= \vec{r}_2 - \vec{r}^{\text{ cm}} = \vec{r}_2 - \frac{m_1\vec{r}_1 + m_2\vec{r}_2}{m_1 + m_2} = \frac{m_1}{m_1 + m_2}(\vec{r}_2 - \vec{r}_1) \end{aligned} \quad (15.15.3)$$

Note that as $m_2 \to \infty$, the second particle becomes the center of mass and $\vec{r}_1^{\text{ cm}} = \vec{r}_1 - \vec{r}_2$ while $\vec{r}_2^{\text{ cm}} = 0$. For particles approaching with a relative velocity $\vec{v}_{\text{relative}}^{\text{initial}} = \vec{v}_2^{\text{initial}} - \vec{v}_1^{\text{initial}}$, a scattering problem requires the determination of $\vec{v}_{\text{relative}}^{\text{final}} = \vec{v}_2^{\text{final}} - \vec{v}_1^{\text{final}}$. Since the force of the first particle on the second $\vec{F}_{12} = -\vec{F}_{21}$,

$$\frac{d^2(\vec{r}_2 - \vec{r}_1)}{dt^2} = \frac{d^2\vec{r}_2}{dt^2} - \frac{d^2\vec{r}_1}{dt^2} = \frac{F_{12}}{m_1} - \frac{F_{21}}{m_2} = \left(\frac{1}{m_1} + \frac{1}{m_2}\right)F_{12} = \frac{1}{\mu}F_{12} \quad (15.15.4)$$

in terms of the *reduced mass*

$$\mu = \left(\frac{1}{m_1} + \frac{1}{m_2}\right)^{-1} = \frac{m_1 m_2}{m_1 + m_2} \tag{15.15.5}$$

The total kinetic energy of the two particles in the center of mass frame equals

$$\begin{aligned} T &= \frac{1}{2}\left(m_1 \left(v_1^{\text{cm}}\right)^2 + m_2 \left(v_2^{\text{cm}}\right)^2\right) \\ &= \frac{1}{2}\left(m_1 \frac{m_2^2}{(m_1+m_2)^2}(\vec{v}_1 - \vec{v}_2)^2 + m_2 \frac{m_1^2}{(m_1+m_2)^2}(\vec{v}_1 - \vec{v}_2)^2\right) \\ &= \frac{1}{2}\mu(\vec{v}_1 - \vec{v}_2)^2 \end{aligned} \tag{15.15.6}$$

while the angular momentum of the system about the center of mass, which is invariant throughout the collision process in the absence of external torques, can similarly be expressed as

$$L = m_1 \vec{r}_1^{\,\text{cm}} \times \frac{d\vec{r}_1^{\,\text{cm}}}{dt} + m_2 \vec{r}_2^{\,\text{cm}} \times \frac{d\vec{r}_2^{\,\text{cm}}}{dt} = \mu(\vec{r}_1 - \vec{r}_2) \times \frac{d(\vec{r}_1 - \vec{r}_2)}{dt} \tag{15.15.7}$$

Defining the *impact parameter*, b, as the distance between the lines tangent to the paths of the two incoming particles in the center of mass frame as $t \to -\infty$, the scattering angle of elastic collisions can be parameterized by the relative, "internal" kinetic energy of motion (Eq. 15.15.6) and the relative angular momentum (Eq. 15.15.7), which evaluates to $m_1|\vec{v}_1|b = m_2|\vec{v}_2|b = \mu|\vec{v}_1 - \vec{v}_2|b$.

For a particle beam incident on a stationary scatterer, the *differential cross section* $d\sigma(\theta,\phi)/d\Omega$ is defined such that $(d\sigma(\theta,\phi)/d\Omega)d\Omega$ is the ratio of the number of scattering events into the solid angle $d\Omega$ about (θ,ϕ) per second to the number of incident particles per unit cross-sectional area per second (the particle flux). Integrating $d\sigma(\theta,\phi)/d\Omega$ over all angles yields the *total cross section*

$$\sigma = \frac{\text{scattered flux}}{\text{incident flux per unit area}} \tag{15.15.8}$$

Since, e.g., a classical scatterer of 0.1 m^2 cross section scatters 1 particle/s for an incident flux of 10 particles/s/m^2, the cross section constitutes an effective scattering area. For ϕ-independent (axially symmetric) scattering and an x-traveling incoming particle beam in the laboratory frame, the differential scattering cross section in the laboratory and center of mass frames differs by a factor $d\Omega_{\text{lab}}/d\Omega_{\text{cm}} = d\cos\theta_{\text{lab}}/d\cos\theta_{\text{cm}}$. This can be obtained from the relationships

$$\begin{aligned} v_1^{\text{final,cm}} \sin\theta_{\text{cm}} &= v_1^{\text{final,lab}} \sin\theta_{\text{lab}} \\ v_1^{\text{final,cm}} \cos\theta_{\text{cm}} + v_{\text{cm}} &= v_1^{\text{final,lab}} \cos\theta_{\text{lab}} \end{aligned} \tag{15.15.9}$$

for the final x and y components of v_1 in the center of mass and laboratory frames, or equivalently,

$$\tan\theta_1^{\text{final, lab}} = \frac{\sin\theta_1^{\text{final, cm}}}{\dfrac{v_{\text{cm}}}{v_1^{\text{final, cm}}} + \cos\theta_1^{\text{final, cm}}} \tag{15.15.10}$$

Lastly, the *impulse approximation* applies to highly energetic particles that maintain a nearly linear path and small relative energy change while scattering. The transverse momentum acquired can be estimated by integrating the transverse force component over this line. If x, y, and b, respectively, denote the coordinate directions parallel and perpendicular to the incoming particle direction and the impact parameter, since for linear motion $F_y = (b/r)\text{F}$ at all x while $2E = pv$ is taken as x independent, the resulting scattering angle is given by

$$\theta \approx \tan\theta = \frac{\Delta p}{p} = \frac{1}{p}\int_{-\infty}^{\infty} F_y(r)dt = \frac{1}{p}\int_{-\infty}^{\infty} \left(\left|\vec{F}(r)\right|\frac{b}{r}\right)\frac{dx}{v} = \frac{b}{2E}\int_{-\infty}^{\infty} \frac{\left|\vec{F}\left(\sqrt{b^2+x^2}\right)\right|}{\sqrt{b^2+x^2}}dx \tag{15.15.11}$$

15.16 MECHANICS OF RIGID BODIES

Since all points on a solid rotating about a point termed the *pivot* possess identical angular velocities, the angular momentum and response to an applied torque are determined by the geometry and density of the object. Considering first the simplified case of a *planar* body in the x–y plane with mass per unit area $\sigma(r, \phi)$ rotating with an angular velocity ω about the z-axis, the velocity of an element $dxdy$ centered at (x, y) equals $\vec{v} = r\omega\hat{e}_\theta = \sqrt{x^2+y^2}\omega\hat{e}_\theta$ resulting in a total angular momentum (where $\rho d^3r = \sigma dA$ is employed for mass distributed over a two-dimensional surface)

$$\vec{L} = \iiint_{V_{\text{body}}} \vec{r} \times (\rho(\vec{r})\,\vec{v})d^3r = \hat{e}_z\omega\iint_{S_{\text{body}}} r^2\sigma(r,\phi)rdrd\phi = \hat{e}_z I\omega \tag{15.16.1}$$

V_{body} and $\rho(r)$ denote the volume and density of the solid, while the *moment of inertia*, I, is defined by

$$I = \iiint_{V_{\text{body}}} \rho(r)r^2d^3r \equiv \iiint_{V_{\text{body}}} r^2dm = \iint_{S_{\text{body}}} r^2\sigma(r,\phi)rdrd\phi \tag{15.16.2}$$

For an isolated particle at $\vec{r}'$ with $|\vec{r}'| = a$, $\rho(\vec{r}) = m\delta^3(\vec{r} - \vec{r}\,')$ and $I = mr'^2 = ma^2$, while for a

$$\begin{array}{ll} \text{Disk or cylinder} & I_{\text{disk}} = \frac{1}{2}ma^2 \\ \text{Thin ring or cylindrical shell} & I_{\text{ring}} = ma^2 \\ \text{Thin rod of length } L\text{, pivot at center} & I_{\text{rod}} = \frac{1}{12}mL^2 \end{array} \tag{15.16.3}$$

The kinetic energy is similarly computed in terms of the moment of inertia according to

$$T = \frac{1}{2}\iiint_{V_{\text{body}}} \rho(\vec{r})v^2 d^3r = \frac{1}{2}\iiint_{V_{\text{body}}} (v(r))^2 dm = \frac{\omega^2}{2}\iint_{S_{\text{body}}} r^2\sigma(r,\phi)r\,dr\,d\phi = \frac{1}{2}I\omega^2 \tag{15.16.4}$$

while the torque is given in terms of the force applied to the body by

$$\vec{\tau} = \iiint_{V_{\text{body}}} \vec{r} \times \vec{F}_{\text{applied}} d^3r \tag{15.16.5}$$

with

$$\vec{\tau} = \frac{d\vec{L}}{dt} = I\frac{d\omega}{dt} = I\alpha \tag{15.16.6}$$

Formulas for linear motion are therefore generally converted into their angular counterparts for such planar objects by replacing m with I and linear by angular quantities.

The computation of the moment of inertia for a thin, laminar body with $\int z^2 dm = 0$ rotating about the z-axis can often be simplified by the *perpendicular axis theorem*

$$I_z = \int (x^2 + y^2)dm = \int (x^2 + z^2)dm + \int (y^2 + z^2)dm = I_x + I_y \tag{15.16.7}$$

so that I_z equals the sum of the moments of inertia about the two perpendicular axes. As well, the *parallel axis theorem* states that if I is calculated about a z-oriented rotation axis C passing through its center of mass with respect to which $\int \vec{r}dm = 0$, the z-component of the moment of inertia, $I_z^{(C+R)}$, about a parallel axis displaced by $\vec{R}$ from C is

$$I_z^{(C+R)} = \int \left[(x - R_x)^2 + (y - R_y)^2\right] dm = \int \vec{r}^{\,2} dm + \int \vec{R}^{\,2} dm - 2\vec{R} \cdot \int \vec{r}dm = I_z^{(C)} + mR^2 \tag{15.16.8}$$

Examples

1. The moment of inertia about the center of a thin circular disk of radius a rotating around the z-axis with $\sigma = m/\pi a^2$ and hence $dm = \sigma \cdot 2\pi r dr$ equals

$$I_z^{(C)} = 2\pi\sigma \int_0^a r^3 dr = 2\pi\sigma \frac{a^4}{4} = \frac{m}{2}a^2 \tag{15.16.9}$$

while if the disk is instead rotated about a point on the rim, $I_z^{(C+a)} = 3ma^2/2$.

2. A cylinder with mass m and radius R rolling without slipping on a plane inclined by an angle α relative to the horizontal can be analyzed in several ways. Designating the distance that the cylinder is displaced *downward* along the plane and the cylinder's rotation angle by x and θ, respectively,

$$L = \frac{1}{2}m\dot{x}^2 + \frac{1}{2}I\omega^2 + mgx\sin\alpha = \frac{1}{2}m\dot{x}^2 + \frac{1}{4}mR^2\dot{\theta}^2 + mgx\sin\alpha \tag{15.16.10}$$

Incorporating the constraint $x = R\theta$ by employing $L' = L + \lambda(R\theta - x)$ in place of L in the Lagrange equations with λ, a Lagrange multiplier leads to the equations

$$\begin{aligned} m\ddot{x} - mg\sin\alpha &= -\lambda \\ \frac{1}{2}mR^2\ddot{\theta} &= \lambda R \end{aligned} \tag{15.16.11}$$

From the first line, λ corresponds to the frictional force. Eliminating λ from the equations above and employing the second derivative of the constraint equation, $R\ddot{\theta} = \ddot{x}$ results in $\ddot{x} = 2g\sin\alpha/3$. The constraint equation can be alternatively employed to eliminate $\dot{\theta}$ in Equation (15.16.10) before applying the Lagrange equation. The upper and lower equation in Equation (15.16.11) also follow directly from $\vec{F}_{\text{total}} = m\vec{a}$ and from equating the torque provided by the frictional force to the time derivative of the angular momentum $L = I_{\text{disk}}\omega$ around a rotation axis at the center of the cylinder. Finally, the stationary point of contact between the cylinder and the plane can be employed as an instantaneous axis around which the cylinder rotates. Around this axis, the torque is $mg\sin\alpha$, while the angular momentum $L = 3mR^2\omega/2$ from the principal axis theorem yielding the above result.

3. Consider next a falling ladder of length L and mass m resting on a frictionless floor extending in the x-direction and a similarly frictionless wall and describing an angle θ with respect to the downward wall (i.e., the $-y$) direction. With the origin at the intersection of the wall and the floor, the ladder's center of mass is located at

$$\vec{r}_{\text{cm}} = \frac{L}{2}\left(\sin\theta\hat{e}_x + \cos\theta\hat{e}_y\right) \tag{15.16.12}$$

(This point moves along the circumference of a circle of radius $L/2$ as the ladder falls as θ also equals the angle between the y-axis and the line from the origin to the center of mass). Accordingly, since $\vec{v}_{\text{cm}} = L\dot{\theta}/2\left(\cos\theta\hat{e}_x - \sin\theta\hat{e}_y\right)$, computing the kinetic energy about the ladder's center of mass yields the Lagrangian

$$\frac{mv_{\mathrm{cm}}^2}{2}+\frac{I\omega^2}{2}-mgy=\frac{mL^2\dot{\theta}^2}{8}+\frac{mL^2\dot{\theta}^2}{24}-\frac{mgL\cos\theta}{2}=\frac{mL^2\dot{\theta}^2}{6}-\frac{mgL\cos\theta}{2} \tag{15.16.13}$$

and hence, $\ddot{\theta}=3g\sin\theta/2L$. If the ladder is initially vertical, since $\ddot{x}_{\mathrm{cm}}=L/2\left(\ddot{\theta}\cos\theta-\dot{\theta}^2\sin\theta\right)>0$ for $3g\cos\theta/2L>\dot{\theta}^2$, while energy conservation implies $T+V=mgL/2=mL^2\dot{\theta}^2/6+mgL\cos\theta/2$ or $\dot{\theta}^2=3g(1-\cos\theta)/L$, the ladder separates from the wall once $\cos\theta=2/3$.

For an *arbitrarily* shaped body rotating with angular velocity $\vec{\Omega}$ about a fixed pivot, designating the vector from the pivot by $\vec{r}$, the velocity at each point of the body is $\vec{\Omega}\times\vec{r}$ so that

$$\vec{L}=\int\vec{r}\times\vec{v}\,dm=\int\vec{r}\times\left(\vec{\Omega}\times\vec{r}\right)dm=\int\left\{r^2\vec{\Omega}-\vec{r}\left(\vec{r}\cdot\vec{\Omega}\right)\right\}dm \tag{15.16.14}$$

In terms of the *inertia tensor* **I** with elements

$$I_{kl}=\int\left(r^2\delta_{kl}-r_kr_l\right)dm \tag{15.16.15}$$

Equation (15.16.14) can be rewritten as

$$L_k=I_{kl}\Omega_l \tag{15.16.16}$$

The kinetic energy is obtained from (since $A\cdot(B\times C)=B\cdot(C\times A)$)

$$T=\frac{1}{2}\int\left(\vec{\Omega}\times\vec{r}\right)\cdot\left(\vec{\Omega}\times\vec{r}\right)dm=\frac{1}{2}\vec{\Omega}\cdot\int\vec{r}\times\left(\vec{\Omega}\times\vec{r}\right)dm=\frac{1}{2}\vec{\Omega}\cdot\vec{L}=\frac{1}{2}\vec{\Omega}\cdot\mathbf{I}\cdot\vec{\Omega} \tag{15.16.17}$$

Example

The inertia tensor of a cube with a pivot at its corner point is evaluated by placing the corner at the origin and orienting the adjacent edges along the coordinate axes. The components of **I** are then by symmetry

$$I_{xx}=I_{yy}=I_{zz}=\int\left(r^2-z^2\right)dm=\frac{m}{L^3}\int_0^L\int_0^L\int_0^L\left(x^2+y^2\right)dxdydz=2\frac{m}{L^3}\frac{L^3}{3}L^2=\frac{2}{3}mL^2$$

$$I_{xy}=I_{yz}=I_{xz}=I_{xz}=I_{yz}=I_{zy}=-\frac{m}{L^3}\int_0^L\int_0^L\int_0^L zydxdydz=-2\frac{m}{L^3}\left(\frac{L^2}{2}\right)^2L=-\frac{1}{4}mL^2 \tag{15.16.18}$$

The general parallel axes theorem, derived as previously, adopts the form

$$I_{kl}^{(C+R)} = m\vec{R}^2\delta_{kl} - mR_kR_l + I_{kl}^{(C)} \tag{15.16.19}$$

Example

Converting the inertia tensor, $I_{kl} = \delta_{kl}\, mL^2/6$, of a cube about its center to **I** relative to the corner at $(L/2, L/2, L/2)$ with this formula reproduces Equation (15.16.18).

For motion around *principal rotation axes* specified by the eigenvectors of the inertia tensor, $\vec{L}$ and $\vec{\Omega}$ are parallel. The corresponding eigenvalues quantify the associated moments of inertia. While $d\vec{L}/dt = \vec{\tau}$ in an inertial laboratory frame, in a frame rotating with the body, from Equation (15.11.4),

$$\left.\frac{dL}{dt}\right|_{\text{body}} + \vec{\Omega} \times \vec{L} = \vec{\tau} \tag{15.16.20}$$

For $\vec{L}$ not along a principal axis, $\vec{L}$ only remains constant if a constant torque is applied. For a coordinate system coinciding with the principal axes, $\vec{L} = (I_1\Omega_1, I_2\Omega_2, I_3\Omega_3)$ and Equation (15.16.20) becomes

$$\begin{aligned} I_1\frac{d\Omega_1}{dt} &= (I_2 - I_3)\Omega_2\Omega_3 + \tau_1 \\ I_2\frac{d\Omega_2}{dt} &= (I_3 - I_1)\Omega_3\Omega_1 + \tau_2 \\ I_3\frac{d\Omega_3}{dt} &= (I_1 - I_2)\Omega_2\Omega_3 + \tau_3 \end{aligned} \tag{15.16.21}$$

which are termed the *Euler equations*. If 3 is a symmetry axis, $I_1 = I_2 = I$, while for zero applied torque,

$$\begin{aligned} I\frac{d\Omega_1}{dt} &= (I - I_3)\Omega_2\Omega_3 \\ I\frac{d\Omega_2}{dt} &= (I_3 - I)\Omega_3\Omega_1 \\ I_3\frac{d\Omega_3}{dt} &= 0 \end{aligned} \tag{15.16.22}$$

Since Ω_3 is time independent, combining the first two equations above yields with $\omega = (1 - I_3/I)\Omega_3$

$$\frac{d^2\Omega_1}{dt^2} = -\omega^2\Omega_1 \tag{15.16.23}$$

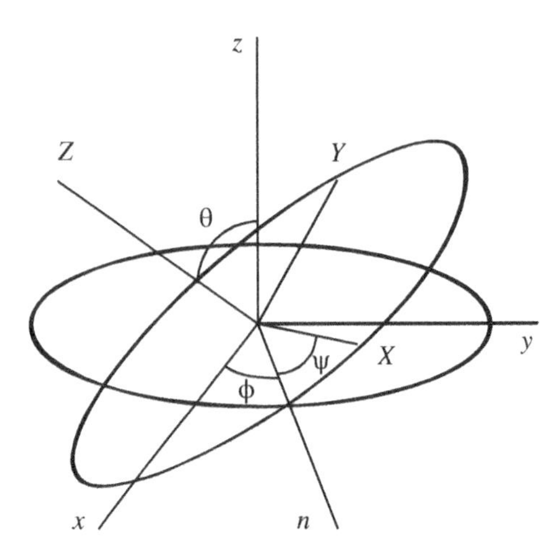

FIGURE 15.3 Euler angles.

Thus, $\Omega_1 = \Omega_T \sin(\omega t + \phi_0)$, $\Omega_2 = \Omega_T \cos(\omega t + \phi_0)$, and $\vec{\Omega}$ rotates or *precesses* around the three axis.

The *Euler angles*, for which one convention is depicted in Figure 15.3, conform to the motion of a top and are defined by a rotation through an angle ϕ about the $z_{\text{laboratory}}$ precession axis followed by a rotation of θ about its x_{top} -axis to incline the z_{top} symmetry axis of the top relative to the $z_{\text{laboratory}}$ precession axis and finally a rotation of ψ around the z_{top} -axis.

The angular momentum components associated with the Euler angles resolved along the body axes, X, Y, Z, are given by, noting that $\dot{\psi}$, $\dot{\phi}$, and $\dot{\theta}$ are oriented along the Z, z and n axis, termed the line of nodes, where $\hat{e}_n = \cos\psi \hat{e}_X - \sin\psi \hat{e}_Y$ and $\hat{e}_z = \sin\theta(\sin\psi \hat{e}_X + \cos\psi \hat{e}_Y) + \cos\theta \hat{e}_Z$,

$$\begin{aligned}
\Omega_X &= \dot{\phi}\hat{e}_X \cdot \hat{e}_z + \dot{\theta}\hat{e}_X \cdot \hat{e}_n + \dot{\psi}\hat{e}_X \cdot \hat{e}_Z = \dot{\phi}\sin\theta\sin\psi + \dot{\theta}\cos\psi \\
\Omega_Y &= \dot{\phi}\hat{e}_Y \cdot \hat{e}_z + \dot{\theta}\hat{e}_Y \cdot \hat{e}_n + \dot{\psi}\hat{e}_Y \cdot \hat{e}_Z = \dot{\phi}\sin\theta\cos\psi - \dot{\theta}\sin\psi \\
\Omega_Z &= \dot{\phi}\hat{e}_Z \cdot \hat{e}_z + \dot{\theta}\hat{e}_Z \cdot \hat{e}_n + \dot{\psi}\hat{e}_Z \cdot \hat{e}_Z = \dot{\phi}\cos\theta + \dot{\psi}
\end{aligned} \tag{15.16.24}$$

The Lagrangian of a top is therefore, where d_{cm} is the distance from the tip to the center of mass,

$$\begin{aligned}
T - V &= \frac{I}{2}\left(\Omega_X^2 + \Omega_Y^2\right) + \frac{I_Z}{2}\Omega_Z^2 - mgd_{\text{cm}}\cos\theta \\
&= \frac{I}{2}\left(\dot{\phi}^2\sin^2\theta + \dot{\theta}^2\right) + \frac{I_Z}{2}\left(\dot{\phi}\cos\theta + \dot{\psi}\right)^2 - mgd_{\text{cm}}\cos\theta
\end{aligned} \tag{15.16.25}$$

The $\hat{e}_z$ and $\hat{e}_Z$ angular momentum components associated with ϕ and ψ are conserved.

A general rotation can be represented as the product of three rotation matrices as

$$R(\phi,\theta,\psi) = e^{-i\psi S_Z^{\text{body}}} e^{-i\theta S_n^{\text{line of nodes}}} e^{-i\phi S_z^{\text{laboratory}}} \tag{15.16.26}$$

Rotating $\hat{e}_x$ and $\hat{e}_y$ an infinitesimal angle $\delta\theta$ around z respectively yields components $\delta\theta$ and $-\delta\theta$ along $\hat{e}_y$ and $\hat{e}_x$. Cyclically, permuting $x \to y \to z \to x$ yields the rotation matrix $\mathbf{R}_{\vec{\Omega}} = e^{-i\Omega\hat{e}_\Omega\cdot\vec{\mathbf{S}}}$ with

$$\vec{S} = i\hat{e}_z \begin{pmatrix} 0 & -1 & 0 \\ 1 & 0 & 0 \\ 0 & 0 & 0 \end{pmatrix}_{\substack{x \to x - y\delta\theta \\ y \to y + x\delta\theta}} + i\hat{e}_x \begin{pmatrix} 0 & 0 & 0 \\ 0 & 0 & -1 \\ 0 & 1 & 0 \end{pmatrix}_{\substack{y \to y - z\delta\theta \\ z \to z + y\delta\theta}} + i\hat{e}_y \begin{pmatrix} 0 & 0 & 1 \\ 0 & 0 & 0 \\ -1 & 0 & 0 \end{pmatrix}_{\substack{z \to z - x\delta\theta \\ x \to x + z\delta\theta}} \tag{15.16.27}$$

From direct multiplication,

$$\left[(S_\Omega)_i, (S_\Omega)_j\right] = i\varepsilon_{ijk}(S_\Omega)_k \tag{15.16.28}$$

Further,

$$-i\left(\vec{\Omega}\cdot\vec{\mathbf{S}}\right)\vec{V} = \begin{pmatrix} 0 & -\Omega_z & \Omega_y \\ \Omega_z & 0 & -\Omega_x \\ -\Omega_y & \Omega_x & 0 \end{pmatrix} \begin{pmatrix} V_x \\ V_y \\ V_z \end{pmatrix} = \begin{pmatrix} \Omega_y V_z - \Omega_z V_y \\ \Omega_z V_x - \Omega_x V_z \\ \Omega_x V_y - \Omega_y V_x \end{pmatrix} = \vec{\Omega} \times \vec{V} \tag{15.16.29}$$

so that $\left(-i\hat{e}_\Omega\cdot\vec{\mathbf{S}}\right)\vec{r} = \hat{e}_\Omega \times \vec{r}$, which together with $\left(-i\hat{e}_\Omega\cdot\vec{\mathbf{S}}\right)^2\vec{r} = \hat{e}_\Omega \times \left(\hat{e}_\Omega \times \vec{r}\right) = \hat{e}_{\vec{\Omega}}\left(\hat{e}_{\vec{\Omega}}\cdot\vec{r}\right) - \vec{r}$ and therefore $\left(-i\hat{e}_\Omega\cdot\vec{\mathbf{S}}\right)^3\vec{r} = -\left(-i\hat{e}_\Omega\cdot\vec{\mathbf{S}}\right)\vec{r}$ yields for the Taylor expansion of the exponential

$$R_{\vec{\Omega}} = e^{-i\Omega\hat{e}_\Omega\cdot\vec{\mathbf{S}}} = \mathrm{I} - i\sin\Omega\,\hat{e}_\Omega\cdot\vec{\mathbf{S}} - (1-\cos\Omega)\left(\hat{e}_\Omega\cdot\vec{\mathbf{S}}\right)^2 \tag{15.16.30}$$

Applying the above equation to a vector $\vec{r}$ results in an expression for the rotated vector $\vec{r}_\Omega$

$$\vec{r}_{\vec{\Omega}} = \vec{r}\cos\Omega + \hat{e}_{\vec{\Omega}} \times \vec{r}\sin\Omega + \hat{e}_{\vec{\Omega}}\left(\hat{e}_{\vec{\Omega}}\cdot\vec{r}\right)(1-\cos\Omega) \tag{15.16.31}$$

termed *Rodrigues' rotation formula*. Here, $\hat{e}_{\vec{\Omega}}\left(\hat{e}_{\vec{\Omega}}\cdot\vec{r}\right)$ coincides with the outer product $\hat{e}_{\vec{\Omega}} \otimes \hat{e}_{\vec{\Omega}}$ applied to the column vector $\vec{r}$.

Infinitesimal rotations unlike finite rotations commute as the noncommuting term after two first-order infinitesimal rotations is second-order and therefore can be neglected. Angular velocities defined in terms of infinitesimal rotations by $\omega_{\Delta\vec{\Omega}} = \lim_{\Delta t\to 0}\left(\left(R_{\Delta\vec{\Omega}} - 1\right)/\Delta t\right)$ therefore add.

15.17 HAMILTON'S EQUATION AND KINEMATICS

The Lagrangian equations provide ND second-order differential equations for a D-dimensional system of N particles. Accordingly, they uniquely specify the evolution of each point in a $2ND$-dimensional *phase space* comprised of the spatial and momentum coordinates of each particle when supplemented by $2ND$ initial conditions. To transform to a formalism explicitly involving phase space variables, a comparison of the Lagrange equation (Eq. 15.5.6) with $\vec{F} = d\vec{p}/dt$ suggests identifying $\partial L/\partial x_i = (\nabla L)_i = -(\nabla V)_i$ with a *generalized force* and $\partial L/\partial \dot{x}$ with a *generalized momentum*, p_i. If $\partial L/\partial x_i = 0$ for some i, the corresponding p_i are conserved (time invariant).

A phase space equation for the p_k and x_k is obtained through a *Legendre transformation* to the *Hamiltonian*

$$H(\vec{p},\vec{x}) = \sum_m p_m \dot{x}_m - L\left(\dot{\vec{x}},\vec{x},t\right) \tag{15.17.1}$$

so that $\partial H/\partial \dot{x}_i = 0$, eliminating the dependence on $\dot{\vec{x}}$. The Lagrange equation then transforms into *Hamilton's equations*:

$$\frac{\partial H}{\partial x_k} = -\frac{\partial L}{\partial x_k} = -F_k = -\dot{p}_k \tag{15.17.2}$$

together with (from the independence of p_i and x_k as a particle can possess any momentum at each spatial point)

$$\frac{\partial H}{\partial p_k} = \dot{x}_k \tag{15.17.3}$$

Since the particle coordinates in phase space evolve with time, $p_k = p_k(t)$ and $x_k = x_k(t)$ and hence

$$\frac{dH}{dt} = \sum_m \left(\dot{p}_m \dot{x}_m + p_m \ddot{x}_m - \underbrace{\frac{\partial L}{\partial x_m}}_{\dot{p}_m} \frac{\partial x_m}{\partial t} - \underbrace{\frac{\partial L}{\partial \dot{x}_m}}_{p_m} \frac{\partial \dot{x}_m}{\partial t} \right) - \left.\frac{\partial L}{\partial t}\right|_{\vec{p},\vec{x}} = -\left.\frac{\partial L}{\partial t}\right|_{\vec{p},\vec{x}} \tag{15.17.4}$$

If L is not explicitly time dependent, H coincides with the conserved energy, e.g., for a single particle

$$H = mv^2 - L = 2T - (T - V) = T + V \tag{15.17.5}$$

Any function of position and momentum variables, $F(x_m, p_m)$ further satisfies the evolution equation

$$\frac{dF}{dt}=\frac{\partial F}{\partial t}+\sum_{m=1}^{N}\left(\frac{\partial F}{\partial x_m}\frac{dx_m}{dt}+\frac{\partial F}{\partial p_m}\frac{dp_m}{dt}\right)=\frac{\partial F}{\partial t}+\sum_{m=1}^{N}\left(\frac{\partial F}{\partial x_m}\frac{\partial H}{\partial p_m}-\frac{\partial F}{\partial p_m}\frac{\partial H}{\partial x_m}\right)\equiv\frac{\partial F}{\partial t}+\{F,H\} \tag{15.17.6}$$

where $\{F, H\}$ is termed the *Poisson bracket* of F and H. Some properties of the Poisson bracket are

$$\{A+B,C\}=\{A,C\}+\{B,C\},\quad \{AB,C\}=A\{B,C\}+\{A,C\}B$$

$$\{x_j,p_k\}=\sum_{m=1}^{N}\left(\frac{\partial x_j}{\partial x_m}\frac{dp_k}{dp_m}-\frac{\partial x_j}{\partial p_m}\frac{dp_k}{dx_m}\right)=\delta_{jk},\quad \{x_j,x_k\}=\{p_j,p_k\}=0$$

$$\{x_j,F\}=\frac{\partial F}{\partial p_j},\quad \{p_j,F\}=-\frac{\partial F}{\partial x_j} \tag{15.17.7}$$

To change the generalized coordinates employed to describe a system, a term $dG(x, X, t)/dt$ containing a new coordinate X can be added to L without altering the Lagrange equations since, specializing for simplicity to one dimension,

$$\begin{aligned}\frac{d}{dt}\left(\frac{\partial}{\partial \dot{x}}\left(L+\frac{dG}{dt}\right)\right)&=\frac{d}{dt}\left(\frac{\partial}{\partial \dot{x}}\left(L+\frac{\partial G(x,X,t)}{\partial t}+\dot{x}\frac{\partial G(x,X,t)}{\partial x}+\dot{X}\frac{\partial G(x,X,t)}{\partial X}\right)\right)\\&=\frac{d}{dt}\left(\frac{\partial L}{\partial \dot{x}}\right)+\frac{d}{dt}\frac{\partial G(x,X,t)}{\partial x}\\&=\frac{\partial}{\partial x}\left(L+\frac{dG(x,X,t)}{dt}\right)\end{aligned} \tag{15.17.8}$$

Two Hamiltonian functions $H(x, p)$ and $\tilde{H}(X,P)$ of different coordinates that refer to the same physical system can therefore be related by

$$L=\dot{x}p-H(x,p)=\dot{X}P-\tilde{H}(X,P)+\frac{dG}{dt}=\dot{X}P-\tilde{H}(X,P)+\frac{\partial G}{\partial t}+\dot{x}\frac{\partial G}{\partial x}+\dot{X}\frac{\partial G}{\partial X} \tag{15.17.9}$$

if the *generating function* $G(x, X, t)$ is determined from the *canonical transformation*

$$\frac{\partial G}{\partial x}=p,\quad \frac{\partial G}{\partial X}=-P,\quad \tilde{H}=H+\frac{\partial G}{\partial t} \tag{15.17.10}$$

The Poisson bracket of any two functions of (x, p) remains invariant when the functions and Poisson bracket are evaluated in (X, P). Canonical transformations are often employed to describe a system in terms of *ignorable coordinates* with zero time derivatives that correspond to constants of the motion.

Example

For a harmonic oscillator $H(x, p) = p^2/2m + m\omega_0^2 x^2/2$, the generating function $G = (m\omega_0 x^2 \cot X)/2$ yields

$$\begin{aligned} P &= -\frac{\partial G}{\partial X} = \frac{m\omega_0 x^2}{2}\csc^2 X & & x = \sqrt{\frac{2P}{m\omega_0}}\sin X \\ & & \Rightarrow & \\ p &= \frac{\partial G}{\partial x} = m\omega_0 x \cot X & & p = \sqrt{2m\omega_0 P}\cos X \end{aligned} \tag{15.17.11}$$

Since the partial derivative of G with respect to time is zero,

$$\tilde{H}(X,P) = H(x,p) = \omega_0 P\left(\cos^2 X + \sin^2 X\right) = \omega_0 P \tag{15.17.12}$$

Hence, $\dot{X} = \omega_0$ and $X = \omega_0 t + \theta_0$ are ignorable, while $P = H/\omega_0 = E/\omega_0$ leading to

$$x = \sqrt{\frac{2E}{m\omega_0^2}}\sin(\omega_0 t + \theta_0) \tag{15.17.13}$$

A further Legendre transformation leads to the X-independent *Hamilton's principal or phase function*

$$S(x,P,t) = G(x,X,t) + PX \tag{15.17.14}$$

with

$$\begin{aligned} \frac{\partial S(x,P,t)}{\partial X} &= \frac{\partial G(x,X,t)}{\partial X} + P = -P + P = 0 \\ \frac{\partial S(x,P,t)}{\partial x} &= \frac{\partial G(x,X,t)}{\partial x} + \frac{\partial (XP)}{\partial x} = p \\ \frac{\partial S(x,P,t)}{\partial P} &= \frac{\partial G(x,X,t)}{\partial P} + \frac{\partial (XP)}{\partial P} = X \\ \bar{H}(X,P,t) &= H(x,p,t) + \frac{\partial S}{\partial t} \end{aligned} \tag{15.17.15}$$

The phase function can be employed to specify a system for which all coordinates and momenta, X, P, are ignorable, while $\bar{H}$ does not depend on time and can be set to zero by redefining the zero of energy. This yields the *Hamilton–Jacobi equation*

$$H(x,p,t) = H\left(x, \frac{\partial S}{\partial x}, t\right) = -\frac{\partial S}{\partial t} \tag{15.17.16}$$

For a time-independent Hamiltonian, H again represents the conserved energy, E, of the system:

$$H\left(x, \frac{\partial S}{\partial x}\right) = E \tag{15.17.17}$$

and therefore, where $W(x, P)$ is termed *Hamilton's characteristic function*,

$$S(x, P, t) = -Et + W(x, P) \tag{15.17.18}$$

Example

For a harmonic oscillator, the Hamilton–Jacobi equation takes the form

$$\frac{1}{2m}\left(\frac{\partial W}{\partial x}\right)^2 + \frac{1}{2}m\omega_0^2 x^2 = E = \frac{1}{2}m\omega_0^2 A^2 \tag{15.17.19}$$

yielding

$$p = \frac{\partial W}{\partial x} = \sqrt{m^2\omega_0^2(A^2 - x^2)} \tag{15.17.20}$$

and hence

$$W(x, P) = \int_{x_0}^{x} p\,dx = \int_{x_0}^{x} \sqrt{m^2\omega_0^2(A^2 - x^2)}\,dx \tag{15.17.21}$$

The *action variable* associated with p (note that the integral corresponds to the area under a circle),

$$2\pi I \equiv \oint p\,dx = 2\int_{-A}^{A} \sqrt{m^2\omega_0^2(A^2 - x^2)}\,dx = 2\pi m\omega_0 \pi A^2 = \frac{2\pi E}{\omega_0} \tag{15.17.22}$$

is then a constant of the motion and, while possessing dimensions of $[MD^2/T]$, can be employed in place of the constant P. The conjugate coordinate is then similarly set to a constant of the motion by

$$X = \text{const} = \frac{\partial S(x, I, t)}{\partial I} = \frac{\partial W(x, I)}{\partial I} - \frac{\partial E}{\partial I}t \tag{15.17.23}$$

The advantage with the above definition of I is that X is now dimensionless and is thus denoted by θ_0, termed an *angle variable*, while $\partial E/\partial I = (1/m\pi A^2)\partial E/\partial\omega_0 = \omega_0$ from Equation (15.17.19). This uniquely describes the motion according to

$$X = \omega_0 t + \theta_0 = \frac{\partial W(x, I)}{\partial I} = \int_{-\sqrt{\frac{2I}{m\omega_0}}}^{x} \left(m\omega_0\sqrt{\frac{2I}{m\omega_0} - x^2}\right)^{-1} dx = \sin^{-1}\left(\sqrt{\frac{m\omega_0}{2I}}x\right) \tag{15.17.24}$$

16

FLUID MECHANICS

While fluids can be exposed to numerous external forces such as pressure, temperature gradients, and electromagnetic forces and can exhibit nonlinearities, multicomponent behavior, and other complex properties, homogeneous fluid motion at low velocities is relatively simply analyzed from Newton's equations and mass conservation. *Steady flow* refers to time-independent fluid motion. *Rotational* flow induces a torque on a small paddle wheel introduced into the fluid, while otherwise the motion can be described by the gradient of a potential function and is termed *irrotational* or *potential flow*. At high velocities, *viscous* (frictional) effects can be substantial. This can generate unstable *turbulent* flow for which a small perturbation can initiate unpredictable changes in the local velocity distribution. Finally, liquid and (particularly low-speed) gas flow can be *compressible*.

16.1 CONTINUITY EQUATION

While fluid motion can be modeled numerically by evolving infinitesimal fluid elements according to Newton's laws of motion, analytic methods introduce continuous density and local velocity distributions, $\rho(\vec{r},t)$ and $\vec{v}(\vec{r},t)$, together with a local scalar pressure, $p(\vec{r},t)$, which is defined by $p = -\vec{F}\cdot\hat{e}_A/A$, where $\vec{F}$ represents the force on a surface element of the fluid of area A and $\hat{e}_A$ is the outward normal to

Fundamental Math and Physics for Scientists and Engineers, First Edition.
David Yevick and Hannah Yevick.

the surface. The unit of pressure $1 \ \text{Pa} = 1 \ \text{N/m}^2$ is termed the *Pascal*. One atmosphere of pressure corresponds to a pressure of $\approx 10^5 \ \text{Pa} \equiv 1 \ \text{bar}$.

Nonviscous, incompressible irrotational, and steady flow can be characterized by continuous *streamlines* directed along the fluid velocity at each point that coincide with the direction of travel of a massless particle. A *tube of flow* is constructed from an arbitrary closed loop in a plane perpendicular to the direction of flow by merging all the streamlines intersected by the loop. Since the local fluid velocity is parallel to the streamline, fluid cannot cross the boundaries of such a tube. Fluid flowing into one end of the tube therefore exits at the opposite end conserving the total mass *flux*, i.e., the mass of fluid flowing through any cross-sectional area of the tube. If the fluid velocity is uniform along each cross section of a tube with varying cross-sectional area, then in terms of two cross sections with areas A_1 and A_2 with normal vectors along the direction of the fluid flow, the *equation of continuity* takes the form

$$\rho_1 v_1 A_1 = \rho_2 v_2 A_2 = \text{constant} \tag{16.1.1}$$

Alternatively, the change of mass within a fixed spatial volume is given by

$$\frac{dM_V}{dt} = \frac{d}{dt}\int_V \rho(\vec{r},t)\, d^3r = \int_V \frac{\partial \rho(\vec{r},t)}{\partial t} d^3r \tag{16.1.2}$$

The partial derivative ensures that $\vec{r}$ is unchanged. If the volume does not contain sources or sinks, this change in mass equals the negative of the integrated mass flux out of the volume

$$\frac{dM_V}{dt} = -\int_{S \subset V} \rho \vec{v} \cdot d\vec{S} = -\int_V \vec{\nabla} \cdot (\rho \vec{v})\, dV \tag{16.1.3}$$

where Gauss's theorem is applied in the second step. Since V is arbitrary,

$$\frac{\partial \rho}{\partial t} = -\vec{\nabla} \cdot (\rho \vec{v}) = -\rho \vec{\nabla} \cdot \vec{v} - \vec{v} \cdot \vec{\nabla} \rho \tag{16.1.4}$$

or

$$\frac{D\rho}{Dt} \equiv \left(\frac{\partial}{\partial t} + \vec{v} \cdot \vec{\nabla}\right)\rho = -\rho \vec{\nabla} \cdot \vec{v} \tag{16.1.5}$$

The *Stokes operator D/Dt*, also often termed the *convective*, *hydrodynamic*, or *substantial derivative*, evaluates to the change in the quantity it acts on per unit time along the system flow. It can also be derived from the multidimensional Taylor expansion

$$\frac{Df}{Dt} = \lim_{\Delta t \to 0} \frac{f(x + v_x \Delta t, y + v_y \Delta t, z + v_z \Delta t, t + \Delta t) - f(x, y, z, t)}{\Delta t} \tag{16.1.6}$$

which is the rate of change in f with time evaluated between $f(\vec{r},t)$ and a second point infinitesimally displaced in both time and in space according to the fluid motion. Indeed, the directional derivative $\vec{v}\cdot\vec{\nabla}f$ for a time-independent function f corresponds to the distance the fluid travels per unit time multiplied by the change per unit distance of f in the direction of the velocity and hence is associated exclusively with the change in f per unit time generated by the fluid motion. If f at a fixed point additionally varies with time, this induces a further time variation that is independent of the fluid motion as evident in Equation (16.1.5).

Since $D\rho/Dt$ represents the change in ρ along the flow, for an incompressible fluid in the absence of sources, $D\rho/Dt=0$ and the continuity equation reduces to $\vec{\nabla}\cdot\vec{v}=0$. Additionally, for steady irrotational flow, $\vec{\nabla}\times\vec{v}=0$. Introducing a potential function, $\vec{v}=-\vec{\nabla}\varpi_v$ then results in Laplace's equation for ϖ_v.

Example

The density of a particular fluid varies with z and t according to $\rho=\rho_0+c_z z+c_t t$. If the fluid is flowing at a velocity of 2.0 m/s in the z-direction, in a frame moving with the fluid, the change of the density with respect to time is $(2(\partial/\partial z)+\partial/\partial t)(\rho_0+c_z z+c_t t)=2c_z+c_t$. If however, the fluid and frame moves at a 90° angle with respect to the z-axis, the density change equals c_t.

16.2 EULER'S EQUATION

Newton's law of motion can be applied to the fluid within each volume element ΔV. For a nonviscous fluid, the force on the fluid element ΔV arises from both external forces $\vec{F}_{\text{external}}=\rho\Delta V\vec{a}_{\text{external}}$ (e.g., for a gravitational field $\vec{a}_{\text{external}}=-g\hat{e}_z$) and from the surrounding pressure. For any constant vector $\vec{c}$ from Gauss's law,

$$\vec{c}\cdot\int_S p\,d\vec{S}=\int_S p\vec{c}\cdot d\vec{S}=\int_V \vec{\nabla}\cdot(\vec{c}p)\,dV=\int_V \vec{c}\cdot\vec{\nabla}p\,dV+\int_V p\,\cancel{\vec{\nabla}\cdot\vec{c}}\,dV=\vec{c}\cdot\int_V \vec{\nabla}p\,dV \tag{16.2.1}$$

which, since $\vec{c}$ is an arbitrary, implies

$$\int_S p\,d\vec{S}=\int_V \vec{\nabla}p\,dV \tag{16.2.2}$$

Since the force exerted by pressure on the surface of an infinitesimal volume ΔV is antialigned to $d\vec{S}$,

$$\vec{F}_{\text{pressure}}=-\int_{S\subset\Delta V} p\,d\vec{S}=-\int_{\Delta V}\vec{\nabla}p\,dV\approx-\vec{\nabla}p\Delta V \tag{16.2.3}$$

Newton's equation of motion then implies that just as an applied force accelerates a particle as it moves along its trajectory, the acceleration of a fluid element along its trajectory from the action of the external forces and internal pressure is described by *Euler's equation* for the fluid velocity field, $\vec{v}(x,y,z,t)$:

$$\begin{aligned}&\rho\Delta V \lim_{\Delta t\to 0}\frac{\vec{v}(x+v_x\Delta t, y+v_y\Delta t, z+v_z\Delta t, t+\Delta t)-\vec{v}(x,y,z,t)}{\Delta t}\\ &\quad=\rho\Delta V\frac{D\vec{v}}{Dt}=\left(-\vec{\nabla}p+\rho\vec{a}_{\text{external}}\right)\Delta V\end{aligned} \tag{16.2.4}$$

Here, the convective derivative incorporates the physical displacement of the fluid element with time.

Example

The velocity field generated by a gravitational field, $\vec{a}_{\text{external}}=-g\hat{e}_z$, equals $\vec{v}=-\hat{e}_z\sqrt{-2gz}$. To obtain this expression, observe that the squared velocity of a fluid element that is initially stationary at $z=t=0$, namely, $v^2=-2gz$, is obtained by equating $\left(\vec{v}\cdot\vec{\nabla}\right)\vec{v}=1/2(\partial v^2/\partial z)\hat{e}_z$ to the gravitational acceleration.

Additionally, in analogy to the motion of an object over a rough surface, two layers of a viscous fluid in relative motion experience a tangential shear force at their interface. The *viscosity* of a linear *Newtonian* fluid moving in the x-direction is thus defined by $\tau=\mu(\partial v_x/\partial y)$ where μ is the (dynamic) viscosity and τ is the shear stress on a layer of fluid or a planar surface with a normal in the direction of the velocity gradient, $\hat{e}_y\partial v_x/\partial y$. Incorporating this viscous force into Euler's equation yields the significantly more complex *Navier–Stokes equation*.

16.3 BERNOULLI'S EQUATION

From the method presented in Section 8.3 together with the relationship $\vec{A}\times\left(\vec{B}\times\vec{C}\right)=\vec{B}\left(\vec{A}\cdot\vec{C}\right)-\vec{C}\left(\vec{A}\cdot\vec{B}\right)$,

$$\begin{aligned}\vec{a}\times\left(\vec{\nabla}_b\times\vec{b}\right)+\vec{b}\times\left(\vec{\nabla}_a\times\vec{a}\right)&=\vec{\nabla}_b\left(\vec{a}\cdot\vec{b}\right)-\vec{b}\left(\vec{a}\cdot\vec{\nabla}_b\right)+\vec{\nabla}_a\left(\vec{a}\cdot\vec{b}\right)-\vec{a}\left(\vec{b}\cdot\vec{\nabla}_a\right)\\ &=\left(\vec{\nabla}_a+\vec{\nabla}_b\right)\left(\vec{a}\cdot\vec{b}\right)-\left(\vec{a}\cdot\vec{\nabla}_b\right)\vec{b}-\left(\vec{b}\cdot\vec{\nabla}_a\right)\vec{a}\end{aligned}$$

$$\vec{a}\times\left(\vec{\nabla}\times\vec{b}\right)+\vec{b}\times\left(\vec{\nabla}\times\vec{a}\right)=\vec{\nabla}\left(\vec{a}\cdot\vec{b}\right)-\left(\vec{a}\cdot\vec{\nabla}\right)\vec{b}-\left(\vec{b}\cdot\vec{\nabla}\right)\vec{a} \tag{16.3.1}$$

Accordingly, for steady-state flow, $\partial \vec{v}/\partial t = 0$, substituting $\vec{v}\cdot\vec{\nabla}\vec{v} = \vec{v}\times(\vec{v}\times\vec{\nabla}) - \vec{\nabla}v^2/2$ in Euler's equation together with $\vec{\nabla}(p/\rho) = \vec{\nabla}p/\rho - p\vec{\nabla}\rho/\rho^2$ and $\vec{a}_{\text{external}} = \vec{f}_{\text{external}} = -\vec{\nabla}u_{\text{external}}$, where $\vec{f}$ and u denote the force *field* and *potential*, yields

$$\vec{\nabla}\left(\frac{1}{2}v^2 + \frac{p}{\rho} + u_{\text{external}}\right) = \vec{v}\times\left(\vec{\nabla}\times\vec{v}\right) - \frac{p\vec{\nabla}\rho}{\rho^2} \quad (16.3.2)$$

For *irrotational, potential flow,* $\vec{\nabla}\times\vec{v} = 0$, while for incompressible fluids, $\vec{\nabla}\rho = 0$ yielding *Bernoulli's equation*

$$\frac{1}{2}v^2 + \frac{p}{\rho} + u_{\text{external}} = \text{constant} \quad (16.3.3)$$

Hence, liquids or gases traveling at high speeds experience reduced pressure. For a gravitational potential $\phi = gz$, Equation (16.3.3) can be more simply derived by considering a pipe or a tube of flow with cross-sectional areas and heights A_1, h_1 and A_2, h_2 and pressures p_1 and p_2 (both directed into the fluid) at its two ends. Displacing the fluid at the lower end by Δz_1 moves the fluid at the opposing, upper, end by Δz_2 where by conservation of mass $\Delta m = \rho\Delta V_{\text{fluid}} = \rho A_1|\Delta z_1| = \rho A_2|\Delta z_2|$ is identical at both ends. Equating the work done by pressure and gravity to the kinetic energy change of the fluid,

$$p_1A_1\Delta z_1 - p_2A_2\Delta z_2 - \Delta mg(h_2 - h_1) = \frac{1}{2}\Delta m\left(v_2^2 - v_1^2\right) \quad (16.3.4)$$

With $\rho A_j\Delta z_j = \Delta m$ for $j = 1, 2$, Equation (16.3.4) reproduces Bernoulli's equation.

Examples

1. For a constant density fluid at rest, Bernoulli's equation implies that $\Delta p = -\rho g\Delta z$. Accordingly, for a compressible media such as air with $\rho = kp$, $p(z) = p_0\exp(-kgz)$.

2. If a ball spins such that its upper surface moves more rapidly opposite the direction of travel than the lower surface, the speed of the air relative to the ball surface is higher and hence the pressure is relatively lower above the ball creating lift. Similarly, an airplane wing is curved such that air traveling above the wing travels farther than the air below the wing. Since the two streams recombine at the end of the wing to avoid a vacuum, the air above the wing again travels faster.

17

SPECIAL RELATIVITY

Special relativity expresses the *invariance of the vacuum speed of light, c, and of physical laws in all inertial frames*, e.g., frames in which external forces are absent. This was first evidenced in the *Michelson–Morley experiment* in which a light beam is divided by a half-silvered mirror into two identical beams, propagating respectively along and perpendicular to the direction of the earth's rotation. After reflection, the beams are recombined, their interference transforming the phase difference generated by the differing travel times along the paths into an intensity signal. It was expected that the phase of the beam propagating in the direction of the earth's rotation would be retarded against the direction of the earth's rotation and advanced in the reverse direction, resulting on average in a longer travel time and hence net phase delay relative to the transverse beam. Rotating the apparatus by 90° would then reverse the sign of this incremental phase difference. The measured interference pattern instead remained constant, indicating that c does not depend on the propagation direction relative to the earth's velocity.

17.1 FOUR-VECTORS AND LORENTZ TRANSFORMATION

The Galilean relativity principle postulates that physical laws are identical for two observers moving at a constant relative velocity. Mathematically, in two dimensions, this law is evident from the invariance of Newton's equations under the forward

Fundamental Math and Physics for Scientists and Engineers, First Edition.
David Yevick and Hannah Yevick.

(direct) and reverse (inverse) coordinate transformations (conventions differ as to which set of equations is termed direct):

$$\begin{aligned} x &= x' + vt' \quad & x' &= x - vt \\ t &= t' \quad & t' &= t \end{aligned} \tag{17.1.1}$$

between the unprimed frame and a primed frame moving at a relative velocity $+v$. The Maxwell equations, which incorporate first derivatives of the individual fields, are not invariant under the above transformation; rather, they are invariant under the Lorentz transformation, Equation (17.1.5). Einstein accordingly postulated that Newton's laws must be modified such that they were similarly invariant under Lorentz transformations. Indeed, the independence of c on the reference frame relates spatial and time intervals such that the time interval Δt between two events in the unprimed frame is different from that of $\Delta t'$ between the same two events in the primed frame. For example, if in a given coordinate frame F light emitted from a source point, x_s, is detected simultaneously at two receivers, $P_{r,\text{right}}$ at $x_s + a$ and $P_{r,\text{left}}$ at $x_s - a$, equally spaced from the source, since light also travels at speed c in a frame F' moving at a velocity v in the $+x$ direction relative to F, in F', the light arrives first at the point, $P_{r,\text{right}}$, moving toward the source and only subsequently arrives at $P_{r,\text{left}}$. These events therefore differ in time by an amount proportional to their physical separation.

To incorporate the constant speed of light in all inertial frames, the Galilean transformation must be modified. Since the Galilean result must however be recovered at small, nonrelativistic velocities $v \ll c$ while the forward and reverse transformations differ only by the choice of reference frame and the transformation should be linear with respect to both space and time, the simplest alternative is

$$x = \gamma(v)(x' + vt') \quad x' = \gamma(v)(x - vt) \tag{17.1.2}$$

For an infinitely short light pulse, $t = x/c$ in F, while $t' = x'/c$ in F'. Combining the equations according to $x = \gamma(1 + v/c)x' = \gamma^2(1 - v^2/c^2)x$ gives

$$\gamma = \frac{1}{\sqrt{1 - \dfrac{v^2}{c^2}}} \tag{17.1.3}$$

To find the time transformation law, observe that

$$\begin{aligned} x' &= \gamma(\gamma(x' + vt') - vt) \\ t &= \gamma t' - \frac{x'}{v\gamma}\left(1 - \left(1 - \frac{v^2}{c^2}\right)^{-1}\right) = \gamma\left(t' + \frac{vx'}{c^2}\right) \end{aligned} \tag{17.1.4}$$

Since F moves at a relative velocity of $-v$ with respect to F', the inverse Lorentz transformation is obtained by interchanging primed by unprimed quantities and v by $-v$ in the above equation. The three-dimensional Lorentz transformation for relative motion in the x-direction is, with $\beta_< \equiv v/c < 1$ and $\gamma_> \equiv (1 - \beta^2)^{-1/2} > 1$,

$$x = \gamma_>(x' + \beta_< ct'),\ y = y',\ z = z',\ ct = \gamma_>(ct' + \beta_< x') \qquad (17.1.5)$$

The quantities x and t are imaginary for $v > c$, indicating that physical velocities cannot exceed c.

An *event coordinate* (x, ct) is defined as the spatial point and c multiplied by the time at which a certain event occurs. The invariance of the speed of light in any inertial frame implies that if a light signal is respectively emitted and detected at $(\vec{x}_s, ct_s)$ and $(\vec{x}_r, ct_r)$ in a frame F, then with $(\Delta\vec{r}, c\Delta t) = (\vec{x}_r - \vec{x}_s, c(t_r - t_s))$ while primed quantities denote the corresponding values in F',

$$(\Delta\tilde{s})^2 \equiv (c\Delta t)^2 - \Delta\vec{r}\cdot\Delta\vec{r} = (\Delta\tilde{s}')^2 = (c\Delta t')^2 - \Delta\vec{r}'\cdot\Delta\vec{r}' = 0 \qquad (17.1.6)$$

More generally, the Lorentz transformation, expressed in terms of space and time intervals,

$$\Delta x = \gamma_>(\Delta x' + \beta_< c\Delta t'),\ \Delta y = \Delta y',\ \Delta z = \Delta z',\ c\Delta t = \gamma_>(c\Delta t' + \beta_< \Delta x') \qquad (17.1.7)$$

similarly preserves the space–time interval $(\Delta s)^2$ between any two event coordinates, which is therefore termed a *relativistic scalar quantity*.

17.2 LENGTH CONTRACTION, TIME DILATION, AND SIMULTANEITY

An object of length Δx_{rest} in its rest frame, F_{rest}, where the positions of its two ends of the object are measured simultaneously with $\Delta t = 0$, possesses a length $\Delta x'$ in F' measured with $\Delta t' = 0$. Inserting $\Delta t' = 0$ into Equation (17.1.7) yields (a common error is to associate identify correctly $\Delta x'$ in Eq. 17.1.7 with $\Delta t' = 0$ as the rest length; however, the observations in F are then taken with $\Delta t \neq 0$)

$$\Delta x' = \frac{\Delta x_{rest,\,\Delta t'=0}}{\gamma_>} \le \Delta x_{rest,\,\Delta t'=0} = \sqrt{1 - \frac{v^2}{c^2}}\Delta x_{rest,\,\Delta t'=0} \qquad (17.2.1)$$

which is termed *Lorentz contraction*. Measurements of the object's front and rear positions that are simultaneous in F however differ in time by $\Delta t'$ in F' according to $\Delta t' + v\Delta x'/c^2 = 0$. The quantity Δx_{rest} is termed the *proper length* since in all frames $(\Delta s)^2 = -(\Delta x_{rest})^2$.

The time interval between two events that occur in the rest frame of a moving object such that a single clock is present at the location of both events and hence $\Delta x' = 0$ is smaller than the *time dilated* interval in F where the events occur at different places. Since

$$\Delta t = \gamma_> \Delta t'_{rest,\,\Delta x'=0} > \Delta t'_{rest,\,\Delta x'=0} \qquad (17.2.2)$$

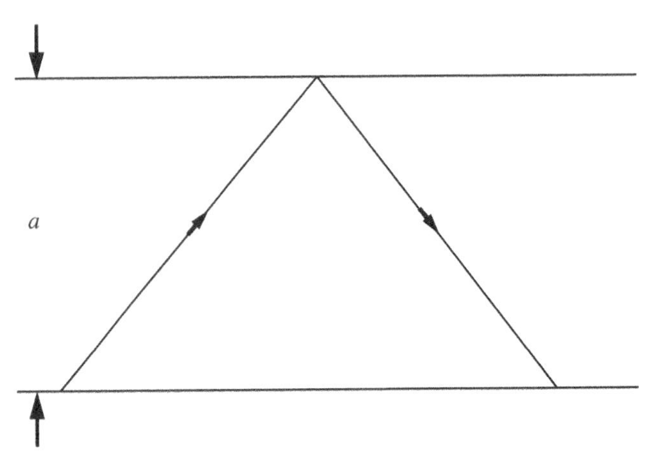

FIGURE 17.1 Light ray in a moving frame.

the quantity $\tau \equiv \Delta t'_{\text{rest}, \Delta x'=0}$, termed the *proper time*, constitutes the smallest time difference between the two events observed in any reference frame. While a spatial volume element is Lorentz contracted by $1/\gamma_>$ in a moving frame, time dilation involves multiplication by $\gamma_>$. Hence, the space–time volume element $d\varsigma = cdtdV$ is invariant and constitutes a scalar under Lorentz transformations.

A second derivation of the above results employs a clock consisting of a vertical light ray that reflects between mirrors along $z = 0$ and $z = a$ in a frame F' traveling at velocity v with respect to frame F. Figure 17.1 shows the resulting path of the light ray as observed from F. The time required for a light beam traveling perpendicular to the mirrors to reflect from the bottom to the top mirror and back in the rest frame is given in the rest frame of the moving clock by $\Delta t'_{\text{rest}, \Delta x'=0} = 2a/c$. However, in the rest frame, F', of the moving clock, in F, the two points at which the beam intersects the bottom mirror are spatially separated such that the beam travels a distance $d = c\Delta t = 2\sqrt{(v\Delta t/2)^2 + a^2} = \sqrt{(v\Delta t)^2 + \left(c\Delta t'_{\text{rest}, \Delta x'=0}\right)^2}$, which reproduces Equation (17.2.2) after solving for Δt (observe that transverse distances are unaffected by longitudinal motion). Further, if a single, stationary point in the moving frame F' passes over an object at rest in a frame F, the length of the object, Δx_{rest}, is $v\Delta t$ as measured in F and $v\Delta t'_{\text{rest}, \Delta x'=0}$ in F'. Consequently, $\Delta x_{\text{rest}, \Delta t=0}/\Delta x' = \Delta t/\Delta t'_{\text{rest}, \Delta x'=0} = \gamma_>$, in agreement with Equation (17.2.1).

Examples

An object at rest with rest length L in a moving frame F' is Lorentz contracted when viewed in a stationary frame F and therefore can seemingly be contained in a smaller region, $R < L$. In F' however, R is instead contracted and therefore cannot enclose the object. However, these statements do not contradict each other as the times at which the front and the back of the object are simultaneously enclosed in R in F are not simultaneous in F'. Similarly, a moving object of rest length L in

F' can drop through a hole of a similar length in frame F in which it is additionally Lorentz contracted. In the frame F' moving along with the object, however, the hole is Lorentz contracted, and hence, the object should not be able to pass through it. Thus, rigidity is not a relativistic invariant as in F' the front end of the object bends downward to enter the hole.

If an object travels to a distant point at a velocity v with respect to a rest frame F and then returns with the same velocity, the time in F advances more rapidly than that experienced by the object as a result of time dilation. conversely, if the moving frame F' is employed as the rest frame, the time in F instead appears retarded during both the outgoing and incoming segments from the transformation inverse to Equation (17.1.7). However, at the midpoint of travel, F' accelerates from $+v$ to $-v$ during which a long period of time elapses at the position of a stationary object in F as observed from F'. While the clocks of F initially appear progressively more retarded at smaller x as viewed from F' with $\Delta t' = 0$ so that $c\Delta t = \beta_{<}\Delta x$ (from the transformation inverse to Eq. 17.1.7 when the velocity is reversed), these clocks instead appear increasingly advanced.

17.3 COVARIANT NOTATION

Just as the manner in which a set of equations transform under rotation or translations can be clarified by recasting the participating quantities as vectors, matrices, or scalars, rewriting expressions in terms of four-dimensional space–time tensor quantities elucidates their transformation properties between different inertial frames. Physical variables that transform according to the standard Lorentz transformation are represented by *contravariant four-vectors* with Greek superscripts, which we here distinguish by column vectors (T denotes transpose) and, when expedient, the subscript C. Thus, the *space–time* four-vector is written as

$$x^{\mu} = \left(x^0, x^1, x^2, x^3\right)_C^T = (ct, x, y, z)_C^T \tag{17.3.1}$$

As well, we introduce *contravariant* quantities with the property that *inner products (contractions) CR or RC can only be formed between a C quantity and a covariant R quantity and yield scalar, frame-independent results*. The *gradient with respect to a contravariant space–time vector is covariant* and is therefore represented below by a row vector and the subscript R since, by the definition of the directional derivative,

$$\Delta f = \frac{\partial f}{\partial x^{\mu}} \Delta x^{\mu} \equiv \left(\partial_{\mu} f\right) \Delta x^{\mu} \tag{17.3.2}$$

where Δf is the change in the scalar function $f(x, t)$ over the displacement Δx^{μ} (the derivative $\partial^{\mu} \equiv \partial / \partial x_{\mu}$ is contravariant). From the two types of vectors, four different matrices or rank two tensors, $A^{\nu\mu}$, A_{μ}^{ν}, A_{μ}^{ν}, and $A_{\nu\mu}$, termed contravariant, mixed, and covariant, respectively, can be defined such that, e.g., the outer product of two contravariant vectors yields a contravariant tensor

$$A^{\mu}B^{\nu} \equiv A_C \otimes B_C \equiv \begin{pmatrix} A_0 \\ A_1 \\ A_2 \\ A_3 \end{pmatrix}_C \otimes \begin{pmatrix} B_0 & B_1 & B_2 & B_3 \end{pmatrix}_C = {}_C\begin{pmatrix} A_0B_0 & A_0B_1 & A_0B_2 & A_0B_3 \\ A_1B_0 & A_1B_1 & \cdots & \\ A_2B_0 & \vdots & \ddots & \\ A_3B_0 & & & \end{pmatrix}_C \tag{17.3.3}$$

Contracting (multiplying), e.g., $A^{\mu\nu}$ with a covariant vector according to $A^{\mu\nu}B_{\nu}$, where a sum is performed for each identical pair of lower and upper indices, results in a contravariant vector, while similarly contracting A^{ν}_{μ} with B_{ν} yields a covariant vector. A matrix corresponding to A^{ν}_{μ} is therefore distinguished with left- and right-hand subscripts R and C, and a dot product (contraction) can only be carried out between an adjacent R and a C subscripted quantity with the two affected subscripts subsequently eliminated. The rank 2 (matrix) *metric tensors* given by

$$g_{\mu\sigma} = {}_R\begin{pmatrix} 1 & 0 & 0 & 0 \\ 0 & -1 & 0 & 0 \\ 0 & 0 & -1 & 0 \\ 0 & 0 & 0 & -1 \end{pmatrix}_R \quad g^{\mu\sigma} = {}_C\begin{pmatrix} 1 & 0 & 0 & 0 \\ 0 & -1 & 0 & 0 \\ 0 & 0 & -1 & 0 \\ 0 & 0 & 0 & -1 \end{pmatrix}_C \tag{17.3.4}$$

are defined such that, e.g., $A^{\mu}g_{\mu\nu}A^{\nu}$ yields the square of the invariant scalar length of the four-vector A^{ν}. As evident from, e.g., direct multiplication, $g_{\mu\nu}g^{\nu\delta} = \delta^{\delta}_{\mu}$ where δ^{δ}_{μ} represents the 4×4 identity matrix for which $x_{\mu} = \delta^{\delta}_{\mu}x_{\delta}$.

A linear transformation $x_i = \sum_j c_{ij}x'_j$ illustrated by the system of two linear equations below can be written in matrix form as, where the subscript x in $\vec{\nabla}'_x$ indicates that $\vec{\nabla}$ must be rearranged to act exclusively on x,

$$\begin{pmatrix} x_0 \\ x_1 \end{pmatrix} = \begin{pmatrix} \partial'_0 x_0 & \partial'_1 x_0 \\ \partial'_0 x_1 & \partial'_1 x_1 \end{pmatrix} \begin{pmatrix} x'_0 \\ x'_1 \end{pmatrix} = \left(\vec{x} \otimes \vec{\nabla}'_x\right)\vec{x}' = \left(\vec{\nabla}' \otimes \vec{x}\right)^T \vec{x}' \tag{17.3.5}$$

From this, the Lorentz transformation or *boost* of Equation (17.1.5) is represented by the tensor

$$L^{\nu}{}_{\mu} = {}_C\begin{pmatrix} \gamma & \beta\gamma & 0 & 0 \\ \beta\gamma & \gamma & 0 & 0 \\ 0 & 0 & 1 & 0 \\ 0 & 0 & 0 & 1 \end{pmatrix}_R \tag{17.3.6}$$

with $x^{\nu} = L^{\nu}{}_{\mu} x'^{\mu}$ and

$$L^{\nu}{}_{\mu} = \frac{\partial x^{\upsilon}}{\partial x'^{\mu}} \tag{17.3.7}$$

A rotation through, e.g., an angle θ around the z-axis is similarly described by

$$R^{\nu}{}_{\mu} = {}^{C}\!\begin{pmatrix} 1 & 0 & 0 & 0 \\ 0 & \cos\theta & -\sin\theta & 0 \\ 0 & \sin\theta & \cos\theta & 0 \\ 0 & 0 & 0 & 1 \end{pmatrix}_{R} \tag{17.3.8}$$

17.4 CASUALITY AND MINKOWSKI DIAGRAMS

Space–time events in a reference frame F can be described by points in four-dimensional *Minkowski* space, a two-dimensional, (ct, x), cross section of which is normally depicted as in Figure 17.2. The path or "world line" of an object with a constant x-velocity v in F is the line, $ct = \beta x$, corresponding to the spatial, x', axis of its rest

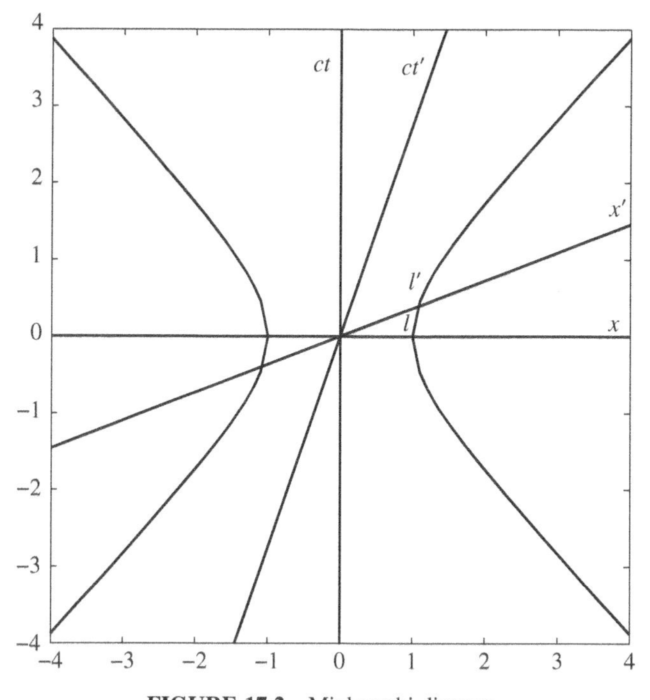

FIGURE 17.2 Minkowski diagram.

frame F'. The ct' axis of F', given by $x' = x - \beta ct' = 0$, similarly corresponds to the line $ct = cx/v = x/\beta$. Consequently, the primed axes possess reciprocal slopes and describe equal angles with respect to the rectangular axes of F. The coordinates (ct', x') in F' of an event given by (ct, x) in F are obtained graphically as the intercepts of the lines through (t, x) parallel to the x' and ct' axes with these axes. Intervals in a Minkowski diagram equal $s^2 = c^2t^2 - x^2$; hence, if an object at rest in the frame F possesses a length l, its length in a frame F' moving at a velocity v relative to F is given by the (larger) value, l', of the point of intersection, r in Figure 17.2, of the x' axis of F' with the hyperbola $s^2 = -l^2$.

An event occurring at the origin of a Minkowski diagram can communicate with a second event situated anywhere along the lines $x = \pm ct$, termed the *light cone*, through the propagation of a light ray. If the coordinates of the second event are situated below the light cone, a frame F' exists for which the x' axis passes through this event. In this frame, both events occur at the same time, and the interval between the events is accordingly termed *space-like*. In frames with respectively smaller and larger velocities than F', the second event, if located in the first quadrant, occurs after and before the event at the origin. As a result, for a space-like interval, unless the frame in which the two events occurred is specified, a conclusion cannot be drawn regarding which event could have caused the other event. An interval from the origin to a second event above the light cone is termed *time-like*. While a frame then exists in which both events occur at the same place, for all frames, the second event follows the first.

17.5 VELOCITY ADDITION AND DOPPLER SHIFT

Classically, if an object moves at a velocity $\vec{v}_{\text{object}}$ relative to an inertial frame, $F_{\text{object frame}}$, that in turn moves with a velocity $\vec{v}_{\text{object frame}}$, relative to a second inertial frame, F_{observer} an observer in the second frame measures an object velocity $\vec{v}_{\text{object, observer}} = \vec{v}_{\text{object}} + \vec{v}_{\text{object frame}}$. However, this would imply the existence of object velocities greater than c as viewed from the second frame. Relativistically, for $\vec{v}_{\text{object}}$ and $\vec{v}_{\text{object frame}}$ along the x-direction, if a particle's coordinates measured in $F_{\text{object frame}}$ change by an amount dx_{object} over an infinitesimal time interval dt_{object}, Lorentz transforming into F_{observer} yields

$$\begin{aligned} dx_{\text{object, observer}} &= \gamma_{\text{object frame}}\left(dx_{\text{object}} + v_{\text{object frame}}\, dt_{\text{object}}\right) \\ dt_{\text{object, observer}} &= \gamma_{\text{object frame}}\left(dt_{\text{object}} + \frac{v_{\text{object frame}}}{c^2} dx_{\text{object}}\right) \end{aligned} \tag{17.5.1}$$

so that

$$v_{x,\text{object,observer}} = \frac{dx_{\text{object,observer}}}{dt_{\text{object,observer}}} = \frac{v_{x,\text{object}} + v_{\text{object frame}}}{1 + \dfrac{v_{x,\text{object}} v_{\text{object frame}}}{c^2}} \tag{17.5.2}$$

which cannot exceed c. Because the form of Equation (17.5.2) is identical to that of

$$\tanh(a+b) = \frac{\sinh(a+b)}{\cosh(a+b)} = \frac{\sinh a \cosh b + \cosh a \sinh b}{\cosh a \cosh b + \sinh a \sinh b} = \frac{\tanh a + \tanh b}{1 + \tanh a \tanh b} \quad (17.5.3)$$

defining the *rapidity* $\chi_{v/c}$ by $v/c = \tanh\chi$ yields $\chi_{\text{object, observer}} = \chi_{\text{object}} + \chi_{\text{object frame}}$. Additionally, from $dz_{\text{object, observer}} = dz_{\text{object}}$,

$$v_{z,\text{object, observer}} = \frac{dz_{\text{object, observer}}}{dt_{\text{object, observer}}} = \frac{v_{z,\text{object}}\sqrt{1 + \dfrac{v^2_{\text{object frame}}}{c^2}}}{1 + \dfrac{v_{x,\text{object}} v_{\text{object frame}}}{c^2}} \quad (17.5.4)$$

If the object travels at an angle θ_{object} with respect to the x-axis (e.g., $\sin\theta_{\text{object}} = v_{z,\text{object}}/v_{\text{object}}$) in frame F_{object}, dividing Equation (17.5.4) by Equation (17.5.2) and then further dividing the numerator and denominator of the right-hand side of the resulting expression by $v_{\text{object}} = \sqrt{v_{z,\text{object}}{}^2 + v_{x,\text{object}}{}^2}$ yield in F_{observer}

$$\tan\theta_{\text{object, observer}} = \frac{\sqrt{1 - \dfrac{v^2_{\text{object frame}}}{c^2}}}{\cos\theta_{\text{object}} + \dfrac{v_{\text{object frame}}}{v_{\text{object}}}} \sin\theta_{\text{object}} \quad (17.5.5)$$

The corresponding formula for a light ray is obtained by setting $v_{\text{object}} = c$ in the above equation.

The *relativistic Doppler effect* (in one spatial dimension) arises when a source in a frame F_{source} emits light with frequency f_{source} while traveling at a velocity v relative to a receiver that is stationary in a frame F_{observer}. Since the phase of a wave, $\phi = \vec{k} \cdot \vec{r} - \omega t$, divided by 2π yields the number of waves between two phase fronts that is identical in any inertial frame, $(\omega/c, \vec{k})_c^T$ comprises a four-vector and therefore $\omega_{\text{observer}} = \gamma(\omega_{\text{source}} + \beta c k_{\text{source}})$. For a light beam with $\omega_{\text{source}} = 2\pi f_{\text{source}} = c k_{\text{source}}$, $\omega_{\text{observer}} = \gamma(1+\beta)\omega_{\text{source}}$, yielding for the frequency detected in F_{observer}

$$f_{\text{observer}} = f_{\text{source}}\sqrt{\frac{1+\beta}{1-\beta}} \quad (17.5.6)$$

17.6 ENERGY AND MOMENTUM

Since the infinitesimal metric $ds = c\,d\tau = c\,dt/\gamma$ reduces for small velocities to $c\,dt$, a velocity four-vector with a spatial component that coincides with γ times the non-relativistic velocity, $\vec{v}_{\text{NR}} = d\vec{r}/dt$, is defined as

$$v^{\mu}=c\frac{dx^{\mu}}{ds}=\gamma\left(c,\vec{v}_{\mathrm{NR}}\right)_{C}^{T} \tag{17.6.1}$$

Multiplying the velocity four-vector by mc then yields the corresponding energy–momentum four-vector (E and p are the *relativistic* energy and momentum), which is also often written without the additional factor of c:

$$p^{\mu}=mcv^{\mu}=\gamma\left(mc^{2},\vec{p}_{\mathrm{NR}}c\right)_{C}^{T}\equiv\left(E,\vec{p}c\right)_{C}^{T} \tag{17.6.2}$$

with $p^{\mu}p_{\mu}=m^{2}c^{4}$, which for a photon with zero rest mass $m_{\gamma}=0$ yields $E=pc$. For small v, a Taylor series expansion yields for p^{0}

$$mc^{2}\left(1-\frac{v^{2}}{c^{2}}\right)^{-\frac{1}{2}}\rightarrow mc^{2}+\frac{1}{2}mv^{2}+\ldots \tag{17.6.3}$$

corresponding to the sum of the classical kinetic energy of the particle and its *rest energy*, which quantifies the maximum energy that can be extracted from a stationary particle of a given mass.

The momentum and energy of an isolated particle can also be obtained by replacing the nonrelativistic free particle Lagrangian $L=T-U=T=mv^{2}/2$ by the Lorentz scalar

$$L_{\mathrm{relativistic}}=-mc^{2}+\frac{1}{2}mv^{2}\rightarrow-mc^{2}\sqrt{1-\frac{v^{2}}{c^{2}}}\approx-\sqrt{p_{\mu}p^{\mu}} \tag{17.6.4}$$

The action is then

$$S=\int_{t_{1}}^{t_{2}}L_{\mathrm{relativistic}}dt=-mc\int_{t_{1}}^{t_{2}}\sqrt{c^{2}-v^{2}}dt=-mc\int_{x^{\mu}(t_{1})}^{x^{\mu}(t_{2})}ds \tag{17.6.5}$$

This yields $\vec{p}=\vec{\nabla}_{v}L_{\mathrm{relativistic}}=m\vec{v}/\sqrt{1-v^{2}/c^{2}}$, and energy $E=\vec{p}\cdot\vec{v}-L_{\mathrm{relativistic}}=mc^{2}/\sqrt{1-v^{2}/c^{2}}$ as above, while consistent with $E^{2}=p^{2}c^{2}+m^{2}c^{4}$, the Hamiltonian is given by $H_{\mathrm{relativistic}}=\sqrt{p^{2}c^{2}+m^{2}c^{4}}$.

Finally, the force four-vector satisfies the relativistic analogue of Newton's equations, namely,

$$f^{\mu}=\frac{dp^{\mu}}{ds}=\frac{\gamma\,dp^{\mu}}{c\;dt}=\gamma\left(\frac{1}{c}\frac{dE}{dt},\frac{d\vec{p}}{dt}\right)_{C}^{T} \tag{17.6.6}$$

For a constant force $\vec{F}$ acting on an initially stationary object in an inertial frame, $d\vec{p}=\vec{F}dt$ from the spatial part of Equation (17.6.6), so that $\vec{F}t=\vec{p}=m\vec{v}/\sqrt{1-v^{2}/c^{2}}$. Solving for $\vec{v}$,

$$m\vec{v} = \frac{\vec{F}t}{\sqrt{1+\left(\frac{Ft}{mc}\right)^2}} \tag{17.6.7}$$

Thus, while the particle momentum increases linearly with time, the velocity approaches c asymptotically for large times.

When the force and velocity are not parallel, $d\vec{p}/dt$ and $d\vec{v}/dt$ are related by

$$\frac{d\vec{p}}{dt} = \frac{d}{dt}\frac{m\vec{v}}{\sqrt{1-\frac{\vec{v}\cdot\vec{v}}{c^2}}} = \frac{m}{\left(1-\frac{\vec{v}\cdot\vec{v}}{c^2}\right)^{\frac{3}{2}}}\left(\frac{d\vec{v}}{dt}\left[1-\frac{\vec{v}\cdot\vec{v}}{c^2}\right] + \vec{v}\left[\frac{d\vec{v}}{dt}\cdot\vec{v}\right]\right) \tag{17.6.8}$$

If $d\vec{v}/dt$ and $\vec{v}$ are parallel, the dot products are replaced by ordinary products yielding

$$\left.\frac{d\vec{p}}{dt}\right|_{\parallel} = m\gamma^{\frac{3}{2}}\frac{d\vec{v}}{dt} \tag{17.6.9}$$

while if they are perpendicular,

$$\left.\frac{d\vec{p}}{dt}\right|_{\perp} = m\gamma\frac{d\vec{v}}{dt} \tag{17.6.10}$$

Further, since $E^2 = c^2p^2 + m^2c^4$,

$$2E\frac{dE}{dt} = 2c^2\vec{p}\cdot\frac{d\vec{p}}{dt} \tag{17.6.11}$$

or, dividing by $2E$,

$$\frac{dE}{dt} = \frac{c^2}{E}\vec{p}\cdot\frac{d\vec{p}}{dt} = \vec{v}\cdot\vec{F} \tag{17.6.12}$$

so that

$$dE = \vec{F}\cdot\vec{v}\,dt = \vec{F}\cdot d\vec{r} = dW \tag{17.6.13}$$

demonstrating that the work–energy theorem $W = \Delta E$ can be applied relativistically.

Relativistic energy–momentum conservation requires that the sum of the four-vectors p^μ of all interacting particles is conserved in collisions if the external force on a system $f^\mu = dp^\mu/dt = 0$; i.e.,

$$\sum_{m\in\text{incoming}} p_m^\mu = \sum_{n\in\text{outgoing}} p_n^\mu \tag{17.6.14}$$

These equations are often solved by contracting with appropriate covariant momentum four-vectors and employing the independence of the resulting scalar quantity on the reference frame.

Examples

If particles of mass m_1 and m_2 are generated by the decay of a particle of mass $m > m_1 + m_2$,

$$\begin{pmatrix} mc^2 \\ 0 \end{pmatrix} = \begin{pmatrix} E_1 \\ cp_1 \end{pmatrix} + \begin{pmatrix} E_2 \\ cp_2 \end{pmatrix} \tag{17.6.15}$$

Hence, $mc^2 = E_1 + E_2$ and $\vec{p}_1 = -\vec{p}_2$. However, $c^2p_1^2 = c^2p_2^2$ implies, since $E^2 - c^2p^2 = m^2c^4$, $E_1^2 - m_1^2c^4 = E_2^2 - m_2^2c^4 = (mc^2 - E_1)^2 - m_2c^4$ so that $E_1 = c^2(m^2 + m_1^2 - m_2^2)/2m$.

For an inelastic collision between a particle with energy E_1 and mass m_1 with a second stationary particle of mass m_2 that generates a particle with mass m and momentum and energy E and p,

$$\begin{pmatrix} E \\ cp \end{pmatrix} = \begin{pmatrix} E_1 \\ cp_1 \end{pmatrix} + \begin{pmatrix} m_2c^2 \\ 0 \end{pmatrix} \tag{17.6.16}$$

Equating the squares, $p_\mu p^\mu$, of both sides of the above equation yields $E^2 - c^2p^2 = m^2c^4 = (E_1 + m_2c^2)^2 - c^2p_1^2 = (E_1 + m_2c^2)^2 - (E_1{}^2 - m_1^2c^4)$ and therefore $m^2c^4 = m_1^2c^4 + m_2^2c^4 + 2m_2c^2E_1$.

Finally, an elastic collision of a particle with mass m_1 with a stationary particle with mass m_2 conserves both particle masses but transforms p_1^μ, p_2^μ to p'^μ_1, p'^μ_2 such that $p_1^\mu + p_2^\mu = p'^\mu_1 + p'^\mu_2$. Since $\vec{p}_2 = 0$,

$$\begin{pmatrix} E_2' \\ c\vec{p}\,'_2 \end{pmatrix} = \underbrace{\begin{pmatrix} E_1 \\ c\vec{p}_1 \end{pmatrix}}_{p_1^\mu} + \underbrace{\begin{pmatrix} m_2c^2 \\ 0 \end{pmatrix}}_{p_2^\mu} - \underbrace{\begin{pmatrix} E_1' \\ c\vec{p}\,'_1 \end{pmatrix}}_{p_3^\mu} \tag{17.6.17}$$

Squaring both sides of this equation results in

$$m_2^2c^4 = \underbrace{m_1^2c^4}_{p_1^\mu p_{1\mu}} + \underbrace{m_2^2c^4}_{p_2^\mu p_{2\mu}} + \underbrace{m_1^2c^4}_{p_3^\mu p_{3\mu}} + \underbrace{2m_2c^2(E_1 - E_1')}_{p_2^\mu(p_{1\mu} - p_{3\mu})} \underbrace{-2(E_1E_1' - p_1p_1', \cos\theta_1)}_{-p_1^\mu p_{3\mu}} \tag{17.6.18}$$

from which, e.g., the scattering angle, θ_1, of particle 1 can be derived after some algebra.

18

ELECTROMAGNETISM

Charged particles give rise to electric and magnetic fields. As time variations in one of these fields induce changes in the second field, *electromagnetism* is divided into the *electrostatics* and *magnetostatics* of time-independent, uncoupled fields and the *electrodynamics* of time-varying field distributions. While electromagnetic fields in media can experience, e.g., direction-dependent or nonlinear forces, only linear, *isotropic* (uniform) media are considered here.

18.1 MAXWELL'S EQUATIONS

Before electrostatics and electrodynamics are reviewed in detail, the heuristic structure of *Maxwell's equations* is introduced in this section with emphasis on units, which are somewhat nonintuitive. The Maxwell's equations describe the mutual coupling of electromagnetic fields as well as their relation to charge and current *sources*. Their underlying structure is partially clarified by considering incompressible fluid entering at a constant rate into the center of a large, filled container. While the fluid can additionally *circulate*, the amount of fluid flowing per unit time through any sphere of surface area $4\pi r^2$ centered at the source (in fact any surface enclosing the source) equals the rate at which the fluid is injected, independent of the fluid properties. Hence, the *radial* component of the velocity vector *averaged* over a sphere with radius r about the origin varies as $1/r^2$. The *force* that the fluid exerts on a standardized

Fundamental Math and Physics for Scientists and Engineers, First Edition.
David Yevick and Hannah Yevick.

test object however additionally depends on the medium properties, particularly the viscosity and the effective cross-sectional dimension of the object. Dividing the force by this dimension converts the *velocity field* into a second, physically relevant, *force field*. By analogy, in electromagnetism, besides the radial fields generated by isolated charges, in atoms and molecules, positive and negative charges exist in close proximity to each other. Applying, e.g., an electric field to a material separates these charges, creating an array of closely separated positive and negative charge pairs or *dipoles*. As each dipole has zero net charge, the *net* field *sourced* by a dipole is zero over macroscopic distances, but its field lines circulate from the positive to the negative charge, possibly generating a further medium and external field-dependent *circulation*.

Maxwell's equations accordingly comprise *source equations* that relate "source," largely medium-independent unmediated fields analogous to the fluid velocity to the charge and current distributions and equations for "force" fields directly associated with forces. Mediated and unmediated quantities cannot occur in the same equation unless a medium-dependent factor such as the conductivity, electric permittivity, or magnetic permeability is present. Otherwise, changing the material would alter some terms in the equation but not others.

Gauss's law relates free, isolated physical charges to the divergence (but not the curl, e.g., rotation, which can be influenced by the presence of charged dipoles at microscopic length scales) of an *electric displacement* source field:

$$\vec{\nabla} \cdot \vec{D}^{(\text{source})} = \rho_{\text{free}} \tag{18.1.1}$$

In SI (international system) units, charge is measured in Coulombs, with the electronic charge

$$q_{\text{electron}} = -e = 1.6 \times 10^{-19} \,\text{Coulombs}\, [Q] \tag{18.1.2}$$

so that the units of electric displacement are Coulombs per meter squared or

$$\left[\vec{D}\right] = \left[\frac{\text{Coulombs}}{\text{meter}^2}\right] = \left[\frac{Q}{D^2}\right] \tag{18.1.3}$$

The charge per unit volume of a material, ρ_{free}, is termed the *charge density*. For a point charge q at $\vec{r}'$, when the volume element d^3r is centered at r', $\rho_{\text{free}}(\vec{r})d^3r = q$, which is expressed as $\rho_{\text{free}}(\vec{r}) = q\delta(\vec{r} - \vec{r}')$. If the charge distribution is confined to a thin wire, multiplying the charge density by the cross-sectional area of the wire yields the charge per unit length, λ while for a surface charge distribution confined to a layer of thickness Δz, multiplying the charge density by Δz yields the charge per unit area, σ; i.e.,

$$\rho d^3r = \sigma \Delta A = \lambda \Delta L = \Delta q \tag{18.1.4}$$

Thus, by superposition, summing an expression over point charges implies a combination of a volume, surface, and line integrals over ρ, σ, and λ, where charges, respectively, form a continuous density distribution, a thin surface layer, and a thin rod.

Moving charges and a time-varying displacement field provide sources for the *magnetic field*, $\vec{H}$, according to *Ampere's law*:

$$\vec{\nabla}\times\vec{H}^{(\text{source})}=\vec{J}_{\text{free}}+\frac{\partial\vec{D}^{(\text{source})}}{\partial t} \tag{18.1.5}$$

The *free current density* $\vec{J}_{\text{free}}$ is defined as the free charge passing through a unit cross-sectional area per unit time (and thus should actually be termed the free current flux). That is, if ρ_{density} *carriers* per unit volume with charge q flow with a velocity $\vec{v}$ in a material, the charge passing through a surface element ΔA oriented perpendicularly to $\vec{v}$ over a time Δt is given by $\Delta Q=q\rho_{\text{density}}v\Delta A\Delta t$ corresponding to $\vec{J}_{\text{free}}=\Delta Q/(\Delta A\Delta t)=q\rho_{\text{density}}\vec{v}$. For currents confined to a depth Δz below the surface of a material or to a thin wire of cross-sectional area A, the *surface current density* $\vec{K}=\vec{J}\Delta z$ and *current* are related by

$$\vec{J}A=\vec{K}L=I\hat{e}_{\vec{l}}\ [\text{Amperes}]=\left[\frac{Q}{T}\right] \tag{18.1.6}$$

where L is a length element measured along the surface over which the current flows in the direction perpendicular to the current and $\hat{e}_{\vec{l}}$ is a unit vector along the wire in the direction that the current flows so that $I\hat{e}_{\vec{l}}dl=Id\vec{l}$ where $d\vec{l}$ is a differential length element along the wire. Inside an integral, the above relationship is expressed as

$$\vec{J}d^3r=\vec{K}dA=Id\vec{l} \tag{18.1.7}$$

As the units of J_{free} are [Amperes/m^2], those of the magnetic field are

$$\left[\vec{H}^{(\text{source})}\right]=\left[\frac{\text{Amperes}}{\text{meter}}\right]=\left[\frac{Q}{DT}\right] \tag{18.1.8}$$

Corresponding to the source fields are force fields that are affected by material properties and are directly related to forces by the *Maxwell force law*:

$$\vec{F}=q\left(\vec{E}^{(\text{force})}+\vec{v}\times\vec{B}^{(\text{force})}\right) \tag{18.1.9}$$

which describes the force on a test charge q moving with velocity $\vec{v}$ in the presence of the *electric field*, $\vec{E}$, and the *magnetic induction*, $\vec{B}$. The units of the electric field and magnetic induction are consequently

$$\begin{aligned}\left[\vec{E}^{(\text{force})}\right]&=\left[\frac{\text{force}}{\text{Columb}}\right]=\left[\frac{MD}{QT^2}\right]\equiv\left[\frac{\text{volts}}{\text{meter}}\right]=\left[\frac{V}{D}\right]\\ \left[\vec{B}^{(\text{force})}\right]&=\left[\frac{\vec{E}^{(\text{force})}}{\vec{v}}\right]=\left[\frac{M}{QT}\right]=\left[\frac{\text{webers}}{\text{meter}}\right]\equiv[\text{telsa}]\end{aligned} \tag{18.1.10}$$

One tesla expressed in cgs units equals 10^4 Gauss. The magnetic field at the surface of a neodymium magnet reaches ≈ 1.25 T, while the earth's magnetic field is ≈ 0.5 G and the field of a refrigerator magnet ≈ 50 G. The electric field for breakdown (sparking) in air is $\approx 10^8$ V/m, while a field of $\approx 10^7$ V/m is required to ionize atoms generating a free conducting electron in, e.g., Si.

Two additional Maxwell's equations then state first that

$$\vec{\nabla}\cdot\vec{B}^{(\text{force})} = 0 \tag{18.1.11}$$

indicating that magnetic charges analogous to electric charges do not exist, while *Faraday's law of induction*

$$\vec{\nabla}\times\vec{E}^{(\text{force})} = -\frac{\partial \vec{B}^{(\text{force})}}{\partial t} \tag{18.1.12}$$

implies that a time-varying magnetic induction induces an electric field.

The motion of carriers in a conductor results from the force provided by the electric and magnetic field and additional forces if present, resulting in an additional equation

$$\vec{J} = \sigma\left(\vec{E}^{(\text{force})} + \vec{v}\times\vec{B}^{(\text{force})} + \text{possible other forces}\right) \tag{18.1.13}$$

In most materials and cases of interest, this reduces to $\vec{J} = \sigma\vec{E}$, which is the local expression of Ohm's law, $V = IR$, where the *conductivity*, σ, possesses units of $(\Omega\text{-cm})^{-1}$. Note that the symbol employed to represent the conductivity is identical to that for surface charge, which can occasionally be misleading. This relationship between force fields and sources enables any *ratio of source and force fields* to be expressed in electrical units, i.e., in terms of those of R, C, and L with the capacitance $C = Q/V$ in Farads or equivalently $[C] = [Q/IR] = [\text{s}/\Omega]$ and inductance L, defined by $V = -LdI/dT$, in Henrys or $[L] = [VT/I] = [IRT/I] = [\Omega\text{-s}]$. In particular, since the units of $\vec{H}$ are those of $\vec{J}$, and therefore $\sigma\vec{E}$, times distance, the ratio

$$Z = \left|\frac{\vec{E}^{(\text{force})}}{\vec{H}^{(\text{source})}}\right| \frac{[\text{volt/m}]}{[\text{Amperes/m}]} = \left|\frac{\vec{E}^{(\text{force})}}{\vec{H}^{(\text{source})}}\right| [\Omega] \tag{18.1.14}$$

termed the *impedance* possesses units of ohms.

The fields $\vec{E}$ and $\vec{D}$ are related by the material-dependent *dielectric permittivity*, ε,

$$\vec{D}^{(\text{source})} = \varepsilon\vec{E}^{(\text{force})} = \varepsilon_0\varepsilon_r\vec{E}^{(\text{force})} \tag{18.1.15}$$

where

$$\varepsilon_0 = 8.85\times10^{-12}\left[\frac{D}{E}\right] = \left[\frac{Q}{D^2}\right]\left[\frac{D}{V}\right] \equiv \left[\frac{C}{D}\right] = \left[\frac{\text{Farads}}{\text{meter}}\right] \tag{18.1.16}$$

is termed the *vacuum permittivity*, while the dimensionless quantity, ε_r, represents the *relative permittivity*, which in the general case can be a spatially varying, nonlinear, anisotropic, and/or time-dependent function of the electric field. The ratio between the magnetic induction and the magnetic field is denoted

$$\vec{B}^{(\text{force})} = \mu \vec{H}^{(\text{source})} = \mu_o \mu_r \vec{H}^{(\text{source})} \tag{18.1.17}$$

in which the *magnetic permeability*, μ, is the product of the dimensionless *relative permeability*, μ_r, and the magnetic permeability of free space, μ_0,

$$\mu_0 = 4\pi \times 10^{-7} \left[\frac{\vec{B}}{\vec{H}}\right] = \left[\frac{\vec{B}}{\vec{E}}\right]\left[\frac{\vec{E}}{\vec{H}}\right] = \left[\frac{T}{D}\right][\Omega] \equiv \left[\frac{L}{D}\right] = \left[\frac{\text{Henrys}}{\text{meter}}\right] \tag{18.1.18}$$

The vacuum permittivity and permeability satisfy the relationship, to be demonstrated later,

$$c_0^2 = \frac{1}{\mu_0 \varepsilon_0} = \frac{D^2}{LC} = \left[\frac{D^2}{T\Omega T\Omega^{-1}}\right] = \left[\frac{D^2}{T^2}\right] \tag{18.1.19}$$

where the vacuum velocity of light, $c_0 \approx 3.0 \times 10^8$ m/s. In a *magnetically active* material μ_r which again can depend in a complicated fashion on $\vec{H}$, differs from unity as electronic or nuclear spins align along or opposite to the direction of an applied magnetic field.

The *differential form* of Maxwell's equations above can be transformed to integral expressions. Integrating Gauss's law over any volume ς

$$\int_{\varsigma} \vec{\nabla} \cdot \vec{D} d^3 r = \int_{\varsigma} \rho_{\text{free}} d^3 r \tag{18.1.20}$$

yields after applying Gauss's theorem, Equation (8.4.1),

$$\oint_{S \subset \varsigma} \vec{D} \cdot d\vec{S} = Q_{\text{free, enclosed}} \tag{18.1.21}$$

where the closed *Gaussian surface* S is described by the boundary of ς, $Q_{\text{free, enclosed}}$ denotes the total free charge enclosed by the surface, and the differential area element $d\vec{S}$ is directed perpendicular to the surface in the *outward* direction from the enclosed volume. Similarly, Equation (18.1.11) for the magnetic induction implies

$$\oint_{S \subset \varsigma} \vec{B} \cdot d\vec{S} = 0 \tag{18.1.22}$$

Integrating Ampere's law, Equation (18.1.5), over a surface that intersects the current source

$$\int_S \left(\vec{\nabla} \times \vec{H}\right) \cdot d\vec{A} = \int_S \left(\vec{J}_{\text{free}} + \frac{\partial \vec{D}}{\partial t}\right) \cdot d\vec{A} \tag{18.1.23}$$

and employing Stokes' theorem, Equation (8.4.2), results in

$$\oint_{C \subset S} \vec{H} \cdot d\vec{l} = I_{\text{free, enclosed}} + \frac{\partial}{\partial t} \int_S \vec{D} \cdot d\vec{A} \tag{18.1.24}$$

in which $I_{\text{free, enclosed}}$ is the total current of free charges that passes through the surface S bounded by C. The direction of $d\vec{l}$ follows the *right-hand rule* that if $d\vec{A}$ points in the direction of one's right thumb, the fingers curl in the direction of positive $d\vec{l}$, e.g., counterclockwise when viewed from the region pointed to by the surface vector.

Any solution of Maxwell's equations must be consistent with charge conservation, i.e., the charge flowing out of a volume equals the negative of the time rate of change of the total charge within the volume,

$$\frac{\partial}{\partial t} Q_{\text{enclosed}} = \frac{\partial}{\partial t} \int_{\varsigma} \rho d^3 r = -\oint_{S \subset \varsigma} \vec{J} \cdot d\vec{A} \tag{18.1.25}$$

as $J dA \cos\theta_{\vec{J}, d\vec{A}}$ corresponds to the current passing through a surface element $d\vec{A}$ oriented at $\theta_{\vec{J}, d\vec{A}}$ with respect to $\vec{J}$. Applying Gauss's theorem to the surface integral yields

$$\frac{\partial}{\partial t} \int_{\varsigma} \rho d^3 r = -\int_{\varsigma} \vec{\nabla} \cdot \vec{J} d^3 r \tag{18.1.26}$$

In the limit that V represents an infinitesimal volume, this yields the *charge continuity equation*

$$\frac{\partial \rho}{\partial t} + \vec{\nabla} \cdot \vec{J} = 0 \tag{18.1.27}$$

This equation can be derived directly by noting that for the current component in, e.g., the x-direction, if $I_x(x)$ is the component of the total current along the x-direction flowing into a region with transverse cross section A and length Δx and $\Delta Q_x(x + \Delta x/2)$ is the change in the total charge in the volume associated with this current component so that $\Delta\rho = \Delta\rho_x + \Delta\rho_y + \Delta\rho_z$,

$$\frac{\partial J_x}{\partial x} = \frac{1}{A\Delta x}\left(I_x(x + \Delta x) - I_x(x)\right) = -\frac{1}{A\Delta x \Delta t} \Delta Q_x \left(x + \frac{\Delta x}{2}\right) = -\frac{\partial \rho_x}{\partial t} \tag{18.1.28}$$

That Maxwell's equations are consistent with charge conservation follows by applying the divergence operator to both sides of Ampere's law, Equation (18.1.5), and applying Gauss's law, Equation (18.1.1), to $\vec{\nabla} \cdot \vec{D}$.

Finally, Faraday's law, Equation (18.1.12), can be rewritten as

$$\oint_{CCS} \vec{E}\cdot d\vec{l} = -\frac{\partial}{\partial t}\int_S \vec{B}\cdot d\vec{A} \tag{18.1.29}$$

In time-invariant systems, Maxwell's equations separate into independent equations for the electric and magnetic fields

$$\begin{aligned} \vec{\nabla}\cdot\vec{D} &= \rho_{\text{free}} \quad & \vec{\nabla}\cdot\vec{B} &= 0 \\ \vec{\nabla}\times\vec{E} &= 0 \quad & \vec{\nabla}\times\vec{H} &= \vec{J}_{\text{free}} \end{aligned} \tag{18.1.30}$$

These are analyzed separately in *electrostatics* and *magnetostatics*.

18.2 GAUSS'S LAW

From the linearity of the divergence and curl operators, which satisfy the relationship $D(\vec{F}_1 + \vec{F}_2) = D\vec{F}_1 + D\vec{F}_2$ for D representing either operator, if μ and ε do not depend on the field strengths, summing the individual solutions of Maxwell's equations for each N sources yields the solution in the presence of all the sources. The fields generated by any charge distribution can consequently be obtained by dividing the distribution into infinitesimal elements and superimposing the vector fields generated by each element through summation or integration. However, for highly symmetric charge distributions, the integral forms of Maxwell's equations can instead be employed directly. That is, knowledge of the integral of a function over an interval does not in general determine the value of the function at each point in the interval. However, if the function exhibits symmetry with respect to displacements, it possesses a constant value over the interval equal to its integral over the interval divided by the interval length. Similarly, if the electric field is invariant with respect to displacements over a Gaussian surface, S, as below, its surface integral over S, which is proportional to the charge enclosed by S, divided by the area, A, of S equals its value at each point on S:

- **Point charge**: The electric field of an isolated point charge q at $\vec{r}'$ in an infinite homogeneous medium is rotationally invariant about the particle position. The field must therefore be constant and radially directed over a Gaussian surface S given by a sphere centered at the point charge with radius r (since any nonradial component can be transformed into its negative through a rotation). Therefore, since $d\vec{S}$ is directed radially outward from $\vec{r}'$ while $A = 4\pi|\vec{r}-\vec{r}'|^2$, Gauss's law yields, in a medium with a dielectric permittivity $\varepsilon = \varepsilon_0\varepsilon_r$,

$$\vec{E}(\vec{r}) = \frac{1}{\varepsilon}\vec{D}(\vec{r}) = \frac{q}{\varepsilon A}\hat{e}_{\vec{r}-\vec{r}'} = \frac{q}{4\pi\varepsilon_0\varepsilon_r|\vec{r}-\vec{r}'|^2}\hat{e}_{\vec{r}-\vec{r}'} \tag{18.2.1}$$

- **Infinite charged line:** The field of an infinite uniform line, e.g., $r=0$ in cylindrical coordinates with a charge per unit length λ is invariant with respect to translations and rotations along the z-axis and reflection in the $z=0$ plane and is therefore cylindrically symmetric and directed in the $\hat{e}_r$ direction. For a cylindrical Gaussian surface, S, a radius r from the charged line terminated by the two planes $z=0, L$, the total enclosed charge equals λL. Since $d\vec{A}$ along the top and bottom faces of S is directed in the $-\hat{e}_z$ and $\hat{e}_z$ directions, $\vec{D}\cdot dA=0$ along these surfaces, while $\vec{D}=D_{\vec{r}}\hat{e}_{\vec{r}}$ over the side of S, with $\vec{r}$ the cylindrical radius vector. Hence, Gauss's law yields, with $A=2\pi rL$,

$$\vec{E}(\vec{r})=\frac{1}{\varepsilon}\vec{D}(\vec{r})=\frac{\lambda L}{\varepsilon A}\hat{e}_{\vec{r}}=\frac{\lambda}{2\pi\varepsilon_0\varepsilon_r r}\hat{e}_{\vec{r}} \qquad (18.2.2)$$

- **Infinite charged plane:** The field a distance z above or below an infinite charged sheet along the $z=0$ plane with uniform surface charge density σ must be invariant with respect to translations in x and y as well as rotation about any line parallel to the z-axis and is consequently directed along sign $(z)\hat{e}_z$. For a Gaussian surface consisting of a box with equal cross-sectional areas A_{charge} along two faces perpendicular to z and that further encloses an area A_{charge} of the charge sheet, $d\vec{A}\cdot\vec{E}=0$ along the sides while $\vec{E}$ and $d\vec{A}$ are parallel on the top and bottom faces. As these faces together possess a total area $A=2A_{\text{charge}}$

$$\vec{E}(\vec{r})=\frac{1}{\varepsilon}\vec{D}(\vec{r})=\text{sign}(z)\frac{\sigma A_{\text{charge}}}{\varepsilon(2A_{\text{charge}})}\hat{e}_z=\text{sign}(z)\frac{\sigma}{2\varepsilon_0\varepsilon_r}\hat{e}_z \qquad (18.2.3)$$

In problems, the sheet, line, or point charges above are often replaced by continuous or multiple-layered planar, cylindrically or spherically symmetric charge distributions.

Examples

1. The field of a point charge together with an infinite line charge is the vector sum of the fields evaluated earlier, while the magnitude of the field of a positive charge sheet with charge density σ situated below a similar sheet with charge density $-\sigma$ equals $\sigma/\varepsilon_0\varepsilon_r$ and zero between and outside the sheets where the fields of the two layers are parallel and antiparallel, respectively.
2. For a charged infinitely thin spherical/cylindrical shell of radius a, a Gaussian surface centered on the center of the shell with a radius $r<a$ (in spherical/cylindrical coordinates, respectively) encloses zero charge, while a surface with $r>a$ encloses the charge of the shell. Consequently, the field inside the shell vanishes, while the field outside coincides with that generated by a point/line charge at the center of the shell with magnitude equal to the total charge/charge per unit length of the shell. *Electromagnetic shielding* additionally employs the property that electric fields outside a hollow metal cavity of any shape generate

large currents that conduct charges to its outer surface until the field inside the metal vanishes. If no charges are present within the cavity, the unique solution to Maxwell's equations inside the cavity is then $\vec{E}=0$ uniformly.

3. The electric displacement for an arbitrary planar charge distribution $\rho(z)$ extending from $-L<z<L$ is obtained in terms of $D_z(-L)$ by applying Gauss's law to a box extending from $-L$ to z with equal cross-sectional areas A_{charge} along the two faces perpendicular to z:

$$\cancel{A_{\text{charge}}}(D_z(z)-D_z(-L))=\int_{-L}^{z}\rho(z')dV'=\cancel{A_{\text{charge}}}\int_{-L}^{z}\rho(z')dz' \tag{18.2.4}$$

Inside a solid, uniformly charged insulating sphere of radius R with total charge Q, the charge density equals $\rho(r)=Q/V=3Q/4\pi R^3$ so that the radial component of D is given by

$$AD_r(r)=4\pi r^2 D_r(r)=\int_0^r 4\pi r'^2\rho(r')dr'=\frac{3Q}{4\pi R^3}\int_0^r 4\pi r'^2 dr'=\frac{Qr^3}{R^3} \tag{18.2.5}$$

Hence, the electric displacement and electric field increase linearly with radius within the charge distribution and decay as $1/r^2$ according to Equation (18.2.1) outside.

18.3 ELECTRIC POTENTIAL

From $\vec{\nabla}\times\vec{E}=0$ for static fields, $\vec{F}=q\vec{E}$ is a conservative force, and a potential energy function U can be defined with $\vec{F}=-\vec{\nabla}U$. Dividing the potential energy of a test charge by its charge yields the *electric potential*, V, with units [volt] = $[MD^2/QT^2]$ related to the electric field by $\vec{F}=q\vec{E}=-\vec{\nabla}U=-q\vec{\nabla}V$ from which $\vec{E}=-\vec{\nabla}V$ or, as an integral expression,

$$V(b)-V(a)=\frac{U(b)-U(a)}{e}=-\frac{1}{e}\int_a^b\vec{F}\cdot d\vec{x}=-\int_a^b\vec{E}\cdot d\vec{x}\ \text{[volt]} \tag{18.3.1}$$

The energy

$$1.0\text{eV}=1.6\times10^{-19}\ [\text{QV = joules}] \tag{18.3.2}$$

gained by an electron by accelerating through a potential of one volt is termed an *electron volt*.

Examples

The potential of an infinite charged sheet along $z = 0$ with charge density σ in the absence of any other external charges or charged boundaries is obtained from

$$V(z) = V(z) - V(0) = -\int_0^z \vec{E}(\vec{r}')\cdot d\vec{l}' = -\frac{\sigma}{2\varepsilon_0\varepsilon_r}\left|\int_0^z dz'\right| = -\frac{\sigma|z|}{2\varepsilon_0\varepsilon_r} \qquad (18.3.3)$$

The absolute value sign arises since $\vec{E}\cdot d\vec{l}$ is positive for both positive and negative z. Hence, the potential slopes downward in both directions away from $z = 0$. A positive test charge "rolls down" this slope as it experiences a constant force away from the sheet.

Similarly, for a point charge for which the electric potential is normally set to zero at infinity,

$$\underbrace{V(\infty)}_{0} - V(r) = -\int_r^\infty \vec{E}(r')\cdot d\vec{r}' = -\frac{q}{4\pi\varepsilon}\int_r^\infty \frac{1}{r'^2}dr' = \frac{q}{4\pi\varepsilon}\frac{1}{r'}\bigg|_r^\infty = -\frac{q}{4\pi\varepsilon_0\varepsilon_r r}$$

$$(18.3.4)$$

As four negative signs are encountered, to verify such a calculation, observe that since a positive charge q repels a positive test charge, the potential must *decrease* toward infinity for the test charge to "roll down," i.e., experience a force away, from the origin.

The potential of a line charge instead varies logarithmically with radius and thus *diverges* both at $r = 0$ and $r = \infty$. A reference potential is therefore specified at a finite distance a in terms of which

$$V(r) - V(a) = -\int_a^r \vec{E}(r')\cdot d\vec{r}' = -\int_a^r \frac{q}{2\pi\varepsilon_0\varepsilon_r r'}dr' = -\frac{q}{2\pi\varepsilon_0\varepsilon_r}\log\left(\frac{r}{a}\right) \qquad (18.3.5)$$

If the source and boundary distributions are not sufficiently symmetric for direct application of Gauss's law, the field can still often be written as a superposition of fields of symmetric charge distributions. Thus, a system containing N discrete point sources yields an electric field

$$\vec{E}(\vec{r}) = \frac{1}{4\pi\varepsilon_0\varepsilon_r}\sum_{m=1}^{N}\frac{q_m}{\left|\vec{r}-\vec{r}_m\right|^2}\hat{e}_{\vec{r}-\vec{r}_m} = \frac{1}{4\pi\varepsilon_0\varepsilon_r}\sum_{m=1}^{N}\frac{q_m}{\left|\vec{r}-\vec{r}_m\right|^2}\frac{\vec{r}-\vec{r}_m}{\left|\vec{r}-\vec{r}_m\right|} \qquad (18.3.6)$$

The vector sum can be avoided by instead adding the *potentials* of each charge

$$V(\vec{r}) = \frac{1}{4\pi\varepsilon_0\varepsilon_r}\sum_{m=1}^{N}\frac{q_m}{\left|\vec{r}-\vec{r}_m\right|} \qquad (18.3.7)$$

and calculating $\vec{E}$ from $\vec{E} = -\vec{\nabla} V$. The equivalence of these methods follows from

$$\left[\vec{\nabla}\frac{1}{|\vec{r}-\vec{r}'|}\right]_i = \hat{e}_i\frac{\partial}{\partial x_i}\left(\sum_{n=1}^{3}(x-x_n')^2\right)^{-\frac{1}{2}} = -\hat{e}_i\frac{(x_i-x_i')}{|\vec{r}-\vec{r}'|^3} = -\frac{1}{|\vec{r}-\vec{r}'|^2}\left[\frac{\vec{r}-\vec{r}'}{|\vec{r}-\vec{r}'|}\right]_i \qquad (18.3.8)$$

Continuous charge distributions can similarly be divided into infinitesimal elements with charge $dQ = \rho d^3r$, yielding the potential

$$V(\vec{r}) = \frac{1}{4\pi\varepsilon_0\varepsilon_r}\int_\varsigma \frac{\rho(\vec{r}')}{|\vec{r}-\vec{r}'|}d^3r' \qquad (18.3.9)$$

and electric field

$$\vec{E}(\vec{r}) = -\vec{\nabla}V(\vec{r}) = \frac{1}{4\pi\varepsilon_0\varepsilon_r}\int_\varsigma \frac{\rho(\vec{r}')(\vec{r}-\vec{r}')}{|\vec{r}-\vec{r}'|^3}d^3r' \qquad (18.3.10)$$

Example

By symmetry, the electric field a distance z above the center of a ring of charge of radius a and charge per unit length λ positioned in the $z = 0$ plane and centered at the origin must point in the z-direction so that $\vec{E} = -\vec{\nabla}V = -\hat{e}_z dV/dz$ with

$$V(\vec{r}) = \frac{1}{4\pi\varepsilon_0\varepsilon_r}\int_\varsigma \frac{\lambda}{|\vec{r}-\vec{r}'|}dl' = \frac{2\pi a\lambda}{4\pi\varepsilon_0\varepsilon_r}\frac{1}{\sqrt{z^2+a^2}} \qquad (18.3.11)$$

yielding

$$\vec{E}(\vec{r}) = -\hat{e}_z\frac{\lambda a}{2\varepsilon_0\varepsilon_r}\frac{\partial}{\partial z}\frac{1}{\sqrt{z^2+a^2}} = \hat{e}_z\frac{\lambda a}{2\varepsilon_0\varepsilon_r}\frac{z}{(z^2+a^2)^{\frac{3}{2}}} \qquad (18.3.12)$$

If the electric field is instead obtained from Equation (18.3.10), as the vector components perpendicular to the z-axis cancel, only the z-component of the field from each charge segment $\lambda dl'$ needs to be evaluated.

The following Octave program graphs a potential function along a two-dimensional plane, illustrating the superposition of the potentials of a separated positive and negative charge:

```
clear all
numberOfPoints = 16;
halfWidth = 10;
chargePosition = 2.5;
```

```
xPositionR = linspace( -halfWidth, halfWidth, ...
  numberOfPoints);
yPositionR = xPositionR;

for outerLoop = 1 : numberOfPoints;
  for innerLoop = 1 : numberOfPoints;
    potential(outerLoop, innerLoop) = 1. / ...
      sqrt( ( xPositionR(outerLoop) - chargePosition ) ^2 ...
      + yPositionR(innerLoop) ^2 ) - 1. / ...
      sqrt( ( xPositionR(outerLoop) + chargePosition ) ^2 ...
      + yPositionR(innerLoop) ^2);
  end
end

mesh( xPositionR, yPositionR, potential);
```

18.4 CURRENT AND RESISTIVITY

Resistance results from the conversion of ordered electron motion into random thermal motion by atomic collisions. This is qualitatively similar to a mechanical system in which a frictional force $F_{\text{friction}} = -c\vec{v}$ yields a terminal velocity proportional to the applied force. Hence, current in a resistor is often conceptualized as fluid flowing through, e.g., a sand-filled pipe. However, unlike, e.g., sliding friction in which doubling the velocity of an object doubles the number of microscopic retarding interactions per unit time, each of which only slightly affects the particle motion, in a resistive material, the number of collisions is approximately independent of time, but the momentum change in a single collision is large and effectively randomizes the particle velocity.

The electric *current*, I, through a surface, S, equals the total charge Q passing through S per unit time. In terms of the component of the *current density vector* (in reality the current flux) $\vec{J}(\vec{r},t) = e\rho_{\text{particle}}(\vec{r},t)\,\vec{v}(\vec{r},t)$, where $\vec{v}(\vec{r},t)$ is the local velocity and $\rho_{\text{particle}}(\vec{r},t)$ is the particle density (e.g., the number of carriers per unit volume), in the direction of $d\vec{S}$

$$I = \frac{dQ_{\text{enclosed}}(t)}{dt} = \int_S \vec{J}(\vec{r},t) \cdot d\vec{S} \tag{18.4.1}$$

The current coincides with the direction of *positive* charge flow; if the current is instead carried by, e.g., electrons with negative charge, the motion of the physical particles opposes the current direction.

While according to Newton's law, the velocity of a free particle in the presence of an electric field increases linearly with time according $\vec{v} = \vec{a}t = q\vec{E}t/m$, unbound charges in most materials exhibit a time-independent *mean or drift velocity*, $\vec{v}_{\text{drift}} = \langle \vec{v} \rangle$ (where the angled brackets indicates an ensemble average over all charges), that varies linearly with the electric field. This linear response originates in the far greater thermal speeds of the charges $v_{\text{thermal}} \approx \langle |\vec{v}| \rangle$, compared to the incremental speed v_{drift} acquired from the electric field over the time interval between energetic collisions that scatter the particles into random directions and thus return their instantaneous average velocity to zero. As the average distance *mean free path*, l_{scatter}, between collisions is approximately velocity independent, the mean *scattering time* between collisions, $\tau_{\text{scatter}} \approx l_{\text{scatter}}/v_{\text{thermal}}$, varies negligibly with $\vec{E}$. The resulting drift velocity $\vec{v}_{\text{drift}} \approx q\vec{E}\tau_{\text{scatter}}/m$, acquired by the charge distribution from the electric field over τ_{scatter}, yields a current

$$\vec{J}\left(\vec{r},t\right) = q\rho_{\text{particle}}\left(\vec{r},t\right)\vec{v}_{\text{drift}} = \frac{q^2\rho_{\text{particle}}\left(\vec{r},t\right)\tau_{\text{scatter}}}{m}\vec{E} \equiv \sigma\vec{E} \tag{18.4.2}$$

in which $\sigma = q^2\rho_{\text{particle}}\tau_{\text{scatter}}/m$ is termed the *conductivity* and its reciprocal, $\rho_{\text{resistivity}} = 1/\sigma$, the *resistivity*. Additionally, $\mu = v_{\text{drift}}/|\vec{E}|$ is labeled the *mobility*.

Applying an electric potential ΔV to constant potential metal contacts or *terminals* covering the two ends of a resistive bar of length L in the z-direction with a constant cross-sectional area, A, generates a z-independent current. From $\vec{J} = \sigma\vec{E}$, the electric field is then uniform within the resistor so that the potential varies linearly with z according to Equation (18.4.2) (this can be visualized by subdividing the resistor into a line of equivalent infinitesimal resistors in parallel with the sum of the identical voltage drops over all resistors equal to ΔV). From $\rho = \vec{\nabla} \cdot \vec{E}$, which only differs from zero at the two end surfaces, charges accumulate at one terminal and are depleted on the opposing terminal providing sources for the electric field as in a capacitor. The *resistance* is computed from $\Delta V = IR$ where ΔV, often simply abbreviated V, is termed the *voltage drop* over the resistor (here the voltage is larger at the resistor terminal positioned opposite to the direction of current flow). With

$$I = \vec{J} \cdot \vec{A} = JA = \sigma EA = \frac{\sigma A \Delta V}{L} \tag{18.4.3}$$

the resistance of the bar is given by

$$R = \frac{\Delta V}{I} = \frac{L}{\sigma A} = \frac{\rho_{\text{resistivity}}L}{A} \tag{18.4.4}$$

When two resistors are connected in *series*, e.g., directly after each other in a circuit as the current is identical through both resistors, the total voltage drop is given by $V_{\text{total}} = V_1 + V_2 = I(R_1 + R_2) = IR_{\text{series}}$ and therefore

$$R_{\text{series}} = R_1 + R_2 \tag{18.4.5}$$

On the other hand, if the two resistors are instead connected in *parallel*, each experiences the same voltage drop so that $I_{total} = I_1 + I_2 = V(1/R_1 + 1/R_2) = V/R_{parallel}$ with

$$\frac{1}{R_{parallel}} = \frac{1}{R_1} + \frac{1}{R_2} \tag{18.4.6}$$

The dissipated *power* in a resistor is obtained by observing that the number of charges per unit time that flow across the resistor equals I/q, each of which loses an energy $q\Delta V$ to heat during transport. The electrical power required to balance this energy loss thus equals

$$P = \Delta VI \equiv VI = I^2R = \frac{V^2}{R} \tag{18.4.7}$$

The resistance of a resistor of a more general shape for which the functional behavior of the electric field or current distribution can be obtained from symmetry considerations is determined in several alternate ways. These are (1) placing positive and negative charges on the contacts and evaluating $\vec{E}$, over some surface S, which should be chosen if possible such that Gauss's law can be applied. (2) The total current I is obtained by integrating $\vec{J} = \sigma\vec{E}$ over S. (3) The potential drop ΔV over the resistor is evaluated by integrating the electric field over the region between the two contacts. (4) R is finally determined from $\Delta V/I$. (1′) Assume a current I between the two contacts. (2′) Find $\vec{J}$ and hence $\vec{E} = \vec{J}/\sigma(\vec{r})$ over a surface S between the contacts, typically by Gauss's law. (3′) Integrate $\vec{E}\cdot d\vec{l}$ between the two contacts to obtain ΔV. (1″) Assume a voltage difference ΔV between the two contacts. (2″) Find $\vec{E}$ from $\vec{E} = -\vec{\nabla}V$. (3″) Find $\vec{J} = \sigma\vec{E}$. (4″) Evaluate $I = \int \vec{J}\cdot d\vec{S}$. The last of these procedures coincides most nearly with the methods appropriate to arbitrary shape electrodes, which form equipotential surfaces although the electric field, current, and charge distributions are typically highly asymmetric.

Example

If a resistive material is situated between two concentric metal spherical shells with radii $r_< < r_>$, a charge Q on the inner sphere and $-Q$ on the outer sphere yields (1) $\vec{E} = Q/4\pi\varepsilon_0 r^2 \hat{e}_r$ for $r_< < r < r_>$ from Gauss's law, (2) $I = \int_S \sigma\vec{E}\cdot d\vec{S} = 4\pi r^2(\sigma Q/4\pi\varepsilon_0 r^2) = \sigma Q/\varepsilon_0$ for any Gaussian surface S, (3) $\Delta V = -\int_{r_>}^{r_<} \vec{E}\cdot d\vec{r} = Q(r_<^{-1} - r_>^{-1})/4\pi\varepsilon_0$, and (4) $R = \Delta V/I = (r_<^{-1} - r_>^{-1})/4\pi\sigma$. Alternatively, (1′) $\vec{J} = I/4\pi r^2 \hat{e}_r$, (2′) $\vec{E} = I/4\pi r^2\sigma\hat{e}_r$, and (3′) $\Delta V = -\int_{r_>}^{r_<} I/(4\pi r^2\sigma)dr = (I/4\pi\sigma)(r_<^{-1} - r_>^{-1})$ or (1″) $V(r) = c_1 + c_2/r, \Delta V = c_2(r_<^{-1} - r_>^{-1})$ from Gauss's law or Poisson equation, (2″) $\vec{E} = c_2/r^2\hat{e}_r$, (3″) $\vec{J} = \sigma\vec{E} = \sigma c_2/r^2\hat{e}_r$, and (4″) $I = \int \vec{J}\cdot d\vec{S} = 4\pi\sigma c_2$.

18.5 DIPOLES AND POLARIZATION

An *electric dipole* with *dipole moment* $\vec{p} \equiv q\vec{d}$ consists of a positive point charge of magnitude q displaced by $\vec{d}$ from a negative charge $-q$. A uniform electric field exerts a torque

$$\vec{\tau} = \frac{d}{2}\left(qE\sin\theta_{\vec{d},\vec{E}} + (-qE)\sin\theta_{-\vec{d},\vec{E}}\right)\hat{e}_{\vec{d}\times\vec{E}} = dqE\sin\theta_{\vec{d},\vec{E}}\hat{e}_{\vec{d}\times\vec{E}} = \vec{p}\times\vec{E} \qquad (18.5.1)$$

on a dipole so that the dipole potential energy relative to $\theta_{\vec{d},\vec{E}} = \pi/2$

$$U\left(\theta_{\vec{d},\vec{E}}\right) - U\left(\frac{\pi}{2}\right) = \int_{\pi/2}^{\theta_{\vec{d},\vec{E}}} \tau(\theta')d\theta' = -pE\cos\theta_{\vec{d},\vec{E}} = -\vec{p}\cdot\vec{E} \qquad (18.5.2)$$

attains a minimum when the dipole and electric field are aligned. While the forces on the two charges are oppositely directed and hence cancel in a uniform field, in a nonuniform field to lowest order, from the definition of the directional derivative, Equation (8.2.2),

$$\left[\vec{F}\right]_i = q\left(E_i\left(\vec{r}+\frac{\vec{d}}{2}\right) - E_i\left(\vec{r}-\frac{\vec{d}}{2}\right)\right) \approx q\left[\left(\vec{d}\cdot\vec{\nabla}\right)\vec{E}\right]_i \qquad (18.5.3)$$

The potential of a dipole for $r \gg d$ is, where Equation (18.3.8) is applied in the last step,

$$\begin{aligned}
V(\vec{r}) &= \frac{q}{4\pi\varepsilon_0\varepsilon_r}\left(\left|\vec{r}-\frac{\vec{d}}{2}\right|^{-1} - \left|\vec{r}+\frac{\vec{d}}{2}\right|^{-1}\right) \\
&= \frac{q}{4\pi\varepsilon_0\varepsilon_r}\left(\left(r^2 - \vec{r}\cdot\vec{d} + \frac{d^2}{4}\right)^{-\frac{1}{2}} - \left(r^2 + \vec{r}\cdot\vec{d} + \frac{d^2}{4}\right)^{-\frac{1}{2}}\right) \\
&= \frac{q}{4\pi\varepsilon_0\varepsilon_r r}\left(\left(1 - \frac{\hat{e}_r\cdot\vec{d}}{r} + \frac{d^2}{4r^2}\right)^{-\frac{1}{2}} - \left(1 + \frac{\hat{e}_r\cdot\vec{d}}{r} + \frac{d^2}{4r^2}\right)^{-\frac{1}{2}}\right) \\
&\approx \frac{q}{4\pi\varepsilon_0\varepsilon_r r}\left(\left(1 + \frac{\hat{e}_r\cdot\vec{d}}{2r} + \ldots\right) - \left(1 - \frac{\hat{e}_r\cdot\vec{d}}{2r} + \ldots\right)\right) \\
&\approx \frac{q\hat{e}_r\cdot\vec{d}}{4\pi\varepsilon_0\varepsilon_r r^2} = \frac{\vec{p}\cdot\vec{r}}{4\pi\varepsilon_0\varepsilon_r r^3} = -\frac{1}{4\pi\varepsilon_0\varepsilon_r}\vec{p}\cdot\vec{\nabla}\left(\frac{1}{r}\right)
\end{aligned} \qquad (18.5.4)$$

For $\vec{d}$ oriented along z so that $\vec{p}\cdot\hat{e}_r = q\vec{d}\cdot\hat{e}_r = p\cos\theta$ in spherical coordinates,

$$\vec{E} = -\vec{\nabla}V = -\frac{1}{4\pi\varepsilon_0\varepsilon_r}\left(\hat{e}_r\frac{\partial}{\partial r}+\hat{e}_\theta\frac{1}{r}\frac{\partial}{\partial\theta}\right)\frac{p\cos\theta}{r^2} = \frac{p}{4\pi\varepsilon_0\varepsilon_r r^3}(2\hat{e}_r\cos\theta+\hat{e}_\theta\sin\theta) \tag{18.5.5}$$

or equivalently (since θ increases *downward* in polar coordinates, $p\hat{e}_z = p\cos\theta\hat{e}_r - p\sin\theta\hat{e}_\theta$),

$$\begin{aligned} E_j = -\hat{e}_j\partial_j V &= -\frac{\hat{e}_j}{4\pi\varepsilon_0\varepsilon_r}\partial_j\left(\sum_{m=1}^{3}\frac{p_m x_m}{r^3}\right) = -\frac{\hat{e}_j}{4\pi\varepsilon_0\varepsilon_r}\sum_{m=1}^{3}p_m\left(\frac{\partial_j x_m}{r^3}+x_m\partial_j\left(\frac{1}{r^3}\right)\right) \\ &= -\frac{\hat{e}_j}{4\pi\varepsilon_0\varepsilon_r}\sum_{m=1}^{3}p_m\left(\frac{\delta_{jm}}{r^3}-\frac{3}{2}x_m\left(\frac{2x_j}{r^5}\right)\right) = \frac{1}{4\pi\varepsilon_0\varepsilon_r}\left(\frac{3(\vec{p}\cdot\hat{e}_{\vec{r}})\hat{e}_{\vec{r}}-\vec{p}}{r^3}\right)_j \end{aligned} \tag{18.5.6}$$

The *polarization*, $\vec{P}$, is defined as the net dipole moment per unit volume such that for n molecules per unit volume with individual dipole moments $\vec{p}$

$$\vec{P} = n\vec{p} \tag{18.5.7}$$

A spatially varying polarization results in an induced *polarization or bound charge*. That is, a sheet of $N(x')$ dipoles with cross-sectional area A and moments $\vec{p} = ed\hat{e}_x$ centered along the plane at $x = x'$ together with a second sheet $N(x'+d)$ centered at $x'+d$ generates a net charge $+qN(x') - qN(x'+d)$ along the plane at $x'+d/2$ from the positive charges of the dipoles at x and the negative charges of the dipoles at $x+d$. Since the two layers together occupy a volume $\varsigma = Ad$ between x' and $x'+d$, the effective charge density inside the region from x' to $x'+d$ equals

$$\rho_{\text{induced}} = \frac{q(N(x')-N(x'+d))}{\varsigma}\cdot\frac{d}{d} = \frac{P(x')-P(x'+d)}{d} \approx -\frac{\partial P(x')}{\partial x'} \tag{18.5.8}$$

Generalizing to three dimensions, the macroscopic electric field, which is related to the force on a test particle and therefore is affected by the local charge density (each dipole is electrically neutral and therefore does not provide a source for the electric displacement except at microscopic scales of the order or smaller than the dipole charge separation), satisfies

$$\vec{\nabla}\cdot\vec{E} = \frac{\rho_{\text{total}}}{\varepsilon_0} = \frac{\rho_{\text{free}}+\rho_{\text{induced}}}{\varepsilon_0} = \frac{1}{\varepsilon_0}\left(\rho_{\text{free}} - \vec{\nabla}\cdot\vec{P}\right) \tag{18.5.9}$$

From $\vec{\nabla}\cdot\vec{D} = \rho_{\text{free}}$,

$$\vec{D} = \varepsilon_0\vec{E} + \vec{P} = \varepsilon\vec{E} \tag{18.5.10}$$

In terms of the *dielectric susceptibility* χ defined by

$$\vec{P} = \varepsilon_0 \chi \vec{E} \tag{18.5.11}$$

the dielectric permittivity adopts the form

$$\varepsilon = \varepsilon_0 (1 + \chi) \tag{18.5.12}$$

In a crystal, the electronic restoring force can be anisotropic (direction dependent) in which case $\vec{P}$ and $\vec{E}$ are generally not collinear and χ and ε constitute *tensor* (e.g., matrix) quantities.

Since

$$\vec{\nabla} \cdot \vec{E} = \vec{\nabla} \cdot \frac{\vec{D}}{\varepsilon} = \frac{1}{\varepsilon} \vec{\nabla} \cdot \vec{D} + \vec{D} \cdot \vec{\nabla} \left(\frac{1}{\varepsilon} \right) = \frac{\rho_{\text{free}}}{\varepsilon} - \vec{E} \cdot \vec{\nabla} \ln \varepsilon \tag{18.5.13}$$

while $\vec{\nabla} \times \vec{D}$ is given by

$$\vec{\nabla} \times \vec{D} = \vec{\nabla} \times \vec{P} = \varepsilon \underbrace{\vec{\nabla} \times \vec{E}}_{0} - \vec{E} \times \vec{\nabla} \varepsilon = a - \frac{\vec{D}}{\varepsilon} \times \vec{\nabla} \varepsilon = -\vec{D} \times \vec{\nabla} \ln \varepsilon \tag{18.5.14}$$

even if $\rho_{\text{free}} = \vec{\nabla} \cdot \vec{D} = 0$, except in a homogeneous medium or a medium with sufficient symmetry that $\vec{\nabla} \times \vec{P} = 0$ throughout space, $\vec{D} = \varepsilon \vec{E} \neq 0$.

Examples

1. An *electret* possesses a permanent polarization vector even in the absence of external fields. For a *cylindrical* electret, with a uniform polarization $\vec{P} = P_z \hat{e}_z$ along its symmetry axis, $\vec{\nabla} \times D = \vec{\nabla} \times \vec{P} = -\hat{e}_\theta (\partial P_z / \partial r)$, which differs from zero along the side of the cylinder. Hence, $\vec{\nabla} \times \vec{P}$ generates the same $\vec{D}$ field as the $\vec{H}$ field resulting from a uniform axial surface current (see Section18.11). Thus, while $\oint_C \vec{E} \cdot d\vec{r} = 0$ around all closed paths C, typically $\oint_{C_{CS}} \vec{D} \cdot d\vec{r} = \int_S \vec{P} \cdot d\vec{S} \neq 0$ if C encloses part of the cylinder's side.
2. For an *infinite* planar dipole layer with surface area $A \to \infty$ and surface charge density $-\sigma$ separated by a distance Δz from a similar layer with surface density σ, $\vec{P} = \hat{e}_z Q_{\text{total}} \Delta z / A \Delta z = \hat{e}_z \sigma A \Delta z / A \Delta z = \sigma \hat{e}_z$, while the potential increase over the distance Δz equals $\Delta V = \sigma \Delta z / \varepsilon_0$. If a second pair of layers is located immediately above this layer, ΔV over the two layers equals $2\sigma \Delta z / \varepsilon_0$. By extension, over an infinite planar electret with $\vec{P} = \sigma \hat{e}_z$ of height L with $\rho_{\text{free}} = 0$, $\Delta V = -L E_z = \sigma L / \varepsilon_0$ so that $\vec{D} = \varepsilon \vec{E} + \vec{P} = 0$ as expected since here both $\vec{\nabla} \times \vec{D} = \vec{\nabla} \times \vec{P} = 0$ by symmetry and $\vec{\nabla} \cdot \vec{D} = 0$.

18.6 BOUNDARY CONDITIONS AND GREEN'S FUNCTIONS

In finite spatial regions, the electric field is determined not only by the enclosed source distribution but also by fields or charges at the boundary. These boundary contributions can be determined through the method of Green's functions.

First, consider an electric field parallel to the surface of a conductor or dielectric (insulator) that occupies the region $z < z_s$. Specifying the coordinate axis such that the $\vec{E} = E_{\|}\hat{e}_y$ while the $\hat{e}_z$ direction is aligned with the surface vector $\hat{e}_n$, for every point infinitesimally above the surface $\vec{x}_s$, i.e., $\vec{x}_s = \vec{x}_{\text{surface}} + \varepsilon\hat{e}_z$ where $\vec{x}_{\text{surface}}$ is a point on the surface and $\varepsilon \ll \Delta z$, from Faraday's law, Equation (18.1.12),

$$\Delta z\left(\vec{\nabla} \times \vec{E}\right)_x = \Delta z\left(\underbrace{\frac{\partial E_z}{\partial y}}_{0} - \overbrace{\frac{\partial E_y}{\partial z}}^{E_{\|}}\right) \approx -\left(E_{\|}(\vec{x}_s) - E_{\|}(\vec{x}_s - \Delta z\hat{e}_z)\right) = -\Delta z\frac{\partial B_x}{\partial t} \tag{18.6.1}$$

which equals zero as $\Delta z \to 0$ since $\vec{B}$ and its time derivatives are finite. Hence, the parallel component of the electric field is continuous at a dielectric interface, while as $\vec{E}$ inside a conductor is zero (otherwise infinite currents would result), $E_{\|}$ vanishes just outside a conductor. Next, Gauss's law, Equation (18.1.1), instead yields with z, the perpendicular ($\perp$) direction, and $\vec{E} = E_{\perp}\hat{e}_z$

$$\Delta z\,\vec{\nabla}\cdot\vec{D} = \Delta z\frac{\partial \varepsilon E_{\perp}(\vec{x}_s)}{\partial z} \approx \varepsilon(\vec{x}_s)E_{\perp}(\vec{x}_s) - \varepsilon(\vec{x}_s - \Delta z\hat{e}_z)E_{\perp}(\vec{x}_s - \Delta z\hat{e}_z) = \Delta z\rho_{\text{free}} = \sigma_{\text{surface}} \tag{18.6.2}$$

for a surface charge density $\sigma_{\text{surface}} = \rho_{\text{free}}/\Delta z$ distributed within a distance Δz below the surface. Hence, $E_{\perp}$ (and $D_{\perp}$ for a free surface charge distribution) is discontinuous across the boundary in the presence of surface charges or a refractive index discontinuity; however, its integral, the electric potential, remains continuous over the boundary. Analogously, the magnetic induction satisfies the boundary conditions

$$B_{\perp}(\vec{x}_s) - B_{\perp}(\vec{x}_s - \Delta z\hat{e}_z) = 0 \tag{18.6.3}$$

while the magnetic field $\vec{H}$ obeys, where $\vec{K}$ is the surface current density,

$$\Delta z\left(\vec{\nabla} \times \vec{H}\right)_x = -\Delta z\frac{\partial H_y(\vec{x}_s)}{\partial z} \approx -\left(H_y(\vec{x}_s) - H_y(\vec{x}_s - \Delta z\hat{e}_z)\right) = \Delta z J_x + \Delta z\frac{\partial D_x}{\partial t} \approx K_x \tag{18.6.4}$$

To determine the electric potential within a bounded volume, ς, Green's theorem, Equation (8.4.5), can be employed with $\phi_1(\vec{r}) = V(r)$ and $\phi_2(\vec{r}) = G(\vec{r}, \vec{r}')$ where the *Green's function* $G(\vec{r}, \vec{r}')$ here satisfies

$$\nabla_{\vec{r}'}^2 G(\vec{r},\vec{r}') = -4\pi\delta(\vec{r}-\vec{r}') \tag{18.6.5}$$

From $\vec{\nabla} \cdot \vec{D} = \rho_{\text{free}}$ and $E = -\vec{\nabla} V$, in a medium of constant permittivity $V(\vec{r})$ further obeys *Poisson equation*

$$\nabla^2 V(\vec{r}) = -\frac{\rho_{\text{free}}(\vec{r})}{\varepsilon_0 \varepsilon_r} \tag{18.6.6}$$

Accordingly, with $d\vec{A} \cdot \vec{\nabla} = dA\hat{e}_n \cdot \vec{\nabla} = dA\partial/\partial n$, where n (or n' below) is the outward normal from ς, after replacing the integral over $V(\vec{r}')\delta(\vec{r}-\vec{r}')$ by $V(\vec{r})$,

$$\begin{aligned} V(\vec{r}) &= \frac{1}{4\pi\varepsilon_0\varepsilon_r}\int_\varsigma G(\vec{r},\vec{r}')\rho_{\text{free}}(\vec{r}')d^3r' \\ &+ \frac{1}{4\pi}\oint_{S\subset\varsigma}\left(G(\vec{r},\vec{r}')\frac{\partial V(\vec{r}')}{\partial n'} - V(\vec{r}')\frac{\partial G(\vec{r},\vec{r}')}{\partial n'}\right)dA' \end{aligned} \tag{18.6.7}$$

Equation (18.6.7) expresses the potential inside a region as the sum of the inhomogeneous solution of Poisson equation associated with the charges within ς and a homogeneous surface integral term, which incorporates the boundary conditions through appropriate source terms on S. That is, the (unprimed) Laplacian operator acting on the Green's functions in the latter term yields zero except on S where $\vec{r} = \vec{r}'$. The first term in the surface integral arises from a normal $\vec{E}$ field component corresponding to a surface charge layer, while the second term is generated by a dipole layer that discontinuously changes the potential at the boundary. Such a layer contributes to the potential within ς through the difference of the Green's functions of its positive and negative sheets that in the limit of zero separation transforms into the normal derivative of the Green's function.

The *Poisson equation possesses unique solutions* since if V_1 and V_2 are solutions with the same sources and boundary conditions, $V_1 - V_2$ satisfies the homogeneous Poisson equation with no sources and zero boundary conditions at the surface and therefore vanishes throughout the problem region. For *Dirichlet boundary conditions*, the potential is specified on the boundary, while the condition $G(\vec{r},\vec{r}') = 0$ is imposed at all boundary points $\vec{r}'$. For *Neumann boundary conditions*, the derivative of the potential is instead specified on S, and $G(\vec{r},\vec{r}')$ is constructed with $\partial G(\vec{r},\vec{r}')/\partial n' = c$ on S, where the value $c = -4\pi/A$ is determined from

$$\begin{aligned} \int_\varsigma \nabla^2 G(\vec{r},\vec{r}')d^3r' &= -4\pi\int_\varsigma \delta(\vec{r}-\vec{r}')d^3r' = -4\pi \\ &= \int_\varsigma \vec{\nabla}\cdot\left[\vec{\nabla} G(\vec{r},\vec{r}')\right]d^3r' \\ &= \int_S d\vec{S}\cdot\left[\vec{\nabla} G(\vec{r},\vec{r}')\right] = \int_S \frac{\partial G(\vec{r},\vec{r}')}{\partial n}dS' = cA \end{aligned} \tag{18.6.8}$$

leading to

$$V(\vec{r}) = \frac{1}{4\pi\varepsilon_0\varepsilon_r}\int_{\varsigma} G(\vec{r},\vec{r}')\rho_{\text{free}}(\vec{r}')d^3r' + \frac{1}{4\pi}\oint_{S\subset\varsigma} G(\vec{r},\vec{r}')\frac{\partial V(\vec{r}')}{\partial n'}dA' + \frac{1}{A}\oint_{S\subset\varsigma} V(\vec{r}')dA' \tag{18.6.9}$$

The last term averages the potential over S but does not affect the electric field. Finally, *mixed boundary conditions* specify the potential and its normal derivative on nonoverlapping regions.

The *method of images* solves a potential problem on a bounded domain D by placing additional sources into a larger domain containing D such that the resulting potential function reproduces the boundary conditions initially specified on the perimeter of D. By uniqueness, the solution to the extended problem then coincides with the desired solution within D.

Example

The potential of a point charge at $(a, 0, 0)$ in the half-space D to the right of a grounded semi-infinite metal plane along at $x = 0$ can be obtained by extending D to all space while introducing a fictitious negative point charge at $(-a, 0, 0)$, yielding

$$V(\vec{r}) = \frac{q}{4\pi\varepsilon_0\varepsilon_r}\left(\frac{1}{|\vec{r}-a\hat{e}_x|} - \frac{1}{|\vec{r}+a\hat{e}_x|}\right) \tag{18.6.10}$$

which satisfies the required zero boundary conditions along $x = 0$ and therefore constitutes the desired unique solution to the problem in D. The charge along the conducting plane at $x = 0$ and axial distance $r_{\|}$ from the origin is then obtained from Equation (18.6.2):

$$\begin{aligned}\frac{\sigma}{\varepsilon_0\varepsilon_r} = E_{\perp}^{+} &= -\left.\frac{\partial V}{\partial x}\right|_{x=0} \\ &= -\frac{q}{4\pi\varepsilon_0\varepsilon_r}\frac{\partial}{\partial x}\left.\left(\frac{1}{\sqrt{r_{\|}^2+(x-a)^2}} - \frac{1}{\sqrt{r_{\|}^2+(x+a)^2}}\right)\right|_{x=0} \\ &= -\frac{2qa}{4\pi\varepsilon_0\varepsilon_r}\frac{1}{\left(r_{\|}^2+a^2\right)^{3/2}}\end{aligned} \tag{18.6.11}$$

The force on the charge at $x = a$ equals that between the charge and its image charge:

$$\vec{F} = -\frac{q^2}{4\pi\varepsilon_0\varepsilon_r}\frac{1}{4a^2}\hat{e}_x \tag{18.6.12}$$

From this expression, the work performed in moving a charge from $x = a$ to infinity equals

$$W = -\int_a^\infty \vec{F}\cdot d\vec{l} = \frac{q^2}{4\pi\varepsilon_0\varepsilon_r}\int_a^\infty \frac{1}{4x^2}dx = \frac{q^2}{16\pi\varepsilon_0\varepsilon_r}\left(-\frac{1}{x}\right)\bigg|_a^\infty = \frac{q^2}{16\pi\varepsilon_0\varepsilon a} \tag{18.6.13}$$

which is half of the potential energy of two charges separated by $2a$, since displacing the source charge displaces the image charge, but the latter motion is not associated with physical work.

Problems often analyzed with image charges include a charge inside a conducting rectangular box, between two parallel conducting planes, or within a wedge with an opening angle π/n for integer $n > 1$. The wedge requires $2n - 1$ image charges, while the other configurations generate infinitely many image charges. The sign of each charge is reversed with respect to its closest neighbors and occupies the position that would be observed visually if the conducting planes are replaced by mirrors. Image charges associated with dielectric boundaries are smaller than the physical charge since only a fraction of the source field penetrates into the dielectric region. Image charges can also be employed for cylindrical and spherical boundaries.

Example

For a charge q inside and at a distance a from the center of a conducting sphere of radius R, the magnitude q' of the image charge and its distance, d, from the center of the sphere are obtained by setting V on the sphere at the two points on the line joining the charges to zero:

$$\begin{aligned} \frac{q}{R-a} &= -\frac{q'}{d-R} \\ \frac{q}{R+a} &= -\frac{q'}{d+R} \end{aligned} \tag{18.6.14}$$

Dividing the two equations and cross multiplying,

$$\begin{aligned} (R-a)(R+d) &= (d-R)(R+a) \\ R^2 + \cancel{(d-a)}R - ad &= -R^2 + \cancel{(d-a)}R + ad \end{aligned} \tag{18.6.15}$$

Hence, $d = R^2/a$, and from Equation (18.6.14), $q' = -qR/a$. When $a \to R$, $d \to R$ as well. The spherical curvature can then be neglected and $q' \to -q$ as for a flat conducting surface.

If the symmetries of the problem coincide with the coordinate directions, the potential can be obtained through separation of variables as discussed in Chapter 13.

Example

An axially symmetric V can be written as $V(r,\theta)=\sum_{l=0}^{\infty}(c_l r^l + d_l r^{-(l+1)})P_l(\cos\theta)$ according to Equation (13.2.5). If a grounded conducting sphere of radius R is placed in a linear electric field $\vec{E}=E_0\hat{e}_z$ for $r \gg R$, $V(\vec{r})=-E_0 z=-E_0 r\cos\theta$ and hence $c_l=-E_0\delta_{l1}$; all other $c_l=0$ to preclude unphysical divergences at infinite radius. The $r^{-(l+1)}$ terms, however, can be nonzero as these approach zero for $r\to\infty$. From the orthogonality of the Legendre polynomials, $V(R)=0$ implies $d_l=E_0R^3\delta_{l1}$ so that $V(r,\theta)=E_0(-r+R^3/r^2)\cos\theta$ and from Equation (18.6.2)

$$\sigma(R,\theta)=-\varepsilon_0\varepsilon_r\frac{\partial V}{\partial r}\bigg|_{r=R}=3\varepsilon_0\varepsilon_r E_0\cos\theta \tag{18.6.16}$$

18.7 MULTIPOLE EXPANSION

The potential far from a localized charge distribution can be recast as a sum of terms decaying as different powers of the distance from the source through the *multipole expansion*. Decomposing $1/|\vec{r}-\vec{r}'|$ into spherical harmonics, for $r>r'$ according to Equation (13.3.39),

$$\begin{aligned}V(\vec{r})&=\frac{1}{4\pi\varepsilon_0\varepsilon_r}\int\frac{\rho(\vec{r}')}{|\vec{r}-\vec{r}'|}d^3r'\\&=\frac{1}{4\pi\varepsilon_0\varepsilon_r}\sum_{l=0}^{\infty}\sum_{m=-l}^{l}\frac{4\pi}{2l+1}\left(\int\rho(\vec{r}')r'^{l}Y_{lm}(\theta',\phi')d^3r'\right)\frac{Y_{lm}(\theta,\phi)}{r^{l+1}}\end{aligned} \tag{18.7.1}$$

The *electric monopole* and *electric dipole* terms are then (recall that $Y_{l-m}=(-1)^m Y_{lm}^*$)

$$\begin{aligned}4\pi\varepsilon_0\varepsilon_r V(\vec{r})&=\frac{1}{r}\int\rho(\vec{r}')d^3r'+\frac{1}{r^2}\left\{\begin{array}{l}\left[\int\rho(\vec{r}')r'\cos\theta' d^3r'\right]\cos\theta+\\ \mathrm{Re}\left\{\left[\int\rho(\vec{r}')r'\sin\theta' e^{-i\phi'}d^3r'\right]\sin\theta e^{i\phi}\right\}\end{array}\right\}+\ldots\\&=\frac{1}{r}\int\rho(\vec{r}')d^3r'+\frac{1}{r^3}\left\{\begin{array}{l}\left[\int\rho(\vec{r}')z' d^3r'\right]z+\\ \mathrm{Re}\left\{\left[\int\rho(\vec{r}')(x'-iy')d^3r'\right](x+iy)\right\}\end{array}\right\}+\ldots\\&=\frac{1}{r}\int\rho(\vec{r}')d^3r'+\frac{1}{r^3}\sum_{i=1}^{3}\left[\int\rho(\vec{r}')x_i' d^3r'\right]x_i+\ldots\end{aligned} \tag{18.7.2}$$

Including the $l=2$ quadrupole term, generated by terms such as $Y_{22} \propto \sin^2\theta e^{2i\phi} \propto x^2/r^2$, yields

$$V(\vec{r}) = \frac{1}{4\pi\varepsilon_0\varepsilon_r}\left(\frac{q}{r} + \frac{\vec{p}\cdot\vec{x}}{r^3} + \frac{1}{2r^5}\sum_{i,j=1}^{3} Q_{ij}x_i x_j + \ldots\right) \tag{18.7.3}$$

with (note from the length dimensions of these expressions that for each term of $V(r)$ to possess the same units requires the dependence on r given above)

$$\begin{aligned} q &= \int \rho(r')dV' \\ p_i &= \int x_i'\rho(r')dV' \\ Q_{ij} &= \int \left(3x_i'x_j' - r'^2\delta_{ij}\right)\rho(r')dV' \end{aligned} \tag{18.7.4}$$

18.8 RELATIVISTIC FORMULATION OF ELECTROMAGNETISM, GAUGE TRANSFORMATIONS, AND MAGNETIC FIELDS

That the Lorentz force law is preserved in all inertial frames implies that the magnetic and electric fields are interrelated much as space and time. However, while space and time form a four-dimensional vector, the six components of the electric and magnetic fields together form a single antisymmetric 4×4 *electromagnetic field tensor*.

To illustrate the physical nature of this relationship, for a line of positive charges with charge density λ moving in the laboratory (rest) frame $F^{(\text{rest})}$ with a velocity $\vec{v} = v\hat{e}_z$ relative to a collinear, oppositely charged stationary background distribution $-\lambda$, the total charge density $\rho^{(\text{rest})} = 0$ so that $\vec{E}^{(\text{rest})} = 0$ while $\vec{B}^{(\text{rest})} = \mu_0\lambda v/2\pi r\hat{e}_\theta$ implying zero force on a *stationary* charge q. However, in a frame $F^{(\text{v})}$ moving with velocity $\vec{v} = v\hat{e}_z$, the length of a segment of the positive charge distribution is expanded by $\gamma_> = \left(1-\beta_<^2\right)^{-1/2}$ with $\beta_< = v/c$ and λ is therefore reduced by a factor $1/\gamma_>$ while the opposite occurs for the negative charge background, yielding a net linear charge density

$$\lambda^{(\text{v})} = \lambda_+ - \lambda_- = \frac{\lambda}{\gamma_>} - \gamma_>\lambda = \frac{\lambda}{\gamma_>}\left(1-\gamma_>^2\right) = \frac{\lambda}{\gamma_>}\left(1 - \frac{1}{1-\beta_<^2}\right) = -\lambda\gamma_>\beta_<^2 \tag{18.8.1}$$

and therefore an electric field (where $c^2 = 1/\mu_0\varepsilon_0$)

$$\vec{E}^{(\text{v})} = -\frac{\lambda\gamma_> v^2}{2\pi\varepsilon_0 rc^2}\hat{e}_r = -\frac{\lambda\mu_0\gamma_> v^2}{2\pi r}\hat{e}_r \tag{18.8.2}$$

that accelerates a positive test charge $q>0$ toward the wire, seemingly contradicting the independence of physical laws on the inertial frame. However, in $F^{(v)}$, the Lorentz-contracted *negative* background charge distribution moves with a velocity $-v\hat{e}_z$ and therefore generates a current $\lambda\gamma_> \vec{v}$. Consequently, $\vec{B}^{(v)} = \gamma_> \vec{B}^{(\text{rest})}$ and the magnetic force on q cancels the electric force as ($c^2 = 1/\mu_0\varepsilon_0$)

$$\vec{F}^{(\text{magnetic},v)} = qv \times \vec{B}^{(v)} = -\frac{q\mu_0\lambda\gamma_> v^2}{2\pi r}\underbrace{\hat{e}_z \times \hat{e}_\theta}_{-\hat{e}_r} = -\vec{F}^{(\text{electric},v)} \tag{18.8.3}$$

Equation (18.8.1) further indicates that $c\rho$ forms the zero component of a relativistic contravariant four-vector $J^\mu \equiv \left(c\rho, \vec{J}\right)_C^T$ for which in this instance, since $\rho^{(\text{rest})} = 0$,

$$c\rho^{(v)} = \gamma_>\left(c\rho^{(\text{rest})} - \beta_< J_z^{(\text{rest})}\right) = -\gamma_> \beta_< J_z^{(\text{rest})} = -\gamma_> \beta_< \lambda v \tag{18.8.4}$$

Indeed, in relativistic notation, the charge continuity equation, Equation (18.1.27), can be rewritten as

$$D_\mu J^\mu \equiv \frac{\partial J^\mu}{\partial x^\mu} = 0 \tag{18.8.5}$$

with $D_\mu = \partial/\partial x^\mu = \left(\partial/\partial ct, \vec{\nabla}_r\right)_R$ the covariant four-vector of Equation (17.3.2).

For static fields, Laplace equation for the scalar electric potential in vacuum in the presence of sources can be written as

$$\nabla^2 V = -\frac{\rho}{\varepsilon_0} = -c^2\mu_0\rho = -c\mu_0 J^0 \tag{18.8.6}$$

while for time-invariant sources in vacuum, $\vec{\nabla} \cdot \vec{B} = \mu_0 \vec{\nabla} \cdot \vec{H} = 0$, implying that $\vec{B}$ can be written in terms of a three-component vector "magnetic" potential $\vec{A}$ as

$$\vec{B} = \vec{\nabla} \times \vec{A} \tag{18.8.7}$$

However, the physical field $\vec{B}$ is then unchanged under the transformation $\vec{A} + \vec{\nabla}\Lambda(\vec{r},t) = \vec{A}'$ (since $\vec{\nabla} \times \vec{\nabla} f = 0$ for any f), where $\Lambda(\vec{r},t)$ is an arbitrary scalar field. If Λ is chosen as the solution of $\nabla^2\Lambda = -\vec{\nabla} \cdot \vec{A}$ (termed the Coulomb gauge), then $\vec{\nabla} \cdot \vec{A}' = 0$ and, after redesignating $\vec{A}'$ as A, the Maxwell's equation $\vec{\nabla} \times \vec{H} = \mu_0 \vec{J}$ yields

$$\vec{\nabla} \times \left(\vec{\nabla} \times \vec{A}\right) = \vec{\nabla}\underbrace{\left(\vec{\nabla} \cdot \vec{A}\right)}_{0} - \vec{\nabla}^2 \vec{A} = \mu_0 \vec{J} \tag{18.8.8}$$

The above formula then comprises three independent equations for each of the three components of $\vec{A}$, each of the form of the Poisson equation, Equation (18.8.14), for the scalar potential with solutions

$$\vec{A}(\vec{r},t)=\frac{\mu_0}{4\pi}\int_{\varsigma}\frac{\vec{J}(\vec{r}',t)}{|\vec{r}-\vec{r}'|}d^3r' \tag{18.8.9}$$

Accordingly, by introducing a four-vector $A^{\mu}=\left(V/c,\vec{A}\right)_C^T$, Laplace equation for both the scalar and vector potentials is obtained from the static (time-invariant) specialization of the covariant *wave equation*

$$\left(\frac{1}{c^2}\frac{\partial^2}{\partial t^2}-\frac{\partial^2}{\partial x^2}-\frac{\partial^2}{\partial y^2}-\frac{\partial^2}{\partial z^2}\right)A^{\mu}=\partial_{\mu}\partial^{\mu}A^{\nu}\equiv\Box A^{\nu}=\mu_0 J^{\nu} \tag{18.8.10}$$

where $\Box$ constitutes the relativistically invariant (scalar) four-dimensional Laplacian or wave operator.

In time-varying systems, the relationship between the potentials and the source distribution depends on the choice of *gauge function.* As $\vec{\nabla}\times\vec{K}=0$ implies that $\vec{K}=\vec{\nabla}\Phi(\vec{r})$ for some scalar function $\Phi(\vec{r})$, from $\vec{\nabla}\times\vec{E}+\partial\vec{B}/\partial t=\vec{\nabla}\times\vec{E}+\partial\left(\vec{\nabla}\times\vec{A}\right)/\partial t=0$, an electric potential can be defined by

$$\vec{E}+\frac{\partial\vec{A}}{\partial t}=-\vec{\nabla}V \tag{18.8.11}$$

Subsequently, $\vec{\nabla}\times\vec{H}-\partial\vec{D}/\partial t=\vec{J}$ in free space yields

$$\begin{aligned}\mu_0\vec{J}&=\vec{\nabla}\times\left(\vec{\nabla}\times\vec{A}\right)-\mu_0\varepsilon_0\frac{\partial}{\partial t}\left(-\vec{\nabla}V-\frac{\partial\vec{A}}{\partial t}\right)\\&=\vec{\nabla}\left(\vec{\nabla}\cdot\vec{A}\right)-\nabla^2\vec{A}+\frac{1}{c^2}\frac{\partial^2\vec{A}}{\partial t^2}+\frac{1}{c^2}\frac{\partial}{\partial t}\vec{\nabla}V\\&=\Box\vec{A}+\vec{\nabla}\left(\vec{\nabla}\cdot\vec{A}+\frac{1}{c^2}\frac{\partial V}{\partial t}\right)\end{aligned} \tag{18.8.12}$$

While Equation (18.8.12) differs from Equation (18.8.10) for time-dependent sources, the *gauge transformation*

$$V'=V-\frac{\partial\Lambda(\vec{r},t)}{\partial t},\quad \vec{A}'=\vec{A}+\vec{\nabla}\Lambda(\vec{r},t) \tag{18.8.13}$$

preserves Equations (18.8.11) and (18.8.7) therefore leaves Maxwell's equations invariant. In the *Coulomb gauge*, $\vec{\nabla} \cdot \vec{A}' = 0$ and therefore

$$\vec{\nabla} \cdot \vec{E}' = -\nabla^2 V' = \frac{\rho}{\varepsilon} \tag{18.8.14}$$

The electric potential, where the primes are omitted, then coincides with the static expression

$$V(r,t) = \frac{1}{4\pi\varepsilon} \int_{\varsigma} \frac{\rho(\vec{r}',t)}{|\vec{r}-\vec{r}'|} d^3r' \tag{18.8.15}$$

which adjusts instantaneously to changes in the charge distribution. However, the observable quantity, namely, the electric field, also depends on $-\partial\vec{A}/\partial t$ with $\vec{A}$ obtained from Equation (18.8.12). With this term included, the effect on $\vec{E}$ of a change in the source distribution propagates outward as expected at the speed of light.

Finally, the relativistically invariant *Lorentz gauge* is defined by[1]

$$\partial_\mu A'^\mu = \frac{1}{c^2}\frac{\partial V'}{\partial t} + \vec{\nabla} \cdot \vec{A}' = 0 \tag{18.8.16}$$

In terms of the contravariant 4-vectors A^μ and $D^\mu = g^{\mu\nu}\partial/\partial x^\nu = \partial/\partial x_\mu = (\partial/\partial(ct), -\vec{\nabla}_{\vec{r}})^T_C$, the source-free Maxwell's equations, Equations (18.8.7) and (18.8.11), are implicit in the definitions of the scalar and vector potentials, Equations (18.8.7) and (18.8.11). In particular, since the antisymmetric component of the outer product generates a cross product according to Equation (5.5.7), $\vec{B} = \vec{\nabla} \times \vec{A}$ can be obtained from the spatial part of the antisymmetrized tensor product of $\vec{\nabla}$ and $\vec{A}$. By associating the gradient operator with the *contravariant* derivative, $D_{A,C}$, where the subscript indicates that the resulting expressions must be algebraically rearranged to act on A, the resulting contravariant tensor additionally contains the electric field within its temporal components (sign conventions differ among authors):

$$\begin{aligned}\mathbf{F}^{\mu\nu} &\equiv D^\mu A^\nu - D^\nu A^\mu \equiv D_{A,C} \otimes A^T - A \otimes D^T_{A,C} \\ &= {}_C\!\left(\begin{array}{c|c} \partial_{ct} V/c & \partial_{ct}\vec{A}^T \\ \hline -\vec{\nabla}V/c & -\partial_i A_j \end{array}\right)_C - {}_C\!\left(\begin{array}{c|c} \partial_{ct} V/c & -\vec{\nabla}^T V/c \\ \hline \partial_{ct}\vec{A} & -\partial_j A_i \end{array}\right)_C \\ &= {}_C\!\left(\begin{array}{c|c} 0 & -\vec{E}^T/c \\ \hline \vec{E}/c & -\varepsilon_{ijk} B_k \end{array}\right)_C\end{aligned} \tag{18.8.17}$$

[1] If $\vec{\nabla} \cdot A \neq 0$ or $\vec{\nabla} \cdot \vec{A} + (1/c^2)\partial V/\partial t \neq 0$ for potentials A, V, the gauge transformation, Equation (18.8.13), yields potentials $\vec{A}'$ and V' that satisfy the Coulomb or Lorenz condition if Λ is identified with the solutions of $\nabla^2\Lambda = -\vec{\nabla} \cdot \vec{A}$ or $\Box\Lambda = \vec{\nabla} \cdot \vec{A} + (1/c^2)\partial V/\partial t$, respectively.

A tensor $\mathbf{T}^{\mu\nu} = A^\mu B^\nu$ is transformed into a moving frame according to $\mathbf{T}'^{\mu\nu} = L^\mu{}_\alpha L^\nu{}_\beta A^\alpha B^\beta$. The A (row) components of $\mathbf{T}$ are thus transformed by multiplying by the Lorentz transformation matrix on the left, while the B (column) components of $A \otimes B^T$ are transformed by multiplying by the transpose of the Lorentz matrix on the right as can be seen by writing out individual components of the product. Hence, transforming to a moving frame in the x-direction,

$$
\begin{aligned}
\left(\mathbf{F}^{lm}\right)^{(v)} &= L^l{}_n L^m{}_s \left(\mathbf{F}^{ns}\right)^{(\text{rest})} = L^l{}_n \left(\mathbf{F}^{ns}\right)^{(\text{rest})} \left(L^m{}_s\right)^T \\
&= {}_C\!\left(\begin{array}{cc|cc} \gamma & -\beta\gamma & 0 & 0 \\ -\beta\gamma & \gamma & 0 & 0 \\ \hline 0 & 0 & 1 & 0 \\ 0 & 0 & 0 & 1 \end{array}\right)_{\not R}
{}_{\not C}\!\left(\begin{array}{cc|cc} 0 & -E_x/c & -E_y/c & -E_z/c \\ E_x/c & 0 & -B_z & B_y \\ \hline E_y/c & B_z & 0 & -B_x \\ E_z/c & -B_y & B_x & 0 \end{array}\right)_{\not C}
{}_{\not R}\!\left(\begin{array}{cc|cc} \gamma & -\beta\gamma & 0 & 0 \\ -\beta\gamma & \gamma & 0 & 0 \\ \hline 0 & 0 & 1 & 0 \\ 0 & 0 & 0 & 1 \end{array}\right)_C
\end{aligned}
\tag{18.8.18}
$$

Since the product of three antisymmetric matrices is antisymmetric, this product can be computed from the results of three block matrix products:

$$
\begin{aligned}
&\frac{1}{c}\begin{pmatrix} \gamma & -\beta\gamma \\ -\beta\gamma & \gamma \end{pmatrix}\begin{pmatrix} 0 & -E_x \\ E_x & 0 \end{pmatrix}\begin{pmatrix} \gamma & -\beta\gamma \\ -\beta\gamma & \gamma \end{pmatrix} = \frac{1}{c}\begin{pmatrix} \gamma & -\beta\gamma \\ -\beta\gamma & \gamma \end{pmatrix}\begin{pmatrix} \beta\gamma E_x & -\gamma E_x \\ \gamma E_x & -\beta\gamma E_x \end{pmatrix} \\
&\quad = \frac{1}{c}\begin{pmatrix} 0 & (-\gamma^2 + \beta^2\gamma^2)E_x \\ (\gamma^2 - \beta^2\gamma^2)E_x & 0 \end{pmatrix} = \begin{pmatrix} 0 & -E_x/c \\ E_x/c & 0 \end{pmatrix} \\
&\begin{pmatrix} 1 & 0 \\ 0 & 1 \end{pmatrix}\begin{pmatrix} 0 & -B_x \\ B_x & 0 \end{pmatrix}\begin{pmatrix} 1 & 0 \\ 0 & 1 \end{pmatrix} = \begin{pmatrix} 0 & -B_x \\ B_x & 0 \end{pmatrix} \\
&\begin{pmatrix} \gamma & -\beta\gamma \\ -\beta\gamma & \gamma \end{pmatrix}\begin{pmatrix} -E_y/c & -E_z/c \\ -B_z & B_y \end{pmatrix}\begin{pmatrix} 1 & 0 \\ 0 & 1 \end{pmatrix} = \begin{pmatrix} -\gamma(E_y/c - \beta B_z) & -\gamma(E_z/c + \beta B_y) \\ \gamma(\beta E_y/c - B_z) & \gamma(\beta E_z/c + B_y) \end{pmatrix}
\end{aligned}
\tag{18.8.19}
$$

yielding generally, where $\|$ denotes the x-direction and $\perp$ the two perpendicular directions,

$$
\begin{aligned}
E_{\|}^{(v)} &= E_{\|}^{(0)} \\
B_{\|}^{(v)} &= B_{\|}^{(0)} \\
\vec{E}_{\perp}^{(v)} &= \gamma\left(\vec{E}_{\perp}^{(0)} + \vec{v} \times \vec{B}_{\perp}^{(0)}\right) \\
\vec{B}_{\perp}^{(v)} &= \gamma\left(\vec{B}_{\perp}^{(0)} - \frac{\vec{v}}{c^2} \times \vec{E}_{\perp}^{(0)}\right)
\end{aligned}
\tag{18.8.20}
$$

Equivalently, for $\vec{v} = v_x\hat{e}_x$, starting from $\vec{E}^{(\mathrm{v})} = -\vec{\nabla}^{(\mathrm{v})}V^{(\mathrm{v})} - \partial_{ct}^{(\mathrm{v})}\left(c\vec{A}^{(\mathrm{v})}\right)$ and $\vec{B}^{(\mathrm{v})} = \vec{\nabla}^{(\mathrm{v})} \times \vec{A}^{(\mathrm{v})}$ and noting that A^{μ} and $\left(\partial_{ct}, -\vec{\nabla}\right)_C^T$ are contravariant 4-vectors, so that, e.g., $\partial_{ct}^{(\mathrm{v})} = \gamma(\partial_{ct} - (v/c)(-\partial_x))$,

$$
\begin{aligned}
E_y^{(\mathrm{v})} &= -\partial_y(\gamma(V-(v/c)cA_x)) - (\gamma(\partial_{ct} + (v/c)\partial_x))cA_y \\
&= -\gamma\left(\partial_{ct}\left(cA_y\right) + \partial_y V\right) - v\left(\partial_x A_y - \partial_y A_x\right) \\
&= \gamma\left(E_y - vB_z\right)
\end{aligned}
\tag{18.8.21}
$$

The increased value of $\vec{E}_{\perp}/\vec{E}_{\parallel}$ in moving frames enhances the radiative coupling of the fields of high-energy particles to transverse electromagnetic waves.

Returning to the remaining Maxwell's source equations, since $F_{00} = 0$, Gauss's law and Ampere's law in free space can be written as, where $-\partial_i \varepsilon_{ijk} B_k = \varepsilon_{jik} \partial_i B_k$,

$$
D_{\mu}\mathbf{F}^{\mu\nu} = \frac{\partial}{\partial x^{\mu}}\mathbf{F}^{\mu\nu} = \left(\partial_{ct} \quad \vec{\nabla}\right)_{\not{R}\not{C}} \left(\begin{array}{c|c} 0 & -\vec{E}^T/c \\ \hline \vec{E}/c & -\varepsilon_{ijk}B_k \end{array}\right)_C = \left(\begin{array}{c} \frac{1}{c}\vec{\nabla}\cdot\vec{E} \\ -\frac{1}{c^2}\frac{\partial E}{\partial t} + \vec{\nabla}\times B \end{array}\right)_C = \mu_0 J^{\nu}
\tag{18.8.22}
$$

In terms of A^{μ}, after changing the order of the mixed partial derivatives, in the Lorentz gauge

$$
\frac{\partial}{\partial x_{\mu}}\frac{\partial A^{\nu}}{\partial x^{\mu}} - \frac{\partial}{\partial x_{\nu}}\underbrace{\frac{\partial A^{\mu}}{\partial x^{\mu}}}_{0} = \mu_0 J^{\nu}
\tag{18.8.23}
$$

Equation (18.8.23) becomes simply, as expected in the Lorentz gauge, Equation (18.8.16), as expected

$$
\Box A^{\nu} = \mu_0 J^{\nu}
\tag{18.8.24}
$$

Since the two source-free Maxwell's equations can be obtained from the remaining Maxwell's equations containing source terms through the transformation $B \to -E/c$, $E/c \to B$, $J, \rho = 0$, the former equations are given by

$$
D_{\mu}\mathbf{G}^{\mu\nu} = \frac{\partial}{\partial x^{\mu}}\mathbf{G}^{\mu\nu} = 0
\tag{18.8.25}
$$

where since

$$
\begin{aligned}
\mathbf{F}_{\mu\nu} &= g_{\mu\alpha}\mathbf{F}^{\alpha\beta}g_{\beta\nu} \\
&= {}_R\left(\begin{array}{c|c} 1 & 0 \\ \hline 0 & -1 \end{array}\right)_{\not{R}\,\not{C}}\left(\begin{array}{c|c} 0 & -\vec{E}/c \\ \hline \vec{E}/c & -\varepsilon_{ijk}B_k \end{array}\right)_{\not{C}\,\not{R}}\left(\begin{array}{c|c} 1 & 0 \\ \hline 0 & -1 \end{array}\right)_R \\
&= {}_R\left(\begin{array}{c|c} 0 & \vec{E}/c \\ \hline -\vec{E}/c & -\varepsilon_{ijk}B_k \end{array}\right)_R
\end{aligned}
\tag{18.8.26}
$$

the dual tensor

$$
\mathbf{G}^{\mu\nu} = \frac{1}{2}\varepsilon^{\mu\nu\alpha\beta}\mathbf{F}_{\alpha\beta} = {}_C\begin{pmatrix} 0 & -B_x & -B_y & -B_z \\ -B_x & 0 & E_z/c & -E_y/c \\ -B_y & -E_z/c & 0 & E_x/c \\ -B_z & E_y/c & -E_x/c & 0 \end{pmatrix}_C = {}_C\left(\begin{array}{c|c} 0 & -\vec{B} \\ \hline \vec{B} & \varepsilon_{ijk}E_k/c \end{array}\right)_C
\tag{18.8.27}
$$

Here, e.g., $\varepsilon^{0123} = \varepsilon^{1230} = 1$ so that $G^{01} = 1/2(F_{23} - F_{32}) = -B_x$, while $G^{12} = 1/2(F_{30} - F_{03}) = E_z$. With the definitions $d\tau = dt\sqrt{1 - dx^2/(cdt)^2} = dt/\gamma$, $p^\mu = (E/c, \vec{p})^T = m\gamma(c, \vec{v})^T$, and $p_\mu = m\gamma(c, -\vec{v})$, the Maxwell force law can be recast in relativistic form as

$$
\frac{dp^\nu}{d\tau} = \frac{q}{m}\mathbf{F}^{\nu\mu}p_\mu
\tag{18.8.28}
$$

Finally, the relativistic action of a particle, Equation (17.6.5), is generalized by observing that in the presence of an electric field, the action $L = T - U = T - qV$. Writing this as a Lorentz invariant quantity leads to

$$
\begin{aligned}
S &= \int_{x^\mu(t_1)}^{x^\mu(t_2)} \left(-mcds - qA_\mu dx^\mu\right) \\
&= \int_{x^\mu(t_1)}^{x^\mu(t_2)} \left(-mcds - qVdt + q\vec{A}\cdot d\vec{r}\right) \\
&= \int_{t_1}^{t_2} \left(-mc^2\sqrt{1 - \frac{v^2}{c^2}} - qV + q\vec{A}\cdot\vec{v}\right)dt = \int_{t_1}^{t_2} Ldt
\end{aligned}
\tag{18.8.29}
$$

Hence, the generalized momentum is given by

$$
\vec{P} = \vec{\nabla}_{\vec{v}}L = \frac{m\vec{v}}{\sqrt{1 - \frac{v^2}{c^2}}} + q\vec{A} = \vec{p} + q\vec{A}
\tag{18.8.30}
$$

while the Hamiltonian is obtained according to

$$\tilde{H}=\vec{v}\cdot\vec{P}-L=\frac{mv^2}{\sqrt{1-\frac{v^2}{c^2}}}+q\vec{A}\cdot\vec{v}+mc^2\sqrt{1-\frac{v^2}{c^2}}-q\vec{A}\cdot\vec{v}+qV=\frac{mc^2}{\sqrt{1-\frac{v^2}{c^2}}}+qV \tag{18.8.31}$$

In the nonrelativistic approximation, Equation (18.8.31) becomes, with the energy computed relative to mc^2,

$$\tilde{H}=\frac{1}{2m}\left(\vec{P}-q\vec{A}\right)^2+qV \tag{18.8.32}$$

To derive the Maxwell force equation from Equation (18.8.32) requires the Hamiltonian equations. Note that $\vec{v}$ includes contributions from both $\vec{P}$ and $\vec{A}$

$$\begin{aligned}\frac{d\vec{r}}{dt}&=\vec{v}=\vec{\nabla}_{\vec{P}}\tilde{H}=\frac{1}{m}\left(\vec{P}-q\vec{A}\right)\\ \frac{d\vec{P}}{dt}&=-\vec{\nabla}_{\vec{r}}\tilde{H}=q\vec{\nabla}_{\vec{r}}\left(\vec{v}\cdot\vec{A}\right)-q\vec{\nabla}_{\vec{r}}V\end{aligned} \tag{18.8.33}$$

Combining these with the identities $d\vec{N}/dt=\partial\vec{N}/\partial t+\vec{v}\cdot\vec{\nabla}\vec{N}$ and since $\vec{\nabla}_{\vec{r}}$ is taken for constant $\vec{P}$ and hence $\vec{v}$, $\vec{v}\times\left(\vec{\nabla}_{\vec{r},A}\times\vec{A}\right)=\vec{\nabla}_{\vec{r},A}\left(\vec{v}\cdot\vec{A}\right)-\left(\vec{v}\cdot\vec{\nabla}_{\vec{r},A}\right)\vec{A}$,

$$m\frac{d^2\vec{r}}{dt^2}=\frac{d\vec{P}}{dt}-q\left(\frac{\partial\vec{A}}{\partial t}+\left(\vec{v}\cdot\vec{\nabla}_{\vec{r}}\right)\vec{A}\right)=\underbrace{-q\left(\vec{\nabla}_{\vec{r}}V+\frac{\partial\vec{A}}{\partial t}\right)}_{q\vec{E}}+\underbrace{q\vec{\nabla}_{\vec{r}}\left(\vec{v}\cdot\vec{A}\right)-q\left(\vec{v}\cdot\vec{\nabla}_{\vec{r}}\right)\vec{A}}_{q\left(\vec{v}\times\left(\vec{\nabla}\times\vec{A}\right)\right)=q\vec{v}\times\vec{B}} \tag{18.8.34}$$

18.9 MAGNETOSTATICS

The magnetic field generated by a time-independent current distribution is derived by applying $\vec{B}=\vec{\nabla}\times\vec{A}$ to Equation (18.8.9). Because $\vec{\nabla}$ acts on $\vec{r}$ but not $\vec{r}'$,

$$\begin{aligned}\vec{B}&=\frac{\mu_0}{4\pi}\vec{\nabla}\times\int_{\varsigma}\frac{\vec{J}_{\text{free}}(r')}{\left|\vec{r}-\vec{r}'\right|}d^3r'\\ &=\frac{\mu_0}{4\pi}\int_{\varsigma}\vec{\nabla}_{\vec{r}}\left(\frac{1}{\left|\vec{r}-\vec{r}'\right|}\right)\times\vec{J}_{\text{free}}(r')d^3r'\\ &=\frac{\mu_0}{4\pi}\int_{\varsigma}\frac{\vec{J}_{\text{free}}(r')\times\left(\vec{r}-\vec{r}'\right)}{\left|\vec{r}-\vec{r}'\right|^3}d^3r'\end{aligned} \tag{18.9.1}$$

For a wire with cross-sectional area A, from Equation (18.1.6), $\vec{J}dV = JAd\vec{l} = Id\vec{l}$, where $d\vec{l}$ points in the direction of the current, yielding the *Biot–Savart law*:

$$\vec{B} = \frac{\mu_0}{4\pi}\int_l d\vec{B} = \frac{\mu_0}{4\pi}\int_l \frac{Id\vec{l}\times(\vec{r}-\vec{r}\,')}{|\vec{r}-\vec{r}\,'|^3} \qquad (18.9.2)$$

The right-hand rule can be applied here in two ways. If the thumb of the right hand points along a current element, its contribution to $\vec{B}$ wraps around in the direction of the fingers of the right hand. For a current loop, if the fingers of the right hand wrap around the loop in the direction of the current, the thumb points in the direction of the contribution to the magnetic field by the loop.

Example

At a position $\vec{r} = (0,0,z)$ above a wire ring positioned along $r' = R$, $z' = 0$ and carrying a counterclockwise current as seen from $z > 0$, $d\vec{l}$ and $\vec{r} - \vec{r}\,'$ are perpendicular while only $B_z \neq 0$ by symmetry. Further, $dB_z = dB\sin\theta_{z,\vec{r}-\vec{r}\,'} = dBR/|\vec{r}-\vec{r}\,'|$ from which

$$\vec{B} = \hat{e}_z\frac{\mu_0}{4\pi}\int_l \frac{I\sin\theta dl}{z^2+R^2} = \hat{e}_z\frac{\mu_0}{2}\frac{RI}{z^2+R^2}\frac{R}{\sqrt{z^2+R^2}} \qquad (18.9.3)$$

In typical problems, the Biot–Savart law is employed to calculate $\vec{B}$ for, e.g., current carrying squares or other symmetric geometric figures. The integral over $d\vec{l}$ can then generally be deduced from that for a single section of the figure. However, for certain symmetric current distributions, $\vec{B}$ can be obtained from the integral form of Ampere's law.

1. *Wire*: The magnetic field of an infinite wire positioned along $r = 0$ in cylindrical coordinates cannot contain a radial component as reversing the current direction and then reflecting through a $z = \text{const}$ plane of symmetry would invert the component while $H_z = 0$ since $d\vec{l} = dl\hat{e}_z$ in Equation (18.9.2). Consequently, $\vec{H}$ is directed in the azimuthal direction according to the right-hand rule explained earlier and

$$\oint_C \vec{H}\cdot d\vec{l} = 2\pi r H_\phi = I \qquad (18.9.4)$$

2. *Plane (current sheet)*: The magnetic field of a uniform current sheet with current per unit length $\vec{K}$ and surface normal $\hat{n}$ is parallel to the sheet since a perpendicular component would change sign if the direction of current flow is reversed. Rotating the sheet 180° about $\hat{n}$ would then reproduce the initial current distribution but with the negative of the perpendicular field. Integrating $\vec{H}\cdot d\vec{l}$ over a rectangular "Amperian"

loop with two sides perpendicular to the surface over which $\vec{H} \cdot d\vec{l} = 0$ and two oppositely directed parallel segments equidistant from and on opposite sides of the surface yields

$$\oint_C \vec{H} \cdot d\vec{l} = 0 + HL + 0 + (-H)(-L) = 2HL = K_J L \tag{18.9.5}$$

The direction $\hat{e}_{\vec{K}} \times \hat{n}$ of the magnetic field is again obtained by placing one's thumb along $\vec{K}$; the direction of the fingers coincides with that of $\vec{H}$ both above and below the current sheet.

3. *Cylinder*: Outside a cylindrical coil of radius R, extending from $z = -\infty$ to $z = +\infty$ formed from a wire with counterclockwise current I as seen looking towards the direction of smaller z and n turns per unit length (or equivalently a cylindrical current sheet with a surface current density $K = nI$) $H_r = 0$ as for a wire. A rectangular Amperian loop with two sides along z of length L, one situated at $r < R$ and the second at $r = \infty$, encloses a current nIL. As $\vec{H} \cdot d\vec{l} \neq 0$ only for the side with $r < R$,

$$\vec{H} = nI\hat{e}_z \tag{18.9.6}$$

4. *Toroid*: A circular Amperian loop of radius r within a toroidal (donut-shaped) coil encloses a current equal to NI, where N represents the total number of turns, so that

$$2\pi r \vec{H} = NI\hat{e}_\phi \tag{18.9.7}$$

Magnetic fields cannot perform work as the power supplied by the magnetic force equals

$$P = \frac{dW}{dt} = F_B \cdot \vec{v} = q(v \times B) \cdot \vec{v} = 0 \tag{18.9.8}$$

Example

A point charge moving with velocity $\vec{v}$ in the x–y plane in the presence of a magnetic field $B\hat{e}_z$ describes a circular orbit obtained by equating the centrifugal force to the magnetic force:

$$qvB = \frac{mv^2}{r} \tag{18.9.9}$$

Since the force is radially directed, no work is performed. The resulting angular frequency of motion $\omega = 2\pi/T$ where $T = 2\pi r/v = 2\pi rm/qBr = 2\pi m/qB$ is termed the (angular) *cyclotron frequency*:

$$\omega_{\text{cyclotron}} = \frac{qB}{m} \tag{18.9.10}$$

18.10 MAGNETIC DIPOLES

For $\vec{r}$ in Equation (18.8.9) far from a localized current distribution, with $1/|\vec{r}-\vec{r}'| \approx (1/r)(1+2r\cdot r'/r^2)^{-1/2}$,

$$\vec{A}(\vec{r}) = \frac{\mu_0}{4\pi}\int_\varsigma \frac{\vec{J}(\vec{r}')}{|\vec{r}-\vec{r}'|}d^3r' = \frac{\mu_0}{4\pi}\int_\varsigma \vec{J}(\vec{r}')\left(\frac{1}{r}+\frac{\vec{r}\cdot\vec{r}'}{r^3}+\ldots\right)d^3r' \tag{18.10.1}$$

Since $\vec{J}=0$ on the surface $S\subset\varsigma$ that bounds ς if ς extends beyond the current distribution while $\vec{\nabla}\cdot\vec{J}=0$ from Equation (18.1.27) and $\vec{J}\cdot\vec{\nabla}x_i = \vec{J}\cdot\hat{e}_i = J_i$, the integral of any component of $\vec{J}$ over V is zero according to

$$\int_\varsigma J_i d^3r = \int_\varsigma \underbrace{\vec{\nabla}\cdot\left(x_i\vec{J}\right)}_{\vec{J}\cdot\vec{\nabla}x_i + x_i\vec{\nabla}\cdot\vec{J}} d^3r - \int_\varsigma x_i \underbrace{\vec{\nabla}\cdot\vec{J}}_{0}\, d^3r = \int_{S\subset\varsigma} x_i \underbrace{\vec{J}}_{0\text{ on }S}\cdot d\vec{S} = 0 \tag{18.10.2}$$

That is, for $\vec{\nabla}\cdot\vec{J}=0$, only the curl of $\vec{J}$ is finite so that the current flux lines form closed paths with equal amounts of integrated current flowing backward and forward along any specified direction. Subsequently, from $\vec{A}\times\left(\vec{B}\times\vec{C}\right) = \vec{B}\left(\vec{A}\cdot\vec{C}\right) - \vec{C}\left(\vec{A}\cdot\vec{B}\right)$,

$$\begin{aligned}\frac{4\pi}{\mu_0}\vec{A}(\vec{r}) &\approx \frac{1}{r^3}\int \vec{J}(\vec{r}')\,\vec{r}\cdot\vec{r}'d^3r' \\ &= \frac{1}{r^3}\int \vec{r}\cdot\vec{J}(\vec{r}')\vec{r}'d^3r' - \frac{1}{r^3}\int \vec{r}\times\left(\vec{r}'\times\vec{J}(\vec{r}')\right)d^3r'\end{aligned} \tag{18.10.3}$$

However,

$$\begin{aligned}0 &= x_i\int_{S\subset\varsigma}\left(x_i'\vec{J}(\vec{r}')x_m'\right)\cdot d\vec{S}' \\ &= x_i\int_\varsigma \vec{\nabla}'\cdot\left(x_i'\vec{J}(\vec{r}')x_m'\right)d^3r' \\ &= \int_\varsigma x_i\sum_k \frac{\partial}{\partial x_k'}\left(x_i'J_k(\vec{r}')x_m'\right)d^3r' \\ &= \int_\varsigma x_i x_i'\sum_k J_k(\vec{r}')\delta_{km}d^3r' + \int_\varsigma x_i x_i'\underbrace{\vec{\nabla}'\cdot J(\vec{r}')}_{0}x_m'd^3r' + \int_\varsigma x_i\sum_k \delta_{ik}J_k(\vec{r}')x_m'd^3r'\end{aligned} \tag{18.10.4}$$

implying

$$\int \vec{J}(\vec{r}')\,\vec{r}\cdot\vec{r}'d^3r' = -\int \vec{r}\cdot\vec{J}(\vec{r}')\vec{r}'d^3r' \tag{18.10.5}$$

The first term on the last line of Equation (18.10.3) consequently equals the negative of the right-hand side of the first line and hence $-4\pi\vec{A}/\mu_0$ so that

$$\frac{4\pi}{\mu_0}\vec{A}(\vec{r}) = -\frac{1}{2r^3}\vec{r}\times\int\vec{r}'\times\vec{J}(\vec{r}')d^3r' = \frac{\vec{m}\times\vec{r}}{r^3} \tag{18.10.6}$$

where the magnetic dipole moment is defined as

$$\vec{m} = \frac{1}{2}\int\vec{r}'\times\vec{J}(\vec{r}')d^3r' \tag{18.10.7}$$

which equals $I\vec{S}$ for a current loop of area s.

The computation of $\vec{B} = \vec{\nabla}\times\vec{A}$ is simplified by noting that the divergence of the $k\vec{r}/r^3$ field of a point charge is zero except at the origin while

$$\left(\vec{m}\cdot\vec{\nabla}\right)\vec{r} = (m_x\partial_x + m_y\partial_y + m_z\partial_z)(x,y,z) = \vec{m} \tag{18.10.8}$$

and

$$\left(\vec{m}\cdot\vec{\nabla}\right)\frac{1}{r^3} = \left(\sum_{i=1}^{3}m_i\partial_i\right)(x^2+y^2+z^2)^{-\frac{3}{2}} = -\frac{3}{2}\sum_{i=1}^{3}\frac{2m_ir_i}{r^5} = -3\frac{\vec{m}\cdot\vec{r}}{r^5} \tag{18.10.9}$$

so that

$$\vec{B} = \vec{\nabla}\times\left(\vec{m}\times\left(\frac{\vec{r}}{r^3}\right)\right) = \vec{m}\underbrace{\vec{\nabla}\cdot\left(\frac{\vec{r}}{r^3}\right)}_{0} - \vec{m}\cdot\vec{\nabla}\left(\frac{\vec{r}}{r^3}\right) = \frac{3(\vec{m}\cdot\vec{r})\vec{r} - r^2\vec{m}}{r^5} \tag{18.10.10}$$

In analogy to an electric dipole in an electric field, the energy of a magnetic dipole in a field $\vec{B}$ is given by $-\vec{m}\cdot\vec{B}$, the torque exerted on the loop equals $\vec{m}\times\vec{B}$, and the force is $\left(\vec{m}\cdot\vec{\nabla}\right)\vec{B}$.

18.11 MAGNETIZATION

If a material contains $N^{(j)}(\vec{r})$ atomic and molecular magnetic dipoles aligned in the jth direction with magnetic moments $\vec{m}^{(j)} = IA\hat{e}_j$ within a volume $\Delta\varsigma$ around the point x, the resulting jth component of the magnetization is defined as

$$\left[\vec{M}(\vec{r})\right]_j \equiv M_j(\vec{r}) = \frac{m^{(j)}N^{(j)}(\vec{r})}{\Delta\varsigma} = \frac{IAN^{(j)}(\vec{r})}{\Delta\varsigma} \tag{18.11.1}$$

Accordingly, consider two stacks of height Δz containing $N^{(z)}(y_0 - \Delta y/2)$ and $N^{(z)}(y_0 + \Delta y/2)$ wire squares carrying a positive, counterclockwise current I with oriented surface vectors $\vec{A} = A\hat{e}_z$ centered at $(x_0, y_0 - \Delta y/2, z_0)$ and $(x_0, y_0 + \Delta y/2, z_0)$, respectively, with $A = \Delta x \Delta y$, where Δx is the square length along x. Since by the right-hand rule, for positive, counterclockwise currents, the current contributions from the stacks at $y_0 - \Delta y/2$ and at $y_0 + \Delta y/2$ are directed in the $-\hat{e}_x$ and $+\hat{e}_x$ direction, respectively, along the $y = y_0$ plane between the two stacks, the total bound current in the $\hat{e}_x$ direction along this plane is

$$I_x^{(\text{bound},z)}(\vec{r}_0) = I\left(N^{(z)}\left(y_0 + \frac{\Delta y}{2}\right) - N^{(z)}\left(y_0 - \frac{\Delta y}{2}\right)\right) \times \underbrace{\frac{A\Delta z}{A\Delta z}}_{\varsigma} \times \frac{\Delta y}{\Delta y} \approx \frac{\partial M_z}{\partial y}\Delta y \Delta z \tag{18.11.2}$$

A similar expression arises from the change with respect to z of the density of y-oriented dipoles except for the presence of an additional minus sign. Summing the two currents and dividing by the surface area $\Delta y \Delta z$ of the dipole layers perpendicular to the x-direction yields

$$\left[\vec{J}_{\text{bound}}\right]_x = \frac{I_x^{(\text{bound},z)} + I_x^{(\text{bound},y)}}{\Delta y \Delta z} = \left(\vec{\nabla} \times \vec{M}\right)_x \tag{18.11.3}$$

Since the above formula is valid under the cyclic transformation $x \to y \to z \to x$,

$$\vec{J}_{\text{bound}} = \vec{\nabla} \times \vec{M} \tag{18.11.4}$$

so that, recalling that J_{free} refers to currents that arise from the macroscopic transport of charges, while $\vec{B}$ is directly associated with the force on a test charge and is therefore sourced by both macroscopic and microscopic currents,

$$\vec{\nabla} \times \vec{B} = \mu_0 \vec{J}_{\text{total}} = \mu_0\left(\vec{J}_{\text{free}} + \vec{J}_{\text{bound}}\right) \tag{18.11.5}$$

and

$$\vec{B} = \mu_0\left(\vec{H} + \vec{M}\right) \tag{18.11.6}$$

In general, $\vec{M} = \chi \vec{H}$, where the *magnetic susceptibility* χ is a tensor (matrix) and field-dependent quantity since the induced magnetic moment in ordered materials is not necessarily aligned with the applied magnetic field. The corresponding *magnetic permittivity tensor*

$$\boldsymbol{\mu} = \mu_0(\mathbf{I} + \boldsymbol{\chi}) \tag{18.11.7}$$

In iron, $|\mu| \sim 10^2 - 10^3$. Although the macroscopic *curl* of $\vec{H}$ is determined by J_{free}, from $\vec{\nabla} \cdot \vec{B} = 0$,

$$\vec{\nabla} \cdot \vec{H} = -\vec{\nabla} \cdot \vec{M} \tag{18.11.8}$$

Example

For a cylindrical magnet of length L with a constant magnetization $\vec{M}$ in the $\hat{e}_z$ direction, from the absence of free currents, $\vec{\nabla} \times \vec{H} = 0$ while $\vec{\nabla} \cdot \vec{H} = \rho_m$ where $\rho_m = -\vec{\nabla} \cdot \vec{M}$ can be interpreted as a uniform positive "magnetic charge" sheet with width $\Delta z \to 0$ and charge density $\sigma_m = \rho_m \Delta z = -\Delta z (dM_z/dz) = M$ at the top of the bar with a corresponding negative charge sheet along its bottom. Hence, $\vec{H}$, which is oriented along $-\hat{e}_z$, is determined by the same formulas as the electric field between two finite area charged sheets and thus approaches zero as $L \to \infty$ while $\vec{B} = \mu_0 (\vec{H} + \vec{M}) \approx \mu_0 \vec{M}$. Alternatively, $J_{\text{bound}} = \vec{\nabla} \times \vec{M}$ is zero except at the cylinder surface where the bound surface current $\left(\vec{K}_{\text{bound}}\right)_\theta = \Delta r \left(\vec{\nabla} \times \vec{M}\right)_\theta = \Delta r (dM_z/dr) \hat{e}_r \times \hat{e}_z = (-M)(-\hat{e}_\theta)$. Hence, the $\vec{B}$ field is approximated by that of a long solenoid with surface current $nI = K = M$, namely, $\vec{B} \approx \mu_0 \vec{M}$, again implying $\vec{H} \approx 0$.

18.12 INDUCTION AND FARADAY'S LAW

Faraday's law states that a time-varying *magnetic flux*, Φ_B, through a surface S bounded by a curve C generates an electric field with an integrated value around C termed the *induced emf* (*electromotive force*), given by

$$\mathcal{V}_{\text{emf}} \equiv \frac{1}{q} \oint_{CCS} \vec{F} \cdot d\vec{l} = \oint_{CCS} \left(\vec{E} + \vec{v} \times \vec{B}\right) \cdot d\vec{l} = -\frac{d}{dt} \int_S \vec{B} \cdot d\vec{A} \equiv -\frac{d\Phi_B}{dt} \tag{18.12.1}$$

where $\vec{E}$ and $\vec{B}$ are measured in the same inertial frame. The positive direction along C is determined by the orientation of the chosen surface vector of S according to the right-hand rule. That is, if a wire loop is placed along C, the magnetic flux change induces a net voltage change of $\mathcal{V}_{\text{emf}}$ around the loop. This produces the same effect as inserting a battery of voltage $\mathcal{V}_{\text{emf}}$ at any point on C such that a *positive* $\mathcal{V}_{\text{emf}}$ corresponds to the positive voltage terminal of this battery positioned further along the positive direction on C than the negative voltage terminal. The minus sign in Equation (18.12.1), commonly referred to as *Lenz's law*, indicates that the induced current generates a magnetic field that counteracts the change in the applied magnetic field, consistent with energy conservation.

Faraday's law can be decomposed into the sum of two contributions, one arising from a changing magnetic field in a region with instantaneously fixed boundaries and the second associated with changes in the boundaries of the region for an instantaneously constant magnetic field, which generates a force on the carriers and hence an emf according to $\vec{F}_{\text{magnetic}} = q\vec{v} \times \vec{B}$. That is, since

$$\frac{d}{dt}\int_{S(t)} \vec{B}(t)\cdot d\vec{S} = \lim_{\Delta t\to 0}\frac{1}{\Delta t}\left[\int_{S(t+\Delta t)} \underbrace{\vec{B}(t+\Delta t)}_{B(t)+\frac{\partial B}{\partial t}\Delta t+\ldots}\cdot d\vec{S} - \int_{S(t)} \vec{B}(t)\cdot d\vec{S}\right]$$

$$= \int_{S(t)} \frac{\partial \vec{B}(t)}{\partial t}\cdot d\vec{S} + \lim_{\Delta t\to 0}\frac{1}{\Delta t}\left[\int_{S(t+\Delta t)} \vec{B}(t)\cdot d\vec{S} - \int_{S(t)} \vec{B}(t)\cdot d\vec{S}\right]$$

$$\equiv \frac{\partial}{\partial t}\int_{S(t)} \vec{B}(t)\cdot d\vec{S}\bigg|_{S} + \frac{d}{dt}\oint_{\Delta S(t)} \vec{B}(t)\cdot d\vec{A}\bigg|_{B} \tag{18.12.2}$$

Equation (18.12.1) can be written schematically as

$$\mathcal{V}_{\text{emf}} = \underbrace{\oint_{C\subset S} \vec{E}\cdot d\vec{l}}_{=-\int_S \frac{\partial \vec{B}}{\partial t}\cdot d\vec{A}\big|_S} + \underbrace{\oint_{C(t)\subset S(t)} \left(\vec{v}\times\vec{B}\right)\cdot d\vec{l}}_{=-\frac{d}{dt}\int_{\Delta S(t)}\vec{B}\cdot d\vec{A}\big|_B} \tag{18.12.3}$$

That the second term in the above equation arises from the change in the surface is verified by noting that, since $\left|\vec{A}\times\vec{B}\right|$ is the area of the parallelogram with sides $\vec{A}$ and $\vec{B}$, if $C(\vec{r},t)$ changes by an amount $d\vec{r}$ at $\vec{r}$ over a time dt, $\vec{v}\times d\vec{l} = \left(d\vec{r}\times d\vec{l}\right)/dt = dA/dt\hat{e}_{\vec{v}\times d\vec{l}}$ where $\hat{e}_{d\vec{A}} = \hat{e}_{\vec{v}\times d\vec{l}}$ is oriented along the normal to the surface element $d\vec{A}$. Hence,

$$\oint_{C(t)} \left(\vec{v}\times\vec{B}\right)\cdot d\vec{l} = -\oint_{C(t)} \left(\vec{v}\times d\vec{l}\right)\cdot\vec{B} = -\frac{d}{dt}\oint_{\Delta S(t)} \vec{B}\cdot d\vec{A}\bigg|_{B} \tag{18.12.4}$$

If the curve C is instead time independent,

$$\oint_{C\subset S} \vec{E}\cdot d\vec{l} = \oint_S \left(\vec{\nabla}\times\vec{E}\right)\cdot d\vec{A} = -\int_S \frac{\partial \vec{B}}{\partial t}\cdot d\vec{A} \tag{18.12.5}$$

Which, as C can represent an arbitrary infinitesimal curve at any point, implies the *Maxwell–Faraday equation*

$$\vec{\nabla} \times \vec{E} = -\frac{\partial \vec{B}}{\partial t} \qquad (18.12.6)$$

Equation (18.12.3) leads to three standard problem types. (1) If the magnitude or direction of the enclosed magnetic field changes through a fixed curve in space or wire loop, the induced emf is evaluated from the first term of Equation (18.12.5) after computing the surface integral of $\vec{B}(t)$. (2) For loops with changing orientation or cross-sectional area at a fixed magnetic field, only the second term in Equation (18.12.3) is present and is evaluated either by performing a line integral of $\vec{v} \times \vec{B}$ or again by calculating the time derivative of the magnetic flux. (3) When both the field and the bounding curve vary with time, both terms in Equation (18.12.3) contribute, and these can be either evaluated separately in their top or equivalent bottom forms in this equation or $-d\Phi_B/dt$ can again be employed.

Examples

1. If the strength of a wire-like filament of magnetic field changes within a curve C in space at rest in an inertial frame, the line integral of the electric field around this curve is identical to the line integral of the $\vec{H}$ field for the equivalent electric current $\vec{J}(\vec{r}) = \partial\vec{B}(\vec{r},t)/\partial t$. That is, $\partial B/\partial t$ can be viewed as a magnetic current that generates a rotational $\vec{E}$ field in space in exactly the same manner that an electric current $\vec{J}(r)$ induces a rotational $\vec{H}$ field.
2. In an *electric AC generator*, a wire is wound N times around the edges of a planar geometrical shape of cross-sectional area A to form N identical connected loops such that the flux through the entire wire therefore equals N times the flux through a single turn. Consequently, rotating the wire in a constant $\vec{B}$ field with an angular frequency ω such that the surface vector $\vec{S}$ is parallel to the magnetic field at $t = 0$ ($\vec{S}$ is perpendicular to the plane of the loop) produces an emf:

$$\mathcal{V}_{emf} = -N\frac{\partial \Phi_B}{\partial t} = -\frac{\partial}{\partial t}(NBA\cos\omega t) = NBA\omega\sin\omega t \qquad (18.12.7)$$

where Φ_B denotes the flux through a single loop of wire.

3. A simple *DC electric generator* consists of a bar moving in the $z = 0$ plane with velocity $v_x\hat{e}_x$ on two parallel rails at $y = 0$ and $y = L$ in a uniform magnetic field $\vec{B} = B\hat{e}_z$. If a resistor R connects the two rails at, e.g., $x = 0$, the flux passing

through the loop formed by the rails, the moving bar, and the resistor, $\Phi_B = BLx$. Then, *from the right part of Equation (18.12.1),* $\mathcal{V}_{\text{emf}} = -\partial\Phi_B/\partial t = -BLv_x$. *However, from the left part of the same equation, the emf can be evaluated directly* from the magnetic force $\vec{F}_{\text{magnetic}} = -qv_xB\hat{e}_y$ on a fictitious positive charge in the rail (which equals the electric force in a frame where the rail is at rest) with an associated emf $\mathcal{V}_{\text{emf}} = +\vec{F}_{\text{magnetic}} \cdot \vec{L}/q = -BLv_x$. The sign indicates that the direction from the positive to negative terminals of the effective battery (generator) voltage is opposite to the positive, here counterclockwise, orientation of the loop (the direction of the right-hand fingers when the thumb points along the surface vector that is aligned with the magnetic field). This generates a *clockwise current* for finite R with an associated $\vec{B}$ field that counteracts the flux change. Here, fictitious positive charges in the bar move in the $-\hat{e}_y$ direction, producing a force on the bar of $\vec{F}^{\,(\text{bar})}_{\text{magnetic}} = -Q^{(\text{bar})}v_yB\hat{e}_x$, with $Q^{(\text{bar})}$ the total charge in the bar; to maintain the motion requires a power $P = -\vec{F}^{\,(\text{bar})}_{\text{magnetic}} \cdot \vec{v} = Q^{(\text{bar})}v_yBv_x = \mathcal{V}_{\text{emf}}I$.

As the magnetic field generated by a time-varying current, I_1, in a wire loop, 1, varies directly with I_1, the resulting magnetic flux $\Phi_{B,2}$ enclosed by a second loop, 2, can be written as

$$\Phi_{B,2} = L_{21}I_1 \tag{18.12.8}$$

where the constant L_{21} is termed the *mutual inductance* with $L_{12} = L_{21}$. The emf in the second loop is then

$$V_2 \equiv \mathcal{V}_{\text{emf},2} = -\frac{d\Phi_{B,2}}{dt} \equiv -L_{21}\frac{dI_1}{dt} \tag{18.12.9}$$

Similarly, a changing current in the first loop alters the magnetic field and therefore the flux contained within the loop itself. The resulting coefficient of *self-inductance* is defined as

$$V_1 \equiv \mathcal{V}_{\text{emf},1} = -\frac{d\Phi_{B,1}}{dt} \equiv -L_{11}\frac{dI_1}{dt} \tag{18.12.10}$$

so that, where the indices are omitted when only loop 1 is present,

$$\Phi_{B,1} = L_{11}I_1 \tag{18.12.11}$$

Examples

To find the (self) inductance of a solenoid (coil) of length l consisting of N turns of cross-sectional area A, since $B = \mu nI$ with $n = N/l$ inside the solenoid, from Ampere's law, $\Phi_{m,1}^{\text{single}}$ turn $= BA = \mu nIA$. Since this flux passes through each of turn of the solenoid, the total enclosed flux $\Phi_{m,1}^{\text{N turns}} = N\mu nIA = \mu n^2 lIA = L_{11}I$, yielding $L \equiv L_{11} = \mu n^2 lA$.

In a *transformer*, two wires are wound N_1 and N_2 times around a common metal core with cross-sectional area A. Since the magnetization $\left|\vec{M}\right| \gg \left|\vec{H}\right|$, the $\vec{B}$ field with $\vec{B} = \mu_0\left(\vec{H} + \vec{M}\right) \approx \mu_0 \vec{M}$ is effectively confined to the metal core. Hence, the $\vec{B}$ field, $\mu n_1 I_1$, generated by a current flowing in the first wire yields a total flux, $\Phi_1 = N_1 \mu n I_1 A$, through the first wire and a flux $\Phi_1 = N_2 \mu n I_1 A$ in the second wire. From $V = -d\Phi/dt$, the ratio of input and output voltages is therefore $V_1/V_2 = N_1/N_2$.

18.13 CIRCUIT THEORY AND KIRCHOFF'S LAWS

Electrical circuits containing resistors, capacitors, and inductors are analyzed by applying *Kirchoff's laws* to a circuit diagram that depicts electronic components linked by wire segments that in turn intersect at junctions. The current in the kth segment is labeled I_k and is assigned an arrow in the assumed direction of current flow (the direction a positive charge moves, which is opposite the motion of electrons).

If the direction of the arrow of I_k is drawn incorrectly, its computed value will be negative. Kirchoff's laws state: (1) From charge conservation, the total current flowing into a junction is zero. Thus, e.g., if at a junction of three wire segments, the current arrows for I_1 and I_2 point toward the junction, while I_3 is directed away from the junction, $I_1 + I_2 + (-I_3) = 0$. (2) The sum of the changes in voltage ΔV around any closed loop equals zero since the voltage is a single-valued function of position. Here, when traversing (i) a generator or a battery in the direction from its negative to its positive terminal (denoted schematically as small and large lines perpendicular to the wire direction), $\Delta V = \mathcal{V}_{\text{emf}}$ (the emf or voltage); (ii) a resistor in the assumed current direction (the direction of the current arrow), $\Delta V = -IR$; (iii) a capacitor in the current direction, $\Delta V = -Q/C$; and (iv) an *inductor* in the current direction, $\Delta V = -LdI/dt$. The sign changes if a component is traversed in the direction opposite to that of the current arrow. Physically, (fictitious) positive charges transported from the negative to the positive terminal of a battery gain an electric potential $\mathcal{V}_{\text{emf}}$. If these charges pass through a resistor, they accumulate on the side of the resistor away from the current direction, which therefore acquires a higher potential than the opposing terminal leading to a voltage change of $\Delta V = -IR$. Positive charges entering a capacitor similarly accumulate at the terminal opposite the direction of the current resulting in a $-Q/C$

voltage change. Finally, an increase in the current yields a counteracting voltage change $-L\partial I/\partial t$ across the inductor which functions as an effective battery with polarity opposing the applied source of emf. In a linear circuit, R, C, and L do not depend on voltage or current, and Kirchoff's laws generate a separate linear equation for each loop and junction. These are often simplified by utilizing circuit symmetries. The standard electrical convention of assigning voltage differences across circuit elements, which will be used below, employs the voltage drop V in place of $-\Delta V$ above so that $V = -L dI/dt$, $V = Q/C$, and $V = IR$.

Since I/q charges per second lose an energy qV in passing through a resistor, the power dissipation $P = VI = I^2R = V^2/R$. Energy is instead stored in a capacitor by displacing electrons between the opposing plates. When the voltage between the two plates equals $V_{\text{intermediate}}$, an energy $qV_{\text{intermediate}}$ is required to transport the subsequent electron. Since this varies linearly with $V_{\text{intermediate}}$, to charge the capacitor to a final voltage V requires an energy given by the total transferred charge times the average potential, $E = QV/2 = CV^2/2$. As the mechanical state of the capacitor is identical before and after charging, for conservation of energy to be valid, this energy can only have been transferred to the electric field. Finally, an energy $dE = LI\,dI/dt$ is expended in transferring I charges per second over a voltage $V = L dI/dt$ in an inductor. Hence, to reach a final current I requires an energy, which is stored in the magnetic field:

$$E = \int_{t_{\text{initial}}}^{t_{\text{final}}} LI\frac{dI}{dt}dt = L\int_0^{I_{\text{final}}} I dI = \frac{1}{2}LI_{\text{final}}^2 \tag{18.13.1}$$

In an alternating current (AC) circuit for which the voltages and currents vary sinusoidally, squared quantities such as $(V(t))^2 = V_0^2 \sin^2(\omega t + \delta)$ or $I(t)V(t) = I_0 V_0 \sin^2(\omega t + \delta)$, when $I(t)$ and $V(t)$ are in phase, are averaged over time by $\sin^2(\omega t + \delta)$ by 1/2.

Examples

Placing a capacitor, resistor, and inductor in *series* (i.e., one after another) with a time-varying voltage source $\mathcal{V}_{\text{emf}}(t)$ yields, from the above equations together with $I = dQ/dt$,

$$\mathcal{V}_{\text{emf}}(t) = V_{\text{capacitor}} + V_{\text{resistor}} + V_{\text{inductor}} = \left(\frac{1}{C} + R\frac{\partial}{\partial t} + L\frac{\partial^2}{\partial t^2}\right)Q(t) \tag{18.13.2}$$

If the inductor is absent, the capacitor is initially uncharged, and a constant voltage source is introduced at $t = 0$ so that $\mathcal{V}_{\text{emf}}(t < 0) = 0$ while $\mathcal{V}_{\text{emf}}(t > 0) = \mathcal{V}_{\text{emf}}$,

$$\mathcal{V}_{\text{emf}} = \frac{Q}{C} + R\frac{dQ}{dt}, \quad Q(0) = 0 \tag{18.13.3}$$

Incorporating the boundary condition through the limits of integration,

$$\int_{Q(0)}^{Q(t)} \frac{dQ'}{C\mathcal{V}_{\text{emf}} - Q'} = \frac{1}{RC}\int_0^t dt' \tag{18.13.4}$$

and therefore

$$-\log\left(\frac{C\mathcal{V}_{\text{emf}} - Q(t)}{C\mathcal{V}_{\text{emf}} - Q(0)}\right) = \frac{t}{RC} \tag{18.13.5}$$

Applying the boundary condition $Q(0) = 0$ and exponentiating,

$$Q(t) = C\mathcal{V}_{\text{emf}}\left(1 - e^{-\frac{t}{RC}}\right) \tag{18.13.6}$$

For times long compared to the characteristic time constant $\tau = RC$, the charge on the capacitor approaches the steady-state value $C\mathcal{V}_{\text{emf}}$. If the capacitor is replaced by an inductor,

$$\mathcal{V}_{\text{emf}}(t) = RI + L\frac{\partial I}{\partial t} \tag{18.13.7}$$

Here R, C, and Q in Equation (18.13.3) are replaced by L, $1/R$, and I, yielding a time constant $\tau = L/R$.

For a charged capacitor connected at $t = 0$ to an inductor, Equation (18.13.2) becomes

$$\frac{d^2Q(t)}{dt^2} = -\frac{1}{LC}Q(t), \quad Q(0) = Q_0, \quad I(0) = \left.\frac{dQ(t)}{dt}\right|_{t=0} = 0 \tag{18.13.8}$$

with the solution $Q(t) = Q_0 \cos(\omega t)$ where $\omega^2 = 1/LC$. The capacitor cannot immediately discharge as the inductor resists a sudden current increase. Instead, the current increases continuously, reaching its maximum value when the capacitor is completely discharged. The inductor now resists a *decreasing* current, and the capacitor therefore charges in the reverse direction. After half a period of oscillation, the capacitor is recharged to its initial value, but with opposite polarity.

Two resistors or inductors in *series* are placed one after the other in a circuit. The same current then passes through both components, while the voltage drops over the two components add, e.g., $\Delta V_{\text{tot}} = IR_1 + IR_2 \equiv IR_{\text{tot, series}}$ or $\Delta V_{\text{tot}} = (L_1 + L_2)\partial I/\partial t \equiv L_{\text{tot, series}}\,\partial I/\partial t$. Therefore,

$$R_{\text{tot,series}} = R_1 + R_2, \quad L_{\text{tot,series}} = L_1 + L_2 \tag{18.13.9}$$

In *parallel*, corresponding terminals of each device are connected, resulting in identical applied voltages. Consequently, $I_{tot} = V/R_{tot,\ parallel} = I_1 + I_2 = V/R_1 + V/R_2$ and

$$\frac{1}{R_{tot,parallel}} = \frac{1}{R_1} + \frac{1}{R_2}, \quad \frac{1}{L_{tot,parallel}} = \frac{1}{L_1} + \frac{1}{L_2} \tag{18.13.10}$$

The total charge over two parallel capacitors is $Q_{tot} = C_{tot}V = C_1V + C_2V$, while for capacitors in series, the charge across each capacitor is identical (the total charge enclosed by a surface that encloses the two adjacent inner capacitor plates and their connecting wire remains equals the static charge residing on the two inner plates independent of V). Hence, $V_{tot} = Q/C_{tot} = Q/C_1 + Q/C_2$ and

$$\frac{1}{C_{tot,series}} = \frac{1}{C_1} + \frac{1}{C_2}, \quad C_{tot,parallel} = C_1 + C_2 \tag{18.13.11}$$

In *linear* AC circuits $Q(t)$, $I(t)$, and $V(t)$, all vary sinusoidally at the same frequency although their phases are retarded or advanced relative to each other by capacitors and inductors. Expressing these as complex *phasor* quantities $\tilde{Q}, \tilde{I}, \tilde{V}$ multiplied by $\exp(i\omega t)$,

$$V(t) = \mathrm{Re}\left(\tilde{V}e^{i\omega t}\right) = L\frac{d}{dt}\mathrm{Re}\left(\tilde{I}e^{i\omega t}\right) = \mathrm{Re}\left(i\omega L\tilde{I}e^{i\omega t}\right) \tag{18.13.12}$$

For a capacitor,

$$I(t) = \mathrm{Re}\left(\tilde{I}e^{i\omega t}\right) = \frac{dQ(t)}{dt} = C\frac{d}{dt}\mathrm{Re}\left(\tilde{V}e^{i\omega t}\right) = \mathrm{Re}\left(iC\omega\tilde{V}e^{i\omega t}\right) \tag{18.13.13}$$

Accordingly, for AC circuits, resistors, capacitors, and inductors can be modeled as generalized resistors with R replaced by the *impedance*, $Z = \tilde{V}/\tilde{I}$, of each component, where

$$\begin{aligned} Z_{resistor} &= R \\ Z_{capacitor} &= \frac{1}{i\omega C} \\ Z_{inductor} &= i\omega L \end{aligned} \tag{18.13.14}$$

After all impedances are combined by applying the formulas for resistors in parallel and series, the output quantity equals the real part of its phasor multiplied by $\exp(i\omega t)$.

Example

The current through a resistor and capacitor placed in series with an alternating voltage source is obtained by combining the impedances of R and C. Accordingly, from

$$I(t) = \mathrm{Re}\left(\frac{\tilde{V}}{Z_{\mathrm{total}}} e^{i\omega t}\right) \tag{18.13.15}$$

the phases satisfy $\arg(I(t)) = \arg(V(t)) - \arg(Z_{\mathrm{total}})$ with

$$Z_{\mathrm{total}} = Z_{\mathrm{resistor}} + Z_{\mathrm{capacitor}} = \frac{-i}{\omega C} + R = \sqrt{R^2 + \left(\frac{1}{\omega C}\right)^2} e^{-i\tan^{-1}\left(\frac{1}{\omega RC}\right)} \tag{18.13.16}$$

.

18.14 CONSERVATION LAWS AND THE STRESS TENSOR

The power supplied to charges within a volume ς per unit time or, equivalently, the work performed on these charges per unit time can be expressed entirely in terms of fields as

$$\begin{aligned} P &= \int_{\varsigma} \vec{F} \cdot \vec{v} d^3 r \\ &= \int_{\varsigma} \rho\left(\vec{E} + \vec{v} \times \vec{B}\right) \cdot \vec{v} d^3 r = \int_V \rho \vec{E} \cdot \vec{v} d^3 r = \int_V \vec{E} \cdot \vec{J} d^3 r \\ &= \int_{\varsigma} \vec{E} \cdot \left(\vec{\nabla} \times \vec{H} - \frac{\partial \vec{D}}{\partial t}\right) d^3 r \end{aligned} \tag{18.14.1}$$

Employing (recall that the subscript E or H denotes the field that the gradient is applied to and $A \cdot (B \times C) = B \cdot (C \times A)$)

$$\begin{aligned} \vec{\nabla} \cdot \left(\vec{E} \times \vec{H}\right) &= \vec{\nabla}_E \cdot \left(\vec{E} \times \vec{H}\right) + \vec{\nabla}_H \cdot \left(\vec{E} \times \vec{H}\right) \\ &= \vec{H} \cdot \left(\vec{\nabla}_E \times \vec{E}\right) - \vec{E} \cdot \left(\vec{\nabla}_H \times \vec{H}\right) \\ &= \vec{H} \cdot \left(\vec{\nabla} \times \vec{E}\right) - \vec{E} \cdot \left(\vec{\nabla} \times \vec{H}\right) \end{aligned} \tag{18.14.2}$$

together with $\vec{\nabla} \times \vec{E} = -\partial \vec{B} / \partial t$ and Gauss's law yields

$$P = -\int_{S \subset \varsigma} \underbrace{\left(\vec{E} \times \vec{H}\right)}_{\vec{S}_P} \cdot d\vec{A} - \frac{d}{dt} \int_{\varsigma} \underbrace{\frac{1}{2}\left(\vec{B} \cdot \vec{H} + \vec{E} \cdot \vec{D}\right)}_{w_E} d^3 r \tag{18.14.3}$$

where the *energy density* w_E is the energy per unit volume contained in the electric field distribution, while the *Poynting vector* $\vec{S}_P$ corresponds to the energy transported by the electric fields through a perpendicular cross section of unit area per unit time (if the Poynting vector points into the volume, P and therefore the work performed on the charges is positive).

The expression for w_E can be reproduced by first calculating the energy in an ideal infinitesimal parallel plate capacitor, which effectively confines the fields to the volume between the plates. With $E = V/d = \sigma/\varepsilon$, where $\sigma = Q/A$, where d and A are the separation and the area of the plates, from the discussion above Equation (18.13.1) and noting that $\vec{D}$ and $\vec{E}$ are parallel,

$$w_e = \frac{1}{2}QV = \frac{1}{2}(\varepsilon AE)(Ed) = \frac{1}{2}E(\varepsilon E)Ad = \frac{1}{2}\vec{E}\cdot\vec{D}\,d^3r \tag{18.14.4}$$

Since in a thin capacitor fringing fields can be neglected, an arbitrary field can be represented as the field generated by an infinite number of infinitesimal thin capacitors positioned throughout space, leading to the electric component in w_E. The magnetic term is similarly obtained by noting that $\vec{H}$ and $\vec{B}$ are similarly confined by a long solenoid. Therefore, any magnetic field can be represented as that formed from an infinite set of solenoids filling space, each directed along a magnetic field line. For each infinitesimal solenoid with n turns per unit length, $L = \mu n^2 lA$, hence from $W = LI^2/2$ and $B = \mu H = \mu nI$,

$$w_m = \frac{1}{2}(\mu nI)(nI)Al = \frac{1}{2}\vec{B}\cdot\vec{H}\,d^3r \tag{18.14.5}$$

As well, the Lorentz force on a continuous isolated charge and current distribution can be written exclusively in terms of electric and magnetic fields according to

$$\vec{F} = \frac{d\vec{p}}{dt} = \int_\varsigma \left(\rho\vec{E} + \vec{J}\times\vec{B}\right)d^3r = \int\left\{\left(\vec{\nabla}\cdot\vec{D}\right)\vec{E} + \left(\vec{\nabla}\times\vec{H} - \frac{\partial D}{\partial t}\right)\times B\right\}d^3r \tag{18.14.6}$$

For a spatially homogeneous medium, $\vec{B} = \mu\vec{H}$ and $\vec{D} = \varepsilon\vec{E}$, yielding

$$\begin{aligned}
\left(\left(\vec{\nabla}\times\vec{H}\right)\times\vec{B}\right)_a &= \underbrace{\varepsilon_{aip}}_{\varepsilon_{ipa}}\left(\varepsilon_{ijk}\partial_j H_k\right)B_p \\
&= \left(\delta_{jp}\delta_{ka} - \delta_{ja}\delta_{kp}\right)\left(\partial_j H_k\right)B_p \\
&= \left(\partial_j H_a\right)B_j - \left(\partial_a H_k\right)B_k \\
&= \partial_j\left(H_a B_j\right) - \underbrace{H_a\partial_j B_j}_{0} - \underbrace{\left(\partial_a H_k\right)B_k}_{\frac{1}{2}\partial_a(H_k B_k)} \\
&= \partial_j\left(H_a B_j - \frac{1}{2}\vec{H}\cdot\vec{B}\,\delta_{aj}\right)
\end{aligned} \tag{18.14.7}$$

and

$$-\left(\frac{\partial \vec{D}}{\partial t}\times\vec{B}\right)_a+\left(\vec{\nabla}\cdot\vec{D}\right)E_a=-\left(\frac{\partial}{\partial t}\left(\vec{D}\times\vec{B}\right)\right)_a+\underbrace{\left(\frac{\vec{D}\times\partial\vec{B}}{\partial t}\right)_a}_{-\left(\vec{D}\times\left(\vec{\nabla}_E\times\vec{E}\right)\right)=\left(\vec{D}\cdot\vec{\nabla}_E\right)\vec{E}-\vec{\nabla}_E\left(\vec{D}\cdot\vec{E}\right)}+\left(\vec{\nabla}\cdot\vec{D}\right)E_a$$

$$=-\left(\frac{\partial}{\partial t}\left(\vec{D}\times\vec{B}\right)\right)_a+\underbrace{D_j\left(\partial_j E_a\right)}_{\partial_j\left(E_aD_j\right)-E_a\left(\partial_j D_j\right)}\underbrace{-D_j\left(\partial_a E_j\right)}_{\frac{1}{2}\partial_a\left(E_jD_j\right)}+\left(\partial_j D_j\right)E_a$$

$$=-\left(\frac{\partial}{\partial t}\left(\vec{D}\times\vec{B}\right)\right)_a+\partial_j\left(E_aD_j-\frac{1}{2}\vec{E}\cdot\vec{D}\,\delta_{aj}\right)$$

(18.14.8)

With $c^2 = 1/\mu\varepsilon$,

$$\frac{d\vec{p}}{dt}+\frac{1}{c^2}\frac{d}{dt}\int_\varsigma\left(\vec{E}\times\vec{H}\right)d^3r=\int_{S\subset\varsigma}\mathbf{T}\cdot d\vec{S} \quad (18.14.9)$$

after applying Gauss's law. The i, j component of the *Maxwell stress tensor tensor* (e.g., matrix) **T**

$$T_{ij}=E_iD_j+B_iH_j-\frac{1}{2}\left(\vec{E}\cdot\vec{D}+\vec{B}\cdot\vec{H}\right)\delta_{ij} \quad (18.14.10)$$

corresponds to the ith component of the momentum crossing a surface perpendicular to the jth coordinate direction per unit time, while the *momentum density* in the electromagnetic fields (the momentum per unit volume) equals $(\vec{E}\times\vec{H})/c^2$. Since the energy transported through a unit cross-sectional area in one second is distributed over a total volume equal to c, comparing with the discussion following Equation (18.14.3) indicates that the momentum in the field satisfies $p = E/c$, as required by relativity. Note that in analogy to the relationship $\vec{F} = -\vec{\nabla}U$ between force and the potential energy of a scalar field, for the vector electromagnetic fields, the force is directly related to the divergence of the Maxwell stress tensor.

Examples

1. In a cylindrical resistor of radius a with a positive terminal at $x = 0$ and a grounded terminal at $x = L$, the terminal voltages differ by IR, yielding an electric field $E = -\Delta V/\Delta x = IR/L\hat{e}_x$ within the resistor and a magnetic field $H = I/2\pi a\hat{e}_\theta$ at its cylindrical outer surface. Integrating the Poynting vector $-\hat{e}_r I^2R/2\pi aL$ over this surface indicates that the work done on the charges inside the resistor and hence

their mechanical energy (which evolves into heat) increases by an amount I^2R per unit time consistent with Equation (18.14.3). Parenthetically, as the boundary conditions insure the continuity of the tangential electric field component at the circular outer surface of the resistor, the Poynting vector is identical infinitesimally inside or outside the resistor. That is, the surface layers of positive and negative charge that accumulate at the high- and low-voltage end contacts, respectively, generate $\vec{E}$ and $\vec{D}$ fields that obey $\vec{E} = \varepsilon_0 \varepsilon_r \vec{D}$ both inside and outside the resistor. However, since both the tangential component of $\vec{D}$ and the normal component of $\vec{E}$ (as a result of the existence of a surface charge layer) are discontinuous along the cylindrical outer surface, $\vec{E}_{\|}$ can remain continuous.

2. For a cylindrical capacitor with a radius R, length L, and plate area $A = \pi R^2$ charging from zero to a charge Q along the lower plate at $z = 0$ in time T, the electric field within the capacitor varies linearly from 0 to $Q/\varepsilon A \hat{e}_z$, while the magnetic field at the outer surface of the capacitor, which remains constant in time, is sourced by the displacement current density $J_{\text{displacement}} = Q/(AT)$ and hence equals $\vec{H} = Q/(2\pi RT)\hat{e}_\theta$. The product of the average value of the Poynting vector over the charging time with $2\pi RLT$ therefore yields the final stored energy $QV/2 = QEL/2$.

The inductance or capacitance of a component can be obtained by evaluating the total energy in the magnetic or electric field and, respectively, equating the result to $LI^2/2$ or $Q^2/2C$.

Examples

For a long cylindrical transmission line of length Λ consisting of a negligibly thin inner cylinder of radius $r_<$ conducting a current I down the cylinder surrounded by a similarly thin sheath of radius $r_>$ carrying a return current $-I$ between the two conductors, $\vec{H} = I/2\pi r\hat{e}_\phi$ and hence

$$\frac{1}{2}\int_{r_<}^{r_>} \vec{B}\cdot\vec{H}d^3r = \frac{1}{2}\int_{r_<}^{r_>} \left(\frac{\mu I}{2\pi r}\right)\left(\frac{I}{2\pi r}\right)2\pi r\Lambda dr = \frac{\mu\Lambda}{4\pi}\log\left(\frac{r_>}{r_<}\right)I^2 = \frac{1}{2}LI^2 \quad (18.14.11)$$

which determines L. The capacitance is similarly obtained by computing the total electric field energy when charges Q and $-Q$ are located on the inner and outer conductors:

$$\frac{1}{2}\int_{r_<}^{r_>} \vec{D}\cdot\vec{E}\,d^3r = \frac{1}{2}\int_{r_<}^{r_>} \left(\frac{Q}{2\pi r\Lambda}\right)\left(\frac{Q}{\varepsilon 2\pi r\Lambda}\right)2\pi r\Lambda dr = \frac{Q^2}{4\pi\varepsilon\Lambda}\log\left(\frac{r_>}{r_<}\right)I^2 = \frac{Q^2}{2C} \quad (18.14.12)$$

The force exerted in a parallel plate capacitor by the positive plate at $z = 0$ on the negative plate with area A at $z = a$ can be evaluated by displacing this plate to

$z = a + \Delta z$ outward. The product of the energy density $\vec{E} \cdot \vec{D}/2$ with the additional volume results in a change in field energy $\Delta E = \sigma^2 A \Delta z / 2\varepsilon$, yielding a force $F_z/A = (-\Delta E/\Delta x)/A = -\sigma^2/2\varepsilon$ per unit area. This result can however also be obtained by integrating the stress tensor over a surface parallel to and between the two capacitor plates joined to a hemisphere in the $z > 0$ half-space at $r = \infty$. As the surface vector in the region between the two plates surrounding the plate at $z = a$ points in the $-\hat{e}_z$ direction,

$$\frac{\vec{F}}{A} = \frac{1}{A}\int_S \mathbf{T} \cdot d\vec{S} = -\mathbf{T} \cdot \hat{e}_z = \vec{E}\left[\vec{D} \cdot (-\hat{e}_z)\right] - \frac{1}{2}\vec{E} \cdot \vec{D}(-\hat{e}_z) = -\frac{\sigma^2}{2\varepsilon}\hat{e}_z \qquad (18.14.13)$$

18.15 LIENARD–WIECHERT POTENTIALS

In the Lorentz gauge, the electric and magnetic fields of time-varying source distributions can be directly determined from the Green's function, $G(\vec{r}-\vec{r}\,', t-t')$, of the wave operator, satisfying

$$\Box G(\vec{r}-\vec{r}\,', t-t') = \delta^3(\vec{r}-\vec{r}\,')\delta(t-t') \qquad (18.15.1)$$

The dependence on $\vec{r}-\vec{r}\,'$ and $t-t'$ instead of $(\vec{r}, t)$ and $(\vec{r}\,', t')$ individually results from the uniformity of space and time. The solution of

$$\Box \phi(\vec{r}, t) = \mu_0 J^i(\vec{r}, t) \qquad (18.15.2)$$

in the absence of boundaries is then

$$\phi(\vec{r}, t) = \phi_0(\vec{r}, t) + \mu_0 \int_S G(\vec{r}-\vec{r}\,', t-t') J^i(\vec{r}\,', t') d^3r' dt' \qquad (18.15.3)$$

Here, $\phi_0(r, t)$ solves the homogeneous equation $\Box \phi_0(\vec{r}, t) = 0$, typified by linear combinations of monochromatic plane waves, $Re\left\{A \exp\left(i\left[\omega t - \vec{k} \cdot \vec{r}\right]\right)\right\}$ with *wave-number* $k \equiv |\vec{k}| = \omega/c$, general nonmonochromatic planar $f(t - \hat{e}_{\hat{k}} \cdot \vec{r}/c)$ solutions or spherical solutions, $f(t \pm r/c)/r$, of $(1/c^2)\partial^2\phi_0(r, t)/\partial t^2 - (1/r)\partial^2 r\phi_0(r, t)/\partial r^2 = 0$.

Heuristically, $G(\vec{r}, t)$ resembles the potential of a point charge of magnitude $q = -\varepsilon$ that exists at $r = 0$ for at a single instant in time. Hence, for $t \approx 0$, $\phi(r, t) \approx \delta(t)/4\pi r$. Comparing to the general radially outgoing solution to the wave equation, $f(t - r/c)/r$, yields

$$G(\vec{r},t)=\frac{1}{4\pi r}\delta\left(t-\frac{r}{c}\right) \tag{18.15.4}$$

To verify that this corresponds to the Green's function, observe that (the last line requires $\delta^3(\vec{r})\delta(t-r/c)=\delta^3(\vec{r})\delta(t)$)

$$\underbrace{\vec{\nabla}}_{\hat{e}_r\frac{\partial}{\partial r}}\left(\frac{\delta(t-r/c)}{r}\right)=\delta(t-r/c)\underbrace{\vec{\nabla}\left(\frac{1}{r}\right)}_{\hat{e}_r\frac{\partial}{\partial r}\frac{1}{r}=-\frac{\hat{e}_r}{r^2}}+\frac{1}{r}\underbrace{\vec{\nabla}\delta(t-r/c)}_{-\frac{\hat{e}_r}{c}\frac{d\delta(t-r/c)}{dt}}$$

$$\underbrace{\vec{\nabla}\cdot}_{\frac{1}{r^2}\frac{\partial}{\partial r}r^2\hat{e}_r\cdot}\vec{\nabla}\left(\frac{\delta(t-r/c)}{r}\right)=-\underbrace{\vec{\nabla}\cdot\left(\frac{\hat{e}_r}{r^2}\right)}_{4\pi\delta^3(\vec{r})}\delta(t-r/c)+\frac{-\hat{e}_r}{r^2}\cdot\frac{-\hat{e}_r}{c}\frac{d\delta(t-r/c)}{dt}$$

$$-\frac{1}{c}\frac{d\delta(t-r/c)}{dt}\frac{1}{r^2}\frac{\partial}{\partial r}r+\frac{1}{rc^2}\frac{d^2\delta(t-r/c)}{dt^2}$$

$$=-4\pi\delta^3(\vec{r})\delta(t)+\frac{1}{rc^2}\frac{d^2\delta(t-r/c)}{dt^2} \tag{18.15.5}$$

Consequently, A^μ in the Lorentz gauge is given by, where $t_{\text{retarded}}=t-\left|\vec{r}-\vec{r}'\right|/c$,

$$\begin{aligned}A^\mu(\vec{r},t)&=\frac{\mu_0}{4\pi}\int_\varsigma d^3r'\int dt'\frac{J^\mu(\vec{r}'(t'),t')}{\left|\vec{r}-\vec{r}'(t')\right|}\delta\left(t-t'-\frac{\left|\vec{r}-\vec{r}'\right|}{c}\right)\\&=\frac{\mu_0}{4\pi}\int_\varsigma d^3r'\frac{J^\mu(\vec{r}'(t_{\text{retarded}}),t_{\text{retarded}})}{\left|\vec{r}-\vec{r}'(t_{\text{retarded}})\right|}\end{aligned} \tag{18.15.6}$$

18.16 RADIATION FROM MOVING CHARGES

The scalar potential of a single charged particle with $\rho(\vec{r},t)=q\delta(\vec{r}-\vec{x}(t))$ is from Equation (18.15.6):

$$\begin{aligned}V(\vec{r},t)&=\frac{\mu_0c^2}{4\pi}\int_\varsigma d^3r'\int dt'\frac{q\delta(\vec{r}'-\vec{x}(t'))}{\left|\vec{r}-\vec{r}'\right|}\delta\left(t-t'-\frac{\left|\vec{r}-\vec{r}'\right|}{c}\right)\\&=\frac{1}{4\pi\varepsilon_0}\int dt'\frac{q}{\left|\vec{r}-\vec{x}(t')\right|}\delta\left(t-t'-\frac{\left|\vec{r}-\vec{x}(t')\right|}{c}\right)\end{aligned} \tag{18.16.1}$$

The zero of the argument of the delta function defines t' implicitly as a function of $\vec{r}$. Since $\delta(f(t)) = \delta(t)/|df(t)/dt|_{t=f^{-1}(0)}$ while, with $\vec{\beta} = v(t')/c$,

$$\frac{d|\vec{r}-\vec{x}(t')|}{dt'} = \frac{d\sqrt{(\vec{r}-\vec{x}(t'))\cdot(\vec{r}-\vec{x}(t'))}}{dt'} = -\frac{(\vec{r}-\vec{x}(t'))}{\sqrt{(\vec{r}-\vec{x}(t'))^2}}\cdot\frac{d\vec{x}(t')}{dt'} = -c\hat{e}_{\vec{r}-\vec{x}(t')}\cdot\vec{\beta}$$

(18.16.2)

Equation (18.16.1) becomes, where $\kappa \equiv 1-\hat{e}_{\vec{r}-\vec{x}(t')}\cdot\vec{\beta}$, $\vec{R} \equiv \vec{r}-\vec{x}(t')$ and the label retarded indicates that t' satisfies $t' = t - |\vec{r}-\vec{x}(t')|/c$,

$$V(\vec{r},t) = \frac{1}{4\pi\varepsilon_0}\left[\frac{q}{\kappa|\vec{r}-\vec{x}(t')|}\right]_{\text{retarded}} = \frac{1}{4\pi\varepsilon_0}\left[\frac{qc}{cR-\vec{v}\cdot\vec{R}}\right]_{\text{retarded}} \tag{18.16.3}$$

and analogously

$$\vec{A}(\vec{r},t) = \frac{\mu_0}{4\pi}\left[\frac{q\vec{v}}{\kappa|\vec{r}-\vec{x}(t')|}\right]_{\text{retarded}} = \frac{\vec{v}}{c^2}V(\vec{r},t) \tag{18.16.4}$$

which are termed the *Lienard–Wiechert potentials.* The factor κ, which approaches unity for nonrelativistic motion, results from the addition at $(\vec{r},t)$ of the potentials generated by sources at $\vec{x}(t')$ along the surface of a collapsing sphere, a distance $|\vec{r}-\vec{x}(t')| = c(t-t')$ from $\vec{r}$ at each earlier time t'. That is, if Δt is the time interval over which the sphere intercepts a body with extent a and traveling with a velocity $v_{\vec{r}-\vec{x}(t')} = \vec{v}\cdot\hat{e}_{\vec{r}-\vec{x}(t')}$ in the direction of $\hat{e}_{\vec{r}-\vec{x}(t')}$, its potential accumulates over a distance $D = a + v_{\vec{r}-\vec{x}(t')}\Delta t = c\Delta t$ implying $\Delta t = a/(c - v_{\vec{r}-\vec{x}(t')})$ and is accordingly enhanced relative to that of a stationary particle by the factor $D/a = 1/\left(1 - v_{\vec{r}-\vec{x}(t')}/c\right)$.

From $\vec{E} = -\partial\vec{A}/\partial t - \vec{\nabla}V$, after involved manipulations, the radiation fields, which unlike static fields decay asymptotically as $1/|\vec{r}-\vec{x}(t')|$, can be isolated. For these fields, the associated Poynting vector varies as $1/|\vec{r}-\vec{x}(t')|^2$, and the total radiated power, computed by integrating the Poynting vector over a sphere of radius $|\vec{r}-\vec{x}(t')|$, remains finite as $|\vec{r}-\vec{x}(t')| \to \infty$. The radiated electric field of a point charge is given by

$$\vec{E}(\vec{r},t) = \frac{q}{4\pi\varepsilon_0 c\kappa^3}\frac{1}{|\vec{r}-\vec{x}(t')|}\left[\hat{e}_{\vec{r}-\vec{x}(t')}\times\left(\left(\hat{e}_{\vec{r}-\vec{x}(t')}-\vec{\beta}\right)\times\dot{\vec{\beta}}\right)\right]_{\text{retarded}} \tag{18.16.5}$$

For a nonrelativistic charge, the time derivative of Equation (18.16.4) yields, omitting terms that contain additional powers of $1/|\vec{r}-\vec{x}(t')|$ or factors of β,

$$\frac{\partial \vec{A}}{\partial t\kappa} = \frac{1}{4\pi\varepsilon_0 c} \frac{q\dot{\vec{\beta}}}{|\vec{r}-\vec{x}(t')|}\Bigg|_{\text{retarded}} \underset{v\ll c}{\longrightarrow} \frac{1}{4\pi\varepsilon_0 c}\frac{q\dot{\vec{\beta}}}{|\vec{r}-\vec{x}(t')|}\Bigg|_{\text{retarded}} \tag{18.16.6}$$

For a coordinate system with its origin near the charge position, the electric field propagates radially far from the source so that the gradient can be approximated by $\vec{\nabla} = \hat{e}_{\vec{r}-\vec{x}(t')}\partial/\partial(|\vec{r}-\vec{x}(t')|)\cdot$, which when acting on a function of $f(t-|\vec{r}-\vec{x}(t')|/c)$ is equivalent to $-(\hat{e}_{\vec{r}-\vec{x}(t')}/c)\partial/\partial t$. Hence, from Equation (18.16.3), noting that only the time derivative of $\vec{v}$ yields an additional factor of $\vec{R}$ that leads to a $1/R$ dependence at large R, where the subscript "retarded" is suppressed:

$$\vec{\nabla} V = -\frac{1}{c}\hat{e}_{\vec{R}}\frac{\partial V}{\partial t} \approx -\frac{qc}{c4\pi\varepsilon_0}\hat{e}_{\vec{R}}\frac{-1}{\left(cR-\vec{v}\cdot\vec{R}\right)^2}\left(-c\dot{\vec{\beta}}\cdot\vec{R}\right) \approx -\frac{q}{4\pi\varepsilon_0 c}\frac{\hat{e}_{\vec{R}}\dot{\vec{\beta}}\cdot e_{\vec{R}}}{|\vec{r}-\vec{x}(t')|} \tag{18.16.7}$$

Alternatively, the Lorentz condition, $\partial V/\partial t = -c^2\,\vec{\nabla}\cdot\vec{A}$, where $\vec{A}$ is given by Equation (18.16.4), yields V:

$$\frac{\partial V}{\partial t} \approx -\frac{qc^2\mu_0}{4\pi}\left(-\frac{1}{c}\hat{e}_{\vec{r}-\vec{x}(t')}\cdot\frac{\partial}{\partial t}\right)\cdot\frac{c\vec{\beta}}{|\vec{r}-\vec{x}(t')|}$$
$$V \approx \frac{qc^2\mu_0}{4\pi}\,\frac{\hat{e}_{\vec{r}-\vec{x}(t')}\cdot\vec{\beta}\left(t-\frac{|\vec{r}-\vec{x}(t')|}{c}\right)}{|\vec{r}-\vec{x}(t')|} \tag{18.16.8}$$

from which, again only retaining the $1/R$ term,

$$\vec{\nabla} V \approx \frac{q}{4\pi\varepsilon_0}\left(-\frac{1}{c}\hat{e}_{\vec{r}-\vec{x}(t')}\cdot\frac{\partial}{\partial t}\right)\frac{\hat{e}_{\vec{r}-\vec{x}(t')}\cdot\vec{\beta}}{|\vec{r}-\vec{x}(t')|} = -\frac{q}{4\pi\varepsilon_0 c}\frac{\left(\hat{e}_{\vec{r}-\vec{x}(t')}\cdot\dot{\vec{\beta}}\right)\hat{e}_{\vec{r}-\vec{x}(t')}}{|\vec{r}-\vec{x}(t')|} \tag{18.16.9}$$

Combining Equations (18.16.6) and (18.16.9) with $\vec{C}-\hat{e}_{\vec{r}-\vec{x}(t')}(\hat{e}_{\vec{r}-\vec{x}(t')}\cdot\vec{C}) \equiv \vec{C}_\perp = -\hat{e}_{\vec{r}-\vec{x}(t')}\times(\hat{e}_{\vec{r}-\vec{x}(t')}\times\vec{C})$ yields, where $\vec{a}$ represents the particle acceleration,

$$\vec{E} = \frac{q}{4\pi\varepsilon_0 c}\left[\frac{\hat{e}_{\vec{r}-\vec{x}(t')} \times \left(\hat{e}_{\vec{r}-\vec{x}(t')} \times \dot{\vec{\beta}}\right)}{|\vec{r}-\vec{x}(t')|}\right]_{\text{retarded}} = -\frac{q}{4\pi\varepsilon_0 c^2}\left[\frac{\vec{a}_\perp(\vec{x}(t'),t')}{|\vec{r}-\vec{x}(t')|}\right]_{\text{retarded}} \tag{18.16.10}$$

Equation (18.16.10) can further be obtained from the field of a stationary point charge initially at the origin that accelerates with uniform acceleration $a\hat{e}_x$ between times t' and $t' + \Delta t = t' + v/a$ and then travels with a constant speed v. Hence, the electric field for $r > c(t - (t' + \Delta t))$ is that of a stationary charge at the origin, while for $r < c$ $(t - t')$ the field coincides with the $1/r^2$ Coulomb field (which is neglected in the above equations) of a point charge moving with a velocity $v\hat{e}_x$. The latter field must coincide with the field of a point charge located at a constant position with respect to a frame moving with velocity v and hence in the nonrelativistic approximation radiates outward with spherical symmetry from the charge's current location. Since the two sets of field lines are displaced by $vt \sin\theta$, where θ is measured from the x-axis (as is easily verified for $\theta = 0, \pi$), continuity of $\vec{E}$ in the region of radius $c\Delta t$ between the two regions requires

$$-\frac{E_\perp}{E_r} = \frac{vt\sin\theta}{c\Delta t} = \frac{v(r/c)\sin\theta}{c(v/a)} = \frac{ar\sin\theta}{c^2} \tag{18.16.11}$$

which together with $E_r = q/4\pi\varepsilon_0 r^2$ reproduces Equation (18.16.10). Note that the distance between the field lines of the initial stationary charge and the field lines of the charge at position vt grows linearly with t and hence with $r = ct$. This additional factor transforms the $1/r^2$ dependence of the static Coulomb field into a $1/r$ radiation field.

Far from the charge, applying the large-distance approximation to $\vec{\nabla}$ in $\vec{\nabla} \times \vec{E} = -\partial\vec{B}/\partial t$,

$$\vec{\nabla} \times \vec{E} \approx -\frac{1}{c}\hat{e}_{\vec{r}-\vec{x}(t')} \times \frac{\partial\vec{E}}{\partial t} = -\frac{\partial\vec{B}}{\partial t} \tag{18.16.12}$$

Integrating with respect to time

$$B = \frac{1}{c}\hat{e}_{\vec{r}-\vec{x}(t')} \times \vec{E} \tag{18.16.13}$$

$\vec{E} \times \vec{B}$ is accordingly oriented along $\hat{e}_{\vec{r}-\vec{x}(t')}$ and the total radiated power is given by the *Larmor formula*, where since the average, 1/3, of $\langle\cos^2\theta\rangle$ over all angles can be computed from

$$\frac{\langle x^2 + y^2 + z^2\rangle}{\langle r^2\rangle} = 1 = 3\frac{\langle x^2\rangle}{\langle r^2\rangle} = 3\langle\cos^2\theta\rangle = 3\left(1 - \langle\sin^2\theta\rangle\right) \tag{18.16.14}$$

integrating the Poynting vector over a spherical surface $\vec{S}$ centered at $\vec{x}(t')$ yields

$$P = \oint_S \vec{S} \cdot d\vec{A} = \oint_S EH\hat{e}_{\vec{r}-\vec{x}(t')} \cdot d\vec{S} = \frac{1}{\mu_0 c}\oint_S E^2 \hat{e}_{\vec{r}-\vec{x}(t')} \cdot d\vec{S}$$

$$= \frac{1}{\mu_0 c}\left(\frac{qa}{4\pi\varepsilon_0 c^2 \left|\vec{r}-\vec{x}(t')\right|}\right)^2 \left|\vec{r}-\vec{x}(t')\right|^2 4\pi\left\langle \sin^2\theta \right\rangle \tag{18.16.15}$$

$$= \left(\frac{1}{\varepsilon_0 \mu_0 c^2}\right)\frac{1}{16\pi^2 \varepsilon_0}\frac{q^2 a^2}{c^3}\frac{8\pi}{3} = \frac{\mu_0}{6\pi}\frac{q^2 a^2}{c}$$

Example

An electron bound by a harmonic restoring force and a velocity-dependent damping force in the presence of an electromagnetic wave with frequency ω is described by

$$\frac{\partial^2 x}{\partial t^2} + \gamma\frac{\partial x}{\partial t} + \omega_0^2 x = \frac{q}{m}\mathrm{Re}\left(E_0 e^{i\omega t}\right) \tag{18.16.16}$$

This equation is solved by adding the homogeneous, zero driving force solution to the inhomogeneous solution incorporating the driving force. The homogeneous solution describes a damped, undriven oscillator and is determined by the initial conditions. The inhomogeneous, steady-state solution is obtained by inserting $x = \mathrm{Re}(x_0 e^{i\omega t})$ into Equation (18.16.16), from which

$$x_0 = \frac{qE}{m\left(\omega_0^2 - \omega^2 + i\gamma\omega\right)} \tag{18.16.17}$$

Equation (18.16.15) then yields for the time-averaged radiated power with $a = \mathrm{Re}(-\omega^2 x_0 \exp(i\omega t))$, where $\omega_0^2 - \omega^2 + i\gamma\omega = \left(\left(\omega_0^2-\omega^2\right)^2 + \gamma^2\omega^2\right)^{1/2} \exp(\phi_0)$ with $\phi_0 = \arctan\left(\gamma\omega/\left(\omega_0^2-\omega^2\right)\right)$ (the result below is also obtained in a somewhat different form by evaluating $\mathrm{Re}\left[\left(\omega_0^2-\omega^2-i\gamma\omega\right)\left(\cos\omega t + i\sin\omega t\right)\right]$)

$$P_{\text{time averaged}} = \frac{1}{4\pi\varepsilon_0}\frac{2}{3}\frac{q^2}{m^2 c^3}\left(\frac{\left(qE\omega^2\right)^2}{\left(\omega_0^2-\omega^2\right)^2 + \gamma^2\omega^2}\right)\overline{\cos^2(\omega t + \phi_0)}$$

$$= \frac{1}{4\pi\varepsilon_0}\frac{1}{3}\frac{q^2}{m^2 c^3}\left(\frac{\left(qE\omega^2\right)^2}{\left(\omega_0^2-\omega^2\right)^2 + \gamma^2\omega^2}\right) \tag{18.16.18}$$

The *Thomson cross section* of an electron is defined as the ratio of the energy radiated by a free electron in the presence of an electric field, obtained by inserting $\vec{a} = e\vec{E}/m$ in Equation (18.16.15), to the incident energy flux (Poynting vector), $E^2/\mu_0 c = c\varepsilon_0 E^2$,

$$\sigma_{\text{Thomson}} = \frac{P}{S} = \frac{8\pi}{3} r_{\text{Thomson}}^2 \tag{18.16.19}$$

in which the *classical electron radius*, denoted by r_{Thompson} or r_e, is by convention the radius for which twice the electric field energy equals the relativistic electron rest energy, e.g.,

$$m_e c^2 = 2\frac{\varepsilon_0}{2}\int E^2 d^3x = \varepsilon_0 \int_{r_{\text{Thompson}}}^{\infty} \left(\frac{e}{4\pi\varepsilon_0 r^2}\right)^2 4\pi r^2 dr = \frac{e^2}{4\pi\varepsilon_0 r_{\text{Thompson}}} \tag{18.16.20}$$

The power radiated by a relativistic electron can be obtained by writing the Larmor formula in covariant form. Recalling that $dt = \gamma_> dt_{\text{rest}} \equiv \gamma_> d\tau$ with $d\tau = \sqrt{(dt)^2 - (dx)^2/c^2}$,

$$P = -\frac{2}{3}\frac{r_{\text{Thompson}}}{mc}\frac{dp_\mu}{d\tau}\frac{dp^\mu}{d\tau} = \frac{2r_{\text{Thompson}}\gamma_>^2}{3mc}\left(\left(\frac{d\vec{p}}{dt}\right)^2 - \frac{1}{c^2}\left(\frac{dE}{dt}\right)^2\right) \tag{18.16.21}$$

However, since $E^2 - c^2p^2 = m^2c^4$,

$$\frac{dE}{d(pc)} = \frac{pc}{E} = \beta_< \tag{18.16.22}$$

and therefore

$$P = \frac{2r_{\text{Thompson}}\gamma_>^2}{3mc}\left(\left(\frac{d\vec{p}}{dt}\right)^2 - \beta_<^2\left(\frac{dp}{dt}\right)^2\right) \tag{18.16.23}$$

In *bremsstrahlung*, (breaking radiation) an electron traveling along a line changes its velocity by $\Delta\nu$ between t' and $t' + \Delta t$. With $\gamma_>^2 = \left(1 - \beta_<^2\right)^{-1}$, a pulse is then generated from $t = t' + R/c$ to $t = t' + \Delta t + R/c$ at a distance R from the electron with a total power

$$P = \frac{1}{4\pi\varepsilon_0}\frac{2e^2}{3m^2c^3}\left(\frac{\Delta p}{\Delta t}\right)^2 = \frac{1}{4\pi\varepsilon_0}\frac{2e^2}{3m^2c^3}\left(1 - \beta_<^2\right)\left(\frac{\Delta p}{\Delta\tau}\right)^2 \tag{18.16.24}$$

and angular distribution for nonrelativistic motion,

$$\frac{dP}{d\Omega} = \frac{1}{\mu_0 c}\left(\frac{q}{4\pi\varepsilon_0}\frac{\Delta v \sin\theta}{\Delta t c^2}\right)^2 \tag{18.16.25}$$

From the relation between the width of a pulse and the width of its Fourier transform, such a pulse exhibits a frequency spread of $\Delta f \approx 1/\Delta t$.

In contrast to bremsstrahlung, in *synchrotron radiation*, a particle changes direction at constant speed so that $dp/dt = 0$. For a highly relativistic electron orbiting in a circle at constant energy $\left|d\vec{p}/dt\right| = \omega\left|\vec{p}\right|$, the first term of Equation (18.16.23) yields

$$P = \frac{1}{4\pi\varepsilon_0}\frac{2}{3}\frac{e^2}{m^2c^3}\gamma_>^2\omega^2\left|\vec{p}\right|^2 \tag{18.16.26}$$

which contains an additional factor of $\gamma_>^2$ compared to linear motion. Since relativistically $\left|\vec{p}\right| = mc\left|\vec{\beta}_<\right|\gamma_>$ while $\omega = v/r = c\beta_</r$ with r the orbit radius,

$$P = \frac{1}{4\pi\varepsilon_0}\frac{2}{3}\frac{e^2c}{r^2}\beta_<^4\gamma_>^4 \tag{18.16.27}$$

For an electron in the constant magnetic field of a synchrotron, $|dp/dt| = q\vec{v}\times\vec{B} = evB = ec\beta_<B$ resulting in, for $\beta_< \approx 1$,

$$P = \frac{1}{4\pi\varepsilon_0}\frac{2}{3}\frac{e^4}{m^2c}\gamma_>^2\beta_<^2B^2 \approx \frac{1}{4\pi\varepsilon_0}\frac{2}{3}\frac{e^4}{m^2c}\gamma_>^2B^2 \tag{18.16.28}$$

The far-field angular distribution of the Poynting vector in the radial direction at large distances emerging from an accelerated relativistic charge is, where $\dot{\vec{\beta}}_\perp = \dot{\vec{\beta}}\sin\theta$ and θ represents the angle between the direction of motion and the unit vector and $\hat{n}$ denotes the direction from the particle to the observation point.

$$\vec{S}\cdot\hat{n} = \frac{c}{\varepsilon_0}\left|\frac{1}{4\pi}\frac{e}{c\kappa^3R}\dot{\vec{\beta}}_\perp\right|^2_{t'=t-\frac{R(t')}{c}} \tag{18.16.29}$$

19

WAVE MOTION

Sinusoidal wave motion with an amplitude-independent period occurs in one dimension for restoring forces that vary as the negative of the departure from equilibrium. In an infinite uniform linear medium in several dimensions, however, for a general displacement, different points in the medium experience unequal restoring forces and therefore accelerate at different rates. As a consequence, the field pattern changes with time. However, for a harmonic perturbation, the curvature of the disturbance is proportional to the negative of its amplitude. For restoring forces that are proportional to the curvature, which generally applies near equilibrium, all points in the medium therefore oscillate with the same frequency, and the shape of the resulting "modal" pattern remains unchanged with time. Since in a linear system a general displacement can be expressed as a sum of such patterns, this chapter is largely concerned with sinusoidal waves and their generalizations.

19.1 WAVE EQUATION

The variation in the pressure of a material with volume at constant entropy is termed its *bulk modulus*:

$$K_{\text{adiabatic}} = -V\left.\frac{dp}{dV}\right|_{S} \tag{19.1.1}$$

Fundamental Math and Physics for Scientists and Engineers, First Edition.
David Yevick and Hannah Yevick.

In a section with equilibrium length $l_{\text{equilibrium}}$ of a gas-filled tube of constant cross section A, the volume occupied by the gas $V = Al_{\text{equilibrium}}$, and hence, $dp = dF/A$ and $dV = Adl$, yielding

$$dF = -K_{\text{adiabatic}} A \frac{dl}{l_{\text{equlibrium}}} \tag{19.1.2}$$

In other words, identical forces are required to extend 1 m of gas by 1 cm and 10 m of gas by 10 cm. In a solid, $K_{\text{adiabatic}}$ is replaced by *Young's modulus* Y.

If x denotes the equilibrium position of each segment of gas and $f(x, t)$ the displacement of this segment from x at time t, the net force $F(x, t)$ on the section of gas with equilibrium position between $x - \Delta x/2$ and $x + \Delta x/2$ equals the sum of the rightward force $F_{\text{right}} = KA\Delta l(x + \Delta x/2)/\Delta x$ arising when the segment with equilibrium position between x and $x + \Delta x$ is lengthened from Δx to $\Delta x + \Delta l$ and a leftward force $F_{\text{left}} = -KA\Delta l(x - \Delta x/2)/\Delta x$ from an elongation of the segment with equilibrium position from $x - \Delta x$ to x:

$$\begin{aligned} F(x,t) &= K_{\text{adiabatic}} \frac{A}{\Delta x}\left[\Delta l\left(x+\frac{\Delta x}{2}\right) - \Delta l\left(x-\frac{\Delta x}{2}\right)\right] \\ &= K_{\text{adiabatic}} A\left[\frac{f(x+\Delta x)-f(x)}{\Delta x} - \frac{f(x)-f(x-\Delta x)}{\Delta x}\right] \\ &= KA\Delta x \frac{\partial^2 f(x,t)}{\partial x^2} \end{aligned} \tag{19.1.3}$$

Thus, if the gas at $x + \Delta x/2$ is compressed or expanded by the same amount as the gas at $x - \Delta x/2$, the rightward and leftward forces balance; otherwise, the element with mass $\rho V = \rho A \Delta x$ accelerates according to

$$ma = \rho A \Delta x \frac{\partial^2 f(x,t)}{\partial t^2} = F = KA\Delta x \frac{\partial^2 f(x,t)}{\partial x^2} \tag{19.1.4}$$

which yields the *scalar wave equation* (note that K/ρ has dimensions of v^2)

$$\frac{\partial^2 f(x,t)}{\partial t^2} - \frac{K}{\rho}\frac{\partial^2 f(x,t)}{\partial x^2} = 0 \tag{19.1.5}$$

In a three-dimensional homogeneous medium, since the above analysis applies to longitudinal displacements in any direction,

$$\frac{\partial^2 f(\vec{r},t)}{\partial t^2} - \frac{K}{\rho}\nabla^2 f(\vec{r},t) = 0 \tag{19.1.6}$$

Two-dimensional transverse waves on an ideal string under tension T situated along the x-axis at equilibrium are similarly analyzed. Denoting by $y = f(x, t)$ and

$\theta(x, t)$ the vertical displacement of the string at x and the angle the tangent to the string describes with respect to the x-axis, the upward force at the right boundary of the segment between $x - \Delta x/2$ and $x + \Delta x/2$ is $T \sin \theta(x + \Delta x/2, t) \approx T \tan \theta(x + \Delta x/2, t) = T\partial f/\partial x|_{x+\Delta x/2}$, while at the left boundary the force (which is negative since the tension at the two ends of any string segment points in opposing directions) equals $-T\partial f/\partial x|_{x-\Delta x/2}$. Summing these and applying Newton's law yields

$$F(x,t) = \rho A \Delta x \frac{\partial^2 f(x,t)}{\partial t^2} = T\Delta x \left[\frac{1}{\Delta x} \left(\left.\frac{\partial f}{\partial x}\right|_{x+\frac{\Delta x}{2}} - \left.\frac{\partial f}{\partial x}\right|_{x-\frac{\Delta x}{2}} \right) \right] = T\Delta x \frac{\partial^2 f(x,t)}{\partial x^2} \tag{19.1.7}$$

Accordingly, Equation (19.1.5) is replaced by

$$\frac{\partial^2 f(x,t)}{\partial t^2} - \frac{T}{\rho A} \frac{\partial^2 f(x,t)}{\partial x^2} = 0 \tag{19.1.8}$$

19.2 PROPAGATION OF WAVES

In terms of the wave velocity $v = \sqrt{T/\rho A}$ or $\sqrt{K/\rho}$, Equations (19.1.5) and (19.1.8) become

$$\frac{\partial^2 f(x,t)}{\partial t^2} - v^2 \frac{\partial^2 f(x,t)}{\partial x^2} = 0 \tag{19.2.1}$$

Factorizing (cf. Eq. 11.5.1)

$$\left(\frac{\partial}{\partial t} - v \frac{\partial}{\partial x} \right) \left(\frac{\partial}{\partial t} + v \frac{\partial}{\partial x} \right) f(x,t) = 0 \tag{19.2.2}$$

yields

$$f(x,t) = g_{\text{right}}(x - vt) + g_{\text{left}}(x + vt) \tag{19.2.3}$$

where the *right and left traveling solutions*, $g_{\text{right}}(x - vt)$ and $g_{\text{left}}(x - vt)$, propagate without distortion with velocity v. The form of these solutions is determined by the *initial* and *boundary conditions*.

While the form of $f(x, t)$ is generally time dependent, the *plane wave solutions* (here generalized to three dimensions)

$$f(\vec{r},t) = A \cos\left(k \left[\hat{e}_{\vec{k}} \cdot \vec{r} - vt \right] + \delta \right) \tag{19.2.4}$$

where $\hat{e}_{\vec{k}}$ denotes a unit vector in the propagation direction, possess an invariant pattern with a linearly varying phase in all symmetry directions and are sometimes termed free space modes. The quantities $\vec{k} = k\hat{e}_{\vec{k}}$ and k are termed the *wavevector* and *wavenumber*, respectively, and possess units of $[k] =$ [radians/m]. In *phasor notation* (the symbol Re is often omitted),

$$f(\vec{r},t) = \mathrm{Re}\left[Ae^{i\delta}e^{\vec{k}\cdot\vec{r}-\omega t}\right] \equiv \mathrm{Re}\left[A_{\mathrm{phasor}}e^{\vec{k}\cdot\vec{r}-\omega t}\right] \tag{19.2.5}$$

Equation (19.2.4) remains invariant when $\vec{r}$ is replaced by $\vec{r} + \hat{e}_k 2\pi/k$; hence, $2\pi/k$ corresponds to the wavelength:

$$\lambda = \frac{2\pi}{k}\left[\frac{\text{meters}}{\text{wavelength}}\right] \tag{19.2.6}$$

Similarly, the invariance under $t \to t + 2\pi/\omega$ relates the angular frequency, ω, with units of [radians/second] = [1/T]; the linear frequency, f, in [Hertz] = [cycles/second] = [1/T] ; and the period, T, by

$$T = \frac{2\pi}{\omega} = \frac{1}{f}\left[\frac{\text{seconds}}{\text{cycle}}\right] \tag{19.2.7}$$

For linear media, the wave frequency does not depend on position; otherwise, the number of waves entering a localized region per unit time would differ from the number leaving, resulting in a divergent or vanishing power density. The velocity of a point of constant phase on the wave, termed the *phase velocity*, can be obtained by considering the point with zero initial phase, $\delta = 0$ in Equation (19.2.4), which translates in time according to $kr - \omega t = 0$ so that

$$v_p = \frac{r}{t} = \frac{\omega}{k} = \frac{2\pi/T}{2\pi/\lambda} = \frac{\lambda}{T} = \lambda f \tag{19.2.8}$$

Alternatively, since a wave advances by a wavelength in one period, its velocity is $v_p = \lambda/T$, or as f wavelengths per second pass a given point each second, the distance traveled per second is $v_p = \lambda f$.

For a one-dimensional mechanical wave, $v(\vec{r},t) = df(\vec{r},t)/dt = \omega A \cos\left(\vec{k}\cdot\vec{r} - \omega t + \delta\right)$, and the time-averaged kinetic energy per unit length therefore equals $\lambda A^2\omega^2 \overline{\cos^2\left(\vec{k}\cdot\vec{r} - \omega t\right)}/2 = \lambda A^2\omega^2/4$ where $\lambda = \rho/A$ represents the mass per unit length. The total energy, which is the sum of the potential and kinetic energy, is conserved and coincides with the maximum kinetic energy $\lambda A^2\omega^2/2$.

19.3 PLANAR ELECTROMAGNETIC WAVES

In a homogeneous medium, the Maxwell's equations can be combined to generate the *electromagnetic wave equation* according to

$$\vec{\nabla}\times\underbrace{\left(\vec{\nabla}\times\vec{E}\right)}_{-\frac{\partial\vec{B}}{\partial t}}=\vec{\nabla}\underbrace{\left(\vec{\nabla}\cdot\vec{E}\right)}_{0}-\nabla^2 E=-\frac{\partial}{\partial t}\vec{\nabla}\times\vec{B}=-\mu\frac{\partial}{\partial t}\underbrace{\vec{\nabla}\times\vec{H}}_{\frac{\partial\vec{D}}{\partial t}}=-\mu\varepsilon\frac{\partial^2\vec{E}}{\partial t^2}=-\frac{1}{c^2}\frac{\partial^2\vec{E}}{\partial t^2}$$

(19.3.1)

with a similar equation for $\vec{H}$. Here vacuum quantities are distinguished by a zero subscript so that $c=c_0/\sqrt{\mu_r\varepsilon_r}\equiv c_0/n$, where $n=\sqrt{\varepsilon_r\mu_r}$ is termed the *dielectric constant* with c_0 the speed of light in free space. The *plane wave solutions* to these equations for propagation in the wavevector direction $\vec{k}$ are, where $\vec{E}_{\text{phasor}}$ and $\vec{H}_{\text{phasor}}$ are complex,

$$\begin{Bmatrix}\vec{E}(\vec{r},t)\\\vec{H}(\vec{r},t)\end{Bmatrix}=\text{Re}\left[\begin{Bmatrix}\vec{E}_{\text{phasor}}\\\vec{H}_{\text{phasor}}\end{Bmatrix}e^{i\left(\vec{k}\cdot\vec{r}-\omega t\right)}\right] \tag{19.3.2}$$

with wavenumber $k=2\pi/\lambda\equiv\left|\vec{k}\right|=k_0 n$ so that $\lambda=2\pi/k=\lambda_0/n$. As, e.g., $\vec{\nabla}\exp(i(\vec{k}\cdot\vec{r}-\omega t))=i\vec{k}\exp(i(\vec{k}\cdot\vec{r}-\omega t))$ acting on Equation (19.3.2), $\vec{\nabla}$ and $\partial/\partial t$ can be replaced by $i\vec{k}$ and $-i\omega$, respectively. Accordingly, $\vec{\nabla}\times\vec{E}=-\partial B/\partial t$ implies

$$\not{i}\vec{k}\times\vec{E}=\mu\left(\not{i}\omega\right)\vec{H} \tag{19.3.3}$$

and $(\vec{k},\vec{E},\vec{H})$, or equivalently, $(\vec{E},\vec{H},\vec{k})$ form a right-handed coordinate system. For $\mu=\mu_0$,

$$Z=\frac{\left|\vec{E}\right|}{\left|\vec{H}\right|}\frac{[\text{volts/meter}]}{[\text{amperes/meter}]}=\frac{\omega\mu}{k}=\frac{c_0\mu_0}{n}=\frac{1}{n}\sqrt{\frac{\mu_0}{\varepsilon_0}}=\frac{1}{n}\left(\frac{4\pi\times10^{-7}}{8.85\times10^{-12}}\right)^{1/2}\Omega=\frac{337}{n}\Omega\equiv\frac{Z_0}{n}\Omega$$

(19.3.4)

in which $Z_0\equiv337\Omega$ is termed the *free space impedance*. As a result,

$$\left|\vec{B}\right|=\mu\left|\vec{H}\right|=\sqrt{\mu\varepsilon}\left|\vec{E}\right|=\frac{\left|\vec{E}\right|}{c} \tag{19.3.5}$$

The energy density in an electromagnetic wave is given by

$$w = \frac{1}{2}\left(\vec{E}\cdot\vec{D}+\vec{B}\cdot\vec{H}\right) = \frac{1}{2}\left(\varepsilon\vec{E}^2 + \underbrace{\sqrt{\varepsilon\mu}}_{1/c}\,\vec{E}\cdot\sqrt{\frac{\varepsilon}{\mu}}\vec{E}\right)$$

$$= \varepsilon E^2 = \varepsilon\left|E_{\text{phasor}}\right|^2\cos^2\left(\vec{k}\cdot\vec{r}-\omega t+\delta_{\text{phasor}}\right) \qquad (19.3.6)$$

resulting in an energy flux equal to the light velocity times the energy density. This coincides with the Poynting vector, given by, for plane waves in media such as lossless dielectrics in which $\vec{E}$ and $\vec{H}$ are in phase,

$$\vec{S} = \vec{E}\times\vec{H} = = \sqrt{\frac{\varepsilon}{\mu}}E^2\hat{e}_k = \frac{\varepsilon}{\sqrt{\varepsilon\mu}}E^2\hat{e}_k = c\varepsilon E^2\hat{e}_k \qquad (19.3.7)$$

The intensity is then defined as the time-averaged power transported across a unit area surface perpendicular to the propagation direction per unit time (the last two expressions again require the absence of a phase shift between $\vec{E}$ and $\vec{H}$):

$$I = \left\langle\left|\vec{E}\times\vec{H}\right|\right\rangle = \frac{1}{2}\text{Re}\left(\vec{E}_{\text{phasor}}\times\vec{H}^{*}_{\text{phasor}}\right) = \left\langle c\varepsilon E^2\right\rangle = \frac{1}{2}c\varepsilon\left|\vec{E}_{\text{phasor}}\right|^2 \qquad (19.3.8)$$

The momentum density and flux then equal the energy density and flux divided by c according to the discussion of Section 18.14, while the radiation pressure, defined as the time-averaged force per unit area on a surface, equals the time average of the momentum flux, I/c. This value is multiplied by two if the wave is reflected rather than absorbed.

19.4 POLARIZATION

A solid medium typified by a flexible and compressible rod can be displaced in one longitudinal and two orthogonal transverse *polarization* directions $\hat{e}_i$. Equation (19.2.3) is then replaced by

$$\vec{f}(x,t) = \sum_{m=1}^{3}\hat{e}_m\left[g_{\text{right},m}(x-vt)+g_{\text{left},m}(x+vt)\right] \qquad (19.4.1)$$

Gases and liquids do not support transverse restoring forces, while for the electromagnetic $\vec{D}$ and $\vec{B}$ source fields in a homogeneous uncharged dielectric medium since $\vec{\nabla}\cdot\vec{D} = \vec{\nabla}\cdot\vec{B} = 0$ implies $\vec{k}\cdot\vec{D} = \vec{k}\cdot\vec{B} = 0$ for each plane wave component, the longitudinal polarization is instead absent.

Unpolarized light typically results from light emission by uncorrelated spatially separated sources. The phase and the polarization of the sum of these individual fields vary rapidly and randomly in both time and space. However, if a collimated beam from an unpolarized source is incident on a random configuration of small particles such as dust, the electrons in the material are accelerated along the direction of the $\vec{E}$ field, which is transverse to the direction of propagation. Hence, perpendicularly scattered light is polarized in a direction perpendicular to both the plane defined by the incident beam and the direction from the scatterers to the observer. Polarized light can be extinguished by an electric field *polarizer* that contains long aligned molecules that conduct along their length in the same manner as miniature wires. The large currents that result induce resistive losses for the electric field components parallel but not perpendicular to the molecular orientation. A mechanical analogue is provided by a set of parallel rigid bars that however in contrast transmit mechanical disturbances that are polarized along rather than perpendicular to the bars.

Mathematically, an electromagnetic wave can be *linearly*, *elliptically*, or *circularly polarized*. For the above wave, the x and y components are often written in terms of the two-component *Jones polarization vector* $\vec{P}$ as

$$\vec{E} = \text{Re}\left[\left|\vec{E}\right| \vec{P} e^{i(kz-\omega t)}\right], \quad \vec{P} = e^{i\delta_x} \begin{pmatrix} \cos\theta \\ \sin\theta e^{i(\delta_y - \delta_x)} \end{pmatrix} \tag{19.4.2}$$

Here, $\delta_y - \delta_x = 0$ corresponds to *linear polarization*; if $\theta = 45°$ and $\delta_y - \delta_x = -\pi/2$, the polarization vector along, e.g., the $z = 0$ plane rotates with time in the negative angular direction as seen from a point with $z > 0$ (the angular direction and therefore the polarization state is however reversed if the electric field is instead represented as E_0 and is therefore termed *left circularly polarized*, while if $\delta_y - \delta_x = \pi/2$, the field is *right circularly polarized.* Otherwise, the endpoint of the polarization vector describes an elliptic path with time, and the field is said to be *elliptically polarized.* A polarizer in the x–y plane with polarization axis along the x-direction transforms a normally (perpendicularly) incident electric field with Jones vector field $\vec{P} = (\cos\theta \exp(i\delta_x), \sin\theta \exp(i\delta_y))$ into a transmitted field $\vec{E} = \text{Re}[|\vec{E}|\hat{e}_x \cos\theta \exp(i(kz - \omega t + \delta_x'))]$ and an unpolarized incident beam into an x-polarized field with half the incident power.

19.5 SUPERPOSITION AND INTERFERENCE

From the linearity of the wave equation, the sum $f_{\text{total}}(\vec{r},t) = c_1 f_1(\vec{r},t) + c_2 f_2(\vec{r},t)$ of two solutions $f_1(\vec{r},t)$ and $f_2(\vec{r},t)$ describes a further solution. This *superposition principle*, however, is often not readily apparent since waves are often observed by energy-sensitive detectors. These record the square of the sum of the waves rather than the sum of the amplitudes of each individual wave.

Superimposing counterpropagating plane wave solutions according to

$$f_1(\vec{r},t) = \text{Re}\left[Ae^{i\delta}e^{-\iota\omega t}\left\{e^{i\vec{k}\cdot\vec{r}} + e^{-i\vec{k}\cdot\vec{r}}\right\}\right] = 2A\cos(\omega t - \delta)\cos\vec{k}\cdot\vec{r} \tag{19.5.1}$$

yields a spatially invariant sinusoidal waveform with a time-varying amplitude. For waves instead incident on two slits at $(x, z) = (d/2, 0)$ and $(-d/2, 0)$ from a coherent or incoherent source at a large negative distance, $(0, -|Z_{\text{source}}|)$ with $|Z_{\text{source}}| \gg d$, the equal amplitude cylindrical waves emerging from the slits possess phases that vary identically in time and are therefore coherent. With I_0, the intensity of a single slit, the intensity at (x, Z) for $Z \gg x, \lambda$ is given by

$$\begin{aligned} I(x) &= I_0\left|e^{ik\sqrt{Z^2+\left(x-\frac{d}{2}\right)^2}+\phi_0} + e^{ik\sqrt{Z^2+\left(x+\frac{d}{2}\right)^2}+\phi_0}\right|^2 \\ &= I_0\left|e^{i\phi_0}\left(e^{ikZ\left(1+\frac{x^2+d^2/4-xd}{Z^2}\right)^{\frac{1}{2}}} + e^{ikZ\left(1+\frac{x^2+d^2/4+xd}{Z^2}\right)^{\frac{1}{2}}}\right)\right|^2 \\ &\approx I_0\left|e^{i\phi_0+ik\left(Z+\frac{x^2+d^2/4}{2Z}\right)}\left(e^{-ik\frac{xd}{2Z}} + e^{ik\frac{xd}{2Z}}\right)\right|^2 \\ &= 4I_0\cos^2\left(\frac{kxd}{2Z}\right) \end{aligned} \tag{19.5.2}$$

For sources located at $\vec{r}_i$ emitting waves with identical temporal frequency ω and polarization, the field amplitude at $\vec{r}$ is given by, suppressing the polarization vector,

$$f(\vec{r},t) = \text{Re}\left[(z_1 + z_2 + \cdots + z_N)e^{-i\omega t}\right] \tag{19.5.3}$$

with

$$z_i = A_i\left(|\vec{r} - \vec{r}_i|\right)e^{i\vec{k}\cdot(\vec{r}-\vec{r}_i)+\delta_i} \tag{19.5.4}$$

In three dimensions, $A(r) \propto 1/r$ since the total radiated flux through any sphere centered on the source point (the product of the energy density, which varies as $(A(r))^2$ with the surface area $4\pi r^2$) is constant. Summing the complex numbers z_i in Equation (19.5.3) is then equivalent to adding N vectors, commonly termed *phasors*, in the complex plane, each of which points from the origin to one of the z_i. If two phasors, z_i and z_j, are proportional and similarly directed, *constructive interference*, $|z_{\text{tot}}| = |z_i| + |z_j|$, results yielding $|z_{\text{tot}}|^2 = |z_1|^2 + |z_2|^2 + 2|z_1||z_2|$, while if these phasors are oppositely directed, the waves are 180° out of phase yielding *destructive interference* for which $|z_{\text{tot}}| = |z_i| - |z_j|$. *Incoherent, unpolarized sources (e.g., sources with polarizations that vary rapidly and randomly with respect to each other over the time scales of the measurement process) as well as orthogonally polarized sources and sources with different frequencies do not interfere* so that $|z_{tot}|^2 = |z_1|^2 + |z_2|^2$. For two equal frequency sources or antennas at $\pm\hat{e}_z(a/2)$ with phase difference φ, the field at $\vec{r}$ with $|\vec{r}| \gg a$ so that the $1/r$ dependence of γ can be neglected at a polar angle $\theta = \theta_{\vec{k},\vec{z}}$ is

$$\begin{aligned}
&\gamma\mathrm{Re}\left[e^{i\left(k\hat{e}_r\cdot\left(\vec{r}-\frac{a}{2}\hat{e}_z\right)-\omega t-\frac{\varphi}{2}\right)}+e^{i\left(k\hat{e}_r\cdot\left(\vec{r}+\frac{a}{2}\hat{e}_z\right)-\omega t+\frac{\varphi}{2}\right)}\right]\\
&\quad=\gamma\mathrm{Re}\left[e^{i(kr-\omega t)}\left(e^{-i\left(\frac{ka\cos\theta}{2}+\frac{\varphi}{2}\right)}+e^{i\left(\frac{ka\cos\theta}{2}+\frac{\varphi}{2}\right)}\right)\right]\\
&\quad=2\gamma\cos(kr-\omega t)\cos\left(\frac{ka\cos\theta}{2}+\frac{\varphi}{2}\right)
\end{aligned} \tag{19.5.5}$$

The time-averaged power attains a maximum when the argument of the second cosine function equals $m\pi$ ($m = 0, 1, \ldots$), which implies from $k = 2\pi/\lambda$

$$a\cos\theta = \left(m - \frac{\varphi}{2\pi}\right)\lambda \tag{19.5.6}$$

As $\Delta l = a\cos\theta$ equals the length difference between the rays from the two sources to the observation point, if the sources radiate in phase, for $\varphi = 0$, constructive interference occurs when $\Delta l = m\lambda$. If instead $\Delta l = (2m+1)\pi$, the contributions interfere destructively, canceling the field.

For N equally spaced sources radiating in phase ($\varphi = 0$) on $z = [-a/2, a/2]$, Equation (19.5.5) is replaced by

$$\begin{aligned}
\gamma' e^{-i\frac{ka\cos\theta}{2}}\sum_{m=0}^{N-1} e^{i\frac{kam\cos\theta}{N-1}} &= \gamma' e^{-i\frac{ka\cos\theta}{2}}\left(\frac{1-e^{i\frac{kaN\cos\theta}{N-1}}}{1-e^{i\frac{ka\cos\theta}{N-1}}}\right)\\
&= \gamma' e^{-i\frac{ka\cos\theta}{2}}\frac{e^{i\frac{Nka\cos\theta}{2(N-1)}}\sin\left(\frac{kaN}{2(N-1)}\cos\theta\right)}{e^{i\frac{ka\cos\theta}{2(N-1)}}\sin\left(\frac{ka}{2(N-1)}\cos\theta\right)}\\
&= \gamma' e^{-i\frac{ka\cos\theta}{2}}\frac{e^{i\left(\frac{ka\cos\theta}{2}+\frac{ka\cos\theta}{2(N-1)}\right)}\sin\left(\frac{kaN}{2(N-1)}\cos\theta\right)}{e^{i\frac{ka\cos\theta}{2(N-1)}}\sin\left(\frac{ka}{2(N-1)}\cos\theta\right)}\\
&= \gamma'\frac{\sin\left(\frac{kaN}{2(N-1)}\cos\theta\right)}{\sin\left(\frac{ka}{2(N-1)}\cos\theta\right)}
\end{aligned} \tag{19.5.7}$$

with $\gamma' \equiv \gamma \exp(i(kr - \omega t))$, which for $N = 2$ reproduces Equation (19.5.5) after employing $\sin(2\varepsilon) = 2 \sin \varepsilon \cos \varepsilon$ with $\varepsilon = (ka \cos \theta)/2$. The maximum relative field amplitude, $N\gamma$, occurs when $X \equiv \pi/2 - \theta \to 0$ so that $\cos \theta = \sin X \approx X$. The subsequent primary maximum is obtained when the denominator of the above expression again approaches zero at $a \cos \theta = (N-1)\lambda$. Between these, $N - 2$ small amplitude subsidiary maxima are located at $a \cos \theta = m\lambda$ with $m = 1, 2, \ldots, N-2$.

In the limit that $N \to \infty$, $\gamma \to 0$ while $N\gamma = \Gamma$,

$$\gamma' \frac{\sin\left(\dfrac{kaN}{2(N-1)} \cos\theta\right)}{\sin\left(\dfrac{ka}{2(N-1)} \cos\theta\right)} \to 2\gamma' N \frac{\sin\left(\dfrac{ka\cos\theta}{2}\right)}{ka\cos\theta} = e^{i(kr-\omega t)} 2\Gamma \frac{\sin\left(\dfrac{ka\cos\theta}{2}\right)}{ka\cos\theta} \tag{19.5.8}$$

corresponding to the diffraction pattern of a wave of amplitude Γ normally incident on a slit of width a consistent with *Huygens' principle*, which regards each point along a wavefront (a line of constant phase) at a given time as the source of a circularly outward propagating wave. Subsequently, these waves interfere constructively along the evolving wavefront. Equivalently, dividing the slit into segments of length dz, each of which contributes an amplitude $(\Gamma/a)dz$, yields for the diffracted field

$$e^{i(kr-\omega t)} \frac{\Gamma}{a} \int_{-\frac{a}{2}}^{\frac{a}{2}} e^{ikz\cos\theta} dz = e^{i(kr-\omega t)} \frac{\Gamma}{a} \frac{e^{i\frac{ka\cos\theta}{2}} - e^{-i\frac{ka\cos\theta}{2}}}{ik\cos\theta}$$

$$= e^{i(kr-\omega t)} 2\Gamma \frac{\sin\left(\dfrac{ka\cos\theta}{2}\right)}{ka\cos\theta} \tag{19.5.9}$$

which possesses a single maximum at $\theta = \pi/2$, zeros at $\cos\theta = \sin\theta' = m\lambda/a$ with $\theta' = 90 - \theta$ the angle measured from the normal to the slit and subsidiary maxima at $\sin\theta' = (2m+1)\lambda/a$. The zeros are located at angles for which the rays from the bottom and top of the slit differ by an integer number of wavelengths. That is, consider a single planar wavefront normally incident on the slit and an outgoing plane wave component traveling at $\theta' \ll \pi/2$. The phase of the outgoing waves at the bottom and top of the slit a long distance from the slit differs by $k\Delta l = (2\pi/\lambda)a \sin\theta'$ where Δl is difference in distance traveled by the two waves. When $\theta' \approx \lambda/a$, the phasor contributions from the different points in the slit are uniformly distributed over all angles from 0 to 2π and therefore sum to zero. This implies *Rayleigh's criterion* that the diffraction patterns of light from two distant sources incident on the same aperture are *resolved* if the angular separation between the two sources exceeds λ/a so that the central maximum of one pattern lies outside the first minimum of the second.

19.6 MULTIPOLE EXPANSION FOR RADIATING FIELDS

For distances $r \gg D$ from a time-dependent charge and current density within a region of size D and for emitted wavelengths, $\lambda = c/f \gg D$, since $|\vec{r}-\vec{r}'| \approx r(1-2\vec{r}\cdot\vec{r}'/r^2 + r'^2/r^2)^{1/2} \approx r - \hat{e}_r\cdot\vec{r}'$,

$$\vec{J}\left(\vec{r}',t-\frac{r}{c}+\frac{\hat{e}_r\cdot\vec{r}'}{c}\right) \approx \vec{J}\left(\vec{r}',t-\frac{r}{c}\right) + \dot{\vec{J}}\left(\vec{r}',t-\frac{r}{c}\right)\frac{\hat{e}_r\cdot\vec{r}'}{c} + \cdots \tag{19.6.1}$$

where $\dot{\vec{J}} \equiv \partial\vec{J}/dt_{\text{ret},0} = \partial\vec{J}/dt$ in which $t_{\text{ret},0} \equiv t - r/c$. Hence, the $1/r$ radiation terms in the vector potential can be approximated by

$$\vec{A}(\vec{r},t) = \frac{\mu_0}{4\pi}\int_\varsigma \frac{\vec{J}\left(\vec{r}',t-\frac{|\vec{r}-\vec{r}'|}{c}\right)}{|\vec{r}-\vec{r}'|}d^3r'$$

$$= \frac{\mu_0}{4\pi r}\int_\varsigma\left(\vec{J}(\vec{r}',t_{\text{ret},0}) + \dot{\vec{J}}(\vec{r}',t_{\text{ret},0})\frac{\hat{e}_r\cdot\vec{r}'}{c}+\cdots\right)\left(\frac{1+\vec{r}\cdot\vec{r}'}{r^2}+\cdots\right)d^3r'$$

$$\approx \underbrace{\frac{\mu_0}{4\pi r}\int_\varsigma \vec{J}(\vec{r}',t_{\text{ret},0})d^3r'}_{\text{electric dipole}} + \underbrace{\frac{\mu_0}{4\pi r}\int_\varsigma \dot{\vec{J}}(\vec{r}',t_{\text{ret},0})\frac{\hat{e}_r\cdot\vec{r}'}{c}d^3r'}_{\text{magnetic dipole and electric quadrupole}} + \cdots \tag{19.6.2}$$

The first, leading, term above is associated with the *electric dipole* moment since a net current within a limited volume must redistribute the charges in direction of the average current and thus alter the electric dipole moment. Mathematically, the current continuity equation $\vec{\nabla}\cdot\vec{J} = -\partial\rho/\partial t$ implies

$$\vec{\nabla}\cdot\left(x_m\vec{J}\right) = \sum_l \frac{\partial}{\partial x_l}(x_m J_l) = \sum_l\left(\delta_{ml}J_l + x_m\frac{\partial J_l}{\partial x_l}\right) = J_m - x_m\frac{\partial\rho}{\partial t} \tag{19.6.3}$$

Further,

$$\int_\varsigma \vec{\nabla}'\cdot\left(x'_m\vec{J}_l(\vec{r}',t_{\text{ret},0})\right)d^3r' = \int_{S\subset\varsigma}\left(x'_m\vec{J}_l(\vec{r}',t_{\text{ret},0})\right)\cdot d\vec{S}' = 0 \tag{19.6.4}$$

since $\vec{J}$ is localized to D and therefore vanishes on any S' situated beyond D. Accordingly,

$$\vec{A}_{\text{electric dipole}}(\vec{r},t) = \frac{\mu_0}{4\pi r}\int_\varsigma \dot{\rho}(\vec{r}',t_{\text{ret},0})\vec{x}'d^3r' = \frac{\mu_0}{4\pi r}\dot{\vec{p}}(t_{\text{ret},0}) \tag{19.6.5}$$

where $\vec{p}$ is the electric dipole moment. Since for a function $f(t_{\text{ret},0})$ of $t_{\text{ret},0}$, $\vec{\nabla} f(t_{\text{ret},0}) = \hat{e}_r \partial_r f(t_{\text{ret},0}) \approx -\hat{e}_r (1/c) \dot{f}(t_{\text{ret},0})$, neglecting terms of order $1/r^2$ that are generated when $\vec{\nabla}$ is applied to $1/r$, with $\vec{p}$ oriented along the z-axis, $\hat{e}_z = \cos\theta \hat{e}_r - \sin\theta \hat{e}_\theta$ (since θ opens downward), $\hat{e}_r \times \hat{e}_\theta = \hat{e}_\phi$, and $p_\perp \equiv p \sin\theta_{\vec{r},\vec{p}}$,

$$\vec{B}_{\text{electric dipole}}(\vec{r},t) = \vec{\nabla} \times \vec{A} = -\frac{1}{c}\hat{e}_r \times \frac{\partial \vec{A}}{\partial t} = -\frac{\mu_0}{4\pi r c}\hat{e}_r \times \ddot{\vec{p}}_{\text{ret},0} = -\frac{\mu_0 \ddot{p}_{\text{ret},0}}{4\pi r c}\hat{e}_{\vec{r}\times\vec{p}} \underbrace{\rightarrow}_{\vec{p}=p\hat{e}_z} \frac{\mu_0 \ddot{p}_{\perp,\text{ret},0}}{4\pi c r}\hat{e}_\phi \tag{19.6.6}$$

Far from the sources,

$$\vec{\nabla} \times \vec{E} = -\hat{e}_r \times \frac{1}{c}\frac{\partial \vec{E}}{\partial t} = -\frac{\partial \vec{B}}{\partial t} \tag{19.6.7}$$

Hence, $\left(\hat{e}_r, \vec{E}, \vec{B}\right)$ form a right-handed coordinate system with $\hat{e}_r \times \vec{E} = c\vec{B}$ so that from $\hat{e}_r \times \hat{e}_\theta = \hat{e}_\phi$,

$$\vec{E}_{\text{electric dipole}} = -c\hat{e}_r \times \vec{B}_{\text{electric dipole}} = -\frac{\mu_0}{4\pi r}\ddot{p}_{\perp,\text{ret},0}\hat{e}_r \times \hat{e}_\phi = \frac{\mu_0}{4\pi r}\ddot{p}_{\perp,\text{ret},0}\hat{e}_\theta \tag{19.6.8}$$

with the corresponding Poynting vector

$$S(t) = \frac{1}{\mu_0}\vec{E} \times \vec{B} = \hat{e}_r \frac{\mu_0}{16\pi^2 c}\frac{\sin^2\theta}{r^2}\left(\ddot{p}(t_{\text{ret},0})\right)^2 \tag{19.6.9}$$

which yields for the total power according to Equation (18.16.14)

$$P_{\text{total, electric dipole}}(t) = \int \vec{S} \cdot d\vec{A} = \left(\frac{2}{3}\right) 4\pi r^2 S = \frac{\mu_0}{6\pi c}\left(\ddot{p}(t_{\text{ret},0})\right)^2 \tag{19.6.10}$$

Equations (19.6.8) and (19.6.10) also follow from Equation (18.16.10) together with Larmor's formula, Equation (18.16.15), for the fields and power radiated by an accelerating point charge with dipole moment $\vec{p} = q\vec{r}$ with respect to a fixed origin.

Example

In a center-fed linear *electric dipole antenna*, an AC source generates a current $I(t, z=0) = I_0 \cos \omega t$ in a vertical wire with cross section A extending from $z = 0$ to $z = D$ and a simultaneous current $I(t, z=0) = -I_0 \cos \omega t$ at z in a second, oppositely directed wire from $z = 0$ to $z = -D$. If the wavelength of the radiation is much larger than D, the current varies as $I(t, z) = \text{sign}(z)(D - z)I_0 \cos(\omega t)/D$. Since $\vec{\nabla} \cdot \vec{J} = (1/A)dI/dz = -\partial\rho/\partial t$, the charge density in the two branches of the antenna after integration $\rho_{\text{upper}} = -\rho_{\text{lower}} = I_0 \sin(\omega t)/\omega AD$ from which the dipole moment $\int_\varsigma \rho(\vec{r}')\vec{r}' d^3r' = \hat{e}_z \int_{-D}^{D} \rho(\vec{r}') z' A dz' = \hat{e}_z \sin(\omega t) I_0 D/\omega$, which as expected is determined by the average separation between the positive and negative charges in the wire. Equation (19.6.10) then yields for the total *time-averaged* radiated power

$$P_{\text{total}} = \frac{\mu_0}{6\pi c}\frac{1}{2}(D\omega I_0)^2 \tag{19.6.11}$$

This result can also be derived directly from

$$\vec{A}(\vec{r},t) = \frac{\mu_0}{4\pi}\int_\varsigma \frac{\vec{J}(\vec{r}', t_{\text{retarded}})}{|\vec{r}-\vec{r}'(t_{\text{retarded}})|} d^3r' \approx \frac{\mu_0}{4\pi r}\int_L I d\vec{l}' = \frac{\mu_0 I_0 D}{4\pi r}\cos\omega\left(t-\frac{r}{c}\right)\hat{e}_z \tag{19.6.12}$$

from which subsequently, with $\hat{e}_z = \hat{e}_r \cos\theta - \hat{e}_\theta \sin\theta$,

$$\vec{B} = \vec{\nabla} \times \vec{A} = -\frac{1}{c}\hat{e}_r \times \frac{\partial \vec{A}}{\partial t} = (-1)^3 \hat{e}_\phi \frac{\mu_0}{4\pi c r} I_0 D\omega \sin\theta \sin\omega\left(t-\frac{r}{c}\right) \tag{19.6.13}$$

together with $\vec{E} = -c\hat{e}_r \times \vec{B}$.

While electric dipole radiation occurs for any nonzero $\vec{J}$ that results in a net redistribution of charges and hence a time-varying electric dipole moment, a time-independent current distribution generates a static nonradiating magnetic field. Hence for *magnetic dipole radiation* $\dot{\vec{J}} \neq 0$. That is, if $\vec{J}$ changes at $t=0$ for a short time interval Δt and then remains stationary in a new configuration, as for the suddenly accelerated electric charge examined in Section 18.16, after a further time t elapses, the magnetic field lines of the initial and modified magnetic field distributions are similarly joined at a distance $r = ct$ by angularly directed magnetic field lines. The spatial displacement of the field lines of the $t<0$ and $t>\Delta t$ distributions varies linearly with r, providing again the additional factor of r that distinguishes radiated and static fields. As the form of the magnetic field from a magnetic dipole is identical to that of the electric field of an electric dipole, while the physical mechanism for the generation of the radiation field is identical, the radiated field formulas differ only by constant factors and appropriate substitutions of $c\vec{B}$ for $\vec{E}$.

Mathematically, magnetic dipole radiation is associated with the antisymmetric pseudovector components (cf. Eq. 5.5.7) of the second term in Equation (19.6.2). Writing the tensor outer product $\dot{j}_a x'_b = \dot{\vec{J}} \otimes \vec{r}'$ as $(\dot{j}_a x'_b + \dot{j}_b x'_a)/2 + (\dot{j}_a x'_b - \dot{j}_b x'_a)/2$ or equivalently $\left(\dot{\vec{J}} \otimes \vec{r}' + \vec{r}' \otimes \dot{\vec{J}}\right)/2 + \left(\dot{\vec{J}} \otimes \vec{r}' - \vec{r}' \otimes \dot{\vec{J}}\right)/2$ and retaining only the antisymmetric term,

$$\left[A_{\text{magnetic dipole}}(\vec{r},t)\right]_a = \frac{\mu_0}{4\pi rc}\hat{e}_b \frac{1}{2}\int_\varsigma \underbrace{\left(\dot{J}_a(\vec{r}')x_b' - \dot{J}_b(\vec{r}')x_a'\right)}_{\varepsilon_{abc}\left(\dot{\vec{J}}\times\vec{x}'\right)_c} d^3r' = -\frac{\mu_0}{4\pi rc}\left(\hat{e}_r \times \dot{\vec{m}}\right)_a \tag{19.6.14}$$

with $\vec{m}$ the magnetic dipole moment pseudovector from which

$$\vec{B}_{\text{magnetic dipole}} = \vec{\nabla} \times \vec{A}_{\text{magnetic dipole}} = \frac{\mu_0}{4\pi rc^2}\hat{e}_r \times \left(\hat{e}_r \times \ddot{\vec{m}}\right) = \frac{1}{c}\hat{e}_r \times \vec{E}_{\text{magnetic dipole}} \tag{19.6.15}$$

(The symmetric term $J_a x_b' + J_b x_a'$ instead yields a four-lobed *electric quadrupole radiation* pattern proportional to the *quadrupole moment tensor* $Q_{ij} = \int_\varsigma \left(3x_i'x_j' - r'^2\delta_{ij}\right)d^3r'$ corresponding to the radiation pattern of an adjacent pair of center-fed linear antennas driven with a 180° relative phase shift.) Accordingly, $\vec{p} \rightarrow \vec{m}/c$, $\left(\vec{E}, c\vec{B}\right)_{\text{electric dipole}} \rightarrow \left(c\vec{B}, -\vec{E}\right)_{\text{magnetic dipole}}$ transforms Equation (19.6.8) into Equation (19.6.15) while leaving $\vec{S} = \vec{E} \times \vec{H}$ invariant and thus, e.g.,

$$P_{\text{total, magnetic dipole}}(t) = \frac{\mu_0}{6\pi c^3}\left(\ddot{m}_\perp(t_{\text{ret},0})\right)^2 \tag{19.6.16}$$

While electric and magnetic dipole and quadrupole radiation fields interfere spatially, the integrated power is the sum of the individual powers.

19.7 PHASE AND GROUP VELOCITY

If the *phase velocity* $v_p(\omega) = x/t = \omega/k$ of a wave in a medium varies with frequency, the propagation speed or *group velocity* of a pulse generally differs from the phase velocity. To illustrate, superimposing two waves with slightly different wavevectors and therefore frequencies yields a modulation envelope that can be interpreted as an infinite train of identical pulses superimposed on a rapidly varying carrier wave:

$$\begin{aligned}\vec{E} &= \vec{E}_0 \text{Re}\left(e^{i((k+\Delta k)z-(\omega+\Delta\omega)t)} + e^{i((k-\Delta k)z-(\omega-\Delta\omega)t)}\right)\\ &= 2\vec{E}_0 \text{Re}\left(e^{i(kz-\omega t)}\cos(\Delta kz - \Delta\omega t)\right)\\ &= 2\vec{E}_0 \cos(kz-\omega t)\cos(\Delta kz - \Delta\omega t)\end{aligned} \tag{19.7.1}$$

The point of zero phase of the wave envelope satisfies $\Delta kz - \Delta\omega t = 0$ so that in the $\Delta\omega$, $\Delta k \rightarrow 0$ limit, employing $\omega = v_p k$ and $k = 2\pi/\lambda$ so that $dk/k = -d\lambda/\lambda$:

$$v_g = \frac{d\omega}{dk} = v_p + k\frac{dv_p}{dk} = v_p - \lambda\frac{dv_p}{d\lambda} \tag{19.7.2}$$

Indeed, if the wavelengths of the two waves differ by $\Delta\lambda$ while the mean wave velocity is v_p,

$$\lambda\frac{dv_p}{d\lambda} = \frac{\lambda}{\Delta\lambda}\cdot\left(\Delta\lambda\frac{dv_p}{d\lambda}\right) = \frac{\lambda}{\Delta\lambda}\Delta v_p \tag{19.7.3}$$

Δv_p corresponds to the relative distance over which the wave with wavelength $\lambda + \Delta\lambda$ travels with respect to the wave with λ in 1 s. If this distance equals $\Delta\lambda$, the point at which the two-component waves interfere constructively and hence the wave envelope is displaced by λ along $-\hat{e}_{\vec{k}}$.

A frequency-dependent phase velocity induces pulse broadening or *dispersion.* A localized pulse is a superposition of an infinite number of waves with different wavevectors, each of which evolves in time with its associated phase velocity. Frequencies that travel at higher and lower speeds, respectively, spread out the leading and trailing edges of a propagating pulse. For a pulse width a with average *carrier wavevector* k_0, expanding ω in a Taylor series around k_0 yields

$$\omega(k) = \omega_0(k_0) + v_g(k_0)(k - k_0) + \frac{\beta(k_0)(k - k_0)^2}{2} + \cdots \tag{19.7.4}$$

so that $v_g(k_0) \equiv \partial\omega/\partial k|_{k_0}$ and $v_g(k) = v_g(k_0) + \beta(k_0)(k - k_0)$ for $k \approx k_0$. The *dispersion parameter* is defined as

$$\beta(k_0) \equiv \left.\frac{\partial^2\omega}{\partial k^2}\right|_{k_0} = \left.\frac{\partial v_g}{\partial k}\right|_{k_0} \quad \left[\frac{D^2}{T}\right] \tag{19.7.5}$$

For an initial (near-Gaussian) pulse width a, since $\Delta x(t=0)\Delta k = a\Delta k \geq 1/2$ from the properties of the Fourier transform or the Heisenberg uncertainty principle, at times for which the pulse spreading far exceeds a, the pulse width $\Delta x(t) = \Delta v_g t \approx \beta t/2a$ from Equation (19.7.5).

19.8 MINIMUM TIME PRINCIPLE AND RAY OPTICS

Wave motion can be modeled classically by evolving *rays*, which are quasiplanar wavefronts generated by a coherent line of sources with a far larger transverse spatial extent than the wavelength and which interfere constructively along the ray path. That is, according to *Babinet's principle*, the field along a plane $x = x'$ results from circular outgoing waves emitted by the field at each point along an adjacent plane $x = x' - L$. A ray is generated when these waves are in phase along $x = x'$ over a distance $\gg \lambda$. In the $\lambda \to 0$ limit, the width of such a beam can therefore be negligible. The phase remains invariant to the lowest order for *any* small but continuous change along a ray trajectory so that waves propagating along all such paths interfere coherently. Hence, a ray path between two points constitutes a local extremum (maximum, minimum, or saddle point) of the phase or equivalently the number of wavelengths. since the phase increment over a length dl of the ray path is $d\phi = kdl$, for a ray between points A and B, if δ denotes any small perturbation of this path,

$$\delta\phi = \delta\int_A^B kdl = 0 \tag{19.8.1}$$

In three-dimensional space, if the ray path is parameterized as $\vec{r}(w)$,

$$\begin{aligned}\delta\int_A^B kdl &= k_0\delta\int_A^B n(\vec{r})d\left(\sqrt{(dx)^2+(dy)^2+(dz)^2}\right)\\ &= k_0\delta\int_A^B \left(n(\vec{r})\sqrt{\left(\frac{dx}{dw}\right)^2+\left(\frac{dy}{dw}\right)^2+\left(\frac{dz}{dw}\right)^2}\right)dw\\ &= k_0\delta\int_A^B L\left(\vec{r},\dot{\vec{r}}\right)dw\end{aligned} \tag{19.8.2}$$

with $\dot{\vec{r}} \equiv d\vec{r}/dw$ and $L(\vec{r},\dot{\vec{r}}) = n(\vec{r})\sqrt{\dot{\vec{r}}^2}$. The Lagrangian equations for the above problem are

$$\frac{d}{dw}\frac{\partial L}{\partial \dot{x}_i} = \frac{\partial L}{\partial x_i} \tag{19.8.3}$$

and, after inserting L,

$$\frac{d}{dw}\left(\frac{n(\vec{r})}{\sqrt{\dot{\vec{r}}^2}}\dot{\vec{r}}\right) = \sqrt{\dot{\vec{r}}^2}\,\vec{\nabla}n(\vec{r}) \tag{19.8.4}$$

The *ray equation* is obtained by returning to the path length variable $dl = \sqrt{\dot{\vec{r}}^2}dw$:

$$\frac{d}{dl}\left(n(\vec{r})\frac{d\vec{r}}{dl}\right) = \vec{\nabla}n(\vec{r}) \tag{19.8.5}$$

19.9 REFRACTION AND SNELL'S LAW

At a planar interface between two dielectric media such that $n = n_1$ for $x < 0$ and $n = n_2$ for $x > 0$, a ray between a point at $(x, z) = (-d_1, 0)$ in medium 1 and a second point at (d_2, L) in medium 2 can be parameterized in terms of a single parameter Z where $(0, Z)$

is the point at which the ray intersects the interface. The extremum condition of the previous section then implies

$$\delta\phi = k_0 \frac{d}{dZ}\left(n_1 \sqrt{d_1^2 + Z^2} + n_2 \sqrt{d_2^2 + (L-Z)^2} \right) = 0 \qquad (19.9.1)$$

or

$$\frac{n_1 Z}{\sqrt{d_1^2 + Z^2}} - \frac{n_2 (L-Z)}{\sqrt{d_2^2 + (L-Z)^2}} = 0 \qquad (19.9.2)$$

Introducing the *angle of incidence*, θ_{inc}, between the ray and the perpendicular to the interface between the two dielectric media yields *Snell's law*

$$n_1 \sin\theta_{1,\text{inc}} = n_2 \sin\theta_{2,\text{inc}} \qquad (19.9.3)$$

Snell's law also follows from the continuity of the wave fronts on both sides of the surface, which implies that the wavelength $\lambda_{\|} \equiv \lambda_z$, and thus, the wavevector component parallel to the direction of the surface, $k_z = \vec{k} \cdot \hat{e}_z = |k| \sin\theta_{\text{inc}} = k_0 n \sin\theta_{\text{inc}}$, must be identical on both sides of the boundary, i.e.,

$$k_{z,1} = \frac{2\pi n_1}{\lambda_0} \sin\theta_{1,\text{inc}} = k_{z,2} = \frac{2\pi n_2}{\lambda_0} \sin\theta_{2,\text{inc}} \qquad (19.9.4)$$

Alternatively, while the wavelength is $\lambda_1 = \lambda_0 / n_1$ below the interface and $\lambda_2 = \lambda_0 / n_2$ below the interface, the distance, λ_z, between successive wavefronts along the interface must be equal so that $\lambda_{z,1} = \lambda_1 / \cos(90 - \theta_{\text{inc},1}) = \lambda_0 / (n_1 \sin\theta_{\text{inc},1}) = \lambda_{z,2} = \lambda_0 / (n_2 \sin\theta_{\text{inc},1})$.

For an electromagnetic wave incident from a higher refractive index medium, 1, onto a second medium, 2, of lower refractive index at an angle of incidence smaller than the *critical angle* defined by

$$\sin\theta_{1,\text{critical inc}} = \frac{n_2}{n_1} \qquad (19.9.5)$$

the direction of travel (e.g., of $\vec{k}$) in medium 2 is more nearly parallel to the interface. As a consequence, the wavelengths along the interface coincide although the wavelength $\lambda_2 = \lambda_0 / n_2$ exceeds $\lambda_1 = \lambda_0 / n_1$. However, at the *critical angle*, the wave in medium 2 is parallel to the interface, while for $\theta_1 > \theta_{1,\text{critical inc}}$ the wavefronts cannot match along the interface. The phases of the wavefronts generated by the incoming wave along the interface at different times then cancel destructively in medium 2 in a manner analogous to the field decay away from the edge of a light beam. Designating by x and z the coordinates perpendicular and parallel to the interface, respectively, since $k_z = k_\| = k_0 n_1 \sin\theta_1$ in both media by continuity while $k^2 = k_{x,2}^2 + k_{z,2}^2 = k_0^2 n_2^2$ in medium 2, $k_{x,2}^2 = k_0^2 n_2^2 - k_{z,2}^2 = k_0^2 n_2^2 - k_{z,1}^2 = k_0^2 n_2^2 - k_0^2 n_1^2 \sin^2\theta_{\text{inc},1} < 0$, and hence,

$$E_2(x,z) = c_1 e^{ik_{z,1}z + \sqrt{k_{z,1}^2 - k_0^2 n_2^2}x - i\omega t} + c_2 e^{ik_{1,z}z - \sqrt{k_{z,1}^2 - k_0^2 n_2^2}x - i\omega t} \tag{19.9.6}$$

The constants c_1 and c_2 depend on the boundary conditions. If an incoming plane wave approaches a lower index interface from below, $c_1 = 0$ since the field must approach zero at $+\infty$. In contrast, both c_1 and c_2 differ from zero within a region $a_1 < x < a_2$ between two higher index dielectric interfaces if electromagnetic waves are incident on both interfaces.

19.10 LENSES

Since the lengths of all radial paths from the center to the edge of an empty, reflecting sphere are identical, electromagnetic waves radiated by a source at the center of the sphere experience equal phase shifts over these paths and hence interfere constructively after a round trip. Therefore, all radiated power returns to the source point. Additionally, since $\lambda f = v$, where the frequency f is identical throughout space, the time delay of a pulse along all rays with equal numbers of wavelengths is identical. Therefore, the temporal shape of a pulse emitted at the source point is preserved after reimaging. In the same manner, all power emitted by a source at one focus of an ellipsoid propagates to the opposite focus without distortion.

A *converging lens* instead focuses light from an object point behind the lens onto an image point in front of the lens through the reduction of the light wavelength in a dielectric medium from λ to λ/n. In particular, the lens is expanded toward its center so that the number of wavelengths accumulated by a ray passing through the lens from the object to the image near its *optical axis*, which is the axis of symmetry perpendicular to the lens, equals those associated with the longer paths traveled by rays incident on the lens at off-center points. If (z, r) represent cylindrical coordinates with respect to the center of the lens at (0, 0) and the optical axis, the distance traveled by a ray emitted from a source point $(-p, 0)$ with $p > 0$ through $(0, r)$ to an image point $(q, 0)$ equals

$$D(r) = \sqrt{p^2 + r^2} + \sqrt{q^2 + r^2} \approx p + q + \frac{r^2}{2}\left(\frac{1}{p} + \frac{1}{q}\right) + \cdots \approx D(0) + \frac{r^2}{2}\left(\frac{1}{p} + \frac{1}{q}\right) \tag{19.10.1}$$

which corresponds to an additional $(D(r) - D(0))/\lambda_0$ wavelengths relative to the straight-line path through (0, 0). Since light propagating a distance Δz in a refractive index medium acquires $(n - 1)\Delta z/\lambda_0$ additional wavelengths compared to vacuum, a lens centered at (0, 0) can equalize the number of wavelengths for all paths through a thickness variation $\Delta z = t(0) - t(r)$ with

$$(n-1)\Delta z = (n-1)(t(0)-t(r)) = \frac{r^2}{2}\left(\frac{1}{p}+\frac{1}{q}\right) \equiv \frac{r^2}{2f} \qquad (19.10.2)$$

leading to the *thin lens equation*

$$\frac{1}{p}+\frac{1}{q}=\frac{1}{f} \qquad (19.10.3)$$

in which $t(r) = t(0) - gr^2$ and $f = 1/(2g(n-1))$ is termed the *focal length*. Parallel rays are generated by a point source at $p = \infty$ and thus are focused at $q = f$, while rays emitted at $p = 2f$ are focused at an equal distance $q = 2f$. The thickness of a *diverging lens* for which $g < 0$ instead increases away from the optical axis so that parallel rays passing through the lens are defocused rather than focused. Thus, $f < 0$ in Equation (19.10.3), implying for $p > 0$ that $q < 0$ so that the source object and its (noninverted) virtual image are situated on the same side of the lens.

A lens can be regarded as a *prism* with a vertex angle that varies with distance from the optical axis. A ray in vacuum incident at an incidence angle θ_{incident} on a triangular prism, where positive angles are directed away from the prism vertex, with refractive index n is refracted according to Snell's law into an angle $\theta_1 = \arcsin(\sin\theta_{\text{incident}}/n)$ inside the prism. If the opening angle of the prism is θ_{prism}, the ray is incident on the second output face of the prism at an angle $\theta_1 - \theta_{\text{prism}}$ leading to $\theta_{\text{ouput}} = \arcsin(n \sin(\theta_1 - \theta_{\text{prism}}))$ with respect to the normal to the output face. As the refractive index of glass is greater for blue light than for red light, as a result of *material dispersion*, blue light is refracted more than red light and emerges from the prism at a larger angle with respect to the direction of the incident beam. While this effect is often utilized in spectroscopy to separate the frequency components of a light beam, in lenses, the frequency-dependent focal length yields a *chromatic distortion* of images.

Examples

If a source is situated at a point $(-p, R)$ with $p > f$ behind a converging lens and away from the optical axis, one of its emitted rays describes a straight line through the center of the lens. Since the image point is located a distance q on the opposite side of the lens, by similar triangles, the height of this "real" image equals $R' = -(q/p)R$ and is therefore inverted and *magnified* by $M = |q/p|$. Objects at large distances ($p \to \infty$ with L finite) are focused to single point on the optical axis, and thus, $M = 0$, while for $p = q = 2f$, $M = 1$. If the source object is positioned a distance $p < f$ from the lens, $q < 0$ from Equation (19.10.2), and the image and the source are located on the same side of the lens. This yields a noninverted, *virtual* image for which $M \to \infty$ as $p \to 0$. Thus, to magnify an object with a single lens, the object is situated such that its virtual image, viewed from the opposing side of the lens, appears at a standard viewing distance, typically $q_{\text{view}} = -0.25$ m. From Equation (19.10.3), $p = fq/(q-f)$ implying $M = |(q_{\text{view}} - f)/f| = 0.25/f + 1$.

In a two-lens *microscope*, an object is placed a distance p_1 slightly *outside* the focal length of a convergent *objective* lens to generate a real image on the opposite side of the lens with $M_{\text{objective}} = |f_{\text{objective}}/(p_1 - f_{\text{objective}})| \gg 1$. This image is then positioned *inside* the focal length of a second convergent lens or *eyepiece* that provides a further magnification $M_2 = |q_2/p_2| = 0.25/f_{\text{eyepiece}} + 1$. In a two-lens *telescope*, the initial object is far beyond the focal distance of the objective lens so that $q_1 \approx f_{\text{objective}}$ yielding a real image much smaller than the original object. For the eyepiece, $p_2 \approx f_{\text{eyepiece}}$ with $p_2 < f_{\text{eyepiece}}$. The final virtual image is then situated a large distance from the front of the telescope similarly to the object distance. Accordingly, $-q_2 \approx p_1 \approx \infty$ resulting in a total magnification $|(q_1/p_1)(q_2/p_2)| \approx q_1/p_2 \approx f_{\text{objective}}/f_{\text{eyepiece}}$.

19.11 MECHANICAL REFLECTION

Both the spatial derivative and the displacement of a wave normally incident on an interface between two dissimilar materials, 1 and 2, are continuous (as these can be obtained from integrating the second derivative terms appearing in the wave equation). Because the velocity and displacement of a sinusoidal wave are proportional, the continuity requirement is typically recast as the equality of the *impedance ratio* Z between the velocity and the displacement, which is independent of the overall wave amplitude, on both sides of the interface. For a single forward propagating wave, $Z_1 = k_1$, while $Z_2 = k_2$ in the two materials. Hence, the incoming wave can only be transmitted without distortion if $k_1 = k_2$. Otherwise, a reflected wave is additionally present with *reversed* velocity relative to the displacement such that the addition of incident and reflected wave yields the same effective impedance at the boundary as the single transmitted wave. For a boundary at $x = 0$ with a wave described by the real part of

$$f(x,t) = \begin{cases} A\left(e^{i(k_1 x - \omega t)} + Re^{-i(k_1 x + \omega t)}\right) & x < 0 \\ ATe^{i(k_2 x - \omega t)} & x > 0 \end{cases} \tag{19.11.1}$$

the equality of impedances, namely,

$$\left.\frac{1}{f}\frac{\partial f}{\partial x}\right|_{0^-} = ik_1 \frac{A(1-R)}{A(1+R)} = \left.\frac{1}{f}\frac{\partial f}{\partial x}\right|_{0^+} = ik_2 \tag{19.11.2}$$

yields the *amplitude reflection and transmission coefficients*, where $T = 1 + R$ by the continuity of f,

$$\begin{aligned} R &= \frac{k_1 - k_2}{k_1 + k_2} = \frac{Z_1 - Z_2}{Z_1 + Z_2} \\ T = 1 + R &= \frac{2k_1}{k_1 + k_2} = \frac{2Z_1}{Z_1 + Z_2} \end{aligned} \tag{19.11.3}$$

For *zero, Dirichlet, boundary conditions*, $f(0, t) = 0$ for all t, associated with an impenetrable barrier such as a fixed end of a string or the closed end of a pipe. This can be implemented by setting $k_2 = i\infty$ in Equation (19.11.1) and hence, $R = -1$ in Equation (19.11.3), suppressing the transmitted field. If an initial waveform $f(x, t = 0)$ is incident from $x < 0$ on such a boundary, its subsequent evolution is identical to that obtained in the region $x < 0$ without the boundary but in the presence of a fictitious left-propagating inverted wave for $x > 0$ defined by $-f(-x, t = 0)$. *Free, Neumann, boundary conditions* at $x = 0$ instead require $df(x, t)/dx|_{x=0} = 0$. These are typified by a rope terminated by a ring sliding on a vertical pole without a transverse frictional restoring force or by a compression wave at the open end of a tube of constant cross-sectional area where from Equation (19.1.2):

$$0 = p - p_{\text{atmospheric}} = \Delta p = -K_{\text{adiabatic}} \frac{\Delta V}{V} = -K_{\text{adiabatic}} \frac{\Delta l}{l} = -K_{\text{adiabatic}} \frac{df(x)}{dx} \quad (19.11.4)$$

The evolution of an incident wave from $x < 0$ reflecting off a free end at $x = 0$ is identical to that obtained in the absence of the boundary by initially including a non-inverted left-propagating wave $f(-x, t = 0)$ over the expanded region $[-\infty, \infty]$. Finally, waveforms in, e.g., a tube bent into a ring with a circumference L much greater than the tube width are subject to *periodic boundary conditions*

$$\begin{aligned} f(x) &= f(x + L) \\ \left.\frac{df(x)}{dx}\right|_x &= \left.\frac{df(x)}{dx}\right|_{x+L} \end{aligned} \quad (19.11.5)$$

In a one-dimensional medium with stationary boundaries at $x = 0, L$, a resonant waveform or *mode* satisfies the boundary conditions at the two endpoints and does not change form with time if the medium and boundary conditions are time independent. For zero boundary conditions at both boundaries, a mode with $n - 1$ *nodes* (zeros) between zero and L is given by $\sin(2\pi x/\lambda_n)\sin(2\pi f_n t + \delta_n)$ with wavelengths $\lambda_n = 2L/n$, $n = 1, 2, \ldots$ and frequencies $f_n = nv/2L$ from $\lambda f = v$. For a zero boundary condition at $x = 0$ and a free boundary condition at $x = L$, the wavelengths instead equal $\lambda_n = 4L/(2n - 1)$, $n = 1, 2, \ldots$, while for periodic boundary conditions, the modes include both $\sin(2\pi x/\lambda_n)\sin(2\pi f_n t + \delta_n)$ and $\cos(2\pi x/\lambda_n)\sin(2\pi f_n t + \delta_n)$ with $\lambda_n = L/n$, $n = 1, 2, \ldots$. The number of modes within a frequency interval Δf is approximated by $2L\Delta f/v$ in all three cases.

19.12 DOPPLER EFFECT AND SHOCK WAVES

For waves propagating from a source to a detector, the recorded frequency depends on the motion of both the detector and source. For example, if the velocity of the source toward the detector equals the wave velocity, all waves arrive simultaneously resulting in a zero observed wavelength and thus infinite frequency. In contrast, if the detector moves toward the source at the wave velocity, the apparent wave velocity doubles, and the observed wavelength thus decreases by a factor of two. Explicitly, an observer moving with a velocity v_{observer} *toward* a source that emits waves of length λ and velocity χ such that $f_{\text{source}} = \chi/\lambda$ intercepts waves approaching at a speed $v = \chi + v_{\text{observer}}$. The time required for the observer to pass one wavelength, or period, which is the inverse of the observed frequency, is then (since $f\lambda = \lambda/T = \chi$)

$$T_{\text{observer}} = \frac{1}{f_{\text{observer}}} = \frac{\lambda}{v} = \frac{\lambda}{\chi + v_{\text{observer}}} = \frac{1}{f_{\text{source}}} \frac{\chi}{\chi + v_{\text{observer}}} \tag{19.12.1}$$

If the source *approaches* the observer at v_{source}, the source moves a distance $v_{\text{source}} T_{\text{source}}$ in one wave period, T_{source}, so that the wavelength detected by the observer is $\lambda_{\text{observer}} = \lambda_{\text{source}} - v_{\text{source}} T_{\text{source}}$ yielding a detected frequency

$$f_{\text{observer}} = \frac{\chi}{\lambda_{\text{observer}}} = \frac{\chi}{\chi T_{\text{source}} - v_{\text{source}} T_{\text{source}}} = f_{\text{source}} \frac{\chi}{\chi - v_{\text{source}}} \tag{19.12.2}$$

When neither the source nor the observer is at rest, the modified frequency associated with the observer motion is given by Equation (19.12.1). However, from Equation (19.12.2), the frequency f_{source} in this equation is additionally altered by the source motion. Hence,

$$f_{\text{observer}} = f_{\text{source}} \frac{\chi + v_{\text{observer}}}{\chi - v_{\text{source}}} \tag{19.12.3}$$

The detected frequency always exceeds the source frequency f_{source} for approaching motion and is less than f_{source} for receding motion as easily verified by listening to a passing siren.

A source with $v > \chi$ emits spherical waves that propagate a distance χt from the location $x = v_{\text{source}} t$ at which they were emitted. A constant phase front termed a *shock wave* is therefore generated that describes a cone with opening angle θ where

$$\sin\theta = \frac{\chi t}{v_{\text{source}} t} = \frac{\chi}{v_{\text{source}}} \tag{19.12.4}$$

For a particle passing through a dielectric medium in which the speed of light is slower than the particle velocity, the corresponding light pulse is termed *Cerenkov radiation.*

19.13 WAVES IN PERIODIC MEDIA

If successive phase fronts of a plane wave in a periodic medium coincide with the periodicity of the medium, waves scattered from equivalent spatial regions superimpose coherently in certain directions leading to refraction. The analysis of this effect is generally performed slightly differently for forward and backward scattering. In forward scattering, consider a wave $\exp(k\sin\theta x + k\cos\theta z)$ incident on successive planes of scatters (each of which can consist of a distribution of point-like scatters or can possess a nonzero thickness in the x-direction) located along planes $x = x_i$ with $x_i = n\Lambda, n = \ldots -2, -1, 0, 1, 2, \ldots$. As is evident by drawing two rays reflecting from successive planes at the points (0, 0) and (0, Λ), for which the additional distance traveled by the lower ray $\Delta l = 2\Lambda\sin\theta$, constructive interference between plans occurs when Δl satisfies the *Bragg reflection* condition

$$\Delta l = 2\Lambda\sin\theta = m\lambda \tag{19.13.1}$$

(a common error is to omit the factor of 2). The refracted beam then forms an angle of 2θ with respect to the incoming beam. Equivalently, refraction occurs when the additional phase increment that the wave experiences in reflecting from the lower plane equals an integer multiple of 2π:

$$\Delta\phi = \Delta k_x\Lambda = 2k\sin\theta\Lambda = \frac{4\pi\Lambda}{\lambda}\sin\theta = 2\pi m \tag{19.13.2}$$

For a plane wave incident at an *incidence* angle θ_{inc} on identical planes of scatters spaced a distance Λ apart in the z-direction, since $\theta_{\text{inc}} + \theta = \pi/2$, $\sin\theta$ is replaced by $sin(\pi/2 - \theta_{\text{inc}}) = \cos\theta_{\text{inc}}$ above and the Bragg condition is generally instead written

$$2\Lambda\cos\theta_{\text{inc}} = m\lambda \tag{19.13.3}$$

Therefore, if $\Lambda = m\lambda/2$, a normally incident wave is reflected.

19.14 CONDUCTING MEDIA

In a homogeneous linear conducting media in which ε, μ, and σ are independent of position and the current density and electric field are related by $\vec{J} = \sigma\vec{E}$, localized charges or electric fields within the conductor generate large currents. These displace charges to the surface after a characteristic relaxation time, τ, which can be determined from the continuity equation

$$\frac{\partial\rho_{\text{free}}^{(q)}}{\partial t} = -\vec{\nabla}\cdot\vec{J} = -\sigma\vec{\nabla}\cdot\vec{E} = -\frac{\sigma}{\varepsilon}\vec{\nabla}\cdot\vec{D} = -\frac{\sigma}{\varepsilon}\rho_{\text{free}}^{(q)} \tag{19.14.1}$$

where $\rho_{\text{free}}^{(q)}$ is the charge density or charge per unit volume so that

$$\rho_{\text{free}}^{(q)}(t) = e^{-\frac{\sigma t}{\varepsilon}}\rho_{\text{free}}^{(q)}(0) \equiv e^{-\frac{t}{\tau}}\rho_{\text{free}}^{(q)}(0) \tag{19.14.2}$$

yielding $\tau = \varepsilon/\sigma = \varepsilon\rho$ with $\rho = 1/\sigma$ the resistivity.

The electric field in a linear conducting medium obeys the wave equation

$$\vec{\nabla}\times\left(\vec{\nabla}\times\vec{E}\right)=\vec{\nabla}\underbrace{\left(\vec{\nabla}\cdot\vec{E}\right)}_{0}-\nabla^2 E=-\frac{\partial}{\partial t}\left(\vec{\nabla}\times\vec{B}\right)$$
$$=-\mu\frac{\partial}{\partial t}\left(\vec{\nabla}\times\vec{H}\right)=-\mu\varepsilon\frac{\partial^2\vec{E}}{\partial t^2}-\mu\sigma\frac{\partial\vec{E}}{\partial t} \tag{19.14.3}$$

with a similar equation for $\vec{B}$. Hence, for a plane wave for which $\left(\vec{E},\vec{B}\right)=\mathrm{Re}\left[\left(\vec{E}_0,\vec{B}_0\right)e^{i\left(\vec{k}\cdot\vec{r}-\omega t\right)}\right]$,

$$k^2=\frac{\omega^2}{c^2}+i\mu\sigma\omega \tag{19.14.4}$$

with $c^2=1/\varepsilon\mu$. Defining $k_0=\omega/c_0$ for a normally incident wave with $\vec{k}=k\hat{e}_z$ where the conductor occupies the half space $z>0$,

$$\begin{Bmatrix}\vec{E}\\\vec{B}\end{Bmatrix}=\mathrm{Re}\left[\begin{Bmatrix}\vec{E}_0\\\vec{B}_0\end{Bmatrix}e^{i\left(k_0\sqrt{\varepsilon_r\mu_r\left(1+i\left(\frac{\sigma}{\omega\varepsilon}\right)\right)}z-\omega t\right)}\right] \tag{19.14.5}$$

inside the conductor. Comparing with the equivalent expression $E_0\exp(i(k_0nz-\omega t))$ for a plane wave in a uniform dielectric medium with $\sigma=0$ yields an effective relative dielectric permittivity

$$\varepsilon_{r,\mathrm{effective}}=n_{\mathrm{effective}}^2=\varepsilon_r\mu_r\left(1+\frac{i\sigma}{\omega\varepsilon}\right) \tag{19.14.6}$$

For $\sigma\ll\omega\varepsilon$, the electric and magnetic fields are nearly in phase as in vacuum, while the decay length or *skin depth* $d_{\mathrm{skin}}=1/\mathrm{Im}(k_0n)$, which is the distance over which the fields fall to $1/e$ of their value, is large. The attenuation is governed by the distance the field travels within a carrier relaxation time and is thus approximately frequency independent if σ does not vary with frequency since (where the resistivity $\rho=1/\sigma[\Omega\mathrm{D}]$ and the impedance $Z=\sqrt{\mu/\varepsilon}[\Omega]$)

$$d_{\mathrm{skin}}=\left[\mathrm{Im}\left(\frac{\omega}{c}\left(1+i\frac{\sigma}{\omega\varepsilon}\right)^{\frac{1}{2}}\right)\right]^{-1}\approx\left[\mathrm{Im}\left(\frac{\omega}{c}\left(1+i\frac{\sigma}{2\omega\varepsilon}+\cdots\right)\right)\right]^{-1}$$
$$\approx\frac{2c\varepsilon}{\sigma}=\frac{2}{\sigma}\left(\frac{\varepsilon}{\mu}\right)^{\frac{1}{2}}=\frac{2\rho}{Z}=2c\tau \tag{19.14.7}$$

For low frequencies, $\sigma\gg\omega\varepsilon$, the large impedance discontinuity at the boundary results in a rapid attenuation of the field below the surface of the conductor with

$$d_{\text{skin}} = \lim_{\omega\to 0}\left[\text{Im}\left(\frac{\omega}{c}\left(1+i\frac{\sigma}{\omega\varepsilon}\right)^{\frac{1}{2}}\right)\right]^{-1} = \left[\left(\frac{\omega}{\not{c}\,\omega\sqrt{\varepsilon\mu}}\right)(\sigma\omega\mu)^{\frac{1}{2}}\text{Im}\underbrace{\sqrt{i}}_{e^{\pi i/4}=\frac{1+i}{\sqrt{2}}}\right]^{-1} = \left(\frac{2}{\sigma\omega\mu}\right)^{\frac{1}{2}} \tag{19.14.8}$$

Further from $\vec{k}\times\vec{E}_0=\omega B_0$ so that $\hat{e}_k\times E_0=(c_0/n_{\text{effective}})B_0$, the, fields are 45° out of phase. For 60 Hz electric current in copper, for which $\sigma\approx 10^6(\Omega\text{ - cm})^{-1}$, $d_{\text{skin}}\approx 1$ cm, which corresponds to the largest dimension of wires typically employed at this frequency. For a microwave signal of frequency 1 MHz (10^6 Hz), d_{skin} is reduced by a factor of $\sqrt{10^6/60}\approx 10^2$, while skin depths of microns are associated with signal frequencies in the gigahertz (1 GHz = 10^9 Hz) region matching the lithographically generated wire sizes of integrated circuits.

19.15 DIELECTRIC MEDIA

In dielectric (insulating) materials, electrons do not move between atomic sites in response to moderate forces. However, an applied field polarizes the individual atoms, generating a bound current. Approximating the electron motion by the response of a damped harmonic oscillator to the Lorentz force of the oscillating electric field yields for the electron displacement $x(t)$

$$m\frac{d^2x}{dt^2}+m\gamma\frac{dx}{dt}+\omega_0^2x=\text{Re}\left(qE_0e^{-i\omega t}\right) \tag{19.15.1}$$

which is solved after introducing $x(t)=\text{Re}(x_0\exp(-i\omega t))$ by

$$x=\text{Re}\left(\frac{qE_0}{m\left(\omega_0^2-\omega^2-i\gamma\omega\right)}e^{-i\omega t}\right) \tag{19.15.2}$$

The current density is then, suppressing Re,

$$J=n_{\text{electron}}qv=\frac{-i\omega n_{\text{electron}}q^2}{m\left(\omega_0^2-\omega^2-i\gamma\omega\right)}E_0e^{-i\omega t}=\sigma E \tag{19.15.3}$$

In vacuum with $\varepsilon_r=\mu_r=1$ in Equation (19.14.6),

$$\varepsilon_r(\omega)=n^2=1+\frac{q^2n_{\text{electron}}}{m\varepsilon_0\left(\omega_0^2-\omega^2-i\gamma\omega\right)} \tag{19.15.4}$$

as also follows from the polarization $\vec{P} = \varepsilon_0 \chi \vec{E} = n_{\text{electron}} \vec{d}$ with $\vec{d} = q\vec{x}$ and $\varepsilon_r = 1 + \chi$. For $\chi \ll 1$,

$$\sqrt{\varepsilon_r} = n_R + in_I = 1 + \frac{q^2 n_{\text{electron}}}{2m\varepsilon_0} \frac{\omega_0^2 - \omega^2 + i\gamma\omega}{\left(\omega_0^2 - \omega^2\right)^2 + \gamma^2\omega^2} \tag{19.15.5}$$

The power in a plane wave propagating in the z-direction dissipates as

$$P(z) = e^{-\alpha z} = e^{gz} = e^{-2k_I z} = e^{-\frac{2\omega n_I}{c_0} z} \tag{19.15.6}$$

where α and g with $\alpha = -g$ are termed loss and gain coefficients, respectively. The loss coefficient reaches a maximum near the resonance at $\omega = \omega_0$, while the real part of the refractive index increases with frequency and therefore decreases with wavelength (since $f\lambda = c$) except for $|\omega - \omega_0| \approx \gamma$. Accordingly, a decreasing functional dependence of $n_R(\lambda)$ on λ, as evident in the smaller refraction of large wavelength red light by a glass prism compared to blue light, is termed normal as opposed to anomalous dispersion.

19.16 REFLECTION AND TRANSMISSION

At an interface between two dielectrics, the absence of surface currents insures the continuity of $\vec{H}$ parallel to the surface, while surface charge layers generate a discontinuity in the perpendicular but not the parallel electric field component (cf. Eqs. 18.6.1 and 18.6.2). For a planar incoming wave normally incident on the interface the impedance ratio $Z = \left|\vec{E}\right| / \left|\vec{H}\right| = \sqrt{\mu/\varepsilon}$ differs in the two media; consequently, these boundary conditions cannot both be fulfilled if only a forward traveling wave is present. However, since $\left(\vec{E}, \vec{B}, \vec{k}\right)$ form a right-handed coordinate system and $\vec{k}$ reverses direction upon reflection, the magnetic fields of backward and forward traveling waves are oppositely oriented relative to the electric field. Accordingly, if both a reflected wave and an incident wave are present, any ratio of electric to magnetic fields can exist at the boundary.

For an interface along the $z = 0$ plane, if $\hat{e}_{\vec{k}}$ and $\vec{E}$ are oriented along $\hat{e}_z$ and $\hat{e}_x$, respectively, from $\hat{e}_{\vec{k}} \times \vec{E} = Z\vec{H}$, the incident and transmitted $\vec{H}$ fields are in the $+\hat{e}_y$ direction for positive E_x, while the reflected $\vec{H}$ field is directed along $-\hat{e}_y$, i.e., $E_x = \pm ZH_y$, where the upper sign applies to the incident and transmitted field and the lower sign to the reflected field. Hence, from the boundary conditions at the interface, where the second and fourth lines are obtained by taking the cross product of $\hat{e}_z$ with the left-hand side of the first and third lines, respectively,

$$\vec{E}_{\parallel,\text{incident}} + \vec{E}_{\parallel,\text{reflected}} = \vec{E}_{\parallel,\text{transmitted}} \Rightarrow E_{x,\text{incident}} + E_{x,\text{reflected}} = E_{x,\text{transmitted}}$$
$$\Rightarrow Z_1\left(H_{y,\text{incident}} - H_{y,\text{reflected}}\right) = Z_2 H_{y,\text{transmitted}}$$
$$\vec{H}_{\parallel,\text{incident}} + \vec{H}_{\parallel,\text{reflected}} = \vec{H}_{\parallel,\text{transmitted}} \Rightarrow H_{y,\text{incident}} + H_{y,\text{reflected}} = H_{y,\text{transmitted}}$$
$$\Rightarrow \frac{1}{Z_1}\left(E_{x,\text{incident}} - E_{x,\text{reflected}}\right) = \frac{1}{Z_2} E_{x,\text{transmitted}} \tag{19.16.1}$$

For $\mu_1 = \mu_2 = \mu_0$, the impedance $Z_m = \sqrt{\mu/\varepsilon_m}, m = 1, 2$. Eliminating $E_{x,\text{transmitted}}$ from the first and last equations above yields $Z_2(E_{x,\text{incident}} - E_{x,\text{reflected}}) = Z_1(E_{x,\text{incident}} + E_{x,\text{reflected}})$. Since $(a-b)/(a+b) = c/d$ implies $a - b = \kappa c$ and $a + b = \kappa d$ from which $a = \kappa(c+d)/2$ and $b = \kappa(d-c)/2$ yielding finally $(d-c)/(d+c) = b/a$ (the same result is obtained if $H_{y,\text{transmitted}}$ is instead eliminated from the second and third equations of Eq. (19.16.1)):

$$R = \frac{E_{x,\text{reflected}}}{E_{x,\text{incident}}} = -\frac{\not{Z}_1 H_{y,\text{reflected}}}{\not{Z}_1 H_{y,\text{incident}}} = \frac{Z_2 - Z_1}{Z_2 + Z_1} = \frac{n_1 - n_2}{n_1 + n_2} \tag{19.16.2}$$

where R is the *amplitude reflection coefficient*. Observe that $\vec{E}$ reverses direction upon reflection if $n_2 > n_1$. The *amplitude transmission coefficient* is then, with $T = 1 + R$,

$$T \equiv \frac{E_{x,\text{ transmitted}}}{E_{x,\text{incident}}} = 1 + \frac{E_{x,\text{reflected}}}{E_{x,\text{incident}}} = 1 + R = \frac{2n_1}{n_1 + n_2} = \frac{Z_2 H_{y,\text{ transmitted}}}{Z_1 H_{y,\text{incident}}} = \frac{n_1 H_{y,\text{ transmitted}}}{n_2 H_{y,\text{incident}}} \tag{19.16.3}$$

The corresponding *intensity reflection and transmission factors*, where the intensity is given by the Poynting vector $\vec{S} = \vec{E} \times \vec{H} = E^2/Z \propto nE^2$, are given by

$$R_{\text{intensity}} = \frac{S_{\text{reflected}}}{S_{\text{incident}}} = \left(\frac{n_1 - n_2}{n_1 + n_2}\right)^2$$
$$T_{\text{intensity}} = \frac{S_{\text{transmitted}}}{S_{\text{incident}}} = \frac{n_2}{n_1}\left(\frac{2n_1}{n_1 + n_2}\right)^2 = \frac{4n_1 n_2}{(n_1 + n_2)^2} = 1 - R_{\text{intensity}} \tag{19.16.4}$$

The impedance of a conducting layer, where $\varepsilon_{\text{effective}} = \varepsilon_0 n_{\text{effective}}^2$,

$$Z = \left|\frac{E}{H}\right| = \sqrt{\frac{\mu}{|\varepsilon_{\text{effective}}|}} = \left(\frac{\mu}{\varepsilon\left|1 + \dfrac{i\sigma}{\omega\varepsilon}\right|}\right)^{\frac{1}{2}} = c\mu\left(1 + \left(\frac{\sigma}{\omega\varepsilon}\right)^2\right)^{-\frac{1}{2}} = c\mu\left(1 + \left(\frac{\mu\sigma c^2}{\omega}\right)^2\right)^{-\frac{1}{2}} \tag{19.16.5}$$

approaches zero as $\sigma/\omega\varepsilon \to \infty$. From $R = (Z_2 - Z_1)/(Z_1 + Z_2)$, the direction of the electric field therefore reverses upon reflection from a conducting plane while that of the magnetic field is preserved since $\vec{E} \times \vec{B}$ is directed along $\vec{k}$, consistent with both the absence of internal electric fields and the zero resistance to the formation of surface currents in a conductor that physically correspond to zero (Dirichlet) and free (Neumann) boundary conditions, respectively.

If the direction of travel (the wavevector) of an incoming plane wave describes an *angle of incidence*, $\theta_{\text{incidence}}$, with respect to the normal to the dielectric interface, only two polarization states of the incident field are preserved upon reflection (with differing reflection coefficients), as can be demonstrated from symmetry arguments. These are termed *transverse electric* (*TE*) and *transverse magnetic* (*TM*) *fields* for which, respectively, the electric field and the magnetic field are perpendicular to the *plane of incidence* that contains both the incident and reflected wavevectors. Any other incident field polarization can be written as a linear combination of the TE and TM solutions and will consequently be altered upon reflection.

For TE polarization, H_y in the third and fourth equations of Equation (19.16.1) is replaced by the component of $\vec{H}$ along the surface, $H_{\parallel} = H_y \cos\theta_{\text{incidence}}$. This corresponds to replacing Z_m in the subsequent derivation by $Z_m/\cos\theta_m$ or equivalently n_m by $n_m \cos\theta_m$ yielding

$$R_{\text{TE}} = \frac{E_{\text{reflected}}}{E_{\text{incident}}} = \frac{n_1\cos\theta_1 - n_2\cos\theta_2}{n_1\cos\theta_1 + n_2\cos\theta_2} = -\frac{H_{\text{reflected}}}{H_{\text{incident}}}$$

$$T_{\text{TE}} = \frac{E_{\text{transmitted}}}{E_{\text{incident}}} = 1 + R = \frac{2n_1\cos\theta_1}{n_1\cos\theta_1 + n_2\cos\theta_2} = \frac{Z_2 H_{\text{transmitted}}}{Z_1 H_{\text{incident}}} = \frac{n_1 H_{\text{transmitted}}}{n_2 H_{\text{incident}}} \tag{19.16.6}$$

Regarding TM polarization, illustrated in Figure 19.1, E_x in the first and second equations is replaced by $E_x \cos\theta$. This is equivalent to substituting $Z_m \cos\theta_m$ for Z_m, or equivalently $n_m/\cos\theta_m$ for n_m in deriving the reflection coefficient from the second and third lines of Equation (19.16.1). Consequently, Equation (19.16.2) is modified according to

$$R_{\text{TM}} \equiv \frac{E_{\text{reflected}}}{E_{\text{incident}}} = \frac{\dfrac{n_1}{\cos\theta_1} - \dfrac{n_2}{\cos\theta_2}}{\dfrac{n_1}{\cos\theta_1} + \dfrac{n_2}{\cos\theta_2}} = \frac{n_1\cos\theta_2 - n_2\cos\theta_1}{n_1\cos\theta_2 + n_2\cos\theta_1} = -\frac{H_{\text{reflected}}}{H_{\text{incident}}}$$

$$T_{\text{TM}} \equiv \frac{E_{\text{transmitted}}}{E_{\text{incident}}} = \left(\frac{\cos\theta_1}{\cos\theta_2}\right)(1+R) = \frac{\cos\theta_1}{\cos\theta_2}\left(\frac{2n_1\cos\theta_2}{n_1\cos\theta_2 + n_2\cos\theta_1}\right)$$

$$= \frac{Z_2}{Z_1}\frac{H_{\text{transmitted}}}{H_{\text{incident}}} = \frac{n_1}{n_2}\frac{H_{\text{transmitted}}}{H_{\text{incident}}} \tag{19.16.7}$$

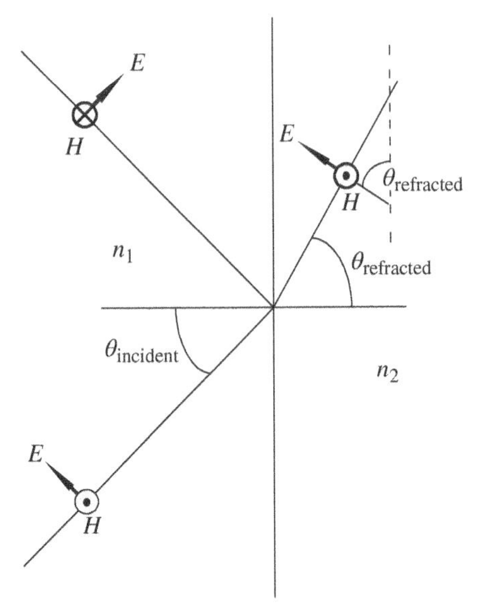

FIGURE 19.1 Refraction at a dielectric interface for TM polarization.

From Snell's law and, e.g., 2 sin a cos b = sin($a + b$) + sin($a - b$), R_{TM} can be rewritten as

$$R_{\text{TM}} = \frac{\frac{n_1}{n_2}\cos\theta_2 - \cos\theta_1}{\frac{n_1}{n_2}\cos\theta_2 + \cos\theta_1} = \frac{\sin\theta_2\cos\theta_2 - \sin\theta_1\cos\theta_1}{\sin\theta_2\cos\theta_2 + \sin\theta_1\cos\theta_1} = \frac{\sin 2\theta_2 - \sin 2\theta_1}{\sin 2\theta_2 + \sin 2\theta_1}$$

$$= \frac{2\sin(\theta_{2,} - \theta_1)\cos(\theta_{2,} + \theta_1)}{2\sin(\theta_2 + \theta_1)\cos(\theta_2 - \theta_1)} = \frac{\tan(\theta_2 - \theta_1)}{\tan(\theta_2 + \theta_1)} \tag{19.16.8}$$

At the *Brewster angle* $\theta_{1,\text{Brewster}} + \theta_{2,\text{Brewster}} = \pi/2$, the denominator of the last line of the above equation becomes infinite, and hence, R_{TM} vanishes (this can be seen directly by noting that the numerator of the first expression for R_{TM} equals zero when $n_1 \cos\theta_2 = n_2 \cos\theta_1$, which after multiplying by $\sin\theta_1 \sin\theta_2$ and employing Snell's law becomes $\sin 2\theta_2 = \sin 2\theta_1$ with a nontrivial solution $2\theta_{1,\text{Brewster}} = 180 - 2\theta_{2,\text{Brewster}}$). From Snell's law,

$$\begin{aligned} n_1 \sin\theta_{1,\text{Brewster}} &= n_2 \sin\theta_{2,\text{Brewster}} = n_2 \sin\left(\frac{\pi}{2} - \theta_{1,\text{Brewster}}\right) \\ &= n_2 \cos\theta_{1,\text{Brewster}} \\ \tan\theta_{1,\text{Brewster}} &= \frac{n_2}{n_1} \end{aligned} \tag{19.16.9}$$

At this angle, the transmitted and reflected wave directions are perpendicular. Hence, a TM incoming field induces electron motion at the surface perpendicular to the direction of the transmitted radiation. Since however the radiation pattern of an oscillating dipole vanishes in the direction perpendicular to the direction of oscillation, the reflected wave vanishes. Accordingly, only the TE-polarized component of an incoming beam is reflected.

19.17 DIFFRACTION

If the effect of polarization on the diffraction or scattering from an aperture is neglected, the *scalar wave (Helmholtz) equation* for a single monochromatic electric or magnetic field component

$$\left(\nabla^2 + k^2\right)\phi = 0 \qquad (19.17.1)$$

can be employed in place of the full Maxwell's equations. The field generated along a plane at $z = Z$ from a pattern $\phi(x, y, z = 0)$ is then most simply obtained by integrating $\gamma \exp\left(\vec{k} \cdot \vec{r} - \omega t\right)$, where γ is an as yet undetermined constant, over all points in this pattern. Since in the *Fresnel approximation*, where primed and unprimed quantities are evaluated at $z = 0$ and $z = Z$, respectively,

$$\vec{k} \cdot \vec{r} = k\sqrt{(x-x')^2 + (y-y') + Z^2} \approx kZ\left(1 + \frac{(x-x')^2 + (y-y')^2}{2Z}\right) \qquad (19.17.2)$$

and consequently

$$\phi(x,y,Z) = \gamma e^{i(kZ-\omega t)} \int_{-\infty}^{\infty}\int_{-\infty}^{\infty} \phi(x',y',z=0) e^{i\frac{k}{2Z}\left((x-x')^2 + (y-y')^2\right)} dx'dy' \qquad (19.17.3)$$

To determine the value of γ, observe that for $\phi(x, y, z = 0) = 1$, the field for $z > 0$ is given by $\exp(i(kz - \omega t))$. Hence, considering the point $\phi(0, 0, Z)$,

$$\frac{1}{\gamma} = \int_{-\infty}^{\infty}\int_{-\infty}^{\infty} e^{i\frac{k}{2Z}\left(x'^2 + y'^2\right)} \underbrace{dx'dy'}_{(x',y') = \sqrt{\frac{2iZ}{k}}(x'',y'')} = \frac{2iZ}{k}\int_{-\infty}^{\infty} e^{-y''^2} dy'' \int_{-\infty}^{\infty} e^{-x''^2} dx'' = \frac{2i\pi Z}{k} = i\lambda Z$$

$$(19.17.4)$$

In the *Fraunhofer diffraction* far-field limit for which $Z \gg x', y'$, $(x - x')^2 \approx x^2 - 2xx'$, $(y - y')^2 \approx y^2 - 2yy'$ leading to

$$\phi(x,y,Z) = -\frac{i}{\lambda Z} e^{i(kZ-\omega t)} e^{ik\frac{x^2+y^2}{2Z}} \int_{-\infty}^{\infty}\int_{-\infty}^{\infty} \phi(x',y',z=0) e^{-i\frac{k}{Z}(xx'+yy')} dx'dy' \quad (19.17.5)$$

Hence, in this limit, apart from a phase factor, the far field is given by the Fourier transform of the near field, indicating that each transverse Fourier component of

$\phi(x', y', z=0)$ propagates at a unique angle with respect to the z-axis and hence intersects the $z=Z$ plane at a single position with the contributions of higher transverse frequencies situated further from the optical axis, $x=y=0$.

Diffraction can also be analyzed with the Green's theorem, Equation (8.4.5). For an aperture cut into an opaque screen at $z'=0$, for $d\vec{S}$ is directed away from the observation point at $\vec{r}$ a large positive distance from the aperture

$$\phi(\vec{r}) = -\oint_{\text{aperture}} \left[G(\vec{r},\vec{r}')\hat{e}_{d\vec{S}} \cdot \vec{\nabla}'\phi - \phi\hat{e}_{d\vec{S}} \cdot \vec{\nabla}' G(\vec{r},\vec{r}')\right] dS' \tag{19.17.6}$$

where in three dimensions The Green's function is given by

$$G(\vec{r},\vec{r}') = -\frac{e^{ik|\vec{r}-\vec{r}'|}}{4\pi|\vec{r}-\vec{r}'|} \tag{19.17.7}$$

since for $\vec{r}'=0$. The Green's function must satisfy $(1/r)d^2/dr^2(rG(r)) + k^2G(r) = \delta(r)$, which is solved by $G(r) \propto \exp(ikr)/r$ outside the singularity and must further approach $1/4\pi r$ as $r \to 0$ where $\exp(ikr)$, and hence, the $k^2G(r)$ term in the differential equation for Green's function can be neglected. If the field equals $\phi(\vec{r})$ inside the aperture and zero outside for large r, the $1/r^2$ terms can be neglected and for a plane wave incoming field ($d\vec{S}$ is in the $-\hat{e}_z$ direction over the aperture and $\vec{\nabla}'|\vec{r}-\vec{r}'| = -\vec{\nabla}'|\vec{r}-\vec{r}'|$)

$$\phi(\vec{r}) \approx \frac{ik}{4\pi}\int_{\text{aperture}} \left[\frac{e^{ik|\vec{r}-\vec{r}'|}}{|\vec{r}-\vec{r}'|}(\hat{e}_z + e_{\vec{r}-\vec{r}'}) \cdot \hat{e}_{d\vec{S}} e^{ikz} dS'\right]_{z'=0} \tag{19.17.8}$$

where θ is the angle to the observation point measured from the z-axis (the presence of the additional factor $1+\cos\theta$ in the above integral compared to the previous derivation results from the smaller number of approximations intrinsic in the formalism). If r is far greater than the aperture width, $|\vec{r}-\vec{r}'| \approx r(1 - 2\vec{r}\cdot\vec{r}'/r^2 + \cdots)^{1/2} \approx r - \vec{r}\cdot\vec{r}'/r$ leading to the *Fraunhofer limit*

$$\phi(\vec{r}) \approx \frac{-i}{2\lambda r}(1+\cos\theta)e^{i(kr-\omega t)}\int_{\text{aperture}} \phi(\vec{r}')e^{-ik\hat{e}_r\cdot\vec{r}'} dS'$$

$$\approx \frac{-i}{2\lambda r}(1+\cos\theta)e^{i(kr-\omega t)}\int_{\text{aperture}} \phi(x',y',z=0)e^{-i\frac{k}{Z}(xx'+yy')} dx'dy' \tag{19.17.9}$$

Babinet's principle states that if an obstacle possesses the same shape as the aperture, summing the fields generated from scattering from an obstacle and from the aperture yields an unscattered field. The radiation pattern generated by the obstacle is therefore the negative of that of its aperture, implying equal cross sections (effective areas) in the two cases. The absorption cross section of a perfectly absorbing disk therefore equals its elastic (nonabsorbing) cross section, yielding a total cross section, σ_{total}, equal to twice its area, i.e., $2A$. The *optical theorem*, which can be generalized to partially absorbing targets, is obtained from the forward scattering amplitude. If A denotes the area of the black (absorbing) disk, from Equation (19.17.9) for a plane wave incident field, $c\exp(i(kz-\omega t))$ and $f(\theta=0)$ is the forward scattering amplitude as defined below

$$\begin{aligned}
\operatorname{Im}(f(\theta=0)) &\equiv \operatorname{Im}\left(r\frac{\text{amplitude of scattered field for } \theta=0 \text{ at } z}{\text{amplitude of unscattered incident field at } z}\right)\\
&= \operatorname{Im}\left(\frac{\frac{-ic}{\lambda}e^{i(kz-\omega t)}A}{ce^{i(kz-\omega t)}}\right)\\
&= -\frac{\sigma_{\text{total}}}{2\lambda}
\end{aligned} \tag{19.17.10}$$

19.18 WAVEGUIDES AND CAVITIES

While the intensity of a localized spherically radiating source in a homogeneous medium decreases as $1/r^2$, a wave confined within an ideal waveguide does not attenuate with distance. Examples are metallic waveguides (transmission lines) for high-frequency electromagnetic waves that do not appreciably penetrate into conductors, acoustic waveguides with nearly incompressible surfaces that supply large restoring forces, and dielectric waveguides utilizing total internal reflection.

An electromagnetic field propagating in a waveguide can be written as a superposition of *modal field* solutions of the electromagnetic wave equation and accompanying boundary conditions. These retain the same field *pattern* with a phase that increases linearly along the waveguide axis. The modal fields of an electromagnetic wave confined to a time-invariant three-, two-, or one-dimensional structure (inside a box, rod, or region between two sheets) composed of conducting or dielectric media thus exhibit the spatial symmetries of the boundaries.

To conceptualize electromagnetic modes, consider a *planar waveguide* consisting of two parallel metallic or dielectric layers separated by a distance d where the z-axis is oriented along the propagation direction of the wave and the x-axis is directed perpendicular to the layers, A single, e.g., upward traveling plane wave incident on the space with wavevector $\vec{k}=(k_x,k_z)=k(\sin\theta,\cos\theta)$ between the two layers is reflected into a direction $(-k_x,k_z)$; the two copropagating waves therefore generate *a* z-dependent interference pattern that is absent at the entrance face of the waveguide. Accordingly, an invariant pattern requires both an upward traveling wave and a downward traveling wave at the entrance face with equal amplitudes. The depletion of the upward traveling wave through reflection into the downward wave is then compensated by the additional upward traveling contribution of the simultaneously reflected downward wave.

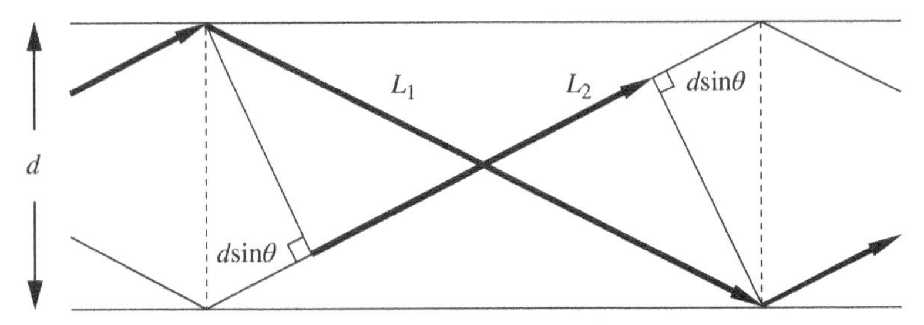

FIGURE 19.2 Modal condition.

As evident from Figure 19.2, a *further* condition is that after an upward traveling wave component reflects from the upper and then the lower boundary, reverting again to an upward traveling wave, it must remain in phase with the upward field component that has still not reflected from the boundary. Since the phase change over a distance Δx is $\Delta\phi = k\Delta x$, the phase difference $k(L_1 - L_2) + 2\delta_r = 2d \sin\theta + 2\delta_r$ between the two rays shown in the figure, where δ_r is the phase change incurred during reflection, must equal $2m\pi$. Hence, modes can only be formed from waves traveling at certain discrete propagation angles. Further, the field *polarization* pattern must be preserved upon multiple reflections. For a planar interface, this requires either TE- or TM-polarized modes that propagate at slightly different angles as in general $\delta_r^{TE} \neq \delta_r^{TM}$.

The modal fields of a hollow waveguide with $\hat{e}_z$ oriented along the propagation direction are expressed as

$$\begin{aligned}\vec{\mathrm{E}}_m(x,y,z) &= \vec{E}_m(x,y)e^{i(k_z z-\omega_m t)}\\ \vec{\mathrm{H}}_m(x,y,z) &= \vec{H}_m(x,y)e^{i(k_z z-\omega_m t)}\end{aligned} \tag{19.18.1}$$

Since $\partial/\partial z$ and $\partial/\partial t$ acting on either field yield multiplicative factors of ik_z or $-i\omega_m$, respectively, the electromagnetic wave equation applied to a mode transforms into the *modal field equation*

$$\left(\nabla_\perp^2 - k_z^2 + \frac{\omega_m^2}{c^2}\right)\begin{Bmatrix}\vec{E}_m\\ \vec{B}_m\end{Bmatrix} = 0 \tag{19.18.2}$$

Dividing the fields into components along and perpendicular to the direction of the z-axis,

$$\vec{\nabla}_\perp \times \underbrace{\left(\vec{\nabla}_\perp \times \vec{E}_\perp\right)}_{-\frac{\partial B_z}{\partial t}\hat{e}_z} = \vec{\nabla}_\perp \underbrace{\left(\vec{\nabla}_\perp \vec{E}_\perp\right)}_{-\frac{\partial E_z}{\partial z} = -ik_z E_z} - \underbrace{\nabla_\perp^2 E_\perp}_{\nabla^2\vec{E}_\perp - \frac{\partial^2 E_\perp}{\partial^2 z^2} = -\frac{\omega^2}{c^2}\vec{E}_\perp + k_z^2\vec{E}_\perp}$$

$$-i\omega\hat{e}_z \times \vec{\nabla}_\perp B_z = \tag{19.18.3}$$

which together with an equivalent manipulation for the magnetic field leads to (as can be deduced from the appearance of a relative factor of $-1/c^2$ in the Maxwell's equations for $\vec{\nabla} \times \vec{E}$ and $\vec{\nabla} \times \vec{B}$)

$$\vec{E}_\perp = i\left(\frac{\omega^2}{c^2} - k_z^2\right)^{-1}\left(-\omega\hat{e}_z \times \vec{\nabla}_\perp B_z + k_z \vec{\nabla}_\perp E_z\right)$$

$$\vec{B}_\perp = i\left(\frac{\omega^2}{c^2} - k_z^2\right)^{-1}\left(\frac{\omega}{c^2}\hat{e}_z \times \vec{\nabla}_\perp E_z + k_z \vec{\nabla}_\perp B_z\right) \tag{19.18.4}$$

Hence, the transverse components of the modal fields are determined by their z-components.

At a conducting waveguide surface, surface charges and currents generate discontinuities in $E_\perp$ and $B_\parallel$, while $E_\parallel|_S = \vec{B}_\perp|_S = 0$. For rectangular geometries, again only TE and TM fields retain their polarization upon reflection and therefore lead to invariant field patterns. Therefore, while the boundary conditions $E_z|_S = 0$ must always be satisfied by a modal field, for a *TE mode*, the electric field polarization at the surface is transverse to the plane of incidence so that $E_z = 0$ while the magnetic field lies in the plane of incidence. For *TM modes* with $B_z = 0$, the magnetic and electric fields are instead polarized transverse and in the place of incidence, respectively. Finally, for a *transverse electromagnetic (TEM) wave*, the z-components of both fields vanish implying $\vec{E} = \vec{B} = 0$ from Equation (19.18.4) unless $\omega_{TEM}^2 = c^2 k_z^2$ and hence, from the modal field equation

$$\nabla_\perp^2 \vec{E}_z = 0 \tag{19.18.5}$$

As the solutions to the Laplace equation are unique, $E_z|_S = 0$ at the boundaries implies $E_z = 0$ within connected regions. TEM modes can however exist within volumes such as those between two parallel wires or cylinders for which the conductors can support different potentials.

For the TM ($B_z = 0$) mode of a rectangular waveguide that extends from zero to L_x in the x-direction and from zero to L_y in y, since $E_\parallel = 0$ at the surface, $E_z = 0$ at the boundary walls and

$$E_z^{(nm,TM)} = E_0 \sin\frac{n\pi x}{L_x}\sin\frac{m\pi y}{L_y}e^{i\left(k_z^{(nm)}z - \omega t\right)} \tag{19.18.6}$$

where from the wave equation

$$k_z^2(nm) = \frac{\omega^2}{c^2} - \pi^2\left(\frac{n^2}{L_x^2} + \frac{m^2}{L_y^2}\right) \equiv \frac{\omega^2}{c^2} - k_z^2(nm, \text{cutoff}) \tag{19.18.7}$$

When ω equals the *cutoff frequency*, $\omega^{(nm,\text{cutoff})} = ck_z^{(nm,\text{cutoff})}$ in Equation (19.18.7), $k_z^{(nm)} = 0$ so that the mode propagates perpendicular to the waveguide axis and therefore exhibits zero group velocity in the direction of the waveguide. For ω below this frequency, the (nm) mode is attenuated.

In the case of TE ($E_z = 0$) polarization, since e.g., on the $x = 0$ surface for which $E_\parallel = 0 \Rightarrow E_y = 0$ and similarly $B_\perp = 0 \Rightarrow B_x = 0$, the Maxwell's equation

$\left(\vec{\nabla}\times B\right)_y = -1/c^2\left(\partial E_y/\partial t\right)$ yields $(i\omega/c^2)E_y = ik_z B_x - \partial_x B_z$ so that $\partial_x B_z = 0$ and by extension $\partial B_z/\partial n|_S = 0$ on the waveguide boundaries. Hence,

$$B_z^{(nm,TE)} = \cos\frac{n\pi x}{L_x}\cos\frac{m\pi y}{L_y}e^{i\left(k_z^{(nm)}z-\omega t\right)} \tag{19.18.8}$$

again implying Equation (19.18.7). The transverse field components follow again from Equation (19.18.4).

From, e.g., $\cos x = (\exp(ix)+\exp(-ix))/2$, the above modes consist of equal amplitude plane waves propagating at angles $\pm\theta_{nm}$ with $\cos\theta_{nm} = k_z^{(nm)}/k$ with respect to the z-axis. Their phase velocity

$$v^{(nm,\text{phase})} = \frac{\omega}{k_z^{(nm)}} = \frac{\omega}{k\cos\theta_{nm}} = \frac{c}{\cos\theta_{nm}} \tag{19.18.9}$$

is obtained by setting $k_z^{(nm)}z-\omega t = 0$ in, e.g., Equation (19.18.8). As θ_{nm} increases, the phase fronts are more widely spaced and therefore propagate more rapidly along the z-axis with $v_{\text{phase}} > c$.

A pulse emitted by a localized source into a waveguide typically propagates in all waveguide modes. Modes with large θ_{nm} are directed away from the waveguide axis and thus propagate more slowly than modes with small θ_{nm}. The resulting pulse spreading is termed *waveguide dispersion*. Quantitatively, information is transferred down the waveguide at the *group velocity*, which is the component $v^{(nm,\text{group})} = c\cos\theta_{nm} = c^2/v^{(nm,\text{phase})}$, of the wave velocity along the waveguide axis. This result can also be derived from $k_z^{2(nm)} = \omega^2/c^2 - k_z^{2(nm,\text{cutoff})}$, which implies $k_z^{(nm)}dk_z^{(nm)} = \omega d\omega/c^2$ and therefore

$$v_{\text{group}} = \frac{\partial\omega}{\partial k_z^{(nm)}} = \frac{c^2 k_z^{(nm)}}{\omega} = \frac{c^2}{v_{\text{phase}}} \tag{19.18.10}$$

In an *electromagnetic cavity*, such as a cube with faces at $x_i = 0, \pi$, each mode possesses a fixed frequency. As an example, a *TE mode with respect to propagation in the z-direction* possesses $\partial_x B_z = 0$ along the $x = 0, \pi$ faces and $\partial_y B_z = 0$ along the $y = 0, \pi$ faces, so that

$$B_z = B_{z0}(c_1\sin m_z z + c_2\cos m_z z)\cos m_x x\cos m_y y \tag{19.18.11}$$

Along the $z = 0, \pi$ planes, $B_z = 0$, leading to $c_2 = 0$. Inserting the resulting expression into the scalar wave equation yields the resonance frequencies

$$\frac{\omega^2_{m_x m_y m_z}}{c^2} = m_x^2 + m_y^2 + m_z^2 \tag{19.18.12}$$

If the cavity is employed as an oscillator, the power loss from the finite metal resistance is as in mechanics generally described by the *Q-factor*

$$Q = \frac{2}{\pi}\frac{\text{energy stored}}{\text{energy lost per cycle}} \tag{19.18.13}$$

on the order of QT where T denotes the period. Since by the uncertainty principle (Δf) $(\Delta t) \backslash \approx 1$, the approximate width of the resonance peak, as can be shown analytically, is given by $\Delta f = f/Q$.

$$1 = \Delta\left(\frac{1}{QT}\right) = \Delta\left(\frac{f}{Q}\right) = \frac{2\pi\Delta\omega}{Q} \tag{19.18.14}$$

20

QUANTUM MECHANICS

The motion of small, light particles is incorrectly described by Newton's laws and instead exhibits the characteristics of wave motion. In particular, the behavior of electrons that are bound to atomic nuclei with characteristic electric binding potentials of the order of volts and that therefore participate in chemical processes is governed by quantum mechanical laws. Consequently, quantum physics derives its relevance from its applicability to the characteristic energy, time, and length scales of chemical and biological processes.

20.1 FUNDAMENTAL PRINCIPLES

If a spatially uniform electron beam of electrons with momentum p passes through a slit of width w and is subsequently recorded by a detector array after a further distance $d \gg w$, each electron within the beam generates a single event at one of the detectors with a location that varies for successive electrons. A histogram of these positions generates a diffraction pattern with an angular half width of $\Delta\theta_{1/2} = h/2pw$, between the maximum and the first zero where *Planck's constant* $h = 6.67 \times 10^{-31}$ J/ms. Since an identical pattern is generated with $\Delta\theta_{1/2} = \lambda/2w$ for, e.g., water or sound waves, this result implies that the particle beam can be described by a plane wave with a *de Broglie wavelength*

Fundamental Math and Physics for Scientists and Engineers, First Edition.
David Yevick and Hannah Yevick.

$$\lambda = \frac{h}{p} \tag{20.1.1}$$

Given the wave nature of particle motion, an electron confined by a time-independent potential, typified by the atomic potential, exhibits time-invariant patterns termed *spatial states* or *eigenmodes* in the same manner that a rubber band fixed in a time-independent manner at both ends supports certain stationary vibrational mode patterns. The nature of these oscillations is clarified through an analogy with a string vibrating in a mode at a frequency f_1 in the presence of a sound wave with a frequency f_2. If the string possesses a further mode at a frequency $f = f_1 + f_2$ or $f = f_1 - f_2$, its interaction with the ambient sound field will couple power between the two modes. Similarly, a monochromatic electromagnetic wave of frequency ω exerts a force $\vec{F} = -e\vec{E}_0 \cos(\omega t + \varphi)$, on an electron. If an electron is initially in a certain modal configuration, the time-dependent dipole moment $\vec{p}(t)$ associated with its displacement from the nucleus interacts with the electric field according to $\Delta U = -\vec{p}(t) \cdot \vec{E}(t)$. This is found to generate a transition to a second atomic state only if a second state exists satisfying the relationship, with "hbar" defined by $\hbar = h/2\pi$,

$$hf = \hbar\omega = E_2 - E_1 \tag{20.1.2}$$

where $E_2 - E_1$ is, e.g., the measured difference in the chemical binding energies of the electron in the two states. Accordingly, an atomic state with energy E oscillates in *time* at a angular frequency

$$\omega = E/\hbar \tag{20.1.3}$$

20.2 PARTICLE–WAVE DUALITY

If a spatially uniform forward propagating beam of electrons with momentum $\vec{p}$ were simply described by a classical wave, $\sigma(\vec{r},t) = A \sin\left(\vec{k} \cdot \vec{r} - \omega t + \delta\right)$, the probability of detecting an electron at a point $\vec{r}$ at time t would be proportional to the power $\propto A^2 \sin^2\left(\vec{k} \cdot \vec{r} - \omega t + \delta\right)$ with $\vec{k} = \vec{p}/\hbar = 2\pi/\lambda$ and $\omega = E(p)/\hbar = 2\pi/T$. Thus, electrons would, e.g., be absent at the planes $\vec{k} \cdot \vec{r} - \omega t + \delta = 2\pi m$, contradicting the uniformity of the measured electron distribution.

To generate a function that describes the beam yet possesses uniform properties, a second field, $\chi(\vec{r},t) = A \cos\left(\vec{k} \cdot \vec{r} - \omega t + \delta\right)$, must accordingly be introduced. The probability of detecting an electron at $(\vec{r},t)$ is then the *square* of the *amplitudes* (the powers) of each field, $\chi^2(\vec{r},t) + \sigma^2(\vec{r},t)$, which remains constant in space and time. The formalism is simplified by postulating a *complex wavefunction*, $\varphi(\vec{r},t) = \chi(\vec{r},t) + i\sigma(\vec{r},t)$, such that for a propagating beam,

$$\varphi(\vec{r},t) = ce^{i\left(\vec{k}\cdot\vec{x}-\omega t\right)} \tag{20.2.1}$$

with complex conjugate $\varphi^*(\vec{r},t) = \chi(\vec{r},t) - i\sigma(\vec{r},t)$. The probability density of the electron occupying a certain point $\vec{r}$ at time t is then

$$P(\vec{r},t) = \varphi^*(\vec{r},t)\varphi(\vec{r},t) \tag{20.2.2}$$

i.e., $P(\vec{r},t)d^3r$ corresponds to the probability of detecting the electron within a volume of size d^3r centered at the position $\vec{r}$ at time t. Consequently, for a *single* electron confined to a volume V_e in N_d spatial dimensions,

$$\int_{V_e} \varphi^*(\vec{r},t)\varphi(\vec{r},t)d^{N_d}r = 1 \tag{20.2.3}$$

As the probability is dimensionless, while the dimension of $d^{N_d}r$ equals $[D^{N_d}]$, the wavefunction possesses units $[1/\mathrm{D}^{N_d/2}]$.

Since $|\vec{p}| = h/\lambda$ implies that $\vec{p} = \hbar\vec{k}$ where $\vec{k}$ is the *wavevector* with magnitude equal to the *wavenumber* $k = |\vec{k}| = 2\pi/\lambda$, while $E = \hbar\omega$, in a homogeneous medium, the wavefunction of a uniform beam of N particles within a volume V_e with momentum $\vec{p}$ and energy E possesses the form

$$\varphi(\vec{r},t) = \sqrt{\frac{N}{V_e}}e^{i\left(\vec{k}\cdot x-\omega t\right)} = \sqrt{\frac{N}{V_e}}e^{i\left(\frac{\vec{p}}{\hbar}\cdot x-\frac{E}{\hbar}t\right)} \tag{20.2.4}$$

While replacing $\vec{k}$ by $-\vec{k}$ reverses the direction of the particle beam, substituting $-\omega$ for ω alters the sign of the particle energy. The latter substitution is unphysical since the energy is uniquely determined by the momentum through, e.g., $E = p^2/2m + U(\vec{r})$, where $U(\vec{r})$ is the potential function.

20.3 INTERFERENCE OF QUANTUM WAVES

For noninteracting particles, if, e.g., two oppositely directed beams with the same frequency are separately described by $\varphi_1(\vec{r},t) = \sqrt{N/V_e}e^{i(kx-\omega t)}$ and $\varphi_2(\vec{r},t) = \sqrt{N/V_e}e^{-i(kx+\omega t)}$, the linear superposition $\varphi(\vec{r},t) = \varphi_1(\vec{r},t) + \varphi_2(\vec{r},t)$ describes a system in which both beams are present and for which the probability of detecting a particle detection at a given point is

$$|\varphi(x,t)|^2 = |\varphi_1(x,t)|^2 + |\varphi_2(x,t)|^2 + 2\mathrm{Re}\left(\varphi_1(x,t)\varphi_2^*(x,t)\right) = \frac{4N}{V_e}\cos^2(kx) \tag{20.3.1}$$

implying that the particles are absent at certain points in space while halfway between these points the probability density equals $4N/V_e$. The overall number of particles however correctly evaluates to $2N$ after integrating over V_e since $\cos^2 x$ averages to 1/2 over any number of periods. In contrast, the probability of detecting a classical particle is the sum of the *probabilities* of finding the particle in each of the individual beams. The classical result is recovered for *incoherent* beams for which the phase of the wavefunctions varies rapidly over the times and distances that characterize the measurement as the interference terms in the probability amplitude then average spatially or temporally to zero.

20.4 SCHRÖDINGER EQUATION

Since wavefunctions superimpose linearly, in a spatially infinite uniform medium with constant potential, U, a general electron wavefunction can be decomposed into a finite or infinite number of plane waves of the form of Equation (20.2.4). Although in spatially varying potentials both $\vec{p}$ and $U(\vec{r},t)$ vary with position, if the potential is approximated by a piecewise constant potential, so that within a small volume of space, ΔV, around a point $\vec{r}_0$ the potential function equals $U(\vec{r}_0,t)$, a general wavefunction can still similarly be expressed locally within ΔV as a superposition of plane waves with all possible momenta. Therefore, a differential equation that for constant potentials possesses only these plane waves as solutions and that enforces the proper boundary conditions at interfaces between the discrete regions generates all physical solutions for general potentials. Noting that the spatial gradient of $\varphi(\vec{r},t) = c\exp(i(\vec{p}\cdot\vec{r}/\hbar - Et/\hbar))$ yields $i\vec{p}/\hbar$ times the wavefunction, while the time derivative similarly leads to a proportionality factor $-iE/\hbar$, *momentum and energy operators* that act on particle wavefunctions can be defined as

$$\vec{p}\,\varphi(\vec{r},t) \rightarrow \vec{p}_{\text{op}}\varphi(\vec{r},t) = -i\hbar\vec{\nabla}\,\varphi(\vec{r},t) \tag{20.4.1}$$

and

$$E\varphi(\vec{r},t) \rightarrow E_{\text{op}}\varphi(\vec{r},t) = i\hbar\frac{\partial}{\partial t}\varphi(\vec{r},t) \tag{20.4.2}$$

(Here as in most treatments of quantum mechanics, the same notation (e.g., $\vec{p}$) is employed for both operators and values).

For each plane wave E and $\vec{p}$ and therefore, the time and spatial dependence are related by $E = p^2/2m + U(\vec{r},t)$. This constraint together with Equations (20.4.1) and (20.4.2) yields the *time-dependent Schrödinger equation*, namely,

$$H_{\text{op}}\varphi \equiv \left(-\frac{\hbar^2}{2m}\nabla^2 + U(\vec{r},t)\right)\varphi(\vec{r},t) = i\hbar\frac{\partial}{\partial t}\varphi(\vec{r},t) = E_{\text{op}}\varphi(\vec{r},t) \tag{20.4.3}$$

where H_{op} is the Hamiltonian of the system, and physical boundary conditions must be imposed to insure a unique wavefunction solution. Since the particle energy and therefore the time derivative of the wavefunction remain finite,

$$\varphi(\vec{r},t+\Delta t)=e^{-\frac{i}{\hbar}\int_t^{t+\Delta t}H_{\text{op}}dt'}\varphi(\vec{r},t')\approx\left(1-\frac{i}{\hbar}\int_t^{t+\Delta t}H_{\text{op}}dt'\right)\varphi(\vec{r},t) \tag{20.4.4}$$

and the absence of divergences in $H_{\text{op}}\varphi(\vec{r},t)$ implies that $\varphi(\vec{r},t)$ is a continuous function of time. Similarly, since the kinetic energy is bounded, excluding unphysical infinite potentials for which a compensating, divergent kinetic energy insures that the total energy remains finite, integrating the Schrödinger equation over position variables establishes the continuity of both $\vec{\nabla}\varphi(\vec{r},t)$ and $\varphi(\vec{r},t)$.

For an energy-conserving system with a time-invariant potential, separating variables with $\varphi(\vec{r},t)=\psi(\vec{r})\chi(t)$ replaces Equation (20.4.3) by the *time-independent Schrödinger equation*

$$H_{\text{op}}\psi(\vec{r})\equiv\left(-\frac{\hbar^2}{2m}\nabla^2+U(\vec{r})\right)\psi(\vec{r})=E\psi(\vec{r}) \tag{20.4.5}$$

while $i\hbar d\chi(t)/dt=E\chi(t)$ so that

$$\varphi(\vec{r},t)=\psi(\vec{r})e^{-i\frac{E}{\hbar}t} \tag{20.4.6}$$

20.5 PARTICLE FLUX AND REFLECTION

The Schrödinger equation preserves the number of particles as left multiplying both sides by $\varphi^*(\vec{r},t)$ yields

$$-\frac{\hbar^2}{2m}\left[\vec{\nabla}\cdot\left(\varphi^*\vec{\nabla}\varphi\right)-\vec{\nabla}\varphi^*\cdot\vec{\nabla}\varphi\right]+\varphi^*U\varphi=i\hbar\varphi^*\frac{\partial}{\partial t}\varphi \tag{20.5.1}$$

In the same manner, multiplying the complex conjugate of the Schrödinger equation

$$(H_{\text{op}}\varphi)^*\equiv\left(-\frac{\hbar^2}{2m}\nabla^2+U(\vec{r},t)\right)\varphi^*(\vec{r},t)=-i\hbar\frac{\partial}{\partial t}\varphi^*(\vec{r},t) \tag{20.5.2}$$

by φ, proceeding as above (which is equivalent to the transformation $\varphi\leftrightarrow\varphi^*$ of Equation (20.5.1)) and subtracting the resulting equation from Equation (20.5.1) leads to

$$-\frac{\hbar^2}{2m}\vec{\nabla}\cdot\left(\varphi^*\vec{\nabla}\varphi-\varphi\vec{\nabla}\varphi^*\right)=i\hbar\left(\varphi^*\frac{\partial}{\partial t}\varphi+\varphi\frac{\partial}{\partial t}\varphi^*\right)=i\hbar\frac{\partial}{\partial t}(\varphi\varphi^*) \tag{20.5.3}$$

In terms of the particle current and density

$$\vec{J} = -\frac{i\hbar}{2m}\left(\varphi^* \vec{\nabla}\varphi - \varphi\vec{\nabla}\varphi^*\right) \tag{20.5.4}$$

and

$$\rho = \varphi\varphi^* \tag{20.5.5}$$

Equation (20.5.3) represents the local particle continuity equation

$$\vec{\nabla}\cdot\vec{J} = -\frac{\partial\rho}{\partial t} \tag{20.5.6}$$

For a uniform beam, Equation (20.2.4), $\vec{J}$ and ρ coincide with $N\vec{v}/V$ and N/V. After integration over any closed volume and applying Gauss's law, Equation (8.4.1)

$$\int_{S\subset V} \vec{J}\cdot dS = -\frac{\partial}{\partial t}\int_V \rho dV \tag{20.5.7}$$

For a particle confined in a region, $\varphi = 0$ on the boundaries. The left-hand side of Equation (20.5.7) then vanishes and the integrated particle density (number of particles) is time invariant.

Example

If a one-dimensional electron beam with $E > U_0$ is incident on a potential step from $x < 0$,

$$U(x) = U_0\theta(x) \tag{20.5.8}$$

where the *Heaviside step function* is defined by

$$\theta(x) = \int_{-\infty}^{x} \delta(x)dx = \begin{cases} 1 & x>0 \\ \frac{1}{2} & x=0 \\ 0 & x<0 \end{cases} \tag{20.5.9}$$

and then with $\hbar k_< = \sqrt{2mE}$, $\hbar k_> = \sqrt{2m(E-U_0)}$

$$\phi(x,t) = \begin{cases} e^{-i\frac{E}{\hbar}t}\left(e^{ik_<x} + Re^{-ik_<x}\right) & x<0 \\ e^{-i\frac{E}{\hbar}t}Te^{ik_>x} & x>0 \end{cases} \tag{20.5.10}$$

With the reflected wave present, both the requirement of wavefunction continuity at $x = 0$

$$1 + R = T \tag{20.5.11}$$

and continuity of the derivative of the wavefunction at $x=0$

$$k_{<}(1-R)=k_{>}T \tag{20.5.12}$$

can be satisfied. Indeed, for a *single* incoming and transmitted plane wave at an interface between two regions with differing $U(x)$, the impedance ratio $(1/\psi)d\psi/dx=ik=i\sqrt{2m(E-U(x))}/\hbar$ between the derivative and the amplitude of the wavefunction cannot be continuous across the interface. If a reflected wave is additionally present, however, its impedance is the negative of that of the incoming wave. Hence, the effective impedance associated with the sum of the reflected and incoming waves can equal that on the transmitted side, as is evident by dividing Equation (20.5.12) by Equation (20.5.11), which equates the impedance ratios on both sides of the boundary. Further, multiplying the corresponding sides of Equations (20.5.11) and (20.5.12) yields

$$k_{<}\left(1-R^2\right)=k_{>}T^2 \tag{20.5.13}$$

which insures the conservation of particle flux at the interface and can therefore be obtained directly from the continuity of Equation (20.5.4) at $x=0$ given (20.5.10). Here, $\hbar k_{<}(1-R^2)$ equals the sum of the momentum component of the incident wave multiplied by the probability of finding the electron in this component (e.g., the particle current) and the corresponding (negative) quantity for the reflected wave.

20.6 WAVE PACKET PROPAGATION

In one dimension, a particle localized within a region $|x|<a$ at $t=0$ can be modeled by a *Gaussian wave packet*

$$\varphi_0(x,t=0)=\frac{1}{\sqrt{2\pi}a}e^{-\frac{x^2}{2a^2}} \tag{20.6.1}$$

with Fourier components

$$\begin{aligned}
\varphi_0(k,t=0)&=\frac{1}{2\pi a}\int_{-\infty}^{\infty}e^{-\frac{x^2}{2a^2}}e^{-ikx}dx\\
&=\frac{1}{2\pi a}e^{-\frac{k^2a^2}{2}}\int_{-\infty}^{\infty}e^{-\frac{1}{2a^2}\left(x+ika^2\right)^2}dx\\
&=\frac{1}{\sqrt{2}\pi}e^{-\frac{k^2a^2}{2}}\int_{-\infty}^{\infty}e^{-x'^2}dx'\\
&=\frac{1}{\sqrt{2\pi}}e^{-\frac{k^2a^2}{2}}
\end{aligned} \tag{20.6.2}$$

Associating the width of $\varphi_0(x)$ and $\varphi_0(k)$ with the standard deviation of the *probability distribution* $|\varphi_0|^2$ implies $\Delta x\approx a/\sqrt{2}$ and $\Delta k\approx 1/\left(\sqrt{2}a\right)$ so that, with $p=\hbar k$,

$$\Delta x\Delta p=\hbar/2 \tag{20.6.3}$$

A Gaussian wave packet propagating with velocity $v = p/m = \hbar k_0/m$ is described at time $t=0$ by $\varphi_{k_0}(k,t=0) = \varphi_0(k-k_0,t=0)$ where the phase of each plane wave wavefunction component evolves in time according to $\varphi_{k_0}(k,t) = \varphi_{k_0}(k,t=0)$ $\exp(-i\omega(k)t)$ with $\omega(k) = E(k)/\hbar$. Defining $k' = k - k_0$ and Taylor expanding $\omega(k)$ in k with the *group velocity* $v_g(k_0) = \partial\omega/\partial k|_{k_0} = \partial E/\partial p|_{p_0}$ and $\omega(k) \approx \omega_0(k_0) + v_g(k_0)k' + (1/2!)\partial v_g/\partial k|_{k_0} k'^2 + \cdots$,

$$\varphi_{k_0}(k,t) = \frac{1}{\sqrt{2\pi}} e^{-\frac{k'^2 a^2}{2}} e^{-i\omega(k)t} \approx \frac{1}{\sqrt{2\pi}} e^{-\frac{k'^2 a^2}{2}} e^{-i\left(\omega_0 t + v_g k' t + \frac{1}{2}\frac{\partial v_g}{\partial k} k'^2 t\right)} \tag{20.6.4}$$

Introducing the variables

$$\begin{aligned} b' &= x - v_g t \\ a'^2 &= a^2 + i\frac{1}{2}\frac{\partial v_g}{\partial k}\bigg|_{k=k_0} t \end{aligned} \tag{20.6.5}$$

and inverse Fourier transforming yield for the propagating wave packet at time t

$$\begin{aligned} \varphi_{k_0}(x,t) &= \frac{e^{-i\omega_0 t}}{2\pi}\int_{-\infty}^{\infty} e^{-\frac{k'^2 a'^2}{2} - ik' v_g t} e^{ikx} dk \\ &= \frac{e^{i(k_0 x - \omega_0 t)}}{2\pi}\int_{-\infty}^{\infty} e^{-\frac{k'^2 a'^2}{2} + ik' b'} dk' \\ &= \frac{e^{i(k_0 x - \omega_0 t)}}{2\pi} e^{-\frac{b'^2}{2a'^2}} \int_{-\infty}^{\infty} e^{-\frac{a'^2}{2}\left(k' - \frac{ib'}{a'^2}\right)^2} dk' \\ &= \frac{e^{i(k_0 x - \omega_0 t)}}{\sqrt{2\pi} a'} e^{-\frac{b'^2}{2a'^2}} \end{aligned} \tag{20.6.6}$$

Accordingly, the wave packet travels at the group velocity, while its width increases linearly with t at large times. That is, since the group velocities of the wavevector components in the packets differ, at large t, the wave packet spreading is governed by this velocity difference multiplied by time.

Example

If the electron wavefunction varies over macroscopic lengths, from the Schrödinger equation with $U = 0$,

$$-\frac{\hbar}{2m_e}\frac{\partial^2\psi(x,t)}{\partial x^2} = -\frac{1.05\times10^{-34}}{2(9.1\times10^{-31})}\frac{\partial^2\psi(x,t)}{\partial x^2} = -\frac{1}{2.09\times10^4}\frac{\partial^2\psi(x,t)}{\partial x^2} = i\frac{\partial\psi(x,t)}{\partial t} \tag{20.6.7}$$

Therefore, an electron wavefunction $\psi(x, t_0) \approx c \exp(-x^2/2(10^{-2})^2)$ with x in meters broadens over second time scales; more generally, the electron probability distribution varies in seconds as approximately 10^4 times the square of its distance variation in meters, as is also evident from $\Delta E \approx (\Delta p)^2/2m_e$, which together with $\Delta x \Delta p \approx \hbar$ implies $\Delta E \approx (\Delta p)^2/2m_e \approx \hbar^2/2m_e(\Delta x)^2$ and therefore from $\Delta t \Delta E \approx \hbar$ (cf. Eq. (20.9.5)) $\Delta t/(\Delta x)^2 \approx 2m_e/\hbar \approx 2 \times 10^4$.

20.7 NUMERICAL SOLUTIONS

The time-dependent Schrödinger equation in a single spatial dimension subject to *initial conditions* given by the values of the wavefunction $\varphi(x_i, t_0)$ on an N point equidistant computational grid $x_i = x_1 + i\Delta x$, $i = 0, 1, 2, ..., N-1$ at an initial time t_0 can be solved numerically for the wavefunction at later times $t_j = t_0 + j\Delta t$ by employing the forward and centered finite difference approximations for the temporal and spatial derivatives, respectively,

$$\left.\frac{\partial \varphi(x,t)}{\partial t}\right|_{x=x_i,t=t_j} \approx \frac{\varphi(x_i, t_{j+1}) - \varphi(x_i, t_j)}{\Delta t}$$

$$\left.\frac{\partial^2 \varphi(x,t)}{\partial x^2}\right|_{x=x_i,t=t_j} \approx \frac{\varphi(x_{i+1}, t_j) - 2\varphi(x_i, t_j) + \varphi(x_{i-1}, t_j)}{(\Delta x)^2} \tag{20.7.1}$$

from which, omitting the sum over the repeated index i′

$$\varphi(x_i, t_{j+1}) \simeq \left(1 + \frac{\Delta t}{i\hbar} H_{ii'}\right)\varphi(x_{i'}, t_j) \tag{20.7.2}$$

where the nonzero elements $H_{ii'}$ of the matrix $\mathbf{H}$ are

$$H_{ii} = \frac{\hbar^2}{m(\Delta x)^2} + V(x_i, t_i)$$

$$H_{ii+1} = H_{ii-1} = -\frac{\hbar^2}{2m(\Delta x)^2} \tag{20.7.3}$$

A program for an initial wavefunction $\varphi(x, 0) = \exp(-x^2/8)$, zero potential and boundary conditions

$$\varphi(x_0, t_j) = \varphi(x_{N+1}, t_j) = 0 \tag{20.7.4}$$

that set the wavefunction equal to zero for all times, t_j, at two points displaced by Δx from the computational window endpoints, is given in atomic units, Sec.21.4, for which $\hbar = m_e = 1$ by

```
clf
hold on
deltaT = .005;
numberOfSteps = 500;
fieldWidth = 2;
computationalWindowWidth = 20;
numberOfGridPoints = 31;

%computational grid
dx = computationalWindowWidth / ( numberOfGridPoints - 1 );
gridPoints = linspace( -computationalWindowWidth / 2, ...
  computationalWindowWidth / 2, numberOfGridPoints );

%initial wavefunction
wavefunction = exp( - gridPoints.^2 ./ ( 2 * fieldWidth^2 ) );
wavefunctionNew = zeros( 1, numberOfGridPoints );

%zero potential
potential = zeros( 1,numberOfGridPoints );

%Hamiltonian matrix elements
diagonalElements = 1 - i * deltaT * ( -1 / dx^2 + potential );
offDiagonalElement = -i * deltaT / ( 2 * dx^2 );

%zero boundary conditions
wavefunction(1) = 0;
wavefunction(numberOfGridPoints) = 0;

for loop = 1 : numberOfSteps
  for innerLoop = 2 : numberOfGridPoints - 1
    wavefunctionNew(innerLoop) = diagonalElements ...
      (innerLoop) * wavefunction(innerLoop) + ...
      offDiagonalElement * (wavefunction(innerLoop + 1) + ...
      wavefunction(innerLoop - 1) );
  end
  wavefunction = wavefunctionNew;
%plotting routines
  if ( rem( loop, 100 ) == 0 )
     plot( abs( wavefunction ), 'g' );
  drawnow;
  end
end
```

More accurate routines, required for rapidly varying potentials, large time steps, or small grid point spacing, are described in, e.g., D. Yevick, *A Short Course in Computational Science and Engineering: C++, Java and Octave Numerical Programming with Free Software Tools*, Cambridge University Press.

20.8 QUANTUM MECHANICAL OPERATORS

While a function transforms the value of its argument into an output value, an operator transforms its function argument into a second, output function. A linear operator satisfies $O(af_1 + bf_2) = aO(f_1) + bO(f_2)$ for any two functions f_1 and f_2. An operator product $AB(f)$ represents $A(Bf)$. Quantum mechanical operators are associative, $A(B + C) = AB + AC$, but not generally commutative, e.g.,

$$(x_{op}p_{op})f = x\left[-i\hbar\frac{d}{dx}f(x)\right] \neq (p_{op}x_{op})f = -i\hbar\frac{d}{dx}[xf(x)] = -i\hbar f + (x_{op}p_{op})f \tag{20.8.1}$$

The *commutator*

$$[A,B] = AB - BA \tag{20.8.2}$$

of two operators, e.g., $[p_{op}, x_{op}] = -i\hbar$, quantifies their degree of noncommutativity. If the equation $Of = g$ possesses a unique solution f for every function g, the inverse operator O^{-1} with $g = O^{-1}f$ exists with $OO^{-1} = O^{-1}O = I_{op}$, where I_{op} represents the identity operator. Functions of operators are defined through power series expansions, e.g., $\exp(O) = \sum_m O^m/m!$ The operator ordering in functions of several noncommuting operators as

$$\exp(A + B) = I_{op} + A + B + \frac{A^2 + AB + BA + B^2}{2!} + \cdots \tag{20.8.3}$$

must however be preserved.

In quantum mechanics, *the result of a physical measurement corresponds to an expectation value of an operator*; e.g., for the average momentum of a state with wavefunction $\varphi(\vec{r}.t)$,

$$\langle \vec{p}_{op} \rangle = \int_V \varphi^*(\vec{r},t)\vec{p}_{op}\varphi(\vec{r},t) = -i\hbar\int_V \varphi^*(\vec{r},t)\,\vec{\nabla}\,\varphi(\vec{r},t) \tag{20.8.4}$$

The *correspondence principle* states that quantum mechanical expectation values such as $\langle x \rangle$ and $\langle p \rangle$ obey classical equations of motion for forces that vary slowly over distances compared to the electron wavelength. For a general time-dependent operator, $O(t)$, typified by, $O(t) = x_{op} + \alpha p_{op}t$, and a Hermitian Hamiltonian H obeying Equation (20.8.8) below, from Equation (20.5.2),

$$\begin{aligned}\frac{d\langle O(t)\rangle}{dt} &= \frac{d}{dt}\int_{-\infty}^{\infty} \psi^*(x,t)O(t)\psi(x,t)dx \\ &= \int_{-\infty}^{\infty}\left[\begin{array}{l}\frac{d\psi^*(x,t)}{dt}O(t)\psi(x,t)+\psi^*(x,t)O(t)\frac{d\psi(x,t)}{dt} \\ +\psi^*(x,t)\frac{dO(t)}{dt}\psi(x,t)\end{array}\right]dx \\ &= \left\langle\frac{dO(t)}{dt}\right\rangle + \frac{1}{i\hbar}\int_{-\infty}^{\infty}\left[-\psi^*(x,t)H_{\text{op}}O(t)\psi(x,t)+\psi^*(x,t)O(t)H_{\text{op}}\psi(x,t)\right]dx \\ &= \left\langle\frac{dO(t)}{dt}\right\rangle + \frac{1}{i\hbar}\left\langle\left[O(t),H_{\text{op}}\right]\right\rangle\end{aligned} \tag{20.8.5}$$

Accordingly, a time-independent operator that commutes with the Hamiltonian remains constant with time and thus corresponds to a physically conserved quantity.

Example

For $O = x_{\text{op}}$, recalling that $[AB, C] = A[B, C] + [A, C]B$, which further implies $[A, f(A)] = 0$,

$$\frac{d\langle x\rangle}{dt} = \frac{1}{i\hbar}\left\langle\left[x_{\text{op}}, H_{\text{op}}\right]\right\rangle = \frac{1}{i\hbar}\left\langle\left[x_{\text{op}}, \frac{p_{\text{op}}^2}{2m}\right]\right\rangle = \frac{i}{2m\hbar}\left\langle p_{\text{op}}\left[p_{\text{op}}, x_{\text{op}}\right] + \left[p_{\text{op}}, x_{\text{op}}\right]p_{\text{op}}\right\rangle = \frac{1}{m}\langle p_{\text{op}}\rangle \tag{20.8.6}$$

so that

$$\begin{aligned}\frac{d^2\langle x\rangle}{dt^2} = \frac{1}{m}\frac{d\langle p_{\text{op}}\rangle}{dt} = \frac{1}{i\hbar m}\left\langle\left[p_{\text{op}}, V(x)\right]\right\rangle &= -\frac{1}{m}\left\langle\left(\frac{d}{dx}\right)V(x) - V(x)\left(\frac{d}{dx}\right)\right\rangle \\ &= -\frac{1}{m}\left\langle\frac{dV(x)}{dx}\right\rangle\end{aligned} \tag{20.8.7}$$

While the expectation values involve integrals over all positions, if a particle is described by a localized wave packet, the force acts approximately at a point and Equation (20.8.7) admits a classical mechanics interpretation.

If an operator O represents a real quantity (a formally complex quantity is typified by the energy of a decaying system described by a non-Hermitian H_{op}), for any function $f(\vec{x})$,

$$\int_V f^*(\vec{r})Of(\vec{r})dV = \left(\int_V f^*(\vec{r})Of(\vec{r})dV\right)^* = \int_V \left(Of(\vec{r})\right)^* f(\vec{r})dV \tag{20.8.8}$$

Defining the *transpose* of the operator O by

$$\int_V f(\vec{r})(Og(\vec{r}))dV = \int_V (O^T f(\vec{r}))g(\vec{r})dV \tag{20.8.9}$$

Equation (20.8.8) implies the Hermitian condition for O

$$O^T = O^* \tag{20.8.10}$$

or equivalently in terms of the *Hermitian conjugate operator* $O^\dagger$

$$O^\dagger \equiv (O^*)^T = O \tag{20.8.11}$$

From the discussion accompanying Equation (12.2.9), a Hermitian operator H_{op} further subject to appropriate Hermitian boundary conditions possesses a complete spectrum of eigenfunctions ψ_n with real eigenvalues E_n satisfying in general

$$H_{\text{op}}(\vec{r},t)\psi_n(\vec{r},t) = E_n(t)\psi_n(\vec{r},t) \tag{20.8.12}$$

that are *orthogonal* and can further by proper normalization be made *orthonormal* such that

$$\int_V \psi_n(\vec{r},t)\psi_m(\vec{r},t)dV = \delta_{nm} \tag{20.8.13}$$

According to Equation (20.4.6), for time-independent Hamiltonians, H,

$$\varphi_n(\vec{r},t) = \psi_n(\vec{r})e^{-i\frac{E_n}{\hbar}t} \tag{20.8.14}$$

so that each eigenfunction preserves its spatial pattern with a phase that increases linearly with time.

For a finite potential, the energy eigenvalue spectrum consists of both a continuum of higher energy states that generally correspond to distorted plane waves that are not confined by the potential and a finite set of discrete bound states, such that for two nondegenerate eigenvalues, $E_{n+1} - E_n > \Delta$, for finite Δ. Accordingly, δ_{nm} in Equation (20.8.13) represents a Kronecker delta function in the discrete part of the spectrum and a Dirac delta function in the continuous spectral region.

That the eigenvectors form a complete set implies that any square-integrable function, i.e., any function that satisfies

$$\int_V |\phi(\vec{r})|^2 dV < \infty \tag{20.8.15}$$

can be written as

$$\phi(\vec{r}) = \sum_n c_n\psi_n(\vec{r}) + \int_{E_{\text{continuous}}} c(E)\psi_E(\vec{r})dE \tag{20.8.16}$$

where in view of Equation (20.8.13)

$$\begin{Bmatrix} c_n \\ c(E) \end{Bmatrix} = \int_V \begin{Bmatrix} \psi_n^*(\vec{r}) \\ \psi_E^*(\vec{r}) \end{Bmatrix} \phi(\vec{r}) dV \tag{20.8.17}$$

and hence

$$\phi(\vec{r},t) = \sum_n c_n \psi_n(\vec{r}) e^{-\frac{i}{\hbar}E_n t} + \int_{E_{\text{continuous}}} c(E) \psi_E(\vec{r}) e^{-\frac{i}{\hbar}Et} dE \tag{20.8.18}$$

For any $\phi(\vec{x})$, where the sum includes an integral over the continuous modal spectrum,

$$\int_V \phi^*(\vec{r}) \phi(\vec{r}) dV = \sum_n c_n^* c_n = \sum_n \int_V \psi_n(\vec{r}') \phi^*(\vec{r}') dV' \int_V \psi_n^*(\vec{r}'') \phi(\vec{r}'') dV''$$

$$= \int_V \phi(\vec{r}'') \int_V \phi^*(\vec{r}') \left(\sum_n \psi_n^*(\vec{r}'') \psi_n(\vec{r}') \right) dV' dV'' \tag{20.8.19}$$

yielding the completeness relation

$$\sum_n \psi_n^*(\vec{r}'') \psi_n(\vec{r}') = \delta^3(\vec{r}'' - \vec{r}') \tag{20.8.20}$$

Further, the probability that a particle in the wavefunction $\phi(\vec{x})$ occupies the eigenstate $\psi_m(\vec{x})$ equals in the position space representation

$$\left| \int_V \psi_m(\vec{r}) \phi(\vec{r}) d^3x \right|^2 = |c_m|^2 \tag{20.8.21}$$

from Equation (20.8.16). If $\phi(\vec{r})$ is normalized to unity, $\sum_m |c_m|^2 = 1$.

20.9 HEISENBERG UNCERTAINTY RELATION

If a wavefunction is an eigenstate of two different Hermitian operators, the physical quantities described by these operators possess definite values. Further, the two operators commute when acting on this eigenfunction, since $O_1 O_2 \psi = \lambda_1 \lambda_2 \psi = \lambda_2 \lambda_1 \psi = O_2 O_1 \psi$. If a complete set of eigenfunctions are eigenstates of both operators, the two operators are fully commuting. However, if two operators commute, an eigenfunction of one operator is only necessarily an eigenfunction of the second operator if the eigenfunction is a *nondegenerate eigenfunction* of the first operator, i.e., only one eigenfunction of the first operator possesses the given eigenvalue.

Otherwise, the eigenfunctions of the second operator are in general linear combinations of the degenerate eigenfunctions of the first operator. In systems with high degrees of symmetry such as the hydrogen atom, several mutually commuting operators $O_1, O_2, \ldots$ termed a *complete set of commuting observables* are required to characterize the system such that a unique state corresponds to each distinct set of eigenvalues $\lambda_1, \lambda_2, \ldots, \lambda_N$ of these operators. These eigenvalues provide the maximum amount of information about a state available at a given time.

Example

In one dimension, eigenfunctions can be degenerate for periodic boundary conditions but not for nonperiodic boundary conditions. Hence, on a one-dimensional circle with circumference L for zero potential, $\exp(ik_n\theta)$ and $\exp(-ik_n\theta)$ with $k_n = 2n\pi/L$ constitute two degenerate eigenfunctions of the Hamiltonian, H_{op}. Further, H_{op} and the *parity operator* defined by $Pf(x) = f(-x)$ commute and form a complete set of observables since, for each k_n, the parity operator eigenfunctions with eigenvalues + 1 and − 1 are unique and are given, respectively, by the symmetric and antisymmetric linear combinations, $\cos(k_n\theta/L)$ and $\sin(k_n\theta/L)$, of the above eigenfunctions.

For noncommuting Hermitian operators, O_1 and O_2, an eigenfunction of O_1 is therefore generally a linear combination of the eigenfunctions of O_2. Repeated measurements of the quantity associated with O_2 yield the eigenvalues of O_2 with probabilities given by the squared moduli of the coefficients in this superposition. The variance of these values about their mean value constitutes the *uncertainty* of the measurement associated with O_2. If O_1 and O_2 are repeatedly applied individually to the same wavefunction, $\psi(x)$, the product of the uncertainties in each operator can be quantified by considering $(\Delta O_1 + i\lambda\Delta O_2)\psi(x)$, with λ real and $\Delta O = O - \langle O \rangle \equiv O - \int_{-\infty}^{\infty} \psi^*(x) O(x) \psi(x) dx$. As $O = \Delta O + \text{const}$ in this expression $[\Delta O_1, \Delta O_2] = [O_1, O_2]$. Minimizing the positive definite quantity

$$\begin{aligned}
f(\lambda) &= \int_{-\infty}^{\infty} [(\Delta O_1 + i\lambda\Delta O_2)\psi(x)]^*(\Delta O_1 + i\lambda\Delta O_2)\psi(x)dx \\
&= \int_{-\infty}^{\infty} \psi^*(x)(\Delta O_1 - i\lambda\Delta O_2)(\Delta O_1 + i\lambda\Delta O_2)\psi(x)dx \\
&= \int_{-\infty}^{\infty} \psi^*(x)\left((\Delta O_1)^2 + i\lambda[\Delta O_1, \Delta O_2] + \lambda^2(\Delta O_2)^2\right)\psi(x)dx \\
&= \left\langle (\Delta O_1)^2 \right\rangle + i\lambda\langle [O_1, O_2] \rangle + \lambda^2 \left\langle (\Delta O_2)^2 \right\rangle
\end{aligned} \tag{20.9.1}$$

according to $df/d\lambda|_{\lambda=\lambda_{\min}} = 0$ results in $\lambda_{\min} = -i\langle[O_1, O_2]\rangle / 2\langle(\Delta O_2)^2\rangle$. Since $f(\lambda) \geq 0$,

$$\left\langle(\Delta O_1)^2\right\rangle - \frac{(-i\langle[O_1, O_2]\rangle)^2}{2\left\langle(\Delta O_2)^2\right\rangle} + \frac{(-i\langle[O_1, O_2]\rangle)^2}{4\left\langle(\Delta O_2)^2\right\rangle} \geq 0 \tag{20.9.2}$$

and

$$\left\langle(\Delta O_1)^2\right\rangle\left\langle(\Delta O_2)^2\right\rangle \geq \frac{1}{4}(-i\langle[O_1, O_2]\rangle)^2 \tag{20.9.3}$$

Setting $O_1 = p_{\text{op}}$, $O_2 = q_{\text{op}}$, $[O_1, O_2] = -i\hbar$ yields the *Heisenberg uncertainty relation*

$$\Delta p \Delta x \geq \frac{\hbar}{2} \tag{20.9.4}$$

where Δx represents the root-mean-squared (r.m.s) value $\langle(\Delta x)^2\rangle^{1/2}$. In a homogeneous medium, the minimum of $\Delta p \Delta x$ is realized by Equation (20.6.3) for a Gaussian wavefunction. Thus, if a particle is localized in a region of extent Δx, the spread in its momentum space wavefunction equals or exceeds $\hbar/2\Delta x$. Employing instead $O_1 = E_{\text{op}} = i\hbar\partial/\partial t$, $O_2 = t$ yields the uncertainty relation

$$\Delta t \Delta E \geq \hbar/2 \tag{20.9.5}$$

indicating that the spread of energies in a wavefunction for a state with lifetime Δt equals or exceeds $\hbar/2\Delta t$. As a consequence, to measure the energy of a system to within a resolution of ΔE requires a time $\geq \hbar/2\Delta E$. Since $E = \hbar\omega$, Equation (20.9.5) is identical to the relationship $\Delta\omega\Delta t \geq 1/2$ between the frequency spread of a pulse and its time duration.

Example

The Fourier transform of a square pulse between $t = -T/2$ and $t = T/2$ with a carrier frequency ω_0 evaluates to

$$\psi(\omega) = \int_{-\frac{T}{2}}^{\frac{T}{2}} e^{-i\omega t} e^{i\omega_0 t} dt = \frac{1}{i(\omega_0 - \omega)}\left(e^{i\frac{(\omega_0-\omega)T}{2}} - e^{-i\frac{(\omega_0-\omega)T}{2}}\right) = \frac{2}{(\omega_0 - \omega)} \sin\left(\frac{(\omega_0 - \omega)T}{2}\right) \tag{20.9.6}$$

The first zeros at $\omega - \omega_0 = \pm 2\pi/T$ correspond to the frequencies for which the phase of the integrand increases linearly by 2π over the integration region. Hence, $\Delta\omega \approx 4\pi/T$ and $\Delta\omega\Delta t = \Delta\omega T \approx 4\pi$. The product exceeds 1/2 since the pulse is non-Gaussian.

20.10 HILBERT SPACE REPRESENTATION

A vector $\vec{V}$ in three-dimensional space is described by an array of three components. In a given orthogonal coordinate system with basis vectors $\hat{e}_j$, these are the coefficients $\vec{V}\cdot\hat{e}_j$ of $\vec{V} = \sum_j \hat{e}_j\left(\vec{V}\cdot\hat{e}_j\right)$. Hence, while the vector is a fixed physical quantity, its representation as a vector with components $\vec{V}\cdot\hat{e}_j$ depends on the choice of coordinate system. In the same manner, a wavefunction can be viewed as a fixed physical quantity analogous to the vector $\vec{V}$ termed a *Hilbert space* (column) vector denoted by a *ket* $|g\rangle$ with as many components as the number of degrees of freedom of the system that it describes. The orthogonal basis that corresponds to $\hat{e}_j$ is then provided by a complete set of eigenfunctions $|\psi_j\rangle$ associated with a Hermitian operator in this system.

To form the expression corresponding to $\vec{V}\cdot\hat{e}_j$, a *bra* $\langle f| \equiv |f\rangle^\dagger = (|f\rangle^*)^T$ representing a row vector formed by Hermitian conjugation is additionally defined such that $\langle f|\, g\rangle$ represents the inner product of the state $|f\rangle$ and $|g\rangle$. An *operator*, which when acting on a function generates a second function, in a Hilbert space acts instead on a ket to generate a second ket; since a ket is represented by a vector, an operator adopts the form of a matrix. The ket generated by the action of the operator O on a ket $|f\rangle$ can be written as either $O|f\rangle$ or $|Of\rangle$. The matrix element of an operator between two states, corresponding in the Schrödinger picture to $\int f^*(\vec{r})Og(\vec{r})dV$, is then written $\langle f|O|g\rangle$, which if O is Hermitian further equals $\langle f|O^\dagger|g\rangle = \langle Of|\, g\rangle$ from the analogous formula $\int f^*(\vec{r})O^\dagger g(\vec{r})dV = \int (Of(\vec{r}))^* g(\vec{r})dV$. A Hilbert space vector $|g\rangle$ is then expressed (represented) in the orthogonal basis $|\psi_j\rangle$ as

$$|g\rangle = \sum_j |\psi_j\rangle\langle\psi_j|\, g\rangle \tag{20.10.1}$$

Since $|g\rangle$ is arbitrary, the identity operator in a Hilbert space is represented by

$$1 = \sum_j |\psi_j\rangle\langle\psi_j| = \sum_j P_j \tag{20.10.2}$$

where the *projection operator* $P_j = |\psi_j\rangle\langle\psi_j|$ is constructed such that $\langle g|P_j$ or $P_j|g\rangle$ yields the j:th component of g in the basis $|\psi_i\rangle$. Additionally, $\langle\psi_n|\,\psi_m\rangle = \delta_{nm}$ implies $P_nP_m = \delta_{nm}P_m$.

The eigenfunctions of the position operator, $O_{\vec{x}} = \vec{x}_{\text{op}}$, are given by

$$\psi_{\vec{x}'}(\vec{x}) = \delta(\vec{x} - \vec{x}') \tag{20.10.3}$$

and are therefore not square integrable. Accordingly, they do not belong to the Hilbert space and are termed generalized states. However, the above definition insures that they are normalized according to

$$\langle \psi_{\vec{x}} | \psi_{\vec{x}'} \rangle = \int_{\times 3} \psi_{\vec{x}}^*(\vec{x}'') \psi_{\vec{x}'}(\vec{x}'') d^3x'' \equiv \int_{-\infty}^{\infty} \int_{-\infty}^{\infty} \int_{-\infty}^{\infty} \psi_{\vec{x}}^*(\vec{x}'') \psi_{\vec{x}'}(\vec{x}'') d^3x'' = \delta(\vec{x} - \vec{x}') \tag{20.10.4}$$

while the closure or completeness relation becomes

$$\int_{\times 3} \psi_{\vec{x}}^*(\vec{x}') \psi_{\vec{x}}(\vec{x}'') d^3x = \delta(\vec{x}' - \vec{x}'') \tag{20.10.5}$$

Often, the eigenstate of an operator is denoted by its eigenvalue so that $\psi_{\vec{x}'}(\vec{x})$ is replaced by $|\vec{x}'\rangle$ and the closure relation is then abbreviated:

$$\int_{\times 3} |\vec{x}\rangle \langle \vec{x}| d^3x = \mathrm{I} \tag{20.10.6}$$

The three-dimensional eigenstates of the momentum operator expressed in position space satisfy

$$-i\hbar \vec{\nabla} \psi_{\vec{p}}(\vec{x}) = \vec{p} \psi_{\vec{p}}(\vec{x}) \tag{20.10.7}$$

which possesses the normalized plane wave solutions

$$\psi_{\vec{p}}(\vec{x}) = \langle \psi | \vec{p} \rangle = (2\pi\hbar)^{-3/2} e^{\frac{i}{\hbar}\vec{p}\cdot\vec{x}} \tag{20.10.8}$$

with

$$\int_{\times 3} \psi_{\vec{p}}^*(\vec{x}) \psi_{\vec{p}}(\vec{x}) d^3x \equiv \delta(\vec{p} - \vec{p}') \tag{20.10.9}$$

The eigenstates of the position operator in momentum space are determined identically from

$$-i\hbar \vec{\nabla}_{\vec{p}} \psi_{\vec{x}}(\vec{p}) = \vec{x} \psi_{\vec{x}}(\vec{p}) \tag{20.10.10}$$

A position space wavefunction $\phi(\vec{x})$ can be expressed in *momentum space* as the coefficient of $\psi_{\vec{p}}(\vec{x})$, which is the Fourier integral,

$$\phi(\vec{p}) = \langle \vec{p} | \phi \rangle = (2\pi\hbar)^{-3/2} \int_{\times 3} e^{-\frac{i}{\hbar}\vec{p}\cdot\vec{x}} \phi(\vec{x}) d^3x \tag{20.10.11}$$

The inverse Fourier transform recasts the momentum space wavefunction into position space

$$\phi(\vec{x}) = \langle \vec{x} | \phi \rangle = (2\pi\hbar)^{-3/2} \int_{\times 3} e^{\frac{i}{\hbar}\vec{p}\cdot\vec{x}} \phi(\vec{p}) d^3p \tag{20.10.12}$$

The normalization of the momentum and position state wavefunctions is identical by Parseval's theorem

$$\begin{aligned}\int_{\times 3} \psi^*(\vec{p})\psi(\vec{p})d^3p &= \int_{\times 3} \psi^*(\vec{p})\left((2\pi\hbar)^{-3/2}\int_{\times 3} \psi(\vec{x})e^{-\frac{i}{\hbar}\vec{p}\cdot\vec{x}}d^3x\right)d^3p \\ &= (2\pi\hbar)^{-3/2}\int_{\times 3}\left(\int_{\times 3} \psi^*(\vec{p})e^{-\frac{i}{\hbar}\vec{p}\cdot\vec{x}}d^3p\right)\psi(\vec{x})d^3x \\ &= \int_{\times 3} \psi(\vec{x})\psi^*(\vec{x})d^3x\end{aligned} \tag{20.10.13}$$

For a system confined to a finite volume, the modal spectrum is discrete and the integrals are replaced by summations. Finally, the classical value of the particle momentum in the momentum representation

$$\langle\vec{p}\rangle = \int_{\times 3} \psi^*(\vec{p})\,\vec{p}\psi(\vec{p})d^3p \tag{20.10.14}$$

corresponds to, in analogy with $\langle\vec{x}\rangle$ in the position representation, the integral over the momentum of each plane wave component weighted by its probability.

20.11 SQUARE WELL AND DELTA FUNCTION POTENTIALS

In an even potential $[P, H_{\text{op}}] = 0$ where P is the parity operator, since

$$P(H_{\text{op}}(x)\psi(x)) = H_{\text{op}}(-x)\psi(-x) = H_{\text{op}}(x)\psi(-x) = H_{\text{op}}P(\psi(x)) \tag{20.11.1}$$

and the eigenfunctions can therefore be simultaneous eigensolutions of P and H_{op}. Even and odd eigenfunctions of P obey $P\psi = \psi$ and $P\psi = -\psi$, respectively, while an arbitrary wavefunction is expressed in terms of its even and odd parts according to

$$\psi(x) = \frac{1}{2}(1+P)\psi(x) + \frac{1}{2}(1-P)\psi(x) = \psi^{\text{even}}(x) + \psi^{\text{odd}}(x) \tag{20.11.2}$$

Example

A particle cannot penetrate the barrier of an *infinite square potential well*

$$U(x) = \begin{cases} 0 & |x| < \dfrac{a}{2} \\ \infty & |x| \geq \dfrac{a}{2} \end{cases} \tag{20.11.3}$$

resulting in the boundary conditions $\psi(-a/2) = \psi(a/2) = 0$. Hence, While the Schrödinger equation admits a plane wave solution inside the well, a single, isolated rightward propagating wave converts through reflection at the square well boundary into a leftward wave. The resulting interference pattern varies with time and therefore does not correspond to an eigenmode. If however the initial field superimposes right and left propagating plane waves with equal amplitudes and phases, as the leftward traveling wave converts into a rightward wave, the rightward wave simultaneously reflects, becoming a leftward wave. At certain discrete wavevectors, the component of the field that is reflected twice is in phase with the unreflected field and the overall field pattern remains unchanged with time.
For a particle energy E, the even solutions of the time-independent Schrödinger equation are, noting that the longest wavelength satisfying the boundary conditions is $2a$,

$$\psi_n(x,t) = \cos\left(\frac{2\pi nx}{2a}\right),\ n = 1,3,5,\ldots \tag{20.11.4}$$

with eigenenergies

$$E_n = \frac{p^2}{2m} = \frac{\hbar^2 k^2}{2m} = \frac{\hbar^2}{2m}\left(\frac{\pi n}{a}\right)^2 \tag{20.11.5}$$

The odd solutions are similarly

$$\psi_n(x,t) = \sin\left(\frac{2\pi nx}{2a}\right),\ n = 2,4,6,\ldots \tag{20.11.6}$$

The lowest-order wavefunction is comprised of plane waves with $k = \pm\pi/a$ implying $\Delta p \approx h/a$, while the spread in position can be approximated by $\Delta x \approx a$. Accordingly, $\Delta x \Delta p \approx h > \hbar/2$.

The *bound state* eigenvalues of a *finite potential well*

$$U(x) = \begin{cases} U_0 & |x| < \dfrac{a}{2} \\ 0 & |x| \geq \dfrac{a}{2} \end{cases} \tag{20.11.7}$$

with $U_0 < 0$ are negative. Hence, the wavefunctions in the outer region satisfy $d^2\psi_n(x)/dx^2 = p_n^2\psi_n(x)$ with $p_n^2 = -2mE_n/\hbar^2 > 0$ for $|x| > a/2$, while in the inner region, $d^2\psi_n(x)/dx^2 = -h_n^2\psi_n(x)$ with $h_n^2 = 2m(E_n - U_0)/\hbar^2 > 0$ so that

$$\psi_n(x) = \begin{cases} A_1 \sin(h_n x) + A_2 \cos(h_n x) & |x| \leq a/2 \\ B_1 e^{p_n x} + B_2 e^{-p_n x} & |x| > a/2 \end{cases} \tag{20.11.8}$$

The coefficients are determined by boundary and continuity conditions. As the even and odd wavefunctions decay exponentially as $|x| \to \infty$,

$$\begin{Bmatrix} \psi_n^{\text{even}}(x) \\ \psi_n^{\text{odd}}(x) \end{Bmatrix} = A \begin{Bmatrix} 1 \\ \text{sign}(x) \end{Bmatrix} e^{-p_n|x|} \tag{20.11.9}$$

while in the inner region,

$$\begin{Bmatrix} \psi_n^{\text{even}}(x) \\ \psi_n^{\text{odd}}(x) \end{Bmatrix} = B \begin{Bmatrix} \cos(h_n x) \\ \sin(h_n x) \end{Bmatrix} \tag{20.11.10}$$

By symmetry, if the boundary conditions $A\psi_n(a/2-\delta) = B\psi_n(a/2+\delta)$ and $A\psi_n'(a/2-\delta) = B\psi_n'(a/2+\delta)$ are imposed at $a/2$, the conditions are simultaneously fulfilled at $-a/2$. Eliminating A and B by dividing the second condition by the first yields a continuity condition for the *logarithmic derivative* (the effective impedance) ψ'/ψ of the wavefunction. After multiplying by $a/2$, the resulting equation contains only the dimensionless quantities $h_n a/2$ and $p_n a/2$:

$$\frac{a}{2}\lim_{\delta\to\infty} \frac{\psi_n'\left(\frac{a}{2}-\delta\right)}{\psi_n\left(\frac{a}{2}-\delta\right)} = h_n \frac{a}{2} \begin{Bmatrix} -\tan\left(\frac{h_n a}{2}\right) \text{even} \\ \cot\left(\frac{h_n a}{2}\right) \text{odd} \end{Bmatrix} = \frac{a}{2}\lim_{\delta\to\infty} \frac{\psi_n'\left(\frac{a}{2}+\delta\right)}{\psi_n\left(\frac{a}{2}+\delta\right)} = -p_n \frac{a}{2} \tag{20.11.11}$$

Additionally, $h_n a/2$ and $p_n a/2$ must be situated on a circle with radius R as

$$\left(\frac{ha}{2}\right)^2 + \left(\frac{pa}{2}\right)^2 = \left(\frac{2m(E_n-U_0)}{\hbar^2} - \frac{2mE_n}{\hbar^2}\right)\frac{a^2}{4} = -\frac{2mU_0}{\hbar^2}\frac{a^2}{4} \equiv R^2 \tag{20.11.12}$$

Plotting this equation as a circle on a graph with $h_n a/2$ as the x-coordinate and $p_n a/2$ as the y-coordinate together with the upper and lower relationships in Equation (20.11.11) yields the bound mode energies from each point of intersection of the curves (the energy eigenvalue is then determined either from $p_n = (-2mE_n/\hbar^2)^{1/2}$ or equivalently from the corresponding value of h_n).

A *delta function potential* $U(x) = -U_0\delta(x)$ alters the wavefunction slope by $-A$ at $x = 0$ resulting in a single symmetric bound mode. That is, for an eigenenergy $E_{\text{bound}} < 0$, the wavefunction possesses the form of Equation (20.11.9) for $x \neq 0$ with logarithmic derivative $+p_{\text{bound}}$ to the left of the origin and $-p_{\text{bound}}$ to the right of the origin. Integrating the Schrödinger equation across $x = 0$ determines p_{bound} and hence the energy $E_{\text{bound}} = -\hbar^2_{\text{bound}}/2m$ of the single bound state according to

$$\frac{1}{\psi_{\text{bound}}(0)}\int_{x=-\delta}^{x=\delta}\left(-\frac{\hbar^2}{2m}\frac{d^2\psi_{\text{bound}}}{dx^2}-U_0\delta(x)\psi_{\text{bound}}\right)dx$$
$$=-\frac{\hbar^2}{2m\psi_{\text{bound}}(0)}\frac{1}{}\frac{d\psi_{\text{bound}}}{dx}\bigg|_{x=-\delta}^{x=\delta}-U_0$$
$$=\frac{2\hbar^2 p_{\text{bound}}}{2m}-U_0$$
$$=\frac{E_{\text{bound}}}{\psi_{\text{bound}}(0)}\int_{x=-\delta}^{x=\delta}\psi_{\text{bound}}(x)dx=0 \tag{20.11.13}$$

20.12 WKB METHOD

The "slowly varying envelope" *WKB method* approximates the solution to the one-dimensional Schrödinger equation by

$$\psi(x)=A(x)e^{i\int_0^x k(x)dx} \tag{20.12.1}$$

where $k(x)=\sqrt{2m(E-U(x))/\hbar^2}$ represents an effective particle wavenumber. For potentials that vary slowly compared to the local particle wavelength, the derivatives of $A(x)$ are far smaller than those of the exponential factor. Thus, after inserting Equation (20.12.1) into $\partial^2\psi/dx^2+k^2(x)\psi=0$,

$$\left(\underbrace{\frac{d^2A}{dx^2}}_{\approx 0}+2ik(x)\frac{dA}{dx}+A\left[-k^2(x)+i\frac{dk(x)}{dx}\right]\right)e^{i\int_0^x k(x')dx'}+Ak^2(x)e^{i\int_0^x k(x')dx'}=0 \tag{20.12.2}$$

The first term can be neglected after which

$$\frac{dA(x)}{A(x)}=-\frac{1}{2}\frac{dk(x)}{k(x)} \tag{20.12.3}$$

implying $A(x)=(k(x))^{-1/2}$. Accordingly, $|A(x)|^2$ reproduces the classical probability of observing a particle within an infinitesimal spatial interval, which varies linearly with the time spent in the interval and hence inversely with the velocity $v=p/m=\hbar k(x)/m$.

Near the *classical turning point*, at which $k(x) \propto E - V \approx 0$, the slowly varying approximation fails as the local particle wavelength is large. For quasi-infinite steplike potentials, the reflected wavefunction cancels the incident wavefunction at each boundary yielding a π phase shift. Otherwise, the potential is approximated by a linear function near the turning point, and, in one of several techniques, the resulting *Airy equation* $d^2\phi/dx^2 + x\phi = 0$ is solved in momentum space, Fourier transformed back to position space, and the method of steepest decent employed to generate

$$\phi(x) = \begin{cases} \dfrac{1}{x^{1/4}} \sin\left(\dfrac{2}{3}x^{3/2} + \dfrac{\pi}{4}\right) & x \to \infty \\ \dfrac{1}{2(-x)^{1/4}} e^{-\frac{2}{3}(-x)^{3/2}} & x \to -\infty \end{cases} \tag{20.12.4}$$

As these expressions coincide with the WKB approximation when the potential is approximated by a linear function, the phase shift upon reflection equals $\pi/2$.

At each point in space inside the turning point, the field is composed of, e.g., an unreflected rightward traveling wave together with a rightward traveling wave that has experienced two reflections (as well as all other multiply reflected waves). For these to interfere constructively and thus preserve the wave pattern, the phase increment after two reflections must coincide with the phase of the unreflected wave up to a multiple of 2π. With $\Delta\theta = k\Delta x$, the resulting *WKB quantization condition* adopts the form

$$\oint k dx = \frac{1}{\hbar}\oint p dx = (n + \mu)2\pi \tag{20.12.5}$$

or equivalently

$$\int_{r_{\text{inner turning point}}}^{r_{\text{outer turning point}}} p dx = \frac{1}{2}(n + \mu)h \tag{20.12.6}$$

where, according to the aforementioned discussion, $\mu = 1/2$ for a potential that varies slowly near the turning point, while $\mu = 1$ for an infinite potential well. If no reflection occurs as for an orbiting electron, $\mu = 0$.

Example

The transmission of particles with energy $E < U_0$ through a finite potential barrier,

$$U(x) = \begin{cases} U_0 & 0 < x < L \\ 0 & \text{elsewhere} \end{cases} \tag{20.12.7}$$

is given in the WKB approximation by

$$|T|^2 \propto e^{-2\int_0^L k(x)dx} \tag{20.12.8}$$

In a complete quantum mechanical analysis, a beam incident from $x<0$ with $E< U_0$, $k_0=\sqrt{2mE}$, and $k=\sqrt{2m(U_0-E)}$ is described by

$$\psi(x) \propto \begin{cases} e^{ik_0x}+Re^{-ik_0x} & x<0 \\ c_1e^{kx}+c_2e^{-kx} & 0\le x\le L \\ Te^{ik_0x} & x>L \end{cases} \tag{20.12.9}$$

From the continuity of the wavefunction and its derivative at $x=L$,

$$\begin{aligned} T &= c_1e^{kL}+c_2e^{-kL} \\ ik_0T &= k\left(c_1e^{kL}-c_2e^{-kL}\right) \end{aligned} \tag{20.12.10}$$

Subsequently eliminating the term proportional to c_1,

$$(k-ik_0)T = 2c_2ke^{-kL} \tag{20.12.11}$$

or

$$|T|^2 = |c_2|^2\left(\frac{4k^2}{k^2+k_0{}^2}\right)e^{-2kL} \tag{20.12.12}$$

For $kL >> 1$, the wavefunction component at $x=0$ resulting from backward reflection from $x=L$ is negligible, and the boundary conditions at $x=0$ are

$$\begin{aligned} 1+R &= c_2 \\ ik_0(1-R) &= -kc_2 \end{aligned} \tag{20.12.13}$$

implying $ik_0(1-(c_2-1)) = -c_2k$ or $c_2 = 2ik_0/(ik_0-k)$ and

$$|T|^2 = \left(\frac{4kk_0}{k^2+k_0^2}\right)^2 e^{-2kL} \tag{20.12.14}$$

If a general potential is approximated by a piecewise constant potential function and the reflected field within each associated internal section neglected, the above formula for the transmission coefficient can be employed within each section. Multiplying the coefficients reproduces the exponential behavior of the WKB result (Eq. (20.12.8)).

20.13 HARMONIC OSCILLATORS

For quadratic potentials, the symmetry of the Hamiltonian with respect to momentum and position operators is similarly manifest in the form of the raising and lowering operators for the eigenfunctions. That is, factoring the Hamiltonian and employing $[p_{op}, x_{op}] = -i\hbar$,

$$\begin{aligned} H_{op} &= \frac{p_{op}^2}{2m} + \frac{m\omega^2}{2} x_{op}^2 \\ &= \hbar\omega\left[\left(\sqrt{\frac{m\omega}{2\hbar}} x_{op} - i\frac{p_{op}}{\sqrt{2m\hbar\omega}}\right)\left(\sqrt{\frac{m\omega}{2\hbar}} x_{op} + i\frac{p_{op}}{\sqrt{2m\hbar\omega}}\right) + \frac{i}{2\hbar}[p_{op}, x_{op}]\right] \\ &\equiv \hbar\omega\left(a^\dagger a + \tfrac{1}{2}\right) \end{aligned} \tag{20.13.1}$$

where the operators a and $a^\dagger$ obey the commutation relations

$$\begin{aligned} &[a, a^\dagger] = 1 \\ &[H_{op}, a] = \hbar\omega[a^\dagger a, a] = \hbar\omega\left(a^\dagger \cancel{[a, a]} + [a^\dagger, a]a\right) = -\hbar\omega a \\ &[H_{op}, a^\dagger] = \hbar\omega[a^\dagger a, a^\dagger] = \hbar\omega a^\dagger \end{aligned} \tag{20.13.2}$$

If χ_n represents an eigenstate with energy E_n, the energy of the eigenstate $a\chi_n$ is $E_n - \hbar\omega$ since

$$H_{op}(a\chi_n) = (aH_{op} + [H_{op}, a])\chi_n = (E - \hbar\omega)\chi_n \tag{20.13.3}$$

while the energy of the eigenstate $a^\dagger\chi_n$ is $E_n + \hbar\omega$. As a and $a^\dagger$ thus transform an eigenstate into the next lower and higher eigenstates, respectively, they constitute lowering and raising operators. Further,

$$a\psi_0(x) = \sqrt{\frac{1}{2m\hbar\omega}}\left(m\omega x + \hbar\frac{d}{dx}\right)\psi_0(x) = 0 \tag{20.13.4}$$

since no state with a lower energy than ψ_0 exists implying

$$\frac{d\psi_0(x)}{\psi_0(x)} = -\frac{m\omega}{\hbar} x dx \tag{20.13.5}$$

Accordingly, the lowest-order wavefunction, after normalizing the probability to unity, equals

$$\psi_0(x) = \left(\frac{m\omega}{\pi\hbar}\right)^{\frac{1}{4}} e^{-\frac{m\omega}{2\hbar}x^2} \tag{20.13.6}$$

Orthogonal higher-order wavefunctions are obtained by applying the raising operator to Equation (20.13.6). Applying $a^\dagger a$ to an eigenfunction yields the order (number) of the state since, from $[a, a^\dagger] = 1$,

$$
\begin{aligned}
(a^\dagger a)\underbrace{a^\dagger a^\dagger \ldots a^\dagger}_{n \text{ times}}\psi_0 &= a^\dagger (aa^\dagger)\underbrace{a^\dagger \ldots a^\dagger}_{n-1 \text{ times}}\psi_0 \\
&= \underbrace{a^\dagger \ldots a^\dagger}_{n \text{ times}}\psi_0 + a^\dagger (a^\dagger a)\underbrace{a^\dagger \ldots a^\dagger}_{n-1 \text{ times}}\psi_0 \\
&\vdots \\
&= n\underbrace{a^\dagger \ldots a^\dagger}_{n \text{ times}}\psi_0 + \underbrace{a^\dagger \ldots a^\dagger}_{n+1 \text{ times}}\underbrace{a\psi_0}_{0} \\
&= n\underbrace{a^\dagger \ldots a^\dagger}_{n \text{ times}}\psi_0
\end{aligned}
\tag{20.13.7}
$$

Then, since $a^\dagger$ and a are Hermitian conjugates and $aa^\dagger = a^\dagger a + 1$,

$$
\begin{aligned}
\int_{-\infty}^{\infty} \left(\underbrace{a^\dagger a^\dagger \ldots a^\dagger}_{m \text{ times}}\psi_0\right)^* \underbrace{a^\dagger a^\dagger \ldots a^\dagger}_{n \text{ times}}\psi_0 \, dV &= \int_{-\infty}^{\infty} \psi_0^* \underbrace{aa \ldots a}_{m \text{ times}}\underbrace{a^\dagger a^\dagger \ldots a^\dagger}_{n \text{ times}}\psi_0 \, dV \\
&= \int_{-\infty}^{\infty} \psi_0^* \underbrace{aa \ldots a}_{m-1 \text{ times}}(aa^\dagger)\underbrace{a^\dagger a^\dagger \ldots a^\dagger}_{n-1 \text{ times}}\psi_0 \, dV \\
&= n \int_{-\infty}^{\infty} \psi_0^* \underbrace{aa \ldots a}_{m-2 \text{ times}}(aa^\dagger)\underbrace{a^\dagger a^\dagger \ldots a^\dagger}_{n-2 \text{ times}}\psi_0 \, dV \\
&= n!\delta_{mn}
\end{aligned}
\tag{20.13.8}
$$

and the orthonormalized wavefunctions are given by

$$
\psi_n = \frac{1}{\sqrt{n!}}\left(a^\dagger\right)^n \psi_0 \tag{20.13.9}
$$

20.14 HEISENBERG REPRESENTATION

In the *Schrödinger formalism* of quantum mechanics operators such as p_{op} and x_{op} are time independent, while the state vectors $|\psi(\vec{x},t)\rangle$ satisfy $i\hbar\partial|\psi(\vec{x},t)\rangle/\partial t = H_{\text{op}}|\psi(\vec{x},t)\rangle$ and therefore depend on time as

$$\left|\psi_n(\vec{x},t)\right\rangle = e^{-\frac{i}{\hbar}H_{\text{op}}t}\left|\psi_n(\vec{x},0)\right\rangle = e^{-\frac{i}{\hbar}E_n t}\left|\psi_n(\vec{x},0)\right\rangle \tag{20.14.1}$$

State vectors in the *Heisenberg picture* instead equal $\left|\psi(\vec{x},0)\right\rangle$ at all t. An operator expectation value, which as a physical quantity must be identical in both formulations, then adopts the form for a Hermitian and time-independent H_{op}:

$$\langle O\rangle = \left\langle \psi(\vec{x},t)\right|O\left|\psi(\vec{x},t)\right\rangle = \left\langle \psi(\vec{x},0)\right|e^{\frac{i}{\hbar}H_{\text{op}}t}Oe^{-\frac{i}{\hbar}H_{\text{op}}t}\left|\psi(\vec{x},0)\right\rangle \tag{20.14.2}$$

Hence, the Heisenberg operator $O_H(t)$ is given by

$$O_H(t) = e^{\frac{i}{\hbar}H_{\text{op}}t}Oe^{-\frac{i}{\hbar}H_{\text{op}}t} \tag{20.14.3}$$

which obeys the time evolution equation

$$\frac{d}{dt}O_H(t) = \frac{i}{\hbar}\left(H_{\text{op}}e^{\frac{i}{\hbar}H_{\text{op}}t}Oe^{-\frac{i}{\hbar}H_{\text{op}}t} - e^{\frac{i}{\hbar}H_{\text{op}}t}Oe^{-\frac{i}{\hbar}H_{\text{op}}t}H_{\text{op}}\right) = \frac{i}{\hbar}\left[H_{\text{op}}, O_H(t)\right] \tag{20.14.4}$$

If the Schrödinger operator O itself depends explicitly on time, the above formula is replaced by

$$\frac{d}{dt}O_H(t) = \frac{i}{\hbar}\left[H_{\text{op}}, O_H(t)\right] + e^{\frac{i}{\hbar}H_{\text{op}}t}\frac{\partial O(t)}{\partial t}e^{-\frac{i}{\hbar}H_{\text{op}}t} = \frac{i}{\hbar}\left[H_{\text{op}}, O_H(t)\right] + \left[\frac{\partial O(t)}{\partial t}\right]_H \tag{20.14.5}$$

Example

For a free particle, $dp_H(t)/dt = (i/\hbar)[p_{\text{op}}^2/2m, p_H(t)] = 0$ so that $p_H(t) = p(0) = p$. Hence, p is conserved consistent with its lack of explicit time dependence and $[p, H] = 0$.

20.15 TRANSLATION OPERATORS

(From this point on, the standard notation that does not distinguish between operators such as p_{op} and the corresponding variables, p, is employed in order to shorten algebraic expressions.) If H is not explicitly time dependent, energy is conserved. The time translation operator, $O_{\Delta t}$,

$$\psi(x,t+\Delta t) = O_{\Delta t}\psi(x,t) = \left(1 + \Delta t\frac{\partial}{\partial t} + \frac{(\Delta t)^2}{2!}\frac{\partial^2}{\partial t^2} + \cdots\right)\psi(x,t) = e^{\Delta t\frac{\partial}{\partial t}}\psi(x,t) = e^{-\Delta t\frac{i}{\hbar}H}\psi(x,t) \tag{20.15.1}$$

commutes with H, while for a position-independent system, H similarly commutes with the momentum and therefore the spatial translation operator

$$\psi(x+\Delta x,t)=O_{\Delta x}\psi(x,t)=\left(1+\Delta x\frac{\partial}{\partial x}+\frac{(\Delta x)^2}{2!}\frac{\partial^2}{\partial x^2}+\cdots\right)\psi(x,t)=e^{\Delta x\frac{i}{\hbar}p}\psi(x,t) \tag{20.15.2}$$

Therefore, for momentum eigenstates, $\psi(x, t) = \exp(ikx)\psi(0, t)$. Finally, if the system is invariant under a rotation about a certain axis, the commutator of H with the rotation operator formed by exponentiating the component of the angular momentum operator about this axis vanishes.

20.16 PERTURBATION THEORY

Assume that an analytically involved Hamiltonian $H' = H^{(0)} + \lambda H$ with $\lambda \ll 1$ differs slightly from a tractable Hamiltonian, $H^{(0)}$, with eigenvalues and eigenstates

$$H^{(0)}\left|\psi_i^{(0)}\right\rangle=E_i^{(0)}\left|\psi_i^{(0)}\right\rangle \tag{20.16.1}$$

satisfying the normalization condition $\left\langle\psi_k^{(0)}\middle|\psi_n^{(0)}\right\rangle=\delta_{kn}$. Approximate eigenstates of H can then be obtained by expanding

$$|\psi_k\rangle=A(\lambda)\left(\left|\psi_k^{(0)}\right\rangle+\lambda\sum_{n\neq k}B_{kn}(\lambda)\left|\psi_n^{(0)}\right\rangle\right) \tag{20.16.2}$$

For $|\psi_i\rangle$ normalized to unity, $A(\lambda = 0) = 1$, while $B_{kn}(\lambda = 0) = 0$. Inserting into $H|\psi_k\rangle = E_k|\psi_k\rangle$ and expanding $B(\lambda)$ and E_k in powers of λ according to

$$\begin{aligned}B_{kn}(\lambda)&=c_n^{(1)}+\lambda c_n^{(2)}+\cdots\\ E_k&=E_k^{(0)}+\lambda E_k^{(1)}+\cdots\end{aligned} \tag{20.16.3}$$

result in

$$\begin{aligned}&\left(H^{(0)}+\lambda H\right)\left(\left|\psi_k^{(0)}\right\rangle+\lambda\sum_{n\neq k}c_n^{(1)}\left|\psi_n^{(0)}\right\rangle+\lambda^2\sum_{n\neq k}c_n^{(2)}\left|\psi_n^{(0)}\right\rangle+\cdots\right)\\ &=\left(E_k^{(0)}+\lambda E_k^{(1)}+\lambda^2E_k^{(2)}+\cdots\right)\left(\left|\psi_k^{(0)}\right\rangle+\lambda\sum_{n\neq k}c_n^{(1)}\left|\psi_n^{(0)}\right\rangle+\lambda^2\sum_{n\neq k}c_n^{(2)}\left|\psi_n^{(0)}\right\rangle+\cdots\right)\end{aligned} \tag{20.16.4}$$

Equating terms proportional to λ and projecting onto $\left\langle\psi_k^{(0)}\right|$,

$$\left\langle \psi_k^{(0)} \middle| \lambda H \middle| \psi_k^{(0)} \right\rangle = \lambda E_k^{(1)} \left\langle \psi_k^{(0)} \middle| \psi_k^{(0)} \right\rangle = \lambda E_k^{(1)} \tag{20.16.5}$$

Thus, to order λ, the deviation of the energy from its unperturbed value is obtained by weighting the perturbation Hamiltonian at each spatial point with the probability of finding the system at the point.

The first-order change of $|\psi_k\rangle$ is found by projecting with $\langle \psi_m^{(0)}|$ for $m \neq k$:

$$\left\langle \psi_m^{(0)} \middle| \lambda H \middle| \psi_k^{(0)} \right\rangle + \lambda c_m^{(1)} E_m^{(0)} \left\langle \psi_m^{(0)} \middle| \psi_m^{(0)} \right\rangle = \lambda E_k^{(0)} c_m^{(1)} \left\langle \psi_m^{(0)} \middle| \psi_m^{(0)} \right\rangle \tag{20.16.6}$$

Hence, for orthonormalized eigenstates,

$$|\psi_k\rangle \approx \left|\psi_k^{(0)}\right\rangle + \lambda \sum_{m \neq k} c_m^{(1)} \left|\psi_m^{(0)}\right\rangle = \left|\psi_k^{(0)}\right\rangle + \lambda \sum_{m \neq k} \frac{\left\langle \psi_m^{(0)} \middle| H \middle| \psi_k^{(0)} \right\rangle}{E_k^{(0)} - E_m^{(0)}} \left|\psi_m^{(0)}\right\rangle \tag{20.16.7}$$

The apparent divergence for closely spaced energy levels is resolved by degenerate perturbation theory. If the first-order correction to the energy is small or zero as for, e.g., an antisymmetric perturbation to a symmetric state, the second-order contribution to the energy, which incorporates the first-order wavefunction change, dominates. Finally, projecting Equation (20.16.4) onto $\left\langle \psi_k^{(0)} \right|$

$$\lambda^2 \sum_{n \neq k} c_n^{(1)} \left\langle \psi_k^{(0)} \middle| H \middle| \psi_n^{(0)} \right\rangle = \lambda^2 E_k^{(2)} \left\langle \psi_k^{(0)} \middle| \psi_k^{(0)} \right\rangle = \lambda^2 E_k^{(2)} \tag{20.16.8}$$

and inserting $c_n^{(1)}$ from Equation (20.16.7) give the second-order energy correction

$$E_k^{(2)} = \sum_{n \neq k} \frac{\left\langle \psi_k^{(0)} \middle| H \middle| \psi_n^{(0)} \right\rangle \left\langle \psi_n^{(0)} \middle| H \middle| \psi_k^{(0)} \right\rangle}{E_k^{(0)} - E_n^{(0)}} \tag{20.16.9}$$

This term contributes negatively to the ground state energy, while the energy shift of other states is in general primarily influenced by the closest energy eigenstate and is negative if this state has a higher energy and is otherwise positive. Accordingly, states tend to repel in second order.

If the exact normalized eigenstate $|\psi_i\rangle$ is approximated by a first-order estimate $|\tilde{\psi}_i\rangle = |\psi_i\rangle + |\delta\psi_i\rangle$, possessing the same symmetries (such as odd or even), a second- or higher-order error is incurred in the estimated energy since for Hermitian H for which $H|\psi_i\rangle = \langle\psi_i|H = E_i$,

$$\begin{aligned} \delta E_i &= \frac{\langle\tilde{\psi}_i|H|\tilde{\psi}_i\rangle}{\langle\tilde{\psi}_i|\tilde{\psi}_i\rangle} - E_i \\ &\approx \frac{E_i + \langle\delta\psi_i|H|\psi_i\rangle + \langle\psi_i|H|\delta\psi_i\rangle}{1 + \langle\psi_i|\delta\psi_i\rangle + \langle\delta\psi_i|\psi_i\rangle} - E_i \\ &\approx \langle\delta\psi_i|H|\psi_i\rangle + \langle\psi_i|H|\delta\psi_i\rangle - E_i(\langle\psi_i|\delta\psi_i\rangle + \langle\delta\psi_i|\psi_i\rangle) = 0 \end{aligned} \tag{20.16.10}$$

Additionally, for the lowest-order eigenstate, $|\psi_1\rangle$,

$$\tilde{E}_1 = \frac{\langle\tilde{\psi}_1|H|\tilde{\psi}_1\rangle}{\langle\tilde{\psi}_1|\tilde{\psi}_1\rangle} = \frac{\sum_{m=1}^{\infty}\langle\tilde{\psi}_1|\psi_m\rangle\langle\psi_m|H|\psi_m\rangle\langle\psi_m|\tilde{\psi}_1\rangle}{\langle\tilde{\psi}_1|\tilde{\psi}_1\rangle}$$

$$= \frac{\sum_{m=1}^{\infty} E_m|\langle\tilde{\psi}_1|\psi_m\rangle|^2}{\langle\tilde{\psi}_1|\tilde{\psi}_1\rangle} > \frac{\sum_{m=1}^{\infty} E_1|\langle\tilde{\psi}_1|\psi_m\rangle|^2}{\langle\tilde{\psi}_1|\tilde{\psi}_1\rangle} = E_1 \tag{20.16.11}$$

Thus, if, e.g., an approximate lowest eigenstate depends on parameters $\alpha_m, m = 1, 2, \ldots$, minimizing the second expression in Equation (20.16.11) (or equivalently $\langle\tilde{\psi}|H|\tilde{\psi}\rangle$ subject to the normalization condition $\langle\tilde{\psi}|\tilde{\psi}\rangle = 1$) with respect to the α_m yields an optimal estimate of the wavefunction and energy within the parameter space. A related result is the *Hellmann–Feynman theorem*, which states that if the Hamiltonian and hence normalized eigenstates of a system depend on a parameter, so that $H(\lambda)|\psi(\lambda)\rangle = E(\lambda)|\psi(\lambda)\rangle$ while $d\langle\psi(\lambda)|\,\psi(\lambda)\rangle/d\lambda = 0$ from $\langle\psi(\lambda)|\,\psi(\lambda)\rangle = 1$,

$$\frac{dE(\lambda)}{d\lambda} = \frac{d}{d\lambda}\langle\psi(\lambda)|H(\lambda)|\psi(\lambda)\rangle$$

$$= \left(\frac{d}{d\lambda}\langle\psi(\lambda)|\right)H(\lambda)|\psi(\lambda)\rangle + \langle\psi(\lambda)|\frac{dH(\lambda)}{d\lambda}|\psi(\lambda)\rangle + \langle\psi(\lambda)|H(\lambda)\left(\frac{d}{d\lambda}|\psi(\lambda)\rangle\right)$$

$$= E(\lambda)\frac{d}{d\lambda}\langle\psi(\lambda)|\psi(\lambda)\rangle + \langle\psi(\lambda)|\frac{dH(\lambda)}{d\lambda}|\psi(\lambda)\rangle$$

$$= \langle\psi(\lambda)|\frac{dH(\lambda)}{d\lambda}|\psi(\lambda)\rangle \tag{20.16.12}$$

Finally, in the *Rayleigh–Ritz method*, $|\tilde{\psi}\rangle$ of Equation (20.16.11) is written as a linear superposition of N approximate eigenstates with unknown coefficients. The partial derivatives of $\tilde{E}_1$ with respect to each of the coefficients are separately set to zero leading to a system of N linear equations; finally, taking inner products of these equations with each of the eigenstates leads to a system of N linear equations.

If the Hamiltonian and boundary conditions of a system are invariant under symmetry transformations, sets of degenerate eigenvalues generally exist resulting in divergent energy denominators in Equations (20.16.7) and (20.16.9) for pairs of these states. An asymmetric perturbation, however, removes this degeneracy. Hence, in *degenerate perturbation theory*, zeroth-order wavefunctions are constructed that replicate this asymmetry, while nondegenerate modes are incorporated as in standard perturbation theory. For two degenerate states $|\psi_j^0\rangle, j = 1,2$ with energy $E^{(0)}$ of a Hamiltonian H, in the presence of an asymmetric perturbation $\lambda H'$, each perturbed wavefunction can be expressed as a linear combination of the ψ_m^0:

$$\psi = \sum_{m=1}^{2} c_m^{(0)}\psi_m^{(0)} \tag{20.16.13}$$

Then to lowest order,

$$\sum_{m=1}^{2}\left(E^{(0)}+\lambda H'-E\right)c_m^{(0)}\psi_m^{(0)}=0 \tag{20.16.14}$$

leading after projecting with $\psi_l^{(0)*}$, $l=1,2$, where $H'_{lm}\equiv\langle\psi_l^{(0)}|H'|\psi_m^{(0)}\rangle$, to the linear equation system

$$\sum_{m=1}^{2}\left(\lambda H'_{lm}+\left(E^{(0)}-E\right)\delta_{lm}\right)c_m^{(0)}=0 \tag{20.16.15}$$

which possesses a nontrivial solution when

$$\begin{vmatrix}\lambda H'_{11}+E^{(0)}-E & \lambda H'_{12}\\ \lambda H'_{21} & \lambda H'_{22}+E^{(0)}-E\end{vmatrix}=0 \tag{20.16.16}$$

The perturbed energies are given by the binomial theorem

$$E_{1,2}=E^{(0)}+\frac{H'_{11}+H'_{22}}{2}\pm\sqrt{\left(\frac{H'_{11}+H'_{22}}{2}\right)^2+H'_{12}H'_{21}} \tag{20.16.17}$$

For each eigenenergy, $c_1^{(0)}$ and $c_2^{(0)}$ are proportional to the elements of the normalized eigenvector.

Time-dependent perturbation theory applies to the time-dependent Schrödinger equation with

$$H(t)=H^{(0)}+\lambda H'(t) \tag{20.16.18}$$

for which a general unperturbed solution with $\lambda=0$ is given in terms of the eigenfunctions of $H^{(0)}$ by

$$\psi^{(0)}(\vec{x},t)=\sum_m c_m^{(0)}e^{-\frac{i}{\hbar}E_m^{(0)}t}\psi_m^{(0)}(\vec{x}) \tag{20.16.19}$$

The effect of the perturbation can then be incorporated through time-dependent coefficients $c_m(t)$:

$$\psi(\vec{x},t)=\sum_m c_m(t)e^{-\frac{i}{\hbar}E_m^{(0)}t}\psi_m^{(0)}(\vec{x}) \tag{20.16.20}$$

Inserting this expression into the time-dependent Schrödinger equation yields

$$\begin{aligned}&i\hbar\sum_m\left(\frac{dc_m(t)}{dt}-\frac{i}{\hbar}\cancel{E_m^{(0)}}c_m(t)\right)e^{-\frac{i}{\hbar}E_m^{(0)}t}\psi_m^{(0)}(\vec{x})\\&=\sum_m c_m(t)e^{-\frac{i}{\hbar}E_m^{(0)}t}\left(\cancel{E_m^{(0)}}+\lambda H'(t)\right)\psi_m^{(0)}(\vec{x})\end{aligned} \tag{20.16.21}$$

Multiplying both sides by $\psi_k^{(0)*}(x)$ and integrating over all space,

$$\begin{aligned}\frac{dc_k(t)}{dt} &= -\frac{i}{\hbar}\lambda\sum_m c_m(t)\left(\int_V \psi_k^{(0)*}(\vec{x})H'(t)\psi_m^{(0)}(\vec{x})dV\right)e^{\frac{i}{\hbar}\left(E_k^{(0)}-E_m^{(0)}\right)t} \\ &\equiv -\frac{i}{\hbar}\lambda\sum_m c_m(t)H'_{km}(t)e^{\frac{i}{\hbar}\left(E_k^{(0)}-E_m^{(0)}\right)t}\end{aligned} \quad (20.16.22)$$

In a system with two states, k and m, and $H' = 2d'\cos\left(\left(E_k^{(0)} - E_m^{(0)}\right)t/\hbar\right)$, Equation (20.16.22) becomes, after neglecting terms that vary rapidly with time as $\exp\left(\pm il\left(E_k^{(0)} - E_m^{(0)}\right)t/\hbar\right)$ with $l = 1, 2$,

$$\begin{aligned}\frac{dc_k(t)}{dt} &= -\frac{i}{\hbar}\lambda d' c_m(t) \\ \frac{dc_m(t)}{dt} &= -\frac{i}{\hbar}\lambda d' c_k(t)\end{aligned} \quad (20.16.23)$$

Combining these reproduces the harmonic oscillator equation. Hence, for $c_m(0) = 1$, $c_k(0) = 0$, the amplitude c_m describes a cosine function, while c_k varies as a sine function. For times small compared to the oscillation period, $h/\lambda d$, the probability of observing the electron in states m and k therefore decreases parabolically as $P_m = P_0 \cos^2(\kappa t) \approx P_0(1 - (\kappa t)^2)$ with $\kappa = \lambda d/\hbar$ and increases linearly, respectively.

More generally, for times sufficiently small that $c_m(t) \approx 1$, the amplitude of the kth state is, where H′ is redefined to include λ,

$$c_k(t) = -\frac{i}{\hbar}\int_0^t H'_{km}(\vec{x},t')e^{\frac{i}{\hbar}\left(E_k^{(0)}-E_m^{(0)}\right)t'}dt' \quad (20.16.24)$$

For

$$H'(x,t) = 2H'(x)\cos(\omega t) = H'(x)\left(e^{i\omega t} + e^{-i\omega t}\right) \quad (20.16.25)$$

the time integral yields

$$c_k(t) = -\frac{i}{\hbar}\left(\frac{e^{\frac{i}{\hbar}\left(E_k^{(0)}-E_m^{(0)}+\hbar\omega\right)t}-1}{\frac{i}{\hbar}\left(E_k^{(0)}-E_m^{(0)}+\hbar\omega\right)} + \frac{e^{\frac{i}{\hbar}\left(E_k^{(0)}-E_m^{(0)}-\hbar\omega\right)t}-1}{\frac{i}{\hbar}\left(E_k^{(0)}-E_m^{(0)}-\hbar\omega\right)}\right)H'_{km} \quad (20.16.26)$$

Again, for $c_m = 1$, $c_k = 0$, in the presence of a static ($\omega = 0$) perturbation, c_k initially increases. However, the relative phases of the two modes evolve in time at a rate proportional to their energy difference. When $t = \pi/\left(E_k^{(0)} - E_m^{(0)}\right)$, the two wavefunctions are 180° out of phase and the coupled amplitude from mode m possesses the opposite sign to the initially transferred amplitude. The amplitude in mode k therefore oscillates sinusoidally. If $\omega \neq 0$, however, the phase of the coupling term additionally varies with time resulting in coupled amplitudes corresponding to the sum and difference of the phase of the perturbation and the wavefunction phase. If for two modes $E_k^{(0)} - E_m^{(0)} \approx \hbar\omega$, the perturbation compensates for the dephasing of the two modes with time in the second term of Equation (20.16.26). That is, when the coupled amplitude is negative as a result of dephasing, the perturbation has changed sign resulting in a net positive coupled amplitude over the entire oscillation cycle. Hence, the second term initially increases linearly with time, while the amplitude of the first term oscillates rapidly and therefore remains small. Accordingly,

$$|c_k(t)|^2 \approx 4\frac{\sin^2\frac{t}{2\hbar}\left(E_k^{(0)} - E_m^{(0)} - \hbar\omega\right)}{\left(E_k^{(0)} - E_m^{(0)} - \hbar\omega\right)^2}|H'_{km}|^2 \tag{20.16.27}$$

For states forming a continuum or quasicontinuum around $E_k^{(0)}$, designating the number of states per unit volume with energy less than E by $\mathrm{N}(E)$ and the *density of states per unit volume* by

$$\mathrm{P}(E) = \frac{d\mathrm{N}(E)}{dE} \tag{20.16.28}$$

yields $\mathrm{P}(E)\Delta E$ for the number of states per unit volume in a region of energy ΔE around E. If the amplitude and frequency of the periodic perturbation change slowly, the coupled modes k are narrowly distributed in energy around $E_k^{(0)} = E_m^{(0)} + \hbar\omega$ compared to the energy scales over which $\mathrm{P}(E)$ and H'_{km} vary significantly. The approximate transition probability to these modes is then

$$\begin{aligned} P(t) &= \sum_k |c_k(t)|^2 \\ &\approx 4|H'_{km}|^2 \int_{-\infty}^{\infty} \frac{\sin^2\frac{t}{2\hbar}\left(E_k^{(0)} - E_m^{(0)} - \hbar\omega\right)}{\left(E_k^{(0)} - E_m^{(0)} - \hbar\omega\right)^2} \mathrm{P}\left(E_k^{(0)}\right) dE_k^{(0)} \\ &\approx 4|H'_{km}|^2 \left(\frac{\pi t}{2\hbar}\right) \mathrm{P}(E_m^{(0)} + \hbar\omega) \end{aligned} \tag{20.16.29}$$

where the integral in the second line requires the formula for the residue at a double pole together with the observation that for a principal part integration, only half of the residue is encircled (cf. Section 10.5), so that, with the substitution $\alpha x = x'$,

$$\int_{-\infty}^{\infty} \frac{\sin^2 \alpha x}{x^2} dx = -\frac{\alpha}{4} \int_{-\infty}^{\infty} \frac{e^{-2ix'} - 2 + e^{2ix'}}{x'^2} dx'$$

$$= \frac{\alpha}{2} \text{Re} \int_{-\infty}^{\infty} \frac{1 - e^{2ix'}}{x'^2} dx'$$

$$= \frac{\alpha}{2} \text{Re} \frac{2\pi i}{2} \left(\frac{1}{(n-1)!} \right) \frac{d^{n-1}}{dx'^{n-1}} \left[x'^2 \frac{1 - e^{2ix'}}{x'^2} \right] \Bigg|_{x=0, n=2} \tag{20.16.30}$$

$$= \frac{\alpha}{2} \text{Re} \frac{2\pi i}{2} \left(-2ie^{2ix'} \right) \Big|_{x'=0}$$

$$= \pi \alpha$$

This yields *Fermi's golden rule* for the initial rate of transitions per unit volume, which is defined as the transition probability per unit time, dP/dt, from a state at $E_m^{(0)}$ to states k near $E_m^{(0)} + \hbar\omega$:

$$R = \frac{2\pi}{\hbar} |H'_{km}|^2 \mathrm{P}\left(E_m^{(0)} + \hbar\omega \right) \tag{20.16.31}$$

An alternative statement of the rule is obtained by identifying

$$\lim_{\alpha \to 0} \frac{\sin^2 \alpha (x - x_0)}{(x - x_0)^2} = \pi \alpha \delta (x - x_0) \tag{20.16.32}$$

which yields for the rate of transition to a single state at $E_m^{(0)} + \hbar\omega$

$$R \approx \frac{2\pi}{\hbar} |H'_{km}|^2 \delta\left(E_k^{(0)} - E_m^{(0)} + \hbar\omega \right) \tag{20.16.33}$$

20.17 ADIABATIC THEOREM

From the discussion following Equation (20.16.26), if the time variation of the potential considerably exceeds the relative period, $T_{\text{beat}} = 1/\Delta f = h/\Delta E$ of a pair of states, the positive amplitude coupled over a half-period $T_{\text{beat}}/2$ between the states will be canceled by the negative amplitude acquired over the subsequent $T_{\text{beat}}/2$ time interval. Hence, if the potential varies more slowly than T_{beat} for all pairs of states and nondegenerate eigenstates remain nondegenerate, the distribution over the system over its eigenstates is preserved.

Inserting the time-dependent system wavefunction written as

$$\psi(t) = \sum_{n=1}^{\infty} a_n(t) \psi_n(t) e^{-\frac{i}{\hbar}\int_0^t E_n(t')dt'} \tag{20.17.1}$$

in which the $\psi_n(t)$ represent eigenfunctions of the instantaneous potential

$$H(t)\psi_n(t) = E_n(t)\psi_n(t) \tag{20.17.2}$$

into the time-dependent Schrödinger equation, $H\psi(t) = \sum_n a_n(t) E_n \psi_n(t) \exp\left(-i\int_0^t E_n(t')dt'/\hbar\right)$ cancels the term in which $i\hbar\partial/\partial t$ acts on the last term of Equation (20.17.1), leading to

$$\sum_{n=1}^{\infty}\left(\frac{da_n(t)}{dt}\psi_n(t) + a_n(t)\frac{d\psi_n(t)}{dt}\right)e^{-\frac{i}{\hbar}\int_0^t E_n(t')dt'} = 0 \tag{20.17.3}$$

For a system initially in a single state m at $t = 0$ that remains nondegenerate during time evolution, projecting onto $\psi_m(t)$ yields

$$\frac{da_m(t)}{dt} = -a_m(t)\langle\psi_m(t)|\frac{d}{dt}|\psi_m(t)\rangle - \sum_{n \neq m} a_n(t)\langle\psi_m(t)|\frac{d}{dt}|\psi_n(t)\rangle e^{-\frac{i}{\hbar}\int_0^t (E_n(t') - E_m(t'))dt'} \tag{20.17.4}$$

Further, differentiating Equation (20.17.2) with respect to time and projecting onto a state $|\psi_m(t)\rangle$ with $m \neq n$ yields, since $\langle\psi_m(t)|dE_n(t)/dt|\psi_n(t)\rangle = dE_n(t)/dt\langle\psi_m(t)|\,\psi_n(t)\rangle = 0$, while by hermiticity $\langle\psi_m(t)|H(d|\psi_n(t)\rangle/dt) = E_m\langle\psi_m(t)|(d|\psi_n(t)\rangle/dt)$,

$$\langle\psi_m(t)|\frac{d}{dt}|\psi_n(t)\rangle = \frac{1}{(E_n - E_m)}\langle\psi_m(t)|\frac{dH}{dt}|\psi_n(t)\rangle \tag{20.17.5}$$

Hence, if $H(t)$ is dominated by low-frequency components such that $H(t) \approx \eta e^{i\omega t}$ with $\eta\omega/(\omega_n - \omega_m) \ll (1/|\psi_n(t)\rangle)d|\psi_n(t)\rangle/dt$ for $n \neq m$, the second term on the right-hand side of Equation (20.17.4) can be neglected relative to the first term, leading to

$$a_m(t) = a_m(0)e^{\frac{i}{\hbar}\int_0^t \langle\psi_m(t')|i\hbar\frac{d}{dt'}|\psi_m(t')\rangle dt'} \tag{20.17.6}$$

Hence, the state amplitude is preserved under an adiabatic transformation, while its phase advances according to the sum of the *geometric phase* appearing in Equation (20.17.6) and the *dynamic phase* of Equation (20.17.1).

21

ATOMIC PHYSICS

While atomic and molecular physics constitutes a fundamental application area of quantum mechanics, the three-dimensional and often asymmetric or nonlinear atomic bonding however introduces considerable mathematical complexity that is generally addressed through involved computational methods. At the same time, knowledge of the properties of the electronic states of highly symmetric potentials, such as that of hydrogen, and the response of these states to external forces can often provide insight into more complex systems.

21.1 PROPERTIES OF FERMIONS

A wavefunction describing a set of identical particles must remain invariant if any two particles are interchanged twice (or, equivalently, the probability distribution must remain invariant upon a single interchange). Hence, e.g., $\psi(\vec{x}_1, \vec{x}_2, \ldots, \vec{x}_N) = c\psi(\vec{x}_2, \vec{x}_1, \ldots, \vec{x}_N)$ with $c^2 = 1$, where $\vec{x}_j$ encompasses all system variables (e.g., spatial and spin) describing the jth particle. From experiment, $c = +1$ for particles with integer spin, termed *bosons*, so that the total wavefunction is symmetric upon interchange of any two particles, while for *fermions* with half-integer spin, $c = -1$ and the total wavefunction is instead antisymmetric under such interchanges. A system such as a nucleus composed of an even number of fermions (protons and neutrons) acts

---

Fundamental Math and Physics for Scientists and Engineers, First Edition.
David Yevick and Hannah Yevick.

as a boson when exchanged with an identical system since the wavefunction is multiplied by −1 upon exchange of each individual subparticle, while it behaves as a fermion for an odd number of subparticles.

The wavefunction of a system of N separately identifiable, noninteracting particles, $1, 2, \ldots, N$, in states $s(1), s(2), \ldots, s(n)$, with $H = \sum_m H_m(x_m)$, and the energy $E = \sum_m E_{s(m)}$, is formed from the product of the individual particle wavefunctions. If the particles are instead identical bosons, with N_l particles in $s(l)$ where $N = \sum_l N_l$, the wavefunction is given by the symmetric product

$$\psi_{s(1),s(2),\ldots,s(N)}(x_1, x_2, \ldots, x_n) = \left(\frac{N_1!N_2!\ldots}{N!}\right)^{\frac{1}{2}} P_m\left(\prod_{k=1}^{N} \psi_{s(k)}(x_m)\right) \tag{21.1.1}$$

where P_m represents *all* permutations of the indices m and the normalization factor is the reciprocal of the number of distinct combinations of the indices. Thus, denoting $a = s(1)$ and $b = s(2)$,

$$\begin{aligned}
&\psi_{a,a,b}(x_1, x_2, x_3) \\
&\quad = \sqrt{\frac{2!}{3!}}(\psi_a(x_1)\psi_a(x_2)\psi_b(x_3) + \psi_a(x_1)\psi_a(x_3)\psi_b(x_2) + \psi_a(x_2)\psi_a(x_3)\psi_b(x_1))
\end{aligned} \tag{21.1.2}$$

where the factor of 2! accounts for the additional three permutations of 1, 2, 3 that yield duplicate wavefunctions. For fermions, each term in the product in Equation (21.1.1) must be multiplied by an additional factor $(-1)^P$, which equals 1 and −1 when the indices m form even or odd permutations of $(1, 2, \ldots, N)$, respectively. Consequently, two particles cannot occupy the same state (e.g., $\psi_a(x_1)\psi_a(x_2)$ equals $-\psi_a(x_1)\psi_a(x_2)$ after interchanging $1 \leftrightarrow 2$ and therefore must vanish as $a = -a$ implies $a = 0$) implying that all $N_i = 1$ in Equation (21.1.1), resulting in a normalization factor $\sqrt{1/N!}$. Since a determinant changes sign upon interchange of any two rows or columns, the wavefunction can then be compactly expressed as a *Slater determinant*

$$\psi = \left(\frac{1}{N!}\right)^{\frac{1}{2}} \begin{pmatrix} \psi_{s(1)}(x_1) & \psi_{s(1)}(x_2) & \cdots & \psi_{s(1)}(x_N) \\ \psi_{s(2)}(x_1) & \psi_{s(2)}(x_2) & \cdots & \psi_{s(2)}(x_N) \\ \vdots & \vdots & \ddots & \vdots \\ \psi_{s(N)}(x_1) & \psi_{s(N)}(x_2) & \cdots & \psi_{s(N)}(x_{N1}) \end{pmatrix} \tag{21.1.3}$$

21.2 BOHR MODEL

In *Rutherford scattering*, heavy, ionized alpha particles (charged helium ions) scatter through Coulomb interactions with positive nuclei. The resulting discovery of a pointlike nuclear structure motivated the *Rutherford model* in which bound electrons describe classical orbits around the nuclei. However, radially accelerating electrons should lose their energy through electromagnetic radiation within microscopic time scales. In the *Bohr model* of an atom, the electron was therefore postulated to occupy a *stationary state*. Electromagnetic radiation with energy

$$hf = \hbar\omega = E_1 - E_2 \tag{21.2.1}$$

would then only be emitted or absorbed only through transitions between two states with energies E_1 and E_2. To justify these assumptions, wavelike electrons are assigned a phase that changes over stationary circular orbits by an integer multiple of 2π, i.e., since $\Delta\phi = k\Delta x$, $\oint kdr = 2\pi rk = 2\pi n$ or $p = m_e v = \hbar k = \hbar n/r$. The radius of the smallest, $n = 1$ hydrogen atom electron orbit is termed the *Bohr radius*, $r = a_B$. The radius, $a_{n,Z,B}$, of the nth orbit for a hydrogen-like atom with a nucleus with Z protons is obtained by equating the Coulomb and centrifugal forces on the electron:

$$\frac{m_e v^2}{a_{n,B}} = \frac{Ze^2}{4\pi\varepsilon_0 a_{n,B}^2} \tag{21.2.2}$$

With $m_e v \equiv n\hbar/a_{n,Z,B}$,

$$a_{n,Z,B} = \frac{4\pi\varepsilon_0}{Ze^2 m_e} m_e^2 v^2 a_{n,Z,B}^2 = \frac{4\pi\varepsilon_0 \hbar^2 n^2}{Ze^2 m_e} \tag{21.2.3}$$

In terms of the dimensionless *fine structure constant* that in all systems of units equals

$$\alpha = \frac{e^2}{4\pi\varepsilon_0 \hbar c} \approx \frac{1}{137.04}\left[Q^2 \frac{D^2 \cdot \text{force}}{Q^2} \frac{1}{T \cdot \text{energy}} \frac{T}{D}\right] = \frac{1}{137.04}\left[\frac{D \cdot \text{force}}{\text{energy}}\right] = \frac{1}{137.04} \tag{21.2.4}$$

and with $hc = 1.24$ eV micron ($\hbar c = 0.395$ eV micron) and $m_e c^2 = 0.5109$ MeV $= 5.109 \times 10^5$ eV,

$$a_{n,Z,B} = \frac{n^2 a_B}{Z} \tag{21.2.5}$$

where the Bohr radius is given by, with $1\ \text{Å} = 10^{-10}$ m (angstrom),

$$a_B = \frac{\hbar}{\alpha m_e c} = 137\left(\frac{\hbar}{m_e c}\right) = 137 \times 3.9 \times 10^{-13}\ \text{m} = 0.53\ \text{Å} \tag{21.2.6}$$

Sinceinaninversesquare-laworbitthekineticenergyis$-1/2$ times the potential energy, the electron binding energy in the nth Bohr orbit is half the potential energy, namely,

$$E_n - E_\infty = -\frac{1}{2}\frac{Ze^2}{4\pi\varepsilon_0\left(\frac{n^2 a_B}{Z}\right)} = -\frac{1}{2}\frac{Ze^2}{4\pi\varepsilon_0\left(\frac{n^2\hbar}{\alpha m_e cZ}\right)}$$
$$= -\frac{1}{2}\frac{Z^2\alpha m_e c^2 \overbrace{\left(\frac{e^2}{4\pi\varepsilon_0\hbar c}\right)}^{\alpha}}{n^2} = -\frac{1}{2}\frac{m_e(\alpha cZ)^2}{n^2} = -13.6\left(\frac{Z}{n}\right)^2 \text{eV} \tag{21.2.7}$$

The quantity $R_y \equiv m_e(\alpha c)^2/2 = \hbar/2a_B^2 m_e = m_e e^4/2(4\pi\varepsilon_0\hbar)^2 = 13.6\text{ eV}$ is termed the *Rydberg energy* and equals half of the *Hartree energy*.

21.3 ATOMIC SPECTRA AND X-RAYS

The atomic spectral lines of an (e.g., thermally) excited atom result from transitions between different quantum states. The emission and absorption frequencies are given in the Bohr model by Equation (21.2.1),

$$f_{12} = \frac{E_1 - E_2}{h} = \frac{Z^2 R_y}{h}\left(\frac{1}{n_1^2} - \frac{1}{n_2^2}\right) \tag{21.3.1}$$

with associated vacuum wavelengths $(\lambda_0)_{12} = c_0/f_{12}$. A transition from the highest $(n_2 = \infty)$ to the lowest $(n_1 = 1)$ energy level in a hydrogen atom yields an ultraviolet with frequency $13.6 \times 1.6 \times 10^{-19}/6.6 \times 10^{-34} = 3.3 \times 10^{15}$ Hz, for which $\lambda_0 = 3.0 \times 10^8/3.3 \times 10^{15} = 9.1 \times 10^{-7}$m or 9.1×10^{-1} μm = 910 nm = 9100 Å (for visible light $\lambda_0 = 0.2 - 0.8$ μm). The inner shell electron binding energies of an atom increases with its atomic number. These electrons can be displaced from their atomic states through collisions with electrons or ions that have been accelerated by an electric field. Subsequent electron transitions to the vacant states generate *X-rays* that decay by interacting with heavy particles. Soft and hard X-ray wavelengths typically range from 1–100 Å to 0.1–1 Å, respectively.

21.4 ATOMIC UNITS

Expressing energy and length in terms of the Hartree energy and the Bohr radius, respectively, removes the explicit dependence of the Schrödinger equation on physical constants. With $x' = x/a_B$,

$$-\frac{\hbar^2}{2m_e a_B^2}\frac{\partial^2\psi}{\partial x'^2} = -\frac{2R_y}{2}\frac{\partial^2\psi}{\partial x'^2} = (E - U)\psi \tag{21.4.1}$$

Substituting $E' = E/2R_y$ and $U' = U/2R_y$ yields Schrödinger equation in *atomic units*:

$$-\frac{1}{2}\frac{\partial^2 \psi}{\partial x'^2} = (E' - U')\psi \tag{21.4.2}$$

The time-dependent Schrödinger equation similarly becomes

$$-\frac{1}{2}\frac{\partial^2 \psi}{\partial x'^2} + U'\psi = i\frac{\partial \psi}{\partial t'} \tag{21.4.3}$$

in terms of the dimensionless variable $t' = t(2R_y/\hbar) = t/t_0$ with $t_0 = \hbar/2R_y = 2.42 \times 10^{-17}$ s. The derivatives in Equation (21.4.3) are of order unity over atomic time and length scales, indicating that electron wavefunctions in an atom are appreciable over distances of the order of the Bohr radius, while its wavefunction varies significantly in phase and possibly amplitude over times of approximately 10^{-17} s.

Atomic units are equivalent to setting $m_e = e = \hbar = 4\pi\varepsilon_0 = 1$, so that charge and mass are expressed in units of the electronic charge, $Q' = Q/e$, and mass, $m' = m/m_e$. Since $a_B = 4\pi\varepsilon_0\hbar^2/e^2 m_e = 1$ and $2R_y = \hbar^2/a_B^2 m_e = 1$, as required, the units of distance and energy then equal the Bohr radius and Hartree energy. Additionally, the time unit $\hbar/2R_y = 1$, while the ratio of the distance and time units, $a_B/(\hbar/2R_y) = \hbar/a_B m_e = e^2/\hbar 4\pi\varepsilon_0 = \alpha c = 1$, demonstrates that the speed of light is $1/\alpha = 137$.

21.5 ANGULAR MOMENTUM

The components of the *angular momentum operator*, $\vec{L} = \vec{r} \times \vec{p}$, obey the commutation relations $[L_i, L_j] = i\hbar\epsilon_{ijk}L_k$ and $\left[\vec{L}, L^2\right] = 0$. Hence, unless $\vec{L} = 0$, simultaneous eigenfunctions of, e.g., both L_z and L_x do not exist, while *spherical harmonic* eigenfunctions $Y_{lm}(\theta, \varphi) = \langle \theta, \varphi | \, l, m \rangle$ of both L^2 and L_z can be constructed with

$$\begin{aligned} L^2|l,m\rangle &= \hbar^2 l(l+1)|l,m\rangle \\ L_z|l,m\rangle &= \hbar m|l,m\rangle \end{aligned} \tag{21.5.1}$$

as described in Section 13.3. If the Hamiltonian depends only on L^2, then $\left[H, \vec{L}\right] = 0$ and angular momentum is conserved.

Example

In the $l = 1$ basis $|1, 1\rangle$, $|1, 0\rangle$, and $|1, -1\rangle$, $\vec{L}$ is represented by

$$L_z = \hbar\begin{pmatrix} 1 & 0 & 0 \\ 0 & 0 & 0 \\ 0 & 0 & -1 \end{pmatrix}, \quad L_x = \frac{\hbar}{\sqrt{2}}\begin{pmatrix} 0 & 1 & 0 \\ 1 & 0 & 1 \\ 0 & 1 & 0 \end{pmatrix}, \quad L_y = \frac{i\hbar}{\sqrt{2}}\begin{pmatrix} 0 & -1 & 0 \\ 1 & 0 & -1 \\ 0 & 1 & 0 \end{pmatrix} \tag{21.5.2}$$

satisfying $[L_i, L_j] = i\hbar\varepsilon_{ijk}L_k$. Thus, for the eigenvector $(2,1,2)^T/\sqrt{9}$ of L_x corresponding to the superposition $(2|1,1\rangle + |1,0\rangle + 2|1,-1\rangle)/\sqrt{9}, l = l_x = 1$. The raising and lowering operators

$$L_+ = L_x + iL_y = \sqrt{2}\hbar\begin{pmatrix} 0 & 1 & 0 \\ 0 & 0 & 1 \\ 0 & 0 & 0 \end{pmatrix}, \; L_- = L_x - iL_y = \sqrt{2}\hbar\begin{pmatrix} 0 & 0 & 0 \\ 1 & 0 & 0 \\ 0 & 1 & 0 \end{pmatrix} \quad (21.5.3)$$

transform $|1,0\rangle = (0,1,0)^T$ into $\sqrt{2}\hbar|1,1\rangle$ and $\sqrt{2}\hbar|1,-1\rangle$, respectively.

21.6 SPIN

A particle wavefunction in general can depend on additional *internal variables* such as *spin* that are associated with its intrinsic structure. The overall wavefunction is then the *product* of the spatial and spin wavefunctions. As first suggested by the *Stern–Gerlach experiment* in which a beam of electrons passing through a nonuniform magnetic field was found to deflect into two beams corresponding to $S_{\vec{B}} = \pm\hbar/2$, a particle with spin s, the component of the spin angular momentum along an applied magnetic field possesses eigenvalues $s, s-1, \ldots, -s$, in analogy to the m angular momentum quantum number. For an electron with spin ½ represented by a two-dimensional vector of unit magnitude with components $|s, s_z\rangle$ given by $|1/2, 1/2\rangle = (1,0)^T$ and $|1/2, -1/2\rangle = (0,1)^T$, the spin operator $\vec{S} = \hbar\vec{\sigma}/2$ where the *Pauli matrices*

$$\sigma_x = \begin{pmatrix} 0 & 1 \\ 1 & 0 \end{pmatrix}, \sigma_y = \begin{pmatrix} 0 & -i \\ i & 0 \end{pmatrix}, \sigma_z = \begin{pmatrix} 1 & 0 \\ 0 & -1 \end{pmatrix} \quad (21.6.1)$$

satisfy the angular momentum commutation relation

$$[\sigma_i, \sigma_j] = 2i\varepsilon_{ijk}\sigma_k \quad (21.6.2)$$

which can be abbreviated as $\vec{\sigma} \times \vec{\sigma} = 2i\vec{\sigma}$ which together with $\sigma_i\sigma_j + \sigma_j\sigma_i = 2\delta_{\text{ij}}$ yields

$$\left(\vec{\sigma}\cdot\vec{A}\right)\left(\vec{\sigma}\cdot\vec{B}\right) = \left(\vec{A}\cdot\vec{B}\right)\mathbf{I} + i\vec{\sigma}\cdot\left(\vec{A}\times\vec{B}\right) \quad (21.6.3)$$

The expectation value of the spin vector $\vec{s}$ in the state, $|\gamma\rangle = (u, d)^T$, is then given by $\langle\gamma|\vec{S}|\gamma\rangle$, i.e., the x spin component $\langle\gamma|S_x|\gamma\rangle = \hbar(u^*d + d^*u)/2$. The spin raising and lowering operators are then obtained from

$$S_+ = S_x + iS_y = \hbar\begin{pmatrix} 0 & 1 \\ 0 & 0 \end{pmatrix}, \; S_- = S_x - iS_y = \hbar\begin{pmatrix} 0 & 0 \\ 1 & 0 \end{pmatrix} \quad (21.6.4)$$

Additionally, the expectation value of $S^2 = S_x^2 + S_y^2 + S_z^2 = \hbar^2 s(s+1)\mathbf{I}$, where $\mathbf{I}$ denotes the $2s+1 \times 2s+1$ unit matrix, for a state of spin s equals $\hbar^2 s(s+1)$.

21.7 INTERACTION OF SPINS

To illustrate the interaction of spins or angular momenta, consider the spins in the lowest state of a helium atom with zero orbital angular momentum. The spins are clearly parallel in the two states $|s_{\text{total}}, s_{\text{total},z}\rangle = |1,1\rangle = |\uparrow\rangle_1|\uparrow\rangle_2$ and $|1,-1\rangle = |\downarrow\rangle_1|\downarrow\rangle_2$ where $|\uparrow\rangle = (1,0)^T$, $|\downarrow\rangle = (0,1)^T$ for which $(S_{1,z} + S_{2,z})|\uparrow\rangle_1|\uparrow\rangle_2 = (S_{1,z}|\uparrow\rangle_1)|\uparrow\rangle_2 + |\uparrow\rangle_1 (S_{2,z}|\uparrow\rangle_2) = \hbar|\uparrow\rangle_1|\uparrow\rangle_2$ while the magnitude of the total spin angular momentum, $\vec{S}_{\text{total}} = \vec{S}_1 + \vec{S}_2$, equals

$$\begin{aligned}
S_{\text{total}}^2 &= S_1^2 + S_2^2 + 2S_1 \cdot S_2 \\
&= S_1^2 + S_2^2 + 2S_{1,z}S_{2,z} + \underbrace{\left(2S_{1,x}S_{2,x} + 2S_{1,y}S_{2,y}\right)}_{\frac{1}{2}[(S_{1+}+S_{1-})(S_{2+}+S_{2-})-(S_{1+}-S_{1-})(S_{2+}-S_{2-})]} \\
&= S_1^2 + S_2^2 + 2S_{1,z}S_{2,z} + S_{1+}S_{2-} + S_{1-}S_{2+}
\end{aligned} \tag{21.7.1}$$

with an expectation value of $\hbar^2(3/4 + 3/4 + 1/2 + 0 + 0) = \hbar^2 s_{\text{total}}(s_{\text{total}} + 1) = 2\hbar$ when applied to $|\uparrow\rangle_1|\uparrow\rangle_2$ and $|\downarrow\rangle_1|\downarrow\rangle_2$, both of which are symmetric upon interchange of the two electron spins. Accordingly, these form a *spin triplet* of $s_{\text{total}} = 1$ states together with the similarly symmetric $m = 0$ combination of spin-up and spin-down states, $|1,0\rangle = (|\uparrow\rangle_1|\downarrow\rangle_2 + |\downarrow\rangle_1|\uparrow\rangle_2)/\sqrt{2}$, which can be obtained by applying the lowering operator to the $s_{\text{total}} = 1$, $m_{\text{total}} = 1$ state according to

$$\begin{aligned}
S_{\text{total},-}|1,1\rangle &= (S_{1,-}|\uparrow\rangle_1)|\uparrow\rangle_2 + |\uparrow\rangle_1(S_{2,-}|\uparrow\rangle_2) \\
&= \sqrt{2}\hbar|1,0\rangle = \hbar\underbrace{\sqrt{(s_{\text{total}} + s_{\text{total},z})(s_{\text{total}} - s_{\text{total},z} + 1)}}_{\sqrt{(1+1)(1-1+1)}}|1,0\rangle
\end{aligned} \tag{21.7.2}$$

The *spatial* wavefunction of the spin triplet states is then antisymmetric such that the overall fermion wavefunction is antisymmetric. The antisymmetric spin wavefunction $|0,0\rangle = (|\uparrow\rangle_1|\downarrow\rangle_2 - |\downarrow\rangle_1|\uparrow\rangle_2)/\sqrt{2}$ thus corresponds to the $m = 0, S = 0$ *spin singlet state* with

$$\begin{aligned}
S_{\text{total}}^2|0,0\rangle &= \frac{\hbar^2}{4}(3+3)|0,0\rangle + \frac{1}{\sqrt{2}}\left(S_{1,-}S_{2,+}|\uparrow\rangle_1|\downarrow\rangle_2 - S_{1,+}S_{2,-}|\downarrow\rangle_1|\uparrow\rangle_2\right) \\
&= \frac{\hbar^2}{4}(6 + 2(+1)(-1))|0,0\rangle + \frac{\hbar^2}{\sqrt{2}}\left(|\downarrow\rangle_1|\uparrow\rangle_2 - |\uparrow\rangle_1|\downarrow\rangle_2\right) \\
&= \hbar^2|0,0\rangle - \hbar^2|0,0\rangle = 0
\end{aligned} \tag{21.7.3}$$

Thus, if two spins are coupled through the interaction of the spin of each electron to the magnetic field of the other electron, $\Delta U = 2c\vec{S}_1 \cdot \vec{S}_2 = c(S_{\text{total}}^2 - S_1^2 - S_2^2)$ and all three spin triplet states are shifted by the same amount in energy relative to the spin singlet. Just as the spin–spin interaction yields a potential that depends on the total spin, if two particles, $i = 1, 2$, each of whose spin and orbital momentum are strongly coupled to form a total particle angular momentum $\vec{j}_i$, the interaction between the $\vec{j}_i$ generates a term $\vec{j}_1 \cdot \vec{j}_2 = c'(J_{\text{total}}^2 - j_1^2 - j_2^2)$ in the Hamiltonian. The total angular momentum values then range from $J_{\text{total}} = j_1 + j_2$ to $J_{\text{total}} = |j_1 - j_2|$ where $2J_{\text{total}} + 1$ associated wavefunctions with differing $m_{\text{total}} = m_1 + m_2$ (generalizing the triplet and singlet states) exist for each value of J_{total}. A wavefunction $|J_{\text{total}}, m_{\text{total}}\rangle$ is comprised of a linear combination of wavefunctions $|j_1, m_1\rangle |j_2, m_2\rangle$ with $|m_1| \le j_1$, $|m_2| \le j_2$ and $m_{\text{total}} = m_1 + m_2$. The *Clebsch–Gordan coefficients* in these superpositions are generally referenced from tables.

21.8 HYDROGENIC ATOMS

The spatial state of hydrogenic atoms is described by products of angular and radial wavefunctions. The kinetic energy of the atom equals the sum of the kinetic energy of the center of mass of the electron and nucleus and the energy about the center of mass. In terms of center of mass and relative coordinates

$$\vec{r}_r = \vec{r}_2 - \vec{r}_1, \; \vec{r}_{cm} = \frac{m_1 \vec{r}_1 + m_2 \vec{r}_2}{m_1 + m_2} \tag{21.8.1}$$

the x_1 derivative in Schrödinger equation, $(T + U)\psi = \left[-\hbar^2/2\left(\nabla_1^2/m_1 + \nabla_2^2/m_2\right) + U\right]\psi = E\psi$, equals

$$\begin{aligned} \frac{\partial^2}{\partial x_1^2} &= \left(\frac{\partial x_{cm}}{\partial x_1}\frac{\partial}{\partial x_{cm}} + \frac{\partial x_r}{\partial x_1}\frac{\partial}{\partial x_r}\right)^2 \\ &= \left(\frac{m_1}{(m_1 + m_2)}\frac{\partial}{\partial x_{cm}} - \frac{\partial}{\partial x_r}\right)^2 \\ &= \frac{m_1{}^2}{(m_1 + m_2)^2}\frac{\partial^2}{\partial x_{cm^2}} - 2\frac{m_1}{(m_1 + m_2)}\frac{\partial^2}{\partial x_{cm}\partial x_r} + \frac{\partial^2}{\partial x_r^2} \end{aligned} \tag{21.8.2}$$

With the corresponding expression for $\partial^2/\partial x_2^2$ obtained with the transformation $x_r \to -x_r$, $1 \leftrightarrow 2$,

$$\frac{1}{m_1}\frac{\partial^2}{\partial x_1^2} + \frac{1}{m_2}\frac{\partial^2}{\partial x_2^2} = \frac{1}{(m_1 + m_2)}\frac{\partial^2}{\partial x_{cm}^2} + \frac{1}{m_r}\frac{\partial^2}{\partial x_r^2} \tag{21.8.3}$$

in which $m_r = m_1 m_2/(m_1 + m_2)$ denotes the reduced mass. Incorporating the y and z derivatives,

$$\left[\frac{\hbar^2}{2(m_1+m_2)}\nabla_{cm}^2 + \frac{\hbar^2}{2m_r}\nabla_r^2 + U(\vec{r}_r)\right]\psi(\vec{r}_{cm},\vec{r}_r) = E\psi(\vec{r}_{cm},\vec{r}_r) \quad (21.8.4)$$

By separation of variables with $\psi(\vec{r}_{cm},\vec{r}_r) = \exp(i\vec{k}_{cm}\cdot\vec{r}_{cm})\psi_r(\vec{r}_r)$, the radial wavefunction obeys

$$\left(-\frac{\hbar^2}{2m_r}\nabla_r^2 + U(r_r)\right)\psi_r(\vec{r}_r) = \left(E - \frac{\hbar^2 k_{cm}^2}{2(m_1+m_2)}\right)\psi_r(\vec{r}_r) \equiv E_r\psi_r(\vec{r}_r) \quad (21.8.5)$$

The Schrödinger equation for the relative motion in spherical coordinates is then, omitting the subscript r,

$$\left(-\frac{\hbar^2}{2m}\nabla^2 + U(r)\right)\psi(\vec{r}) = \left(-\frac{\hbar^2}{2m}\left(\frac{\partial^2}{\partial r^2} + \frac{2}{r}\frac{\partial}{\partial r} - \frac{L^2}{\hbar^2 r^2}\right) + U(r)\right)\psi(\vec{r}) = E\psi(\vec{r}) \quad (21.8.6)$$

with L^2 here given by $\hbar^2$ multiplied by Equation (13.3.5). Separating variables according to $\psi(\vec{r}) = R_{nl}(r)Y_{lm}(\theta,\varphi)$ with n the *radial quantum number* yields

$$\left(-\frac{\hbar^2}{2m}\left(\frac{\partial^2}{\partial r^2} + \frac{2}{r}\frac{\partial}{\partial r} - \frac{l(l+1)}{r^2}\right) + U(r)\right)R_{nl}(r) = E_{nl}R_{nl}(r) \quad (21.8.7)$$

For the $l = m = 0$ ground state of a hydrogenic atom, the above equation becomes with $m_e \approx m_r$

$$\left(\frac{\partial^2}{\partial r^2} + \frac{2}{r}\frac{\partial}{\partial r} + \frac{2m_e Ze^2}{\hbar^2 4\pi\varepsilon_0 r}\right)R_{00}(r) = \left(\frac{\partial^2}{\partial r^2} + \frac{2}{r}\frac{\partial}{\partial r} + \frac{2Z}{a_B r}\right)R_{00}(r) = -\frac{2m_e E_{00}}{\hbar^2}R_{00}(r) \quad (21.8.8)$$

The terms containing $1/r$ cancel for $R_{00}(r) = 2(Z/a_B)\exp(-rZ/a_B)$, which is normalized according to $\int_V (R_{00}(r))^2 dV = 1$, yielding an energy eigenvalue $E_{00} = -\hbar^2 Z^2/2m_e a_B^2 = -Z^2 R_y$. The general radial eigenfunctions possess the form $R_{nl} = \exp(-rZ/na_B)r^l p_{nl}(r)$ where $p_{nl}(r)$ is a polynomial of order $n-l-1$. Accordingly, only the $l=0$ wavefunctions that do not experience a centrifugal repulsion differ from zero at the origin. The $m_z = 0$ wavefunctions are elongated (for $l \neq 0$) along the z-axis, while for larger values of $|m_z|$, for fixed l, the direction of elongation tilts toward the $x-y$ plane.

21.9 ATOMIC STRUCTURE

From the properties of the angular wavefunctions, for each atomic radial quantum number n, the angular quantum number l varies from 1 to n, while for each l, the azimuthal quantum number m varies from $-l$ to l. For a hydrogen atom, energy levels with the same value of n are degenerate with energies given by the Bohr model as a result of the particular symmetry of the $1/r$ potential. Any perturbation that does not also vary as $1/r$ therefore reduces the energy level degeneracy. Similarly, in classical mechanics, elliptical planetary orbits in a $1/r$ potential are stationary; forces that do not vary as $1/r^2$ induce precession.

At zero temperature, atomic electrons occupy the lowest available energy levels subject to the Fermi exclusion principle that only one spin-up and one spin-down electron can be present in each atomic energy level. In all atoms except hydrogen, electron–electron repulsion removes the $1/r$ symmetry, and small l states typically possess a lower energy than large l states with the same n. In chemical notation, the distribution of electrons in an atom is abbreviated by nl^{2m} where the $l = 0, 1, 2, \ldots$ states are designated instead as $l = s, p, d, \ldots$. The order in which these states are filled is given by

$$1s^2\,2s^2\,2p^6\,3s^2\,3p^6\,4s^2\,3d^{10}\,4p^6\,5s^2\,4d^{10}\,5p^6\,6s^2\left(5d^{10}\right)^2 4f^{12}\left(5d^{10}\right)^8 6p^6 \ldots \tag{21.9.1}$$

21.10 SPIN–ORBIT COUPLING

The period, T, of a charge q moving at a constant speed v in a circular orbit of radius R equals $2\pi R/v$, yielding an effective current (charge per unit time) $I = q/T = qv/2\pi R$ and magnetic moment, where $\vec{A}$ is the area vector,

$$\vec{\mu} = IA\hat{e}_A = \frac{q}{2\pi R}\left(\pi R^2\right)\hat{e}_R \times \vec{v} = \frac{q\vec{L}}{2m} \equiv \gamma\vec{L} \tag{21.10.1}$$

where γ is termed the *gyromagnetic ratio*. The resulting interaction between the particle and a magnetic field is described by $H = -\vec{\mu}\cdot\vec{B}$.

In contrast to a spinless charge, an electron possesses a magnetic moment $\vec{\mu}_e = \gamma_e\vec{S} = (-g_e e/2m_e)\vec{S} = -g_e\mu_B\vec{S}/\hbar$ where the electron *g-factor* $g_e \approx 2$ as a result of relativistic and quantum mechanical effects and $\mu_B = e\hbar/2m$ is labeled the *Bohr magneton.* From the Heisenberg equation of motion, $\vec{S}$ accordingly evolves in time as

$$\frac{d\vec{S}}{dt} = \frac{1}{i\hbar}\left[\vec{S}, H_{\text{spin-field}}\right] = -\frac{1}{i\hbar}\gamma_e\left[\vec{S}, \vec{S}\cdot\vec{B}\right] \tag{21.10.2}$$

Since

$$B_j\left[S_i, S_j\right] = \frac{\hbar^2}{4}B_j\left[\sigma_i, \sigma_j\right] = i\frac{\hbar^2}{2}\varepsilon_{ijk}B_j\sigma_k = i\hbar\left(\vec{B}\times\vec{S}\right)_i \tag{21.10.3}$$

Equation (21.10.2) can be rewritten as (with $q = -e < 0$)

$$\frac{d\vec{S}}{dt} = \gamma_e \vec{S} \times \vec{B} = \vec{\mu}_e \times \vec{B} \tag{21.10.4}$$

The spin of an electron thus *precesses* (circles) around the magnetic field direction with an angular velocity proportional to the field strength $\omega = -\gamma_e B = g\omega_{\text{cyclotron}}/2$ with the *cyclotron frequency* $\omega_{\text{cyclotron}} = eB/m_e$.

In *paramagnetic resonance*, atoms are placed in a constant z-directed magnetic field while a weak sinusoidal magnetic field is applied in a perpendicular direction. The sinusoidal component can be decomposed into the sum of right- and left-handed circularly polarized magnetic fields. If the angular velocities of one of these matches that of the spin vector precession, e.g., for a hydrogen atom $-\gamma_e B_z$, it appears as a small static magnetic field in the frame of the rotating spin. Therefore, the spin is rotated in the plane defined by its instantaneous direction and the z-axis, yielding a transition between the low energy state in which the spin is aligned along the applied DC magnetic field and an antialigned state. The resulting transition radiation can be directly measured.

The magnetic interaction energy of an electron in an atom moving with a velocity $\vec{v}$ in the electric field of the other atomic particles is—where the additional factor of ½ arises from a relativistic effect termed *Thomas precession*—

$$\Delta E = -\frac{1}{2}\vec{\mu}_e \cdot \vec{B} = -\frac{1}{2}\left(\frac{e}{2m_e} g\vec{S}\right) \cdot \frac{\vec{v} \times \vec{E}}{c^2} \tag{21.10.5}$$

For a hydrogen-like atom with $g = 2$,

$$\Delta E = -\frac{1}{2}\left(\frac{eg}{2m_e}\vec{S}\right) \cdot \frac{(m_e\vec{v}) \times \vec{r}}{m_e c^2}\left(\frac{Ze}{4\pi\varepsilon_0 r^3}\right) = \frac{Z}{8\pi\varepsilon_0 r^3}\frac{e^2}{m_e^2 c^2}\vec{S} \cdot \vec{L} \equiv c_{\text{spin-orbit}}\frac{1}{r^3}\vec{S} \cdot \vec{L} \tag{21.10.6}$$

This *spin–orbit coupling* term can be evaluated by noting that since $\vec{J} = \vec{S} + \vec{L}$,

$$\vec{S} \cdot \vec{L} = \frac{1}{2}(J^2 - L^2 - S^2) \tag{21.10.7}$$

requiring the specification of an atom state by its simultaneous j, l, and s quantum numbers. With the notation ${}^{2S+1}L_J$ for a hydrogen atom, the $1s$ ground state is denoted ${}^2S_{1/2}$, while the first group of degenerate excited states consists of the ${}^2S_{1/2}$ $n = 2$, $l = 0$ and $n = 2$, $l = 1{}^2P_{1/2}$, and ${}^2P_{3/2}$ states. The spin–orbit interaction raises and lowers the energies of the ${}^2P_{3/2}$ and ${}^2P_{1/2}$ states, respectively, but does not affect the ${}^2S_{1/2}$ state energy for which $\vec{S} \cdot \vec{L} = 0$. However, the relativistic correction to the electron kinetic energy,

$$\sqrt{p^2c^2+m^2c^4}-mc^2-\frac{p^2}{2m}=mc^2\left(1+\frac{1}{2}\frac{p^2}{m^2c^2}+\frac{1}{2!}\left(\frac{1}{2}\right)\left(-\frac{1}{2}\right)\left(\frac{p^4}{m^4c^4}\right)+\cdots\right)$$
$$-mc^2-\frac{p^2}{2m}$$
$$\approx-\frac{p^4}{8m^3c^2} \tag{21.10.8}$$

for hydrogen-like atoms, which can be evaluated in first-order perturbation theory, restores the degeneracy of the $^2S_{1/2}$ and $^2P_{1/2}$ states, while the $^2P_{3/2}$ level possesses a slightly higher energy. A further quantum electrodynamics correction, the *Lamb shift*, then removes the degeneracy of the $^2S_{1/2}$ and $^2P_{1/2}$ by a minute amount.

Example

For $L=0$ hydrogenic states, $J=S$, while

$$\frac{1}{2}(J^2-L^2-S^2)\begin{Bmatrix}\left|j=l+\frac{1}{2},l,\frac{1}{2}\right\rangle\\ \left|j=l-\frac{1}{2},l,\frac{1}{2}\right\rangle\end{Bmatrix}=\frac{\hbar^2}{2}\begin{Bmatrix}\left[\left(l+\frac{1}{2}\right)\left(l+\frac{3}{2}\right)-l(l+1)-\frac{1}{2}\cdot\frac{3}{2}\right]\left|l+\frac{1}{2},l,\frac{1}{2}\right\rangle\\ \left[\left(l-\frac{1}{2}\right)\left(l+\frac{1}{2}\right)-l(l+1)-\frac{1}{2}\cdot\frac{3}{2}\right]\left|l-\frac{1}{2},l,\frac{1}{2}\right\rangle\end{Bmatrix}$$
$$=\frac{\hbar^2}{2}\begin{Bmatrix}l\left|l+\frac{1}{2},l,\frac{1}{2}\right\rangle\\ -(l+1)\left|l-\frac{1}{2},l,\frac{1}{2}\right\rangle\end{Bmatrix} \tag{21.10.9}$$

Thus, in first-order perturbation theory, the spin–orbit-induced energy change is $\delta E_j=\langle\psi_{n,j,l,1/2}|H_{\text{so}}|\psi_{n,j,l,1/2}\rangle=c_{\text{spin-orbit}}\hbar^2L(l)\langle R_{nl}|(1/r^3)|R_{nl}\rangle/2$ where $L(l)=l$ for $J=l+1/2$ and $-(l+1)$ for $J=l-1/2$.

Finally, *hyperfine structure* arises from the magnetic field of the nuclear magnetic dipole moment, resulting in a perturbation $e\left(\vec{L}+2\vec{S}\right)\cdot\vec{B}/2m$, where $\vec{B}$ is proportional to the nuclear spin, $\vec{S}_{\text{nuclear}}$. For $L=0$ states, the interaction is proportional to $\vec{S}\cdot\vec{S}_{\text{nuclear}}$, which as in the spin–orbit case can be written as

$$S_{\text{total}}(S_{\text{total}}+1)-S(S+1)-S_{\text{nuclear}}(S_{\text{nuclear}}+1)\text{ with }\vec{S}_{\text{total}}=\vec{S}+\vec{S}_{\text{nuclear}}$$

21.11 ATOMS IN STATIC ELECTRIC AND MAGNETIC FIELDS

The change in an atomic Hamiltonian induced by an external z-directed electric field is given by

$$\Delta H'=\Delta U=-\vec{F}\cdot\vec{r}=e\vec{E}\cdot\vec{r}=e\mathrm{E}_z z \tag{21.11.1}$$

An unperturbed nondegenerate wavefunction is $|\psi_m\rangle$ either symmetric or antisymmetric with respect to reflection in the $x-y$ plane. However, in both cases, the product of $\Delta H'$ with the symmetric squared wavefunction in the formula for the first-order energy shift $\delta E_m^{(1)} = \langle \psi_m^{(0)} | H' | \psi_m^{(0)} \rangle = 0$ with Equation (21.11.1) after integration over all space. Since $\delta E = -\vec{p} \cdot \vec{E}$, with $\vec{p}$ as the dipole moment, this further establishes that an atom in a nondegenerate state cannot possess a permanent dipole moment. However, second-order perturbation theory incorporates the asymmetric distortion in the wavefunction induced by the electric force. The magnitude of the contributions to the energy shift from above and below the $z=0$ plane then differs, resulting in a ground state *Stark shift*:

$$\delta E^{(2)} = e^2 E_z^2 \sum_{m>0} \frac{\left| \left\langle \psi_0^{(0)} \middle| z \middle| \psi_m^{(0)} \right\rangle \right|^2}{E_0^{(0)} - E_m^{(0)}} \tag{21.11.2}$$

If $\psi_m^{(0)}$ is degenerate with a second state, $\psi'^{(0)}_m$, with opposite parity under reflection, the first-order energy shift of the state does not vanish (as the sum of a symmetric and an antisymmetric function possesses neither symmetry), yielding a permanent dipole moment. That is, $\delta E_m^{(1)} = \left\langle c_1 \psi_m^{(0)} + c_2 \psi'^{(0)}_m \middle| H' \middle| c_1 \psi_m^{(0)} + c_2 \psi'^{(0)}_m \right\rangle = 2\mathrm{Re}\left[c_1^{*} c_2 \left\langle \psi_m^{(0)} \middle| z \middle| \psi'^{(0)}_m \right\rangle \right] \neq 0$ since the product of $\psi_m^{(0)}$ with z possesses the same parity under reflection as $\psi'^{(0)}_m$. The coefficients c_i are obtained from degenerate perturbation theory.

Example

For the $n=2$ levels of the hydrogen atom, $\psi_{200} = R_{20}(r)|0, 0\rangle$ with angular dependence $Y_{00} = 1/\sqrt{4\pi}$, $\psi_{211} = R_{21}(r)|1, 1\rangle$, ψ_{21-1} with angular variation $Y_{1\pm1}(\theta, \phi) \propto \sin\theta \exp(\pm i\phi) = (x \pm iy)/r$, and $\psi_{210} \sim Y_{10}(\theta, \phi) \propto \cos\theta = z/r$ are degenerate. An electric field reduces the spherical symmetry of the $1/r$ potential to a two-dimensional cylindrical rotational symmetry, resulting in a twofold degeneracy. Indeed, superimposing to leading order in $|\vec{E}|$ the angularly invariant $l=m=0$ state with the $\cos\theta$ variance of the $l=1$, $m=0$ state with a 0 or 180° relative phase shift generates orthogonal wavefunctions with a maximum toward or against the direction of the electric field. These are, respectively, lowered and raised in energy relative to the two $|m|=1$ states, which instead vary as $\sin\theta$ and therefore do not change the average electron position in the field direction to first order. Mathematically, for $m' \neq m$, the matrix elements of $e\mathrm{E}_z z$ equal zero since from $[e\mathrm{E}_z z, L_z] = 0$ (as L_z only contains x and y derivatives),

$$\begin{aligned} 0 &= \left\langle \psi_{n,l,m} \middle| [z, L_z] \middle| \psi_{n',l',m'} \right\rangle \\ &= \left\langle \psi_{n,l,m} \middle| (zL_z - L_z z) \middle| \psi_{n',l',m'} \right\rangle \\ &= (m' - m) \left\langle \psi_{n,l,m} \middle| z \middle| \psi_{n',l',m'} \right\rangle \end{aligned} \tag{21.11.3}$$

For the two $m=0$ states, direct integration yields $\langle\psi_{200}|z|\psi_{210}\rangle = \langle\psi_{210}|z|\psi_{200}\rangle = -3a_B$, resulting in the eigenvalue equation in degenerate perturbation theory:

$$-3e\mathrm{E}_z a_B \begin{pmatrix} 0 & 1 \\ 1 & 0 \end{pmatrix} \begin{pmatrix} c_1 \\ c_2 \end{pmatrix} = \delta E \begin{pmatrix} c_1 \\ c_2 \end{pmatrix} \tag{21.11.4}$$

The antisymmetric and symmetric eigenvectors are accordingly $c_1 = -c_2 = 1/\sqrt{2}$ with eigenvalue $3e\mathrm{E}_z a_B$ and $c_1 = c_2 = 1/\sqrt{2}$ with eigenvalue $-3e\mathrm{E}_z a_B$.

If the coupling energy of two spins or angular momentum components, $\vec{A}_1$ and $\vec{A}_2$, sufficiently exceeds that of the individual spins with an external $\vec{B}$ field, the field interaction must be evaluated between coupled angular momenta eigenstates while in the opposite case, the levels are eigenstates of the individual components. In a constant magnetic field $B\hat{e}_z$, atomic energy levels undergo *Zeeman splitting* into closely spaced groups of levels that differ by the projections of the spin and angular momentum onto the field axis. The magnetic field interaction energy is generally far smaller than the spin–orbit interaction energy, so that the energy level shift of the state $|j, m_j\rangle$ is determined to first order by the matrix element (the factor of 2 before $\vec{S}$ results from $g_e \approx 2$ for electrons):

$$\begin{aligned} \frac{e}{2m_e}\langle j, m_j|\left(\vec{L}+2\vec{S}\right)\cdot\vec{B}|j, m_j\rangle &= \frac{eB}{2m_e}\langle j, m_j|(J_z+S_z)|j, m_j\rangle \\ &= \frac{eB}{2m_e}\left(\hbar m_j + \langle j, m_j|S_z|j, m_j\rangle\right) \end{aligned} \tag{21.11.5}$$

The matrix element of S_z in the above expression can be evaluated by inserting the expressions for $|j, m_j\rangle$ in terms of the eigenstates $|l, m_l, s, m_s\rangle$. A simplified semiclassical procedure models $\vec{L}$ and $\vec{S}$ as vectors that precess rapidly around $\vec{J}$. After time averaging, only the component of $\vec{S}$ in the direction of $\vec{J}$ remains, and $\langle j, m_j|S_z|j, m_j\rangle$ can be replaced by

$$\begin{aligned} \langle j, m_j|S_z|j, m_j\rangle &= \langle j, m_j|\hat{e}_z\cdot\left(\vec{J}\frac{S\cdot J}{J^2}\right)|j, m_j\rangle \\ &= \hbar m_j\langle j, m_j|\frac{J^2+S^2-L^2}{2J^2}|j, m_j\rangle \\ &= \hbar m_j\frac{j(j+1)+\frac{3}{4}-l(l+1)}{2j(j+1)} \end{aligned} \tag{21.11.6}$$

Each unperturbed energy level is therefore split into a set of levels that differ according to m_j:

$$\Delta E^{(1)}_{\text{magnetic},j} = \frac{eB\hbar}{2m_e} m_j \left(1 + \frac{j(j+1) + \frac{3}{4} - l(l+1)}{2j(j+1)}\right) \tag{21.11.7}$$

The Hamiltonian for an electron in an electromagnetic field is obtained with $\vec{p} \to -i\hbar\vec{\nabla}$ in Equation (18.8.32) when the interaction of the electron spin with the field is neglected:

$$i\hbar \frac{\partial \psi(\vec{r},t)}{\partial t} = \left(\frac{1}{2m_e}\left(-i\hbar\vec{\nabla} + e\vec{A}(\vec{r},t)\right)^2 - eV(\vec{r},t)\right)\psi(\vec{r},t) \tag{21.11.8}$$

Equation (21.11.8) is preserved under the gauge transformation, $V(\vec{r},t) \to V(\vec{r},t) + \partial\Lambda(\vec{r},t)/\partial t$, $\vec{A}(\vec{r},t) \to \vec{A}(\vec{r},t) - \vec{\nabla}\Lambda(\vec{r},t)$ if

$$\psi(\vec{r},t) \to \psi(\vec{r},t) e^{i\frac{e}{\hbar}\Lambda(r,t)} \tag{21.11.9}$$

as evident from

$$\begin{aligned} &\left(-i\hbar\vec{\nabla} + e\vec{A}(\vec{r},t) - e\vec{\nabla}\Lambda(r,t)\right)\left[\psi(\vec{r},t)e^{i\frac{e}{\hbar}\Lambda(r,t)}\right] = e^{i\frac{e}{\hbar}\Lambda(r,t)}\left(-i\hbar\vec{\nabla} + e\vec{A}(\vec{r},t)\right)\psi(\vec{r},t) \\ &\left(i\hbar\frac{\partial}{\partial t} + eV(\vec{r},t) + e\frac{\partial\Lambda(r,t)}{\partial t}\right)\left[\psi(\vec{r},t)e^{i\frac{e}{\hbar}\Lambda(r,t)}\right] = e^{i\frac{e}{\hbar}\Lambda(r,t)}\left(i\hbar\frac{\partial}{\partial t} + eV(\vec{r},t)\right)\psi(\vec{r},t) \end{aligned} \tag{21.11.10}$$

In a constant magnetic field $B = B\hat{e}_z$ and a gauge for which $\vec{A} = (0, Bx, 0)$ (recall $\vec{B} = \vec{\nabla} \times \vec{A}$ for static fields)

$$H\psi(\vec{r},t) = \left(\frac{1}{2m_e}\left(p_x^2 + p_y^2 + p_z^2 + 2eBxp_y + e^2B^2x^2\right)\right)\psi(\vec{r},t) = E\psi(\vec{r},t) \tag{21.11.11}$$

Since $[H, p_y] = [H, p_z] = 0$, simultaneous eigenfunctions of p_y, p_z and H exist. If p_y and p_z are constants of the motion, $-i\hbar\partial\psi/\partial y = p_y\phi$ and $-i\hbar\partial\psi/\partial z = p_z\phi$ with $\vec{p} = \hbar\vec{k}$. Consequently, setting

$$\psi(\vec{r}) = e^{i(k_y y + k_z z)}\phi(x) \tag{21.11.12}$$

yields

$$\frac{1}{2m_e}\left(-\hbar^2\frac{d^2}{dx^2}+e^2B^2\left(x+\frac{\hbar k_y}{eB}\right)^2\right)\phi(x)=\left(E-\frac{\hbar^2k_z^2}{2m_e}\right)\phi(x)\equiv E'\phi(x) \quad (21.11.13)$$

If the particle motion is confined to the $x-y$ plane, $k_z=0$ and the above equation describes a harmonic oscillator with equilibrium position $x_B=-\hbar k_y/eB$, implying $k_y=-eBx_B/\hbar$ and potential, $m\omega^2x'^2/2$, given by $e^2B^2x'^2/2m_e$, with $x'=x-x_B$. Hence, the eigenenergies are with integer N

$$E'_N=\frac{\hbar eB}{m_e}\left(N+\frac{1}{2}\right)=\hbar\omega_{\text{cyclotron}}\left(N+\frac{1}{2}\right) \quad (21.11.14)$$

with $\omega_{\text{cyclotron}}$ as the cyclotron frequency (Equation (18.9.10)). Semiclassically, equating magnetic and centrifugal forces, $evB=m_ev^2/r$, yields $p=m_ev=erB$. The WKB condition, $\oint krd\theta=\oint(p/\hbar)rd\theta=2(N+1/2)\pi$, then implies $r^2=(N+1/2)\hbar/eB$. As the energy in a circular orbit equals twice the kinetic energy, $E_N=p^2/m_e=r^2(eB)^2/m_e=(N+1/2)\hbar eB/m_e$.

The nonrelativistic *Pauli equation* additionally describes the interaction of electron spin with the magnetic field. In an explicitly gauge-invariant form, the equation is written as

$$-i\hbar\frac{\partial\varphi}{\partial t}=\left[\frac{1}{2m}\left(\vec{\sigma}\cdot\left(\vec{p}-\frac{e}{c}\vec{A}\right)\right)^2+eU\right]\varphi \quad (21.11.15)$$

where φ represents a two-component spinor. Applying the formula $(\vec{\sigma}\cdot\vec{A})(\vec{\sigma}\cdot\vec{B})=\vec{A}\cdot\vec{B}+i\vec{\sigma}\cdot(\vec{A}\times\vec{B})$ together with, where the subscript indicates the expression upon which the operator acts, $\vec{p}_A\times\vec{A}=-i\hbar\vec{B}$ and $(\vec{p}\times\vec{A}+\vec{A}\times\vec{p})\varphi=(\vec{p}_A\times\vec{A}-\vec{A}\times\vec{p}_\varphi+\vec{A}\times\vec{p}_\varphi)\varphi=-i\hbar\vec{B}\varphi$ yields the terms of Equation (21.11.8) together with the additional coupling term, Equation (21.10.5), with $g=2$ and without the relativistic factor of ½.

21.12 HELIUM ATOM AND THE H_2^+ MOLECULE

The Fermi exclusion principle generates an effective spin–spin interaction among electrons in an atom or molecule. For example, the Hamiltonian of a two-electron helium atom for which $Z=2$ is given by

$$H=H_0(r_1)+H_0(r_2)+H_I(\vec{r}_1-\vec{r}_2)=\frac{p_1^2+p_2^2}{2m_e}-\frac{Ze^2}{4\pi\varepsilon_0}\left(\frac{1}{r_1}+\frac{1}{r_2}\right)+\frac{e^2}{4\pi\varepsilon_0}\frac{1}{|\vec{r}_1-\vec{r}_2|} \quad (21.12.1)$$

The last term describes the Coulomb interaction potential $H_I(\vec{r}_1 - \vec{r}_2)$ between electrons. For $H_I = 0$, the total wavefunction factors into a product of the two individual electron wavefunctions. In the ground state, both electrons occupy the lowest-order spatial state, yielding a symmetric spatial and therefore antisymmetric spin wavefunction $\psi_0(\vec{r}_1, \vec{r}_2) = \psi_{100}(\vec{r}_1)\psi_{100}(\vec{r}_2)|1,0\rangle_-$, with $|1,0\rangle_-$ the antisymmetric spin singlet state. In the absence of H_I, the noninteracting electrons together possess a binding energy $E_{\text{binding}} = -2(mc^2\alpha^2 Z^2/2) = -108.8$ eV, while the lowest-order correction associated with H_I

$$\langle \psi_0(\vec{r}_1, \vec{r}_2)|H_I(\vec{r}_1 - \vec{r}_2)|\psi_0(\vec{r}_1, \vec{r}_2)\rangle = \int \rho_0(\vec{r}_1)\left(\int H_I(\vec{r}_1 - \vec{r}_2)\rho_0(\vec{r}_2)d^3r_1\right)d^3r_2 \tag{21.12.2}$$

where the electron density, $\rho_0(\vec{r}) = |\psi_{100}(\vec{r})|^2$, neglects the distortion of each electron wavefunction resulting from the presence of the other electron

In the lowest He excited state, the electrons occupy different states, yielding for the total wavefunction, where the upper and lower entries are associated with spin singlet and triplet states, respectively,

$$\psi_{\left\{\substack{\text{singlet} \\ \text{triplet}}\right\}}(\vec{r}_1, \vec{r}_2) = \frac{1}{\sqrt{2}}\Big(\psi_{100}(\vec{r}_1)\psi_{210}(\vec{r}_2)$$

$$\pm \psi_{100}(\vec{r}_2)\psi_{210}(\vec{r}_1)\Big)\left\{\begin{matrix} |0,0\rangle_{\text{spin singlet}} \\ |1,1 \text{ or } 0 \text{ or} -1\rangle_{\text{spin triplet}} \end{matrix}\right\} \tag{21.12.3}$$

Evaluating the resulting interaction energy by employing the symmetry upon interchange of $\vec{r}_1$ and $\vec{r}_2$,

$$\Delta E_{\left\{\substack{\text{singlet} \\ \text{triplet}}\right\}} = \frac{e^2}{4\pi\varepsilon_0}\iint \frac{|\psi_{100}(\vec{r}_1)|^2|\psi_{210}(\vec{r}_2)|^2 \pm \psi_{100}^*(\vec{r}_1)\psi_{210}^*(\vec{r}_2)\psi_{210}(\vec{r}_1)\psi_{100}(\vec{r}_2)}{|\vec{r}_1 - \vec{r}_2|}d^3r_1 d^3r_2 \tag{21.12.4}$$

Designating the *direct* and *exchange interaction* terms above by G and H, since $S_{\text{total}}^2 = s_1^2 + s_2^2 + 2\vec{s}_1 \cdot \vec{s}_2$ implies $2\vec{s}_1 \cdot \vec{s}_2/\hbar^2 = \vec{\sigma}_1 \cdot \vec{\sigma}_2/2 = S_{\text{total}}(S_{\text{total}} + 1) - 3/2$, which is $1/2$ for triplet and $-3/2$ for singlet states, Equation (21.12.4) can be written as

$$\Delta E = G - \frac{1}{2}(1 + \sigma_1 \cdot \sigma_2)H \tag{21.12.5}$$

Hund's rule postulates that states with largest total spin are the most spatially symmetric and therefore typically possess the lowest energies. This spin correlation energy in certain substances considerably exceeds the spin–orbit interaction leading to a large magnetic response.

In the H_2^+ ion with only a single bound electron, the electron wavefunction in the electric field of protons at $\pm \vec{R}/2$ satisfies, where the $1/R$ term arises from the Coulomb repulsion of the protons,

$$\left[-\frac{\hbar^2}{2m_e}\nabla_r^2 - \frac{e^2}{4\pi\varepsilon_0}\left(\frac{1}{\left|\vec{r}-\vec{R}/2\right|^2}+\frac{1}{\left|\vec{r}+\vec{R}/2\right|^2}-\frac{1}{R}\right)\right]\psi\left(\vec{r},\vec{R}\right)=E\psi\left(\vec{r},\vec{R}\right) \tag{21.12.6}$$

For large R, the electron is bound to a single atom with a binding energy of E_B, while for $R \to 0$, the $1/R$ term diverges. The resulting Hamiltonian is symmetric with respect to reflection in the perpendicular plane bisecting to the line joining the nuclei so that the spatial wavefunction symmetry with respect to this transformation depends on the spin state of the two neutrons. The Ritz variational method estimates the electronic ground state energy with a trial wavefunction composed from a symmetric or antisymmetric superposition of hydrogenic wavefunctions:

$$\psi_{a,s} = c\left(e^{-\frac{1}{a_B}\left|r-\frac{R}{2}\right|} \pm e^{-\frac{1}{a_B}\left|r+\frac{R}{2}\right|}\right) \tag{21.12.7}$$

where a_B denotes the Bohr radius. The nuclear separation with the highest binding energy is found by minimizing $\langle\psi_{a,s}|H|\psi_{a,s}\rangle/\langle\psi_{a,s}|\,\psi_{a,s}\rangle$ with respect to the dimensionless parameter R/a_B. The result indicates that *only the binding energy of the symmetric spatial state possesses a minimum as a function of R leading to molecular bonding and a stable ion configuration.*

A molecular wavefunction is a product of nuclear and electronic spin and spatial wavefunctions that can typically be separated into *vibrational, rotational*, and *electronic* states such that summing the energies of the individual excitations approximates the total molecular energy. That is, if the potential between the two nuclei is approximated by a parabolic function near its minimum, the nuclear motion forms a series of equally spaced vibrational energy levels typically separated by a few electron volts. The rotational motion of the molecule is associated with a term in the Hamiltonian of the form $H = L^2/2I$, yielding a spectrum of rotational energy levels with $\Delta E_l = \hbar^2 l(l+1)/2I$.

Example

At low temperatures, the H_2 molecule occupies its symmetric electronic and vibrational ground states. As the nuclear spins are half-integral, the total wavefunction is antisymmetric with respect to interchange of the nuclei. If the nuclei occupy the nondegenerate $s_{\text{total, nuclear}} = 0$ odd spin state, their orbital momentum must accordingly be even, while the three degenerate $s_{\text{total, nuclear}} = 1$ spin states require odd orbital momentum. Thus, if H_2 molecules are equally distributed among the spin states through randomizing collisions at low temperature, odd orbital angular momentum states will be three times as prevalent as even states.

21.13 INTERACTION OF ATOMS WITH RADIATION

The electromagnetic interaction of an electron with an electric field $\mathrm{E}_z\hat{e}_z\cos(kx-\omega t)$ possessing a time-averaged energy density

$w=\langle\vec{E}\cdot\vec{D}+\vec{B}\cdot\vec{H}\rangle/2=\varepsilon_0|\mathrm{E}_z|^2/2$ contributes a term

$$H' = e\mathrm{E}_z z\cos\omega t \tag{21.13.1}$$

to the Hamiltonian. If the electron is initially in an eigenstate $|n\rangle$ with energy E_n and a second eigenstate $|m\rangle$ exists with energy $E_m=E_n+\hbar\omega_{mn}$, transitions between the two states are generated through the component $e\mathrm{E}_z z e^{-i\omega t}/2$ of H' if $\omega=\omega_{mn}$. Time-dependent perturbation theory approximates the resulting transition rate by

$$\frac{2\pi}{\hbar}\frac{e^2\mathrm{E}_z^2}{4}|\langle m|z|n\rangle|^2\mathrm{P}(E_n+\hbar\omega_{mn}) \tag{21.13.2}$$

in which the density of states function per unit energy and volume $\mathrm{P}(E_n+\hbar\omega_{mn})\to\delta(E_m-E_n-\hbar\omega_{mn})$ for discrete states.

If the electromagnetic energy per unit volume in a frequency interval $\Delta\omega$ about ω_{mn} equals $w(\omega_{mn})\Delta\omega$, then in the corresponding energy interval $(w(\omega_{mn})\Delta\omega/\Delta E)\Delta E=(w(\omega_{mn})/\hbar)\Delta E$. Thus, replacing $\varepsilon_0\mathrm{E}_z^2(\omega_{mn})\mathrm{P}(E_n+\hbar\omega_{mn})/2$ in Equation (21.13.2) by $w(\omega_{mn})/\hbar$, the absorption rate in an unpolarized field where the average field energy in each polarization direction is one-third the total energy (a dot product is implicit in the squared matrix element below),

$$\frac{\pi e^2}{3\varepsilon_0\hbar^2}|\langle m|\vec{r}|n\rangle|^2 w(\omega_{mn}) \tag{21.13.3}$$

Alternatively, in the Coulomb gauge, a photon in a plane wave state can be represented as

$$\vec{A}(\vec{r},t)=\Xi\sum_{i=1,2}\left(a_i\hat{e}_i e^{i(\vec{k}\cdot\vec{r}-\omega t)}+a_i^*\hat{e}_i e^{-i(\vec{k}\cdot\vec{r}-\omega t)}\right)\equiv\Xi\left(\vec{a}e^{i(\vec{k}\cdot\vec{r}-\omega t)}+\vec{a}^*e^{-i(\vec{k}\cdot\vec{r}-\omega t)}\right) \tag{21.13.4}$$

Here, Ξ and $\hat{e}_i$ are a normalization factor and orthogonal unit vectors perpendicular to the wavevector $\vec{k}$. With $(A\times B)\cdot(C\times D)=(A\cdot C)(B\cdot D)-(A\cdot D)(B\cdot C)$ and $k^2/\varepsilon_0\mu_0=c_o^2k^2=\omega^2$, the energy density

$$\begin{aligned}u&=\frac{\varepsilon_0}{2}\left(-\frac{\partial\vec{A}}{\partial t}\right)^2+\frac{1}{2\mu_0}\left(\vec{\nabla}\times\vec{A}\right)^2\\&=\Xi^2\left(\frac{\varepsilon_0\omega^2}{2}\vec{a}\cdot\vec{a}^*+\frac{1}{2\mu_0}\left(\vec{k}\times\vec{a}\right)\cdot\left(\vec{k}\times\vec{a}^*\right)+c.c.\right)\\&=\Xi^2\left(\varepsilon_0\omega^2\,\vec{a}\cdot\vec{a}^*+c.c.\right)\end{aligned} \tag{21.13.5}$$

from which the total energy in a volume V for the field of Equation (21.13.4) equals $2\varepsilon_0\omega^2|a|^2\Xi^2V$. (Note that since for a plane wave $\vec{B}\times\vec{H}=\vec{E}\times\vec{D}$, the two terms above coincide.) With $|a|^2=1$ and Ξ specified such that u yields the energy of N photons with energy $\hbar\omega$ within V,

$$\vec{A}\left(\vec{r},t\right)=\sqrt{\frac{\hbar N}{2\varepsilon_0\omega V}}\sum_{i=1,2}\left(a_i\hat{e}_i e^{i\left(\vec{k}\cdot\vec{r}-\omega t\right)}+a_i^*\hat{e}_i e^{-i\left(\vec{k}\cdot\vec{r}-\omega t\right)}\right) \tag{21.13.6}$$

In a full description, since the first and second term above increase and decrease the energy of the system, a_i^* and a_i are interpreted as operators a_i^+ and a that increase and decrease the number of photons in the polarization state $\hat{e}_i$ by unity. The wave function for a single-photon state with polarization $\hat{e}$

$$\vec{A}_{\vec{k},\hat{e}}=\sqrt{\frac{\hbar}{2\omega\varepsilon_0 V}}\hat{e}e^{-i\left(\vec{k}\cdot\vec{r}-\omega t\right)} \tag{21.13.7}$$

is then obtained by applying the modified version of Equation (21.13.6) to a state with zero photons. The photon emission rate in a transition from a state $|m\rangle$ to $|n\rangle$ resulting from the interaction term $e\vec{A}\cdot\vec{p}/m$ in the Hamiltonian is accordingly

$$\frac{2\pi}{\hbar}\frac{e^2}{m_e^2}\frac{\hbar}{2\omega\varepsilon_0 V}\left|\langle n|e^{-i\vec{k}\cdot\vec{r}}\hat{e}\cdot\vec{p}\,|m\rangle\right|^2 \tag{21.13.8}$$

Electronic transitions typically occur between states separated by a few electron volts. The resulting optical wavelength $\lambda=h/p=hc/E$ where, if λ and E are expressed in microns and electron volts,

$$\lambda=\frac{6.63\times10^{-34}[\mathrm{J\cdot s}]}{E[\mathrm{eV}]}3\times10^8\left[\frac{\mathrm{m}}{\mathrm{s}}\right]\frac{1}{1.6\times10^{-19}}\left[\frac{\mathrm{J}}{\mathrm{eV}}\right]10^6\left[\frac{\mu\mathrm{m}}{\mathrm{m}}\right]=\left(\frac{1.24}{E[\mathrm{eV}]}\right)[\mu\mathrm{m}] \tag{21.13.9}$$

Hence, a transition between two states differing by 1 eV results in a photon of 1.24 μm wavelength, which greatly exceeds the several angstrom spatial extent of the electronic wavefunction. Hence, $\vec{k}\cdot\vec{r}\approx a_B/\lambda\ll1$ in Equation (21.13.8). The *electric dipole approximation* accordingly replaces $\exp\left(-i\vec{k}\cdot\vec{r}\right)\approx1-i\vec{k}\cdot\vec{r}+\cdots$ and retains only the leading nonzero term, which, since $\langle n|m\rangle=0$, is

$$\langle n|e^{-i\vec{k}\cdot\vec{r}}\hat{e}\cdot\vec{p}\,|m\rangle\approx\langle n|\hat{e}\cdot\vec{p}\,|m\rangle=\frac{im_e}{\hbar}\hat{e}\cdot\langle n|\left[H,\vec{r}\right]|m\rangle=im_e\frac{E_n-E_m}{\hbar}\hat{e}\cdot\langle n|\,\vec{r}\,|m\rangle \tag{21.13.10}$$

Consequently, Equation (21.13.8) becomes, with $\omega^2=(E_n-E_m)/\hbar$,

$$\frac{2\pi}{\hbar}\frac{\hbar}{2\omega\varepsilon_0 V}\frac{e^2}{m_e^2}\omega^2\left|\langle n|\hat{e}\cdot\vec{r}|m\rangle\right|^2=\frac{2\pi e^2\hbar\omega}{\hbar\,2\varepsilon_0 V}\left|\langle n|\hat{e}\cdot\vec{r}|m\rangle\right|^2 \tag{21.13.11}$$

Since $\hbar\omega/V$ is the energy density for a single photon in a volume V, the above expression coincides with Equation (21.13.3) after converting between energy and frequency densities.

21.14 SELECTION RULES

In transitions between states of isolated hydrogen atoms, the dipole matrix element of the previous section factors into the product of a radial and an angular matrix element:

$$\langle \psi' | \hat{e} \cdot \vec{r} | \psi \rangle = \int_0^{\infty} r^3 R_{nl}^*(r) R_{n'l'}(r) dr \int Y_{l'm'}^*(\theta, \varphi) \hat{e} \cdot \hat{r} Y_{lm}(\theta, \varphi) d\Omega \tag{21.14.1}$$

Since the components x, y, and z all act as $P_1(\cos\theta) = \cos\theta$ with respect to the corresponding axis, the dipole term is a combination of first-order $l = 1$ spherical harmonics. Indeed, noting that $Y_{lm}(\theta, \varphi) \propto \exp(im\varphi)$,

$$\begin{aligned} \hat{e} \cdot \hat{r} &= \hat{e}_x \sin\theta \cos\varphi + \hat{e}_y \sin\theta \sin\varphi + \hat{e}_z \cos\theta \\ &= \frac{\sin\theta}{2} \left[\left(\hat{e}_x - i\hat{e}_y \right) e^{i\varphi} + \left(\hat{e}_x + i\hat{e}_y \right) e^{-i\varphi} \right] + \hat{e}_z \cos\theta \end{aligned} \tag{21.14.2}$$

And therefore, $\hat{e} \cdot \hat{r}$ is a combination of Y_{11} Y_{10} and Y_{1-1} terms and can through the addition of angular momentum only induce transitions from l to $l' = l, l \pm 1$ states. The transverse polarization of the photon further implies that, if $\hat{e}_z$ is associated with the photon propagation direction, $\hat{e}$ is confined to the $x-y$ plane. Hence, only transitions from a state with azimuthal quantum number m to state with $m' = m + 1$ or $m' = m - 1$ are allowed leading to the outgoing forward-traveling radiation that is, respectively, right-circularly polarized (i.e., described by $(\hat{e}_x - i\hat{e}_y)/\sqrt{2}$ in Equation (21.13.6)) and left-circularly polarized. The projection (helicity) of the unit photon angular momentum along the z-axis of the emitted photon then equals -1 for $\Delta m = +1$ (from $\text{Re}\left(\left(\hat{e}_x - ie_y \right) \left(\cos\omega t - i \sin\omega t \right) \right) = \hat{e}_x \cos\omega t - \hat{e}_y \sin\omega t$) and $+1$ for $\Delta m = -1$, so that the total angular momentum of the atom–photon system is preserved. Further, since the dipole operator is odd, the final and initial atomic states must possess opposite parity. The $(-1)^l$ parity of $Y_{lm}(\theta, \varphi)$ thus excludes $\Delta l = 0$ transitions, while the spin of the final state is unchanged from that of the initial states as a result of the spin independence of the electric dipole interaction. Accordingly, the *atomic dipole transition selection rule* takes the form

$$\Delta l = \pm 1, \Delta s = 0 \tag{21.14.3}$$

For containing more than one electron, the dipole moment is composed of a sum of the dipole moments of the individual electrons, while the atomic parity is not determined only by the total angular momentum, l, as a consequence of which $\Delta l = 0$

transitions are also permitted. The $l = 0$ to $l' = 0$ transition, however, is always forbidden since the photon possesses unit angular momentum. Weaker spectral lines are associated with, e.g., electric quadrupole and magnetic dipole transitions associated with the $\vec{k} \cdot \vec{r}$ term in the expansion of the exponential in Equation (21.13.8) that, respectively, induce parity, preserving $\Delta l = \pm 2, \pm 1, 0$ and $\Delta l = \pm 1, 0$ transitions with no parity change. As well, a small $\vec{S} \cdot \vec{B}$ atomic interaction term leads to $\Delta s \neq 0$ transitions.

21.15 SCATTERING THEORY

In center of mass coordinates, the general wavefunction solution of Schrödinger equation describing a particle in an incoming z-propagating uniform monoenergic particle beam scattering from a central potential of the form $U(r)$ is given by

$$\psi(r,\theta) = e^{ikz} + \frac{f(\theta,\varphi)}{r} e^{ikr} = e^{ikr\cos\theta} + \frac{f(\theta,\varphi)}{r} e^{ikr} \tag{21.15.1}$$

Since $m = 0$ for the incoming, azimuthally symmetric beam, a φ dependence is only present for spin-dependent potentials that couple states with differing m. Excluding these, separating $\psi(r, \theta)$ into a product of radial and angular functions results in

$$\psi(r,\theta) = \sum_{l=0}^{\infty} c_l R_l(kr) P_l(\cos\theta) \tag{21.15.2}$$

with

$$\frac{1}{r}\frac{d^2}{dr^2}(rR_l(kr)) + \left(\frac{2m}{\hbar^2}(E - U(r)) - \frac{l(l+1)}{r^2}\right) R_l(kr) = 0 \tag{21.15.3}$$

From the central result, Equation (13.5.20), for a plane wave,

$$e^{ikr\cos\theta} = \sum_{l=0}^{\infty} i^l (2l+1) j_l(kr) P_l(\cos\theta) \to \frac{1}{kr} \sum_{l=0}^{\infty} i^l (2l+1) \sin\left(kr - \frac{\pi l}{2}\right) P_l(\cos\theta) \tag{21.15.4}$$

If $U(r) \to 0$ and therefore $2m(E - U(r))/\hbar^2 \to k^2$ faster than $1/r$ as $r \to \infty$, multiplying Equation (21.15.3) by r similarly yields in the $r \to \infty$ limit, where c_l and δ_l are determined from the exact solution,

$$R_l(kr) = \frac{c_l}{r}\sqrt{\frac{2}{\pi}} \sin\left(kr + \delta_l - \frac{\pi l}{2}\right) \tag{21.15.5}$$

For a general solution to describe an incoming plane wave together with an elastically scattered radially outgoing wave, according to Equation (21.15.1), the radially incoming part ($\sim \exp(-ikr)$) of each term in the sum of Equations (21.15.2) and (21.15.5) must coincide with that of Equation (21.15.4), which implies

$$c_l = \frac{1}{k}\sqrt{\frac{\pi}{2}}(2l+1)i^l e^{i\delta_l} \tag{21.15.6}$$

From

$$\begin{aligned} 2i^l \sin(kr - \pi l/2) &= -i \cdot i^l\left(\exp(ikr)(-i)^l - \exp(-ikr)i^l\right) \\ &= i\left((-1)^l \exp(-ikr) - \exp(ikr)\right) \end{aligned} \tag{21.15.7}$$

the asymptotic field can be rewritten as

$$\psi(r,\theta) = \frac{i}{2kr}\sum_{l=0}^{\infty}(2l+1)\left[(-1)^l e^{-ikr} - e^{2i\delta_l}e^{ikr}\right]P_l(\cos\theta) \tag{21.15.8}$$

which is the sum of an incoming plane wave solution (Equation (21.15.4)) and a scattered field with

$$f(\theta) = \frac{1}{2ik}\sum_{l=0}^{\infty}(2l+1)\left(e^{2i\delta_l} - 1\right)P_l(\cos\theta) \tag{21.15.9}$$

The scattering cross section, $d\sigma$, into a solid angle $d\Omega$ is defined by $d\sigma = |f(\theta)|^2 d\Omega$ with $d\Omega = \sin\theta d\theta d\varphi$. The orthogonality of the $P_l(\cos\theta)$, Equation (13.2.26), then yields the total cross section

$$\sigma = 2\pi\int_0^{\pi}|f(\theta)|^2 \sin\theta d\theta = \frac{4\pi}{k^2}\sum_{l=0}^{\infty}(2l+1)\sin^2\delta_l \tag{21.15.10}$$

If $U(r) = 0$ for $r > a$, only the *s-wave* ($l = 0$) amplitude of a slow incoming particle with $\lambda = 2\pi/k \gg a$ and $\hbar^2 k^2/2m \ll U(r)$ is significantly altered by scattering since, writing Equation (21.15.3) in terms of the dimensionless variable $r' = kr$, U is nonzero only in the region $r' \ll 1$ for which $-l(l+1)/r'^2 \gg 1$. That is, for $l \neq 0$, the angular momentum term in the Hamiltonian excludes the partial wavefunction from the vicinity of the scattering center.

The spatial wavefunction for two identical scattering particles is either symmetric or antisymmetric upon interchange. In a center of mass system, substituting $\theta \to \pi - \theta$ interchanges the particles so that

$$\psi_{\left\{\substack{\text{symmetric} \\ \text{antisymmetric}}\right\}}(r) = e^{ikz} \pm e^{-ikz} + \frac{e^{ikr}}{r}\left(f(\theta) \pm f(\pi - \theta)\right) \tag{21.15.11}$$

For two spin ½ particles, the top sign applies to the $s_{\text{total}} = 0$, symmetric spatial wavefunction state, and the lower sign to the $s_{\text{total}} = 1$, antisymmetric wavefunction state. The differential cross section is then

$$d\sigma_{\left\{\substack{\text{symmetric} \\ \text{antisymmetric}}\right\}} = |f(\theta) \pm f(\pi-\theta)|^2 d\Omega \tag{21.15.12}$$

Since identical but unpolarized fermions possess three times as many $s_{\text{total}} = 1$ as $s_{\text{total}} = 0$ spin states,

$$\begin{aligned} d\sigma_{\text{unpolarized}} &= \frac{1}{4} d\sigma_{\text{symmetric}} + \frac{3}{4} d\sigma_{\text{antisymmetric}} \\ &= \left(|f(\theta)|^2 + |f(\pi-\theta)|^2 - \frac{1}{2}\left[f(\theta) f^*(\pi-\theta) + f^*(\theta) f(\pi-\theta) \right] \right) d\Omega \end{aligned} \tag{21.15.13}$$

If the two particles are distinguishable, the interference term is absent, yielding instead

$$d\sigma_{\text{distinguishable}} = \left(|f(\theta)|^2 + |f(\pi-\theta)|^2 \right) d\Omega \tag{21.15.14}$$

A scattering event accompanied by a change in the internal state of the system, such as a change in the occupied energy levels of the particles, their chemical composition or number, is termed *inelastic*. Each accessible change forms a separate *reaction channel*. For a spherically symmetric central force, inelastic collisions do not alter the particle angular momentum but do modify the channel amplitudes in Equation (21.15.8):

$$\psi(r,\theta) = \frac{i}{2kr} \sum_{l=0}^{\infty} (2l+1) \left[(-1)^l e^{-ikr} - S_l e^{ikr} \right] P_l(\cos\theta) \tag{21.15.15}$$

The difference $\psi - \exp(ikz)$ of the scattered and the incoming waves then adopts the form

$$f(\theta) = \frac{1}{2ik} \sum_{l=0}^{\infty} (2l+1)(S_l - 1) P_l(\cos\theta) \tag{21.15.16}$$

yielding for the cross section associated with the outgoing field, termed the *elastic cross section*,

$$\sigma_{\text{elastic}} = \frac{\pi}{k^2} \sum_{l=0}^{\infty} (2l+1) |S_l - 1|^2 \tag{21.15.17}$$

while the *inelastic cross section* refers to the sum of the amount of power lost in each channel,

$$\sigma_{\text{inelastic}} = \frac{\pi}{k^2} \sum_{l=0}^{\infty} (2l+1) \left(1 - |S_l|^2 \right) \tag{21.15.18}$$

The total cross section is the sum of these two cross sections,

$$\sigma_{\text{tot}} = \sigma_{\text{elastic}} + \sigma_{\text{inelastic}} = \frac{\pi}{k^2}\sum_{l=0}^{\infty}(2l+1)\left(2 - S_l - S_l^*\right) \tag{21.15.19}$$

Setting $\theta = 0$ results in the *optical theorem*

$$\text{Im}(f(0)) = \text{Im}\left[\frac{1}{2ik}\sum_{l=0}^{\infty}(2l+1)(S_l - 1)\right] = \text{Re}\left[\frac{1}{2k}\sum_{l=0}^{\infty}(2l+1)(1 - S_l)\right] = \frac{k}{4\pi}\sigma_{\text{tot}} \tag{21.15.20}$$

The *Born approximation* for scattering from a localized time-independent potential distribution neglects momentum transfer to the incoming particle. Assuming a stationary scatterer (otherwise, particle velocities must be replaced by relative velocities and masses by effective masses), the golden rule matrix element between initial and final states with momenta p_i and p_f is given by, for wavefunctions normalized to one particle per unit volume,

$$\left\langle \psi_{\vec{p}_f} \middle| H'(x) \middle| \psi_{\vec{p}_i} \right\rangle = \frac{1}{V}\int_V U(x) e^{\frac{i}{\hbar}(\vec{p}_i - \vec{p}_f)} d^3x \equiv U(\vec{p}_i - \vec{p}_f) \tag{21.15.21}$$

The density of continuum plane wave states can be obtained by imposing periodic boundary conditions along the sides of a cube of volume V for which the allowed values of k_α are spaced by $2\pi/L$ along each coordinate direction and the momentum space volume occupied by a state therefore equals $V_p = \hbar^3 V_k = \hbar^3(2\pi/L)^3 = h^3/V$. Relativistically, $E^2 - c^2p^2 = m^2c^4$ implies $EdE = c^2pdp$ where $E = \gamma mc^2$ and $\vec{p} = \gamma m\vec{v}$ so that $p/E = v/c^2$ where $\gamma = (1 - v^2/c^2)^{1/2}$. Hence, if P($E$) and $P(p)$ denote the density of states per unit energy interval and volume and the density of states per unit momentum space volume,

$$\text{P}(E)dE\,d\Omega = \underbrace{\text{P}(p)}_{1/V_p}\frac{c^2p}{E}dE\,d\Omega = \frac{1}{h^3}\frac{p^2}{v}dE\,d\Omega \tag{21.15.22}$$

with a scattering cross section per unit solid angle from Fermi's golden rule (Equation (20.16.31)) (after division by the incoming particle velocity to change the normalization from one particle per unit volume to one incident particle per square meter per second):

$$\frac{d\sigma}{d\Omega} = \frac{m^2\gamma^2}{4\pi^2\hbar^4}\left|U(\vec{p}_i - \vec{p}_f)\right|^2 \tag{21.15.23}$$

This formula is generally accurate when the phase change over the spatial extent, D, of the perturbing potential on the incoming wavefunction,

$$\left(k_{\text{potential}} - k\right)D = \frac{D}{\hbar}\left(\sqrt{2m\left(\frac{p^2}{2m} + U\right)} - p\right) \ll 2\pi \tag{21.15.24}$$

and therefore the distortion of the incoming plane wave wavefront is small.

Example

To find $U(p)$ in the Born approximation for the nonrelativistic scattering of a charge Q_1 by a second charge Q_2, note that the potential energy $U(r) = Q_1Q_2/4\pi\varepsilon_0 r$ solves the Poisson equation $\nabla^2 U(\vec{r}) = -Q_1Q_2\delta(\vec{r})/\varepsilon_0$. Fourier transforming both sides of this equation and integrating each term within the integral twice by parts yield the desired matrix element, $U(\vec{p}-\vec{p}')$, according to

$$\int_{\times 3} e^{\frac{i}{\hbar}(\vec{p}_i-\vec{p}_f)\cdot\vec{r}}\left(\frac{\partial^2}{\partial x^2}+\frac{\partial^2}{\partial y^2}+\frac{\partial^2}{\partial z^2}\right)U(r)dV = -\frac{(\vec{p}_i-\vec{p}_f)^2}{\hbar^2}U(\vec{p}_i-\vec{p}_f) = -\frac{Q_1Q_2}{\varepsilon_0} \tag{21.15.25}$$

Employing $|\vec{p}_i-\vec{p}_f|^2 = p^2(2-2\cos\theta) = 4p^2\sin^2(\theta/2)$ then reproduces the classical *Rutherford scattering* result:

$$\frac{d\sigma}{d\Omega} = \frac{Q_1^2Q_2^2}{4\pi^2}\left(\frac{m}{\varepsilon_0 p^2}\right)^2 \frac{1}{16\sin^4\frac{\theta}{2}} \tag{21.15.26}$$

22

NUCLEAR AND PARTICLE PHYSICS

Nuclear and particle physics generally involves time and energy scales far removed from atomic binding energies (volts) and reaction times (>femtoseconds = 10^{-15}s). Despite this, classical and quantum methods often provide an adequate framework for understanding the underlying features of the relevant processes.

22.1 NUCLEAR PROPERTIES

A nucleus, denoted by ${}_Z^AX$, is characterized by the number of its protons that is termed the *atomic number*, Z, and the *mass number*, $A = N + Z$, corresponding to the total number of nucleons (neutrons and protons). *Isotopes* refer to atoms with the same number of protons but different number of neutrons. For example, terrestrial carbon contains 98.6% ${}_6^{12}C$, 1.1% ${}_6^{13}C$, and small amounts of ${}_6^{11}C$ and ${}_6^{14}C$. Similarly, hydrogen exists in three isotopes, atomic hydrogen ${}_1^1H$, deuterium ${}_1^2H$, and tritium ${}_1^3H$. The *atomic weight* of an atom equals the mass of its isotopes weighted according to their average natural abundance where the isotopic mass is approximated by the sum of the masses of its component particles with

$$\begin{aligned} m_{\text{proton}} &= 1.00728 \text{ amu} = 938.3 \text{ MeV}/c^2 = 1.672 \times 10^{-27} \text{ kg} \\ m_{\text{neutron}} &= 1.00867 \text{ amu} = 939.6 \text{ MeV}/c^2 = 1.675 \times 10^{-27} \text{ kg} \\ m_{\text{electron}} &= 0.00055 \text{ amu} = 0.51 \text{ MeV}/c^2 = 9.11 \times 10^{-31} \text{ kg} \end{aligned} \tag{22.1.1}$$

Fundamental Math and Physics for Scientists and Engineers, First Edition.
David Yevick and Hannah Yevick.

The fraction of neutrons to protons in nuclei generally increases with atomic number. The *nuclear binding energy*

$$B = \left(Zm_{\text{proton}} + Nm_{\text{neutron}} - m_{\text{nucleus}}(Z,N)\right)c^2 \tag{22.1.2}$$

per particle grows rapidly with Z until $Z \approx 25$ and then remains nearly constant until finally decreasing roughly linearly for $Z > 75$. In fact, particularly stable nuclei exist when either or both of Z and N equal *magic numbers* 2, 8, 20, 28, 50, 82, 126 as the binding energy of a neutron (the energy required to remove a single neutron from the nucleus) among the isotopes of a single element increases when this condition is satisfied. The binding energy is also larger for even numbers of neutrons suggesting the presence of a *pairing interaction*. Scattering experiments with spin-polarized neutrons indeed indicate that the force between nucleons is attractive within a radius of about ≈ 2 fm (1 fm $= 10^{-15}$ m) and repulsive inside a core region extending for a few tenths of a Fermi. The approximate nuclear radius

$$r_{\text{nuclear}} \approx \left(1.2 \times 10^{-15}\ \text{m}\right)A^{\frac{1}{3}} \tag{22.1.3}$$

is determined through collisions with accelerated ions that possess sufficient energy to surmount the nuclear Coulomb barrier. The $\sqrt[3]{A}$ dependence suggests that the nuclear density does not vary with A. The *liquid drop model* of nuclear structure accordingly first includes the implied linear dependence of the binding energy on A. A second negative contribution proportional to the surface area results from the reduced number of nucleons with which nucleons at the surface can bind. A Coulomb contribution proportional to the inverse of the nuclear radius multiplied by the square of the nuclear charge models the potential energy of a uniform spherical charge distribution. Finally, a "symmetry energy" accommodates the exclusion principle, as a result of which neutrons that are in excess of the number of protons must occupy higher nuclear states, decreasing the binding energy. This contribution can be estimated by modeling the nucleus as a Fermi liquid contained within a volume $(1.3A)$ fm^3 yielding $p_{N,\text{Fermi}} = \hbar/1.3\ \text{fm}(9\pi N/4A)$ for the neutron Fermi momentum according to Equation (23.15.4) with a similar expression for protons. The average energy of the nucleons then approximates the observed symmetry energy. Combining all four contributions,

$$E_{\text{binding}} = a_{\text{volume}}A - a_{\text{surface}}A^{\frac{2}{3}} - a_{\text{Columb}}Z^2A^{-\frac{1}{3}} - a_{\text{symmetry}}\frac{(Z-N)^2}{A^2} \tag{22.1.4}$$

With suitable coefficients, Equation (22.1.4) accurately predicts nuclear binding energies. Alternatively, the *nuclear shell model* employs the quantum mechanical energy levels of an empirical nuclear potential. When the number of neutrons suffices to fill a level, the binding energy increases, increasing the nuclear stability in the same manner that closed orbitals enhance the stability of noble gases. However, spin and other interactions are required to reproduce the observed magic numbers.

The nuclear angular momentum corresponding to the vector sum of the angular momenta of the nucleons is an integer or half-integer. The proton and neutron magnetic moments are, respectively, $2.79\mu_{\text{nuclear}}$ and $-1.91\mu_{\text{nuclear}}$ where the magnitude of the *nuclear magneton*

$$\mu_{\text{nuclear}} = \frac{e\hbar}{2m_{\text{proton}}} = 5.05 \times 10^{-27}\,\text{J/T} \tag{22.1.5}$$

equals approximately 1/2000 of the Bohr magneton as a consequence of the greater proton mass. Hence, an applied magnetic field splits an atomic level into *hyperfine* levels with different nuclear spin orientations. In *nuclear magnetic resonance spectroscopy*, a low-frequency (typically radio) wave induces transitions between these levels, leading to increased absorption. In the presence of a magnetic field gradient, the resonance frequency becomes position dependent, enabling *magnetic resonance imaging*.

22.2 RADIOACTIVE DECAY

As nucleons are less strongly bound in large atoms and isotopes, these often spontaneously radiate or decompose into more stable fragments. Such processes typically release, in order of penetrating ability, (i) ${}^{4}_{2}\text{He}$ nuclei (*alpha rays*); (ii) electrons or positively charged electron *antiparticles*, e^{+}, termed *positrons* (*beta rays*); and (iii) high-energy photons (*gamma rays*). Defining the *Q-value* of the reaction ${}^{A}_{Z}X \rightarrow {}^{A-4}_{Z-2}Y + {}^{4}_{2}\text{He}$ as

$$Q = (m_X - m_Y - m_\alpha)c^2 \tag{22.2.1}$$

alpha particles are emitted with a discrete spectrum of Q-values, indicating that the reacting nucleons belong to different energy levels. The emission process is therefore modeled by the formation and subsequent tunneling of an alpha particle inside the nucleus through a potential barrier. High-energy alpha particles tunnel more rapidly and therefore possess shorter decay time constants. In beta decay, an electron or positron is instead emitted with a continuous spectrum of energies from zero to a maximum kinetic energy. This requires the existence of two new particles, the (electron) *neutrino*, υ_e, and *antineutrino*, $\bar{\upsilon}_e$, with zero charge, negligible mass, a spin of ½, and a minute cross section with matter. These particles receive momentum and energy from the scattering process according to either

$$\begin{aligned} {}^{A}_{Z}X &\rightarrow {}^{A}_{Z-1}Y + e^{+} + \upsilon_e \\ {}^{A}_{Z}X &\rightarrow {}^{A}_{Z+1}Y + e^{-} + \bar{\upsilon}_e \end{aligned} \tag{22.2.2}$$

Finally, gamma rays are generated when a nucleus decays to an excited state and subsequently emits a photon with a characteristic energy spectrum in a further transition to a stable state.

In a homogeneous radioactive sample, the decay probability is identical and time independent for every atom; hence, the decay rate (the number of decays per unit time) is proportional to the number of decaying atoms. As a result, the number of radioactive nuclei decreases with time according to

$$\frac{dN}{dt} = -\lambda N \tag{22.2.3}$$

where λ is termed the *decay constant* so that $N = N(t = 0)\exp(-\lambda t)$. The *decay rate* or *activity*

$$R = \left|\frac{dN}{dt}\right| = \lambda N(t=0)e^{-\lambda t} \tag{22.2.4}$$

is expressed in *becquerel*, which is one decay per second, or *curies*, 3.7×10^{10} decays per second. The *half-life* is the time over which half of a given sample decays so that $N(t=0)e^{-\lambda t_{1/2}} = N(t=0)/2$ or

$$t_{1/2} = \frac{\ln 2}{\lambda} = \frac{0.693}{\lambda} \tag{22.2.5}$$

After M half-lives, a fraction $1/2^M$ of the original sample remains.

22.3 NUCLEAR REACTIONS

A high-energy particle incident on a nucleus X can change its composition, generating a new nucleus Y in a nuclear reaction denoted by $A_{\text{incoming}} + X \rightarrow B_{\text{outgoing}} + Y$ or equivalently $X(A_{\text{incoming}}, B_{\text{outgoing}})Y$. The *reaction energy* generated is related to the individual particle rest energies by

$$Q = (M_A + M_X - M_B - M_Y)c^2 \tag{22.3.1}$$

If the incoming and outgoing particles as well as the nuclei X and Y are identical, the scattering is termed *elastic* if $Q = 0$ and *inelastic* otherwise. Processes with $Q > 0$ and $Q < 0$ are termed *exothermic* and *endothermic*, respectively, while the kinetic energy of A_{incoming} required for the reaction to occur is called the *threshold* energy. Nuclear reactions conserve relativistic energy and momentum and preserve the number of nucleons.

22.4 FISSION AND FUSION

Since the particle binding energy increases as a function of Z for $Z < 25$ and decreases with Z for $Z > 75$, energy can be released either by the *fusion* of two small nuclei into a larger nucleus or the *fission* (splitting) of a heavy nucleus. Fission can generate a *chain reaction*, as when ^{235}U absorbs a slow neutron thus becoming larger and more unstable. As the liquid drop model predicts, the nucleus then splits into two roughly equal size nuclei, releasing excess neutrons. These neutrons generate further reactions in the presence of a sufficient, *critical mass* of uranium. While this normally yields a nuclear explosion, if the neutron energy is properly controlled, the process can be instead employed for power generation. At high temperatures and densities, small nuclei can overcome large Coulomb barriers to produce fusion. In a star, e.g., the *proton–proton cycle* produces a helium nucleus, electron, and neutrino from four protons:

$$\begin{aligned} {}_1^1\mathrm{H} + {}_1^1\mathrm{H} &\rightarrow {}_1^2\mathrm{H} + e^+ + \nu \\ {}_1^1\mathrm{H} + {}_1^2\mathrm{H} &\rightarrow {}_2^3\mathrm{H}^e + \gamma \\ {}_2^3\mathrm{H}^e + {}_2^3\mathrm{H}^e &\rightarrow {}_2^4\mathrm{H}^e + {}_1^1\mathrm{H} + {}_1^1\mathrm{H} \end{aligned} \tag{22.4.1}$$

22.5 FUNDAMENTAL PROPERTIES OF ELEMENTARY PARTICLES

High-energy collisions in particle accelerators identified numerous elementary particles that complement the common stable particles, namely, the electron, neutrino, proton, neutron, and photon. While the unstable particles possess short lifetimes that severely limit practical application, analysis of their interactions led to models of physical forces in which *strong forces* mediated by *gluons* bind subatomic particles termed *quarks*. The strong force increases with distance precluding the observation of isolated quarks. *Electromagnetic forces* result from interactions with "virtual" photons, while *weak forces*, which lead to processes such as beta decay, are transferred by the W and Z bosons. It is predicted by the *electroweak theory* that the electromagnetic and weak forces are identical at high energies. Finally, *gravitational forces* are mediated by gravitons. That is, from the uncertainty principle, energy conservation can be violated over a time proportional to the inverse of the magnitude of the energy violation. In a simple model, while, e.g., individual gluons are confined to a single nucleon, a proton can release a massive "virtual" neutral pion, π^0, composed of a quark and an antiquark that can transfer energy and momentum to a second nucleon. Such a pion of mass M_π exists for only time given by $\Delta t \approx \hbar/(2M_\pi c^2)$ and cannot propagate further than $c\Delta t$. Hence, this force component is only substantial for distances of the order of the nuclear radius. In contrast, the graviton and photon are massless, leading to long range $1/r^2$ forces.

Particles can be classified as, first, the six *leptons*, namely, the *electron*, *muon*, *tau lepton*, and their corresponding neutrinos that interact primarily through electroweak forces. The lepton masses are 0.511, 105, and 1,784 MeV, respectively, while the neutrinos are assumed massless. Next, six spin ½ and baryon number 1/3 *quarks* can bind through strong interactions to form *hadrons* such as protons and neutrons. These quarks are the up, down, charmed, strange, and top and bottom with masses of 360, 350, 1500, 540 MeV, and 173 and 5 GeV, respectively (GeV = billion electron volts). Every quark possesses one of three *colors*, red, green, and blue, while hadrons cannot contain three quarks of the same type as a result of Fermi statistics. The up, charmed, and top quarks possess a fractional charge of $2e/3$, while the charge of the remaining quarks is $-e/3$. Since in relativity theory, $E^2 - c^2p^2 = m^2c^4$, implying $E = \pm\sqrt{p^2c^2 + m^2c^4}$, every particle possesses a corresponding *antiparticle* with the negative energy and opposite values of internal quantum properties such as charge. While these negative energy states are normally filled, a high-energy excitation can excite a particle from a negative to a positive energy state, leaving a "vacancy" associated with an antiparticle. A collision between a particle and its antiparticle refills this vacancy, releasing approximately twice the particle rest mass. Hadrons are subdivided into *mesons* that contain one quark and one antiquark and therefore possess zero or unit spin and spin ½ or 3/2 b*aryons* with 3 quarks that constitute fermions. Thus, e.g., a π^+ meson is formed from an up and an antidown quark, while a neutron is composed of one up and two down quarks.

Some properties and *conservation rules* of particles and particle interactions are:

1. Conservation of electron-lepton, muon-lepton, and tau-lepton numbers where, e.g., the electron leptons, namely, electrons and electron neutrinos, and electron antileptons are assigned electron-lepton numbers of 1 and −1.
2. Conservation of baryon number: All baryons possess baryon number 1 and all antibaryons −1. The proton is absolutely stable because it comprises the smallest mass baryon.
3. Conservation of strangeness, charm, topness, and bottomness in *strong but not weak* interactions: For example, charm quantum numbers are 1 and −1 for a charmed quark and antiquark, respectively.
4. Absence of colored baryons: The quarks in a baryon are either red, green, and blue or antired, antigreen, and antiblue, while a meson contains a quark and an antiquark of the same color.

Being comprised of two quarks, mesons form an octagonal pattern termed the *eightfold way* when plotted as functions of charge and strangeness. As up, down, and strange quarks possess charges of $2e/3$, $-e/3$, and $-e/3$, two positively charged mesons, the π^+ and K^+ mesons with strangeness 0 and 1, can be formed as $u\bar{d}$ and $u\bar{s}$. The strangeness 1 and −1 uncharged K^0 and $\bar{K}^0$ mesons are then formed as $d\bar{s}$ and $\bar{d}s$. An additional three (η, η', and π^0) mesons with strangeness 0 are composed of the three neutral quark pairs $u\bar{u}$, $d\bar{d}$, and $s\bar{s}$. Displaying the charge and strangeness of each particle as (x, y) pairs yields, after including the π^- and K^- antiparticles,

	charge = −1	charge = 0	charge = 1
strangeness = −1		K^0	K^+
strangeness = 0	π^-	η, η', π^0	π^+
strangeness = 1	K^-	$\bar{K}^0$	

(22.5.1)

For spin 1/2 baryons, the corresponding diagram is

	charge = −1	charge = 0	charge = 1
strangeness = 0		n	p
strangeness = −1	Σ^-	Λ^0, Σ^0	Σ^+
strangeness = −2	Ξ^-	Ξ^0	

(22.5.2)

23

THERMODYNAMICS AND STATISTICAL MECHANICS

While the equations of motion for even a few interacting particles can only be solved numerically, systems containing a large number of particles can be characterized *statistically* in terms of average, macroscopic properties such as temperature, energy, and pressure. Thermodynamics relates these quantities based on a few underlying physical assumptions, while *statistical mechanics* instead proceeds from the principle that a system with many degrees of freedom transitions randomly between all accessible configurations consistent with its macroscopic properties.

23.1 ENTROPY

In contrast to the description of a deterministic mechanical or quantum mechanical configuration, the behavior of a statistical mechanical system is characterized by the average properties of a large number of independent realizations or replicas, termed a *statistical ensemble*. These can be multiple identical systems at a given time or a single system sampled repeatedly over time intervals longer than the *correlation time* required to eliminate its dependence on the previously sampled state. The properties of an ensemble depend on whether energy and/or particles are exchanged with its environment. The energy of an *isolated* system is determined within an accuracy ΔE that is, in theory, inversely proportional to the measurement time implying that all system configurations within this energy interval, termed a *microcanonical* ensemble, are equally probable.

Fundamental Math and Physics for Scientists and Engineers, First Edition.
David Yevick and Hannah Yevick.

If a system instead exchanges energy/particles with its environment, only the average energy/particle density is quantifiable. Assuming no additional prior system information or measured experimental parameters beyond these constraints implies that in an optimal description, the system can occupy *any* possible state subject to these constraints with state occupation probabilities that insure maximum randomness. In a *canonical* ensemble, energy is exchanged with the external environment so that only the mean energy, quantified through the *temperature*, can be measured. A system that exchanges both heat and particles (as exchanging particles necessarily alters the energy) is described by a *grand canonical ensemble* in which the mean number of particles is quantified through the *chemical potential.*

The particles of a system further obey classical or quantum mechanical *statistics.* For a system of *distinguishable* or "identifiable" parts such as isolated atoms in a spatial lattice or a gas of dissimilar particles, a realization in which, e.g., an atom A is in state 1 and B in state 2 differs from a realization in which A is in 2 and B in 1. These two states are instead identical for *indistinguishable* components such as atoms in a monatomic gas. For *Fermi–Dirac* statistics, each state can contain at most one particle, while for *Bose–Einstein* statistics, any number of particles can be present in a state.

The degree of randomness of a system is quantified by its *entropy* or equivalently *information* content. To illustrate, consider a system of distinguishable noninteracting particles each occupying one of two equally likely states with energies 0 and E. Adding a particle to the system increases its randomness by unity, as one additional element of information becomes uncertain, while doubling the number of states (since a system state is specified by that of the additional particle together with the state of the remainder of the system). Therefore, the entropy is proportional to the number of such particles or equivalently to the logarithm of the number, 2^N, of equally probable states. In general, a system composed of two distinguishable subsystems, with N and N' states, possesses NN' states, while the entropy, S, is given by the sum of the entropies of the two subsystems; i.e.,

$$S(NN') = S(N) + S(N') \tag{23.1.1}$$

which coincides with the properties of the logarithm function. Mathematically, if $N' = 1 + \delta$, from a Taylor expansion, noting that the entropy of a single state system equals zero,

$$\begin{aligned} S(N(1+\delta)) &= \cancel{S(N)} + N\delta \left.\frac{dS(N)}{dN}\right|_N \\ S(N) + S(1+\delta) &= \\ \cancel{S(N)} + \underbrace{S(1)}_{0} + \delta \left.\frac{dS(N)}{dN}\right|_{N=1} &= \end{aligned} \tag{23.1.2}$$

Accordingly,

$$\frac{dS(N)}{dN} = \frac{1}{N}\left.\frac{dS}{dN}\right|_{N=1} \equiv \frac{\Xi}{N} \tag{23.1.3}$$

verifying that the entropy is proportional to the logarithm of the total number of accessible states. In information theory, the constant $\Xi = 1/\log_e 2$ yielding the information $I = \log_2 N$, while in statistical mechanics, $\Xi = k$. Hence, $S = k \ln N$ where the *Boltzmann constant* $k = 1.38 \times 10^{-23}$ J/°K.

Consider next a three-state system for which the probability of states of energy E and $2E$ is ¼, while that of a state with energy 0 is ½. The 50% of particles in state 0 contribute $k \ln 2 = -k \ln(1/2)$ to the entropy as these effectively populate a subsystem with two equally likely outcomes, while the contribution of each of the other remaining particles equals $-k \ln(1/4)$. Weighting each of these by its probability of occurrence p_n yields, in the general case of N states with occupation probabilities p_m,

$$S = -k \sum_{m=1}^{N} p_m \ln p_m \tag{23.1.4}$$

23.2 ENSEMBLES

As noted in the previous section, a *microcanonical ensemble* models a fully isolated system that cannot exchange heat or particles with its environment and therefore possesses an energy that can be determined according to quantum mechanics within a range $\Delta E \approx \hbar/\Delta t$ where Δt denotes the measurement time. The randomness of the ensemble is maximized subject to the lack of any prior knowledge of the microscopic condition of the system by assigning an equal occupation probability to each state with a total energy within the interval $[E, E + \Delta E]$ and zero probability to all other states. Consider an ensemble of distinguishable atoms with energy levels $\{E_1, E_2, \ldots, E_m\}$ that are occupied on average by $\{N_1, N_2, \ldots, N_m\}$ particles (to insure that all $N_i \gg 1$, the N_i can also be interpreted either as the number of states of the entire ensemble with an *average* energy within each range $[E_i, E_{i+1}]$ or as the *total* number of atoms in state i in the entire ensemble). The number of unique system configurations subject to these constraints equals the product of the ${}_N C_{N_1}$ distinct ways, with N as the total number of particles, to distribute N_1 electrons into the $i = 1$ state(s), with the ${}_{N-N_1} C_{N_2 N_2}$ ways to distribute N_2 of the remaining electrons into the $i = 2$ state(s), etc., yielding the multinomial distribution

$$P(N_1, N_2, \ldots, N_m) = \frac{N!}{N_1!\cancel{(N-N_1)!}} \frac{\cancel{(N-N_1)!}}{N_2!(N-N_1-N_2)!} \cdots = \frac{N!}{N_1! N_2! \ldots N_m!} \tag{23.2.1}$$

where

$$\begin{gathered} \sum_i N_i = N \\ \sum_i N_i E_i = E \end{gathered} \tag{23.2.2}$$

The occupation probability of the *individual* energy states in the microcanonical ensemble, for which the total energy is fixed, can be obtained from the condition that a small energy-conserving perturbation of the most likely distribution does not alter P to the lowest order. Considering, e.g., a subset of three states, L, j, H, with energies $E_j - \Delta E_L$, E_j, $E_j + \Delta E_H$, of the m states of this system, if the change in energy, $N_{H\to j}\Delta E_H = a\Delta E_L \Delta E_H$, resulting from displacing $a\Delta E_L$ electrons from H to j is compensated by displacing of $a\Delta E_H$ electrons (approximated by an integer) from L to j, this condition takes the form

$$P = \frac{N!}{N_1!N_2!\ldots(N_L - a\Delta E_H)!\ldots(N_j + a(E_H + E_L))!\ldots(N_H - a\Delta E_L)!\ldots N_m!} = \frac{N!}{N_1!N_2!\ldots N_m!} \tag{23.2.3}$$

If the displacement is small compared to the number of particles in each of the three states,

$$\frac{(N_j + a)!}{N_j!} = (N_j + a)(N_j + a - 1)\ldots(N_j + 1) \approx (N_j)^a \tag{23.2.4}$$

Hence, multiplying Equation (23.2.3) by the product of both denominators,

$$(N_L)^{a\Delta E_H}(N_H)^{a\Delta E_L} = (N_j)^{a(\Delta E_H + \Delta E_L)} \tag{23.2.5}$$

and therefore,

$$\left(\frac{N_L}{N_j}\right)^{a\Delta E_H}\left(\frac{N_H}{N_j}\right)^{a\Delta E_L} = 1 \tag{23.2.6}$$

Taking the logarithm

$$a\Delta E_H \ln\left(\frac{N_L}{N_j}\right) = -a\Delta E_L \ln\left(\frac{N_H}{N_j}\right) \tag{23.2.7}$$

or

$$\frac{1}{\Delta E_L}\ln\left(\frac{N_L}{N_j}\right) = \frac{1}{\Delta E_H}\ln\left(\frac{N_j}{N_H}\right) \equiv \beta \tag{23.2.8}$$

implying in both cases the *Boltzmann condition*

$$\frac{N_j}{N_k} = e^{-\beta(E_j - E_k)} \tag{23.2.9}$$

A *heat reservoir* consists of an extended body with far more quantum mechanical modes than the system. Consequently, energy transferred from the system does not measurably alter the modal occupation probabilities in the reservoir and hence its temperature. A group of systems in thermal contact with either a reservoir or equivalently with a large number of identical systems constitutes a *canonical ensemble*, in which the system energy fluctuates with time. In a *grand canonical ensemble*, systems exchange both heat and *particles* with an adjoining reservoir or a multitude of identical systems through, e.g., porous membranes. If in a grand canonical ensemble of K systems $K_{i,j}$ systems possess particle numbers from $[N_i, N_i + \Delta N]$ within a fixed energy interval $[E_j, E_j + \Delta E]$ (or simply N_i particles in state E_j),

$$\sum_{i,j} K_{i,j} = K$$
$$\sum_{i,j} K_{i,j} E_j = E \qquad (23.2.10)$$
$$\sum_{i,j} K_{i,j} N_i = N$$

the state occupation probability is obtained by maximizing the entropy from Equation (23.2.1)

$$\frac{S}{k} = \ln P(K_{i,j}) = K \ln K - \sum_{i,j} K_{i,j} \ln K_{i,j} \qquad (23.2.11)$$

subject to the constraints of Equation (23.2.10). Introducing the constraint equations through Lagrange multipliers (cf. Section 8.5) implies

$$\frac{\partial}{\partial K_{k,l}}\left(\frac{S}{k} - \alpha\left(\sum_{n,m} K_{n,m} N_n - N\right) - \beta\left(\sum_{n,m} K_{n,m} E_m - E\right) - \gamma\left(\sum_{i,m} K_{n,m} - K\right)\right) = 0 \qquad (23.2.12)$$

with the solution

$$-\ln M_{k,l} - 1 - \alpha N_k - \beta E_l - \gamma = 0 \qquad (23.2.13)$$

or

$$K_{i,j} = c e^{-\beta E_j - \alpha N_i} \qquad (23.2.14)$$

The probability of occupation of a state is then

$$p(E_j, N_i) = \frac{K_{i,j}}{K} = \frac{e^{-\beta E_j - \alpha N_i}}{\sum_{i,j} e^{-\beta E_j - \alpha N_i}} \qquad (23.2.15)$$

The values of β and α are found from

$$\bar{E} = \sum_{k,l} E_k p(E_k, N_l) = \sum_k E_k p(E_k) = \frac{\sum_{k,l} E_k e^{-\beta E_k - \alpha N_l}}{\sum_{k,l} e^{-\beta E_k - \alpha N_l}}$$

$$\bar{N} = \sum_{k,l} N_l p(E_k, N_l) = \sum_l N_l p(N_l) = \frac{\sum_{k,l} N_l e^{-\beta E_k - \alpha N_l}}{\sum_{k,l} e^{-\beta E_k - \alpha N_l}} \tag{23.2.16}$$

where $\bar{E}$ and $\bar{N}$ represent the *average* energy and number of particles. In the canonical ensemble, N does not vary and α is therefore absent in the formalism, reproducing Equation (23.2.9). Indeed, in the derivation of Equation (23.2.9), the subsystem consisting of the state or states at energy $E_j + \Delta E_H$ exchanges energy with the remainder of the system, which therefore functions as a heat reservoir. As will be demonstrated presently, $\beta = 1/kT$ in terms of the temperature T, where kT corresponds to the average increase in energy if an additional one-dimensional harmonic oscillator component is added to the system. Analogously, $\alpha = \mu/kT$ in which the *chemical potential*, μ, corresponds to the average work (up to an additive constant) required to displace an additional particle from infinity to a location at rest within the system (note that α must be dimensionless since N is dimensionless). Both T and μ are time or ensemble averaged and therefore time-independent quantities.

23.3 STATISTICS

A system of two distinguishable particles occupying a single spatial state with spins + and − possess four distinct spin states, (+,+), (+,−), (−,+) and (−,−), yielding $S = k \ln 4 = 2k \ln 2$ or twice the entropy of a single spin. If the particles are instead indistinguishable and obey Bose–Einstein statistics, (+,−) and (−,+) are identical and $S = k \ln 3$, while for Fermi–Dirac statistics, (+,+) and (−,−) (for which two electrons are present in the same space × spin state) are additionally excluded. Hence, the probability of configurations that contain more than 1 particle in a given energy state, e.g., the probability of either (+,+) or (−,−), is enhanced for Bose–Einstein particles (here 2/3) and lowered for Fermi–Dirac particles (0/1) relative to that of classical distinguishable particles (2/4). The classical result can however be partially corrected by multiplying by the reciprocal, 1/2!, of the number of permutations of two identical particles. The resulting estimate $(2\,!)^2/2! = 2$ is, respectively, one less and one more than the Bose–Einstein and Fermi–Dirac enumerations.

To determine the distribution functions for Bose–Einstein and Fermi–Dirac particles, note first that Equation (23.2.14) can be obtained by maximizing $\ln P$ with P given by Equation (23.2.1), subject to the constraints of Equation (23.2.2), as this leads to, after introducing Lagrange multipliers α and β,

$$\frac{\partial}{\partial N_l}\left[N \ln N - N - \sum_i N_i \ln(N_i) + N - \alpha\left(\sum_m N_m - N\right) - \beta\left(\sum_m N_m E_m - E\right)\right] = 0 \tag{23.3.1}$$

The Bose–Einstein state occupation probability as a function of energy follows similarly by first evaluating the number of ways that N identical particles can be distributed among different energy levels such that N_i particles occupy the M_i energy levels with energy E_i (or alternatively with energies between E_i and $E_{i+1} = E_i + \Delta E_i$). For a given E_i, the number of distinct configurations coincides with the number of ways that N_i similar objects can be separated with $M_i - 1$ partitions. This in turn equals the number of different arrangements of a set of N_i and $M_i - 1$ identical objects of two different types. Incorporating all states of different energies,

$$P(N_1, M_1, N_2, M_2, \ldots) = \prod_i \frac{(M_i + N_i - 1)!}{N_i!(M_i - 1)!} \tag{23.3.2}$$

Again, maximizing the logarithm of this function subject to

$$\begin{aligned} \sum_i N_i &= N \\ \sum_i N_i E_i &= E \end{aligned} \tag{23.3.3}$$

yields for large N_i and M_i

$$\frac{\partial}{\partial N_l}\left[\begin{array}{l}(N_i + M_i)\ln(N_i + M_i) - (N_i + M_i) - N_i \ln(N_i) - M_i \ln(M_i) \\ + N_i + M_i - \alpha\left(\sum_i N_i - N\right) - \beta\left(\sum_i N_i E_i - E\right)\end{array}\right] = 0 \tag{23.3.4}$$

and therefore, where α and β are found from Equation (23.3.3),

$$\ln\left(\frac{N_l + M_l}{N_l}\right) = \alpha + \beta E_l \tag{23.3.5}$$

or, in terms of the *occupation probability* of a state of energy E_i,

$$n_l \equiv \frac{N_l}{M_l} = \frac{1}{e^{\beta E_l + \alpha} - 1} \tag{23.3.6}$$

If $N_i \ll M_i$ for all states i, P can be instead approximated by

$$P(N_1, M_1, N_2, M_2, \ldots) \approx \frac{M_i^{N_i}}{\prod_i N_i!} \tag{23.3.7}$$

resulting in the Maxwell–Boltzmann distribution function

$$n_i \equiv \frac{N_i}{M_i} = ce^{-\beta E_i} \tag{23.3.8}$$

after maximization subject to fixed E and N. Indeed, since $N_i \ll M_i$ implies $\beta E_i + \alpha \gg 1$, Equation (23.3.6) reduces to Equation (23.3.8) in this limit.

For *Fermi–Dirac* statistics, N indistinguishable particles occupy the M energy levels such that each state only contains zero or one electron. The number of possible realizations for state i is therefore ${}_{M_i}C_{N_i}$, e.g., the number of ways that N_i heads can appear in M_i coin flips. Thus,

$$P(N_1, M_1, N_2, M_2, \ldots) = {}_{M_1}C_{N_1}\, {}_{M_2}C_{N_2} \ldots \tag{23.3.9}$$

and

$$\ln P = \sum_i [M_i \ln M_i - N_i \ln N_i - (M_i - N_i) \ln (M_i - N_i)] \tag{23.3.10}$$

Maximizing this function as above leads to

$$\ln\left(\frac{M_l - N_l}{N_l}\right) = \alpha + \beta E_l \tag{23.3.11}$$

or equivalently

$$n_l = \frac{N_l}{M_l} = \frac{1}{e^{\beta E_l + \alpha} + 1} \tag{23.3.12}$$

Again, in the limit that the probability of finding a particle in a given state is small compared to unity, $\exp(\beta E_i + \alpha) \gg 1$ for all i, reproducing the Maxwell–Boltzmann distribution function.

23.4 PARTITION FUNCTIONS

The macroscopic thermodynamic properties of a system such as the energy or pressure as well as microscopic properties such as the state occupation probabilities can be obtained from a single *partition function* and its derivatives. Since the total energy associated with the mth energy level equals its eigenenergy ε_m multiplied by the number of electrons occupying the state, the *grand canonical partition function* of this state (the argument V of Z_m enters implicitly through the energy level positions and is often suppressed) is first defined as

$$Z_m(\beta, \alpha, V) = \sum_{\text{all occupation numbers } n_i} e^{-(\beta \varepsilon_m + \alpha) n_i} \tag{23.4.1}$$

which equals

$$Z_m = \begin{cases} \displaystyle\sum_{n_i=0}^{1} e^{-(\beta\varepsilon_m + \alpha) n_i} = 1 + e^{-(\beta\varepsilon_m + \alpha)} & \text{Fermi-Dirac (FD)} \\ \displaystyle\sum_{n_i=0}^{\infty} e^{-(\beta\varepsilon_m + \alpha) n_i} = \frac{1}{1 - e^{-(\beta\varepsilon_m + \alpha)}} & \text{Bose-Einstein (BE)} \end{cases} \tag{23.4.2}$$

In the classical limit, both expressions approach $1+\exp(-(\beta\varepsilon_m+\alpha))$. The mean number of particles in state m is then obtained from

$$\bar{n}_m = \frac{\displaystyle\sum_{i=0}^{\substack{1(\text{FD})\text{ or}\\ \infty(\text{BE})}} n_i e^{-(\beta\varepsilon_m-\alpha)n_i}}{\displaystyle\sum_{i=0}^{\substack{1(\text{FD})\text{ or}\\ \infty(\text{BE})}} e^{-(\beta\varepsilon_m-\alpha)n_i}} = -\frac{1}{\beta}\frac{\partial \ln Z_m}{\partial \varepsilon_m}\bigg|_{\alpha,\beta} \tag{23.4.3}$$

leading to Equations (23.3.6) and (23.3.12) in the form

$$n_m = \begin{cases} \dfrac{e^{-(\beta\varepsilon_m+\alpha)}}{1+e^{-(\beta\varepsilon_m+\alpha)}} & \text{Fermi-Dirac} \\[2ex] \dfrac{e^{-(\beta\varepsilon_m+\alpha)}}{1-e^{-(\beta\varepsilon_m+\alpha)}} & \text{Bose-Einstein} \end{cases} \tag{23.4.4}$$

In analogy to Equation (23.4.1), the system *grand canonical partition function*, Z, can be defined by (the argument V is again generally suppressed):

$$\begin{aligned} Z(\beta,\alpha,V) &= \sum_{i=\{n_1,n_2,\ldots\}} e^{-\beta E_i-\alpha N_i} \\ &= \sum_{\substack{\text{all permissible } l\\ \text{for level } m}} e^{-\beta\left(\sum_{\text{all energylevels } m} n_l\varepsilon_m\right)-\alpha n_l} \\ &= \sum_{\text{all } n_i=0}^{\substack{1(\text{FD})\text{ or}\\ \infty(\text{BE})}} e^{-\beta(\varepsilon_1 n_1+\varepsilon_2 n_2+\ldots)-\alpha(n_1+n_2+\ldots)} \end{aligned} \tag{23.4.5}$$

However, since $\exp(a+b)=\exp(a)\exp(b)$, these can be recast as the product

$$Z=\prod_{m=1}^{\infty} Z_m \tag{23.4.6}$$

While, e.g., Equation (23.4.3) still applies with Z_m replaced by Z, the average energy of the entire system and the probability of a state with a set of occupation numbers collectively indexed by i leading to a total energy E_i is designated $p(E_i, N_i)$:

$$\bar{E}=\sum_i E_i p(E_i,N_i)=\frac{\sum_i E_i e^{-\beta E_i-\alpha N_i}}{\sum_i e^{-\beta E_i-\alpha N_i}}=\frac{\partial \ln Z}{\partial \beta} \tag{23.4.7}$$

The average number of particles is similarly given by

$$\bar{N}=\sum_i N_i p(E_i,N_i)=\frac{\sum_i N_i e^{-\beta E_i-\alpha N_i}}{\sum_i e^{-\beta E_i-\alpha N_i}}=\frac{\partial \ln Z}{\partial \alpha} \tag{23.4.8}$$

If a system only exchanges heat with its environment, all members of the ensemble possess the same total number of particles $N_i=\sum_l n_l$. Since then

$$p(E_i)=\frac{e^{-\beta E_i}}{\sum_i e^{-\beta E_i}} \tag{23.4.9}$$

the average energy becomes

$$\bar{E}=\sum_i E_i p(E_i)=\frac{\sum_i E_i e^{-\beta E_i}}{\sum_i e^{-\beta E_i}}=\frac{\partial}{\partial \beta}\left(\ln \sum_i e^{-\beta E_i}\right)\equiv\frac{\partial \ln z}{\partial \beta} \tag{23.4.10}$$

where (one or more of the arguments of z are often omitted)

$$z(\beta,V,N)=\sum_i \exp(-\beta E_i) \tag{23.4.11}$$

is termed the *canonical partition function*. If each member of the ensemble is further isolated from all adjacent members, all states possess the same energy, E_i, or their energies fall into the same infinitesimal range $[E_i, E_i+\Delta E_i]$, and thus, all terms in Z or z are equal. The *microcanonical partition function* $\Omega(E, V, N)$, where one or more of the arguments are again often suppressed, thus corresponds to the number of states with energies between E_j and $E_{j+1}=E_j+\Delta E$. If the *total* number of states with energies less than E is denoted as $\mathrm{N}_{\mathrm{total}}(E, V, N)$,

$$\Omega(E,V,N)=\frac{\partial \mathrm{N}_{\mathrm{total}}(E,V,N)}{\partial E}\Delta E \tag{23.4.12}$$

For the grand canonical partition function, the entropy is obtained from

$$S=-k\sum_i p(E_i,N_i)\ln p(E_i,N_i)=-k\sum_i p(E_i,N_i)\ln\frac{e^{-\beta E_i-\alpha N_i}}{Z}=k\beta\bar{E}+k\alpha\bar{N}+k\ln Z \tag{23.4.13}$$

An identical calculation with $\{\exp(-\beta E_i), z\}$ replacing $\{\exp(-\beta E_i-\alpha N_i), Z\}$ yields $S=k\beta\bar{E}+k\ln z$. Hence, $\ln z(\beta,N)=\ln Z(\beta,\alpha)+\alpha\bar{N}$, which from Equation (23.4.8) constitutes a Legendre transform of $\ln Z$ with respect to α as $\partial \ln z/\partial N = 0$. Similarly, $S = k \ln \Omega$ implies $\ln \Omega(E,N) = \ln Z(\beta,\alpha) + \alpha\bar{N} + \beta\bar{E}$. Since $\Omega(E, N)$ corresponds to the number of states with energy between E and $E + \Delta E$, if E_0 denotes the ground state energy,

$$z(\beta,N) = \sum_{\text{all states } i} e^{-\beta E_i} = \sum_m e^{-\beta(E_0 + m\Delta E)}\Omega(E_0 + m\Delta E,N) \equiv \sum_{E_i} e^{-\beta E_i}\Omega(E_i,N) \tag{23.4.14}$$

That is, the canonical partition function counts the number of states at each E_i weighted by the relative probability of occurrence of a single state of energy E_i. The product of the rapidly decreasing and increasing functions $\exp(-\beta E)$ and $\Omega(E, N)$ possesses a pronounced maximum at the energy characterizing the microcanonical ensemble. Similarly, grouping all system realizations with the same total number of particles N_i into an inner sum recasts Equation (23.4.5) as

$$Z(\beta,\alpha) = \sum_{N_i} e^{-\alpha N_i} z(\beta, N_i) \tag{23.4.15}$$

The partition functions for indistinguishable particles in the classical limit for which $\exp(-\beta\varepsilon_r - \alpha) \ll 1$ follow from Equation (23.4.6). In this limit, from Equation (23.4.4), $\bar{n}_i \approx \exp(-\beta\varepsilon_r - \alpha)$, and therefore, (i) $N = \sum_l n_i = \sum_i \exp(-\beta\varepsilon_i - \alpha) = \exp(-\alpha) z_i$ where z_i is the single particle canonical partition function so that (ii) $\alpha = \ln(z_i/N)$. For, a classical system with total particle number, N, recalling that $\ln(1 + \varepsilon) \approx \varepsilon$, $\ln z = \alpha N + \ln Z \approx \alpha N + \sum_i \ln(1 + \exp(-\beta\varepsilon_i - \alpha)) \approx \alpha N + \sum_i \exp(-\beta\varepsilon_i - \alpha)$, and hence, substituting (i) and (ii), $\ln z = N \ln(z_i/N) + N = N \ln z_i - N \ln N + N$, Which yields $z = z_i^N/N!$, from Simpson's rule. The $1/N!$ factor corrects for overcounting states that differ only by particle exchange. Such a factor does not appear, e.g., for *internal degrees of freedom* such as the vibrational or rotational excitations of atoms or molecules that are localized to a single particle and therefore distinguishable.

23.5 DENSITY OF STATES

If periodic boundary conditions are imposed at the edges of a three-dimensional homogeneous (i.e., uniform) cubic solid with sides of length L, the confined eigenmodes, given by $\exp(\pm i2\pi lx/\mathrm{L}) \exp(\pm i2\pi mx/\mathrm{L}) \exp(\pm i2\pi nx/\mathrm{L}) \exp(-i\omega_{lmn}t)$, are comprised wavevectors spaced by $\Delta k \equiv 2\pi/L$ in each coordinate direction. Hence, a single mode occupies a volume $(\Delta k)^3 = 8\pi^3/L^3$ in three-dimensional wavevector space, while more generally, in a $2N_d$-dimensional phase space in N_d dimensions, $(\Delta p)^{N_d}(\Delta x)^{N_d} = h^{N_d}$. The number of states $\mathrm{N}_{\text{total}}(k)$ with wavevectors less than k equals the volume of a sphere of radius k divided by the phase space volume of a state. For electron wavefunctions, multiplying by two to incorporate the spin polarizations for each spatial state,

$$\mathrm{N}_{\text{total}}(k) = \underbrace{2}_{\text{polarizations}} \frac{L^3}{(2\pi)^3} \frac{4\pi}{3} k^3 \tag{23.5.1}$$

This result with an appropriate polarization factor applies to any system described by a wave equation. Since $k = \sqrt{2mE}/\hbar$,

$$\mathrm{N}_{\text{total}}(E) = \frac{V}{\hbar^3} \frac{1}{3\pi^2} \left(\sqrt{2mE}\right)^{3/2} \tag{23.5.2}$$

in which $V = L^3$ is the sample volume. If zero boundary conditions are instead applied, the eigenmodes, $\sin n\pi x/L \sin m\pi y/L \sin p\pi z/L \exp(-i\omega_{nmp}t)$, are formed from the sum of equal positive and negative traveling wave components in each coordinate direction. Adjacent k values are then spaced by $\Delta k = \pi/L$ along each coordinate direction, yielding $(\Delta \mathrm{W})^3 = \pi^3/L^3$. However, Equation (23.5.1) still applies as, e.g., both $\sin(kx)$ and $\sin(-kx)$ describe the same field after normalization and thus only 1/8 of the sphere of $\vec{k}$ values contributes to $\mathrm{N}_{\text{total}}(E)$.

For a spinless particle in N_d dimensions, or with $N_d = 3N$ for N particles in 3 dimensions, the corresponding phase space possesses $3N$ spatial dimensions and Equation (23.5.2) becomes

$$\mathrm{N}_{\text{total}}(E) = 2\xi \left(\frac{L}{2\pi}\right)^{N_d} \frac{(2mE)^{\frac{N_d}{2}}}{\hbar^{N_d}} \tag{23.5.3}$$

where the constant ξ is given by Equation (8.1.5). This yields the *density of states*, which represents the number of states per unit volume:

$$\mathrm{P}(E) = \frac{1}{V} \frac{\partial \mathrm{N}_{\text{total}}(E)}{\partial E} \propto E^{N_d/2-1} \approx E^{N_d/2} \tag{23.5.4}$$

Finally, in a system of K one-dimensional harmonic oscillators with average energy E, the motion of each oscillator, i, describes an ellipse in phase space with $p^2/2m + kx^2/2 = E$. Hence, $\mathrm{N}_{\text{total}}(E)$ is proportional to the volume of $2K$-dimensional phase space confined within a hyperellipse of average radius E and is therefore to E^K without a V or L^K volume dependence (as momentum and position variables are functionally equivalent).

23.6 TEMPERATURE AND ENERGY

The *temperature* of a system is defined as the change in its energy per unit change in entropy (at fixed volume and number of particles):

$$T = \left.\frac{\partial E}{\partial S}\right|_{V,N} \tag{23.6.1}$$

While the entropy and energy are proportional to the system size and are thus termed *extensive*, the temperature does not depend on V and is labeled *intensive*.

> **Example**
>
> In a system of N spins with levels at 0 and E, if a single spin is excited to E (at an unknown location) when all other spins are in the zero state, $\Delta S \propto k \ln N$ is large and hence $T \approx 0$ for large N, while when all spins are randomly oriented, a spin change does not affect S, implying $T = \infty$.

From the dependence of the entropy on energy and volume, the energy and pressure can be related to temperature. Thus, e.g., the microcanonical entropy for N free particles with $3N$ degrees of freedom, from the discussion of the previous section, where ϖ represents a constant factor,

$$S(E,V) = k \ln \Omega(E,V) \approx k \ln \left(\varpi V^N E^{\frac{3N}{2}}\right) = Nk\left(\ln V + \frac{3}{2}\ln E + \varpi'\right) \tag{23.6.2}$$

implies

$$\frac{1}{T} = \left.\frac{\partial S}{\partial E}\right|_V = \frac{3}{2}\frac{Nk}{E} \tag{23.6.3}$$

or, in terms of the number of moles n with $Nk = nR$ in which the *ideal gas constant* $R = 8.31$ J/(mol · K),

$$E = \frac{3}{2}NkT = \frac{3}{2}nRT \tag{23.6.4}$$

Additionally, if a system performs work by displacing a surface element A a distance Δx against an inward pressure, its decrease in energy is $\Delta E = -F\Delta x = -(F/A)A\Delta x = -p\Delta V$ with

$$p = -\left.\frac{\partial E(S,V)}{\partial V}\right|_S = \frac{\left.\frac{\partial S}{\partial V}\right|_E}{\left.\frac{\partial S}{\partial E}\right|_V} = T\left.\frac{\partial S}{\partial V}\right|_E \tag{23.6.5}$$

according to Equation (7.1.7). Inserting Equation (23.6.2) for S yields the *ideal gas law*

$$p = \frac{NkT}{V} \tag{23.6.6}$$

If further, e.g., magnetic or electric fields with strengths $\chi^{(i)}$ are present, in analogy with $\Delta E = -\vec{F} \cdot \Delta\vec{x}$, *generalized forces* are defined by

$$f^{(i)} = -\left.\frac{\partial E}{\partial \chi^{(i)}}\right|_{S,V} \tag{23.6.7}$$

In the canonical ensemble, the ideal gas law follows by applying Equation (23.4.10) to either $z = z_i^N/N!$ for N indistinguishable particles or from $z = z_i^N$ for distinguishable particles where z_i denotes the canonical partition function for a single particle confined to a cubic region. With zero boundary conditions so that $p_i = \hbar k_i = \hbar \pi n_i/L = hn_i/2L$ and applying the classical approximation in the last step,

$$z_i = \sum_{n_x, n_y, n_z = 0}^{\infty} e^{-\frac{\beta \pi^2 \hbar^2}{8mL^2}\left(n_x^2 + n_y^2 + n_z^2\right)} = \left(\sum_{n_x=0}^{\infty} e^{-\frac{\beta h^2 n^2}{8mL^2}}\right)^3 \approx \left(\int_0^{\infty} e^{-\frac{\beta h^2 n^2}{8mL^2}} dn\right)^3 = \left(\frac{L}{h}\int_{-\infty}^{\infty} e^{-\frac{p^2}{2kTm}} dp_x\right)^3 \tag{23.6.8}$$

Performing the integration according to $\int_{-\infty}^{\infty} \exp(-x^2)dx = \sqrt{\pi}$,

$$z_i = \left(\frac{L}{h}\sqrt{2\pi mkT}\right)^3 = \frac{V}{h^3}(2\pi mkT)^{\frac{3}{2}} \tag{23.6.9}$$

More generally, Equation (23.6.4) follows from the *equipartition theorem*, which if the energy for a degree of freedom J (associated with either a momentum or a position coordinate) possesses the form, in which where the index $i = 1, 2, \ldots J-1, J+1, \ldots$ does *not* include J,

$$E = E'(\alpha_i) + c\alpha_J^2 \tag{23.6.10}$$

the coordinate J contributes $\bar{\varepsilon}_J = kT/2$ to the average system energy, as evident from Equation (23.4.10) with

$$z_J \propto \int_{-\infty}^{\infty} e^{-c\beta\alpha_J^2} d\alpha_J = \sqrt{\frac{\pi}{\beta c}} \tag{23.6.11}$$

Examples

For a single harmonic oscillator with $E = p^2/2m + kx^2/2$, $\bar{\varepsilon} = \bar{\varepsilon}_p + \bar{\varepsilon}_x = kT$. For a diatomic molecule with three translational and two degrees of rotational freedom, the corresponding average energy $\bar{\varepsilon} = 5kT/2$.

From the equipartition theorem, the thermal energy of a solid with N atoms and hence $3N$ degrees of freedom expressed in normal coordinates equals $E = \sum_{i=1}^{3N}\left(p_i^2/2m + kx_i^2/2\right)$, so that $\bar{E} = 3NkT$, termed the *law of Dulong and Petit.* At low temperatures however many oscillators occupy their ground states invalidating the Boltzmann distribution. The energy can then instead be obtained from the single harmonic oscillator canonical partition function:

$$z_i = \sum_{m=0}^{\infty} e^{-\beta\left(m+\frac{1}{2}\right)\hbar\omega} = e^{-\beta\frac{\hbar\omega}{2}} \sum_{m=0}^{\infty} e^{-\beta m\hbar\omega} = \frac{e^{-\beta\frac{\hbar\omega}{2}}}{1 - e^{-\beta\hbar\omega}} \tag{23.6.12}$$

where Equation (4.1.24) for the sum of a geometric series has been applied. Accordingly,

$$\bar{E} = -3N\frac{\partial \ln z_i}{\partial \beta} = 3N\left(\frac{1}{2}\hbar\omega + \frac{\hbar\omega e^{-\beta\hbar\omega}}{1-e^{-\beta\hbar\omega}}\right) = 3N\hbar\omega\left(\frac{1}{2} + \frac{1}{e^{\beta\hbar\omega}-1}\right) \tag{23.6.13}$$

The quasiclassical result $\bar{E} \approx 3N(\hbar\omega/2 + kT)$ is recovered when $kT \gg \hbar\omega$ or, in terms of the *Einstein temperature* $T_E = \hbar\omega/k$, $T \gg T_E$, for which the denominator of the second term approaches $\beta\hbar\omega$.

23.7 PHONONS AND PHOTONS

Small-frequency, long-wavelength lattice vibrations termed *acoustic phonons* result from both *shear* and *compression waves*, which are respectively associated with transverse and longitudinal displacements of successive planes of atoms. Since a crystal with N atoms thus possesses $3N$ acoustic modes, each occupying a k-space volume $(\Delta k)^3 = (2\pi)^3/V$, the maximum crystal wavevector is found by setting the total number of acoustic modes with $k < k_{\max}$ equal to $3N$:

$$3N = \mathrm{N}_{\text{total}}(k_{\max}) = \underbrace{3}_{\text{polarizations}} \frac{V}{(2\pi)^3}\frac{4}{3}\pi k_{\max}^3 \tag{23.7.1}$$

As $\omega/k = v = c_{\text{sound}}$, with c_{sound} the sound velocity, the corresponding maximum or cutoff frequency is

$$\omega_{\max} = 2\pi c_{\text{sound}}\left(\frac{3N_{\text{total}}}{4\pi V}\right)^{\frac{1}{3}} \tag{23.7.2}$$

The energy spectrum, i.e., the energy present between ω and $\omega + d\omega$, then follows from Equation (23.6.13), where $\theta(x)$ denotes the Heaviside step function, which equals unity for positive x and zero for $x < 0$,

$$E(\omega,T)d\omega = \underbrace{3}_{\substack{\text{number of}\\ \text{polarizations}}} \cdot \underbrace{\theta(\omega_{\max}-\omega)}_{\substack{\text{occupation probability}\\ \text{of a state with frequency } \omega}} \cdot \underbrace{\hbar\omega\left(\frac{1}{2} + \frac{1}{e^{\frac{\hbar\omega}{kT}}-1}\right)}_{\substack{\text{energy of excitation}\\ \text{with frequency } \omega}} \cdot \underbrace{\frac{4\pi}{(2\pi c_{\text{sound}})^3}\omega^2}_{\substack{\text{density of states } \mathrm{P}(\omega)/3\\ \text{for one polarization}}} V d\omega \tag{23.7.3}$$

The total acoustic phonon energy is obtained by integrating Equation (23.7.3) up to the cutoff frequency:

$$\bar{E}(T) = \frac{3V\hbar}{2\pi^2 c_{\text{sound}}^3}\int_0^{\omega_{\max}}\left\{\frac{1}{2} + \frac{1}{e^{\frac{\hbar\omega}{kT}}-1}\right\}\omega^3 d\omega \tag{23.7.4}$$

The second term in the integral is approximately constant until $\hbar\omega \approx kT$ and then falls rapidly with ω. Its contribution to $\bar{E}(T)$ can therefore be approximated by integrating ω^3 from 0 to $kT/\hbar$ and is therefore proportional to T^4 from which $C_v \equiv \partial E/\partial T|_V \propto T^3$. At room temperature $\hbar\omega \ll kT$ so that $\exp(\hbar\omega/kT) \approx 1 + \hbar\omega/kT$, yielding the law of Dulong and Petit in the form $E = 3NkT + \text{constant}$ after integration from Equation (23.7.2).

Thermal radiation from electromagnetic modes in a cavity is termed *black body radiation*. The spectrum follows from Equation (23.7.3), by substituting the Bose–Einstein state occupation probability $n(\omega) = 1/(\exp(\hbar\omega/kT) - 1)$ for $\theta(\omega_{\max} - \omega)$, replacing the excitation energy by $\hbar\omega$ and the three acoustic polarizations by two electromagnetic polarizations. Accordingly,

$$\varepsilon(\omega, T)d\omega = 2 \cdot \frac{1}{e^{\frac{\hbar\omega}{kT}} - 1} \cdot \hbar\omega \cdot \underbrace{\frac{1}{2\pi^2 c^3}\omega^2}_{\mathrm{P}(\omega)/2} d\omega \tag{23.7.5}$$

where $\varepsilon(\omega, T) = E(\omega, T)/V$ represents the *energy density* of the electric field in the cavity. For $kT \gg \hbar\omega$, Equation (23.7.5) approaches $kT\mathrm{P}(\omega)d\omega$ yielding an energy of $kT/2$ for each degree of freedom in a unit volume, termed the *Rayleigh–Jeans law*. The energy density then initially increases with ω as ω^3, attaining a maximum at $\approx 3kT/\hbar$ (termed *Wien's displacement law*). The total radiated energy varies as T^4, similar to phonons, and is expressed as $4\sigma T^4/c$, with the *Stefan–Boltzmann constant* $\sigma = 5.67 \times 10^{-8}\,\mathrm{W/m^2 \cdot K^4}$.

23.8 THE LAWS OF THERMODYNAMICS

Thermodynamics can be derived from four basic principles that apply to thermally isolated equilibrium systems.

1. Two systems in thermal equilibrium with a third system are in mutual thermal equilibrium
2. Every system possesses an *internal energy*, E, which is time independent for a thermally isolated, constant volume system in equilibrium. The internal energy increases as heat, Q, is added to the system and decreases when work, W, is performed *by* the system, according to

$$E_{\text{final}} - E_{\text{initial}} = Q_{\text{absorbed by system}} - W_{\text{performed by system}} \tag{23.8.1}$$

   This effectively corresponds to a restatement of energy conservation. Heat and energy are often measured in *calories* (or kilocalories) with 1 cal equal to 4.184 J, the approximate amount of energy required to raise the temperature of 1 g (kg) of water by 1°.
3. A thermodynamic system further possesses an *entropy*, S, with the property that in any process the entropy change of an isolated system, $\Delta S \geq 0$. Further,

for infinitesimal *quasistatic* processes that remain near an equilibrium state at all times,

$$S_{\text{final}} - S_{\text{initial}} = \frac{\Delta Q_{\text{absorbed by system}}}{T} \tag{23.8.2}$$

The second law possesses numerous equivalent reformulations, such as the impossibility of extracting and transforming heat into work from a source with a spatially uniform temperature.

4. As $T \to 0$, S approaches a constant value that is determined by the microscopic nature of the system (e.g., for 5 electrons distributed among the ground states of 10 atoms, the entropy $S = -k\sum_{i=1}^{10} p_i \ln p_i$ approaches $5k \ln 2$ since for each atom $p_i = 1/2$).

To relate the entropy and energy of a system, consider its average energy:

$$E\left(\vec{\xi}\right) = \sum_m p_m E_m\left(\vec{\xi}\right) \tag{23.8.3}$$

where E_m and p_m are the energies and occupation probabilities of each quantum state and $\vec{\xi}$ represents the collection of macroscopic parameters such as volume and electric field upon which the system energy depends. Denoting by $f^{(l)}$ the *generalized force* associated with the variable $\xi^{(l)}$ and further defining ΔQ and ΔW as the heat flowing into and the work performed by the system, respectively, a transformation to an infinitesimally different system configuration changes the energy of the system by

$$\Delta E = \underbrace{\sum_m \Delta p_m E_m}_{\Delta Q} + \underbrace{\sum_m p_m \Delta E_m}_{-\Delta W} = \Delta Q + \sum_l \sum_m p_m \frac{\partial E_m}{\partial \xi^{(l)}} \Delta \xi^{(l)} = \Delta Q - \underbrace{\sum_l f^{(l)} \Delta \xi^{(l)}}_{\Delta W} \tag{23.8.4}$$

Here, $\Delta W = -\sum_m p_m \Delta E_m$ results from changes in the individual energy levels, ΔE_m, induced by changes in system variables such as displacements of a container wall in contrast to $\Delta Q = \sum_m \Delta p_m E_m$, which arises from changes to the occupation probabilities for fixed energy levels. While ΔE according to Equation (23.8.3) depends only on the initial and final states, ΔQ and ΔW are affected by the nature of the transition between these two states. For example, a gas that is first cooled, decreasing its pressure, performs less work upon expanding to a larger volume than if it is first expanded and then cooled to reach the same final state. In a slow, *quasistatic* process, however, in which the system remains infinitesimally close to an equilibrium state, the work, W, is uniquely defined by the initial and final states according to $W = \sum_l \int_{\text{initial}}^{\text{final}} f^{(l)} d\xi^{(l)}$ so that the heat transferred equals $\Delta E + W$. Further, in an *adiabatic* process, the system is thermally isolated and consequently, $\Delta Q = 0$ in Equation (23.8.4). (This does not require $\Delta p_m = 0$, since if a system is composed of nonidentical subsystems whose energy levels become coincident at some point in time during the process, their occupation probabilities can equilibrate, altering Δp_m.) As the difference in energy of the initial and final states equals $E_{\text{initial}} - E_{\text{final}} = \Delta W$ in an adiabatic process, such processes are *reversible* as undoing the displacement returns the system to its original energy state. For a nonadiabatic process, the amount of heat supplied equals the sum of the work performed by the system $W = \sum_l \int_{\text{initial}}^{\text{final}} f^{(l)} d\xi^{(l)}$ and the energy difference between the initial and final states measured in an adiabatic process. Lastly, from the third law of thermodynamics Dividing ΔQ by the temperature yields an entropy change $\Delta S = \Delta Q/T = (1/T) \sum_{mEm} \Delta p_m$, which depends uniquely on the initial and final states.

Example

To construct an *absolute temperature* scale, a system is placed in contact with a heat reservoir at a temperature T_i and a predetermined amount of heat ΔQ_i is added, resulting in $\Delta S_1 = \Delta Q_i / T_i$. If the system then evolves into a final state (by changing values of variables such as pressure) with a temperature, T_f, through an adiabatic $\Delta S = 0$ process, the overall change in energy is identical to that generated by first performing an adiabatic process to raise the temperature to T_f and then adding $\Delta Q_f = T_f \Delta S_2$ until the same final state is reached. From $\Delta S_1 = \Delta S_2$, the initial and final temperatures are related by $T_f / T_i = \Delta Q_f / \Delta Q_i$.

23.9 THE LEGENDRE TRANSFORMATION AND THERMODYNAMIC QUANTITIES

The laws of thermodynamics together with the definitions of ΔW and ΔS indicate that in a *thermally isolated* system with fixed volume, the system energy is a unique function of the entropy and volume (assuming the absence of other generalized forces than the pressure). However, for a system held at constant pressure and/or temperature, the volume and/or entropy fluctuates with time, motivating the introduction of redefined "energies" that depend on the constant variables. Since $T = \partial E/\partial S|_V$ while $p = -\partial E/\partial V|_S$, to replace, e.g., S by T as an independent variable, the *Helmholtz free energy*, F, is defined as a *Legendre transformation* corresponding to the y-intercept of the tangent to the curve for the energy as a function of S:

$$F(T,V) = E(S,V) - \frac{\partial E(S,V)}{\partial S} S = E(S,V) - TS \tag{23.9.1}$$

Since if E varies linearly with S, the y-intercept F is independent of S, changes in F only arise from variations in the slope, T. Thus, F is determined by T rather than S. Indeed, from Equation (23.9.1),

$$dF = dE - SdT - TdS = -pdV - SdT \tag{23.9.2}$$

confirming that F depends only on T and V. The resulting partial derivative pair

$$\left.\frac{\partial F}{\partial V}\right|_T = -p$$
$$\left.\frac{\partial F}{\partial T}\right|_V = -S \tag{23.9.3}$$

are termed *equations of state*. *Maxwell's relations* follow from the mixed partial derivative relation $\partial^2/\partial x\partial y = \partial^2/\partial y\partial x$, which gives, e.g.,

$$-\left.\frac{\partial S}{\partial V}\right|_T = \frac{\partial^2 F}{\partial V\partial T} = \frac{\partial^2 F}{\partial T\partial V} = -\left.\frac{\partial p}{\partial T}\right|_V \tag{23.9.4}$$

Similarly, p can replace V as a variable in systems held at fixed pressure by constructing the *enthalpy*

$$H(S,p) = E + pV \tag{23.9.5}$$

If both T and p are held constant, the *Gibbs free energy* is given by

$$G(T,p) = H - TS = E - TS + pV \tag{23.9.6}$$

The quantities $\{E, F, G, H\}$ and $\{p, V, S, T\}$ are termed *thermodynamic potentials* and *variables*, respectively. The potentials are independent of the process employed to reach a given equilibrium state of a system. While p and T are *intensive*, i.e., independent of the system dimensions, all other variables and potentials are *extensive* quantities that scale with system size. Intensive quantities are typically denoted by small letters except where confusion results (such as between T and time, t). The thermodynamic relationships are summarized in the *thermodynamic square* of Figure 23.1, which can be recalled through the mnemonic "to verify fundamental thermodynamics, get powerful help from square's extremities" where "to" and "from" indicate the direction of the diagonal arrows.

Each thermodynamic potential, situated at the middle of a side, is a function of the variables at its two adjacent corners. Arrows pointing toward and away from a corner are associated with a negative and positive sign, respectively. Thus, $dF = -pdV - SdT$ (as well as $F = E - TS$) follows from the arrow pointing from S to T, implying $\partial F/\partial T|_V = -S$ and $p = -\partial F/\partial V$. The Maxwell relation

$$\partial T/\partial V|_S = -\partial p/\partial S|_V \tag{23.9.7}$$

is obtained by starting with T and cycling counterclockwise around two corners to generate $\partial T/\partial V|_S$. Similarly, cycling clockwise two corners from p yields $\partial p/\partial S|_V$.

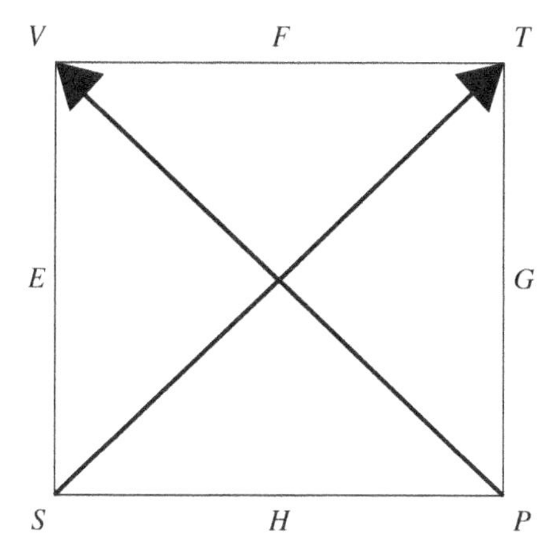

FIGURE 23.1 Thermodynamic square.

Since the arrows toward the starting points T and p are oppositely directed, these two expressions possess opposite signs in Equation (23.9.7).

Derivatives of thermodynamic quantities are often expressed in terms of those of other quantities, in particular the Gibbs free energy $G(T, p)$ as T and p are often most readily controlled and thus amenable to measurements. Here, C_p with

$$\frac{\partial^2 G}{\partial T^2} = -\left.\frac{\partial S}{\partial T}\right|_p = -\frac{C_p}{T} \tag{23.9.8}$$

and $\kappa_{\text{isothermal}}$ and $\kappa_{\text{adiabatic}}$ defined by

$$\frac{\partial^2 G}{\partial p^2} = \left.\frac{\partial V}{\partial p}\right|_T = -V\kappa_{\text{isothermal}}, \quad -\left.\frac{1}{V}\frac{\partial V}{\partial p}\right|_S = \kappa_{\text{adiabatic}} \tag{23.9.9}$$

(where again $\partial G/\partial p|_T = V$ in $\kappa_{\text{adiabatic}}$) are termed the (extensive) *specific heat at constant pressure* and the *isothermal compressibility* and *adiabatic compressibility*, respectively. Both compressibilities and the further *volume coefficient of expansion*, α,

$$\frac{\partial^2 G}{\partial T \partial p} = \left.\frac{\partial V}{\partial T}\right|_p = V\alpha \tag{23.9.10}$$

are intensive as doubling the system volume doubles the volume variation induced by a pressure change. The *specific heat at constant volume*, C_V, is then defined analogously to Equation (23.9.8), namely (note that $\partial G/\partial T|_p = -S$),

$$\left.\frac{\partial S}{\partial T}\right|_V = \frac{C_V}{T} \tag{23.9.11}$$

which can be eliminated in favor of the other quantities since, from Equation (7.1.11),

$$C_V = \left.\frac{\partial S}{\partial T}\right|_p + \left.\frac{\partial S}{\partial p}\right|_T \left.\frac{\partial p}{\partial T}\right|_V = \left.\frac{\partial S}{\partial T}\right|_p + \left.\frac{\partial V}{\partial T}\right|_p \left.\frac{\partial p}{\partial T}\right|_V = \left.\frac{\partial S}{\partial T}\right|_p + \left.\frac{\partial V}{\partial T}\right|_p \frac{\left.\frac{\partial V}{\partial T}\right|_p}{\left.\frac{\partial V}{\partial p}\right|_T} = C_p - \frac{\alpha^2 VT}{\kappa_{\text{isothermal}}} \tag{23.9.12}$$

For an ideal gas, $pV = nRT$ implies both $dV/V|_p = dT/T|_p$ and therefore $\alpha = 1/T$, while $dV/V|_T = -dp/p|_T$ similarly yields $\kappa_{\text{isothermal}} = 1/p$. Hence, $C_p - C_V = R$, while $E = 3RT/2$ implies $C_V = 3R/2$.

To express derivatives of thermodynamic variables in terms of typically convenient quantities, partial derivatives at fixed values of the thermodynamic potentials are written as quotients of partial derivatives of the potentials that can be evaluated with the thermodynamic square. For example,

$$\left.\frac{\partial T}{\partial V}\right|_E = -\frac{\left.\frac{\partial E}{\partial V}\right|_T}{\left.\frac{\partial E}{\partial T}\right|_V} = -\frac{T\left.\frac{\partial S}{\partial V}\right|_T - p}{T\left.\frac{\partial S}{\partial T}\right|_V} \tag{23.9.13}$$

Derivatives of the entropy at fixed p or V are then expressed as temperature derivatives through the Maxwell relations by manipulations such as (cf. Eq. 7.1.9)

$$\left.\frac{\partial S}{\partial V}\right|_p = \frac{\left.\frac{\partial S}{\partial T}\right|_p}{\left.\frac{\partial V}{\partial T}\right|_p} \tag{23.9.14}$$

The temperature derivatives of the entropy yield C_p and C_V, while the remaining partial derivatives are transformed into volume derivatives, according to, e.g.,

$$\left.\frac{\partial p}{\partial T}\right|_V = -\frac{\left.\frac{\partial V}{\partial T}\right|_p}{\left.\frac{\partial V}{\partial p}\right|_T} \tag{23.9.15}$$

After replacing the volume derivatives by κ and α, Equation (23.9.12) is finally employed to eliminate C_V.

23.10 EXPANSION OF GASES

From $pV = nRT$, if an ideal gas expands *isothermally* (at constant T),

$$p_2 - p_1 = nRT\int_{V_1}^{V_2} \frac{1}{V} dV = nRT \log\left(\frac{V_2}{V_1}\right) \tag{23.10.1}$$

Under *adiabatic* expansion, with zero heat transfer to the surroundings,

$$\left.\frac{\partial V}{\partial p}\right|_S = -\frac{\left.\frac{\partial S}{\partial p}\right|_V}{\left.\frac{\partial S}{\partial V}\right|_p} = -\frac{\left.\frac{\partial S}{\partial T}\right|_V \left.\frac{\partial V}{\partial T}\right|_p}{\left.\frac{\partial p}{\partial T}\right|_V \left.\frac{\partial S}{\partial T}\right|_p} = -\frac{\frac{C_V}{T}}{-\frac{\left.\frac{\partial V}{\partial T}\right|_p}{\left.\frac{\partial V}{\partial p}\right|_T}} \frac{\left.\frac{\partial V}{\partial T}\right|_p}{\frac{C_p}{T}} = \frac{C_V}{C_p}\left.\frac{\partial V}{\partial p}\right|_T = -\frac{C_V}{C_p}\frac{nRT}{p^2} = -\frac{C_V}{C_p}\frac{V}{p} \tag{23.10.2}$$

Accordingly,

$$\int_{V_1}^{V_2} \frac{1}{V} dV = \log\left(\frac{V_2}{V_1}\right) = -\frac{C_V}{C_p}\int_{p_1}^{p_2} \frac{1}{p} dp = -\frac{C_V}{C_p}\log\left(\frac{p_2}{p_1}\right) = -\log\left(\frac{p_2}{p_1}\right)^{\frac{C_V}{C_p}} \tag{23.10.3}$$

so that

$$p^{\frac{C_V}{C_p}} V = \text{constant} \tag{23.10.4}$$

This result can also be obtained by noting that $dS = 0$ implies $dE = -pdV$ with $dE = 3RdT/2 = C_V dT$, while $pV = nRT$ simultaneously implies $pdV + Vdp = nRdT$. Eliminating dT,

$$pdV + Vdp = -\frac{RpdV}{C_V} \tag{23.10.5}$$

or

$$pdV\underbrace{(C_V + R)}_{C_p} + VC_V dp = 0 \tag{23.10.6}$$

Quasistatic, reversible isothermal expansion in an insulated container is typified by action against a slowly moving piston. Employing the Maxwell relation for $\partial S/\partial V|_T$,

$$\left.\frac{\partial E}{\partial V}\right|_T = T\left.\frac{\partial S}{\partial V}\right|_T - p = T\left.\frac{\partial p}{\partial T}\right|_V - p = -T\frac{\left.\frac{\partial V}{\partial T}\right|_p}{\left.\frac{\partial V}{\partial p}\right|_T} - p = \frac{T\alpha}{\kappa_{\text{isothermal}}} - p \tag{23.10.7}$$

For an ideal gas, $T\alpha = p\kappa_{\text{isothermal}} = 1$ (cf. the derivation in Eq. 23.9.12), implying $\partial E/\partial V|_T = 0$ as required since the energy only depends on temperature from $E = 3kT/2$. However, in an imperfect, *van der Waals gas*, the finite size of the molecules limits the accessible volume, while the pressure is reduced by the attractive forces between molecules. This results in an approximate equation of state for one mole of a gas:

$$p = \frac{RT}{V-b} - \frac{a}{V^2} \tag{23.10.8}$$

Inserting into the second equation of Equation (23.10.7), namely, $T\partial p/\partial T|_V - p$, yields $\partial E/\partial V|_T = a/V^2$. Hence, as the confining volume is increased, the effective molecular attraction is reduced, which augments the energy.

Irreversible processes leading to an entropy gain in a thermally isolated system are typified by *free expansion* in which opening a valve enables a gas to expand suddenly from a volume V_i to V_f. Since the gas is isolated and does not perform work against a force, $\Delta Q = \Delta W = 0$. From $\Delta E = \Delta Q - \Delta W$, energy is conserved and

$$\left.\frac{\partial T}{\partial V}\right|_E = -\frac{\left.\frac{\partial E}{\partial V}\right|_T}{\left.\frac{\partial E}{\partial T}\right|_V} = -\frac{1}{C_V}\left(\frac{T\alpha - p\kappa_{\text{isothermal}}}{\kappa_{\text{isothermal}}}\right) \tag{23.10.9}$$

which equals zero for an ideal gas and $-a/C_V V^2$ for a van der Waals gas.

Finally, in Joule–Thomson expansion, a gas passes through a porous material inside a thermally isolated cylinder such that the pressure differs on the two sides of the cylinder while $\Delta Q = 0$. If a volume V_1 of gas enters the cylinder at pressure p_1 and exits as V_2 at p_2, the energy change equals the net work performed on the gas; i.e., after integrating $dE = -pdV$ from V_1 to V_2,

$$E_2 - E_1 = p_1 V_1 - p_2 V_2 \tag{23.10.10}$$

Hence, the enthalpy $H(S, p) = E + pV$ remains constant, consistent with the system being determined by the time-independent entropy and pressure. Accordingly,

$$\left.\frac{\partial T}{\partial p}\right|_H = -\frac{\left.\frac{\partial H}{\partial p}\right|_T}{\left.\frac{\partial H}{\partial T}\right|_p} = -\frac{T\left.\frac{\partial S}{\partial p}\right|_T + V}{T\left.\frac{\partial S}{\partial T}\right|_p} = \frac{T\left.\frac{\partial V}{\partial T}\right|_P - V}{C_p} = \frac{V(T\alpha - 1)}{C_p} \tag{23.10.11}$$

For an ideal gas, $T\alpha = 1$ and Equation (23.10.11) vanishes. Although $\Delta Q = 0$, the process is not quasistatic as the entropy increases according to

$$\left.\frac{\partial S}{\partial p}\right|_H = -\frac{\left.\frac{\partial H}{\partial p}\right|_S}{\left.\frac{\partial H}{\partial S}\right|_p} = -\frac{V}{T} \tag{23.10.12}$$

23.11 HEAT ENGINES AND THE CARNOT CYCLE

An engine extracts energy from heat to perform work while its inverse mode of operation produces a refrigerator or heat pump that employs work to extract heat energy. The most efficient engine utilizes quasistatic processes and is reversible and cyclic, in that the entropy and energy revert to their original values after a stroke. If the engine subsystem extracts heat Q_h from a heat bath at T_h and subsequently evolves to T_c through a thermally isolated, adiabatic process with zero entropy change in which work is performed and then discards a smaller amount of heat Q_c to a bath at T_c before evolving again adiabatically to its initial state, zero net entropy for the entire system after change a cycle implies

$$\Delta S_{\text{total}} = \Delta S_c + \Delta S_h = \frac{Q_c}{T_c} - \frac{Q_h}{T_h} = 0 \tag{23.11.1}$$

or

$$Q_c = Q_h \frac{T_c}{T_h} \tag{23.11.2}$$

Since the energy also remains invariant after a cycle, the engine performs an amount of work

$$W = Q_h - Q_c = Q_h\left(1 - \frac{T_c}{T_h}\right) \equiv Q_h \eta \tag{23.11.3}$$

where η is termed the *efficiency*. In a refrigerator, the work source instead transfers heat from the colder to the warmer source (this is also the principle of operation of a *heat pump*, which however instead exploits the heat created by the process). Work is thus performed on the refrigerator rather than by the engine, reversing the direction of heat flow. Since all variables earlier therefore change sign, Equation (23.11.3) remains invariant although for a refrigerator the expression is generally rewritten as

$$W = Q_h - Q_c = Q_h\left(1 - \frac{T_c}{T_h}\right) = Q_c \frac{T_h}{T_c}\left(1 - \frac{T_c}{T_h}\right) = Q_c\left(\frac{T_h}{T_c} - 1\right) \tag{23.11.4}$$

A cyclic *Carnot engine* can be formed from a gas container with a movable piston that is alternately placed in contact with a high and a low temperature source. The gas is initially at a high temperature T_h and occupies a large volume. In the first step (cf. Figure 23.2), the gas contacts a low-temperature heat source at T_c and is compressed isothermally and quasistatically, transferring $\Delta Q = \int T_c dS = T_c \Delta S$ to the cold source while work $\Delta W = \int p dV = \int (nRT_c/V) dV$ is performed against it. Next, the gas is thermally isolated and compressed adiabatically ($\Delta S = 0$) with $\Delta Q = \int T dS = 0$ while the piston does additional work on the gas that can be computed from the adiabatic law (23.10.4). The system is then placed in contact with the reservoir at T_h and expands isothermally. Finally, the container is again isolated and the gas expands adiabatically until it returns to its initial state.

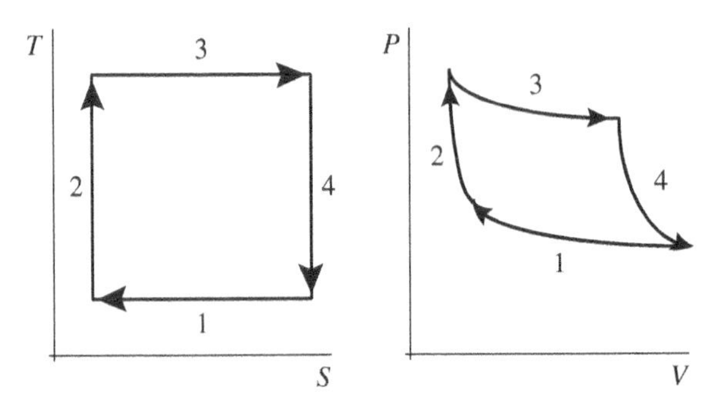

FIGURE 23.2 Carnot cycle.

23.12 THERMODYNAMIC FLUCTUATIONS

The properties of fluctuations in a statistical mechanical system depend on which of the thermodynamic variables E, S, T or p, are held constant at its boundaries. For two distinguishable ideal gas systems 1 and 2 in thermal contact, i.e. that can exchange energy but otherwise isolated, the microcanonical partition function of the overall system is formed from the product of the individual partition functions. If the total energy of the two systems and the energy of system 1 subject to a fluctuation of magnitude ΔE from its equilibrium energy E_1 equal E and $E_1 + \Delta E$, respectively,

$$\Omega_{\text{total}}(E, E_1 + \Delta E) = \Omega_1(E_1 + \Delta E)\Omega_2(E - E_1 - \Delta E)$$

$$= e^{\ln(\Omega_{\text{total}}(E, E_1)) + \left(\frac{\partial \ln(\Omega_1(E_1))}{\partial E_1} + \frac{\partial \ln(\Omega_2(E - E_1))}{\partial E_1}\right)\Delta E + \frac{1}{2}\left(\frac{\partial^2 \ln(\Omega_1(E_1))}{\partial E_1^2} + \frac{\partial^2 \ln(\Omega_2(E - E_1))}{\partial E_1^2}\right)(\Delta E)^2 + \ldots} \tag{23.12.1}$$

The maximum of $\ln \Omega_{\text{total}}(E_1, E_{\text{total}})$ with respect to E_1 yields the system equilibrium point at which

$$\frac{\partial \ln \Omega_1(E_1)}{\partial E_1} + \frac{\partial \ln \Omega_2(E - E_1)}{\partial E_1} = \frac{\partial \ln \Omega_1(E_1)}{\partial E_1} - \frac{\partial \ln \Omega_2(E - E_1)}{\partial (E - E_1)} = 0 \tag{23.12.2}$$

Since $S = -k \ln \Omega$ and $T = \partial E / \partial S$, the two systems possess equal temperatures. Further, if the two systems are composed of ideal gases, writing the exponential function as $\exp\left(-(\Delta E)^2 / 2\overline{\Delta E^2}\right)$ and recalling that in the microcanonical ensemble, $\Omega(E, V) = V^N E^{3N/2}$ from Equation (23.5.3) with $E = 3NkT/2$,

$$\frac{1}{\overline{\Delta E^2}} \approx -\frac{3}{2}\left[N_1 \frac{\partial}{\partial E_1}\left(\frac{1}{E_1}\right) + N_2 \frac{\partial}{\partial E_2}\left(\frac{1}{E_2}\right)\right] = \frac{2}{3}\frac{1}{(kT)^2}\frac{N_1 + N_2}{N_1 N_2} \tag{23.12.3}$$

If $N_1 \ll N_2$, $\overline{\Delta E^2} = 3(kT)^2/2$, which also follows directly from the Boltzmann distribution since the density of states is proportional to $\sqrt{E}$ while $\bar{E} = 3kT/2$,

$$\overline{E^2} - \bar{E}^2 = \frac{\int_0^\infty E^{\frac{5}{2}} e^{-\frac{E}{kT}} dE}{\int_0^\infty E^{\frac{1}{2}} e^{-\frac{E}{kT}} dE} - \bar{E}^2 = \frac{(kT)^{\frac{7}{2}} \Gamma\left(\frac{7}{2}\right)}{(kT)^{\frac{3}{2}} \Gamma\left(\frac{3}{2}\right)} - \left(\frac{3}{2} kT\right)^2 = (kT)^2 \left(\frac{5}{2} \cdot \frac{3}{2} - \left(\frac{3}{2}\right)^2\right) \tag{23.12.4}$$

For a system in contact with a heat reservoir at temperature and pressure $T_{\text{res}} = T_{\text{sys}}$ and $p_{\text{res}} = p_{\text{sys}}$, as the equilibrium state of a system maximizes the sum of the entropy of the system with its surroundings, for fluctuations around the equilibrium state, $\Delta S_{\text{tot}} = S_{\text{sys}} + S_{\text{res}} < 0$. A fluctuation that augments the energy and volume of the system by ΔE and ΔV oppositely impacts the reservoir, for which $\Delta E_{\text{res}} = -\Delta E$ and $\Delta V_{\text{res}} = -\Delta V$, changing its entropy by

$$\Delta S_{\text{res}} = \frac{\Delta Q_{\text{res}}}{T_{\text{res}}} = -\frac{\Delta E + p_{\text{res}} \Delta V}{T_{\text{res}}} \tag{23.12.5}$$

The total entropy variation is therefore given by

$$\Delta S_{\text{sys}} + \Delta S_{\text{res}} = \Delta S_{\text{sys}} - \frac{\Delta E + p_{\text{res}} \Delta V}{T_{\text{res}}} = -\frac{\Delta E - T_{\text{res}} \Delta S_{\text{sys}} + p_{\text{res}} \Delta V}{T_{\text{res}}} = -\frac{\Delta G_{\text{sys}}}{T_{\text{sys}}} < 0 \tag{23.12.6}$$

Accordingly, at equilibrium G is a minimum in a system at fixed T and p. If the system volume is fixed, the ΔV terms above are absent and F is instead minimized; if such a system is thermally isolated, E attains a minimum, while for a thermally isolated system at constant pressure, H is minimized. These considerations follow naturally by associating the fluctuating variables with the arguments of the relevant thermodynamic potential.

Suppressing the subscript on p_{res} and t_{res}, the fluctuation probability is given by

$$e^{\frac{\Delta S_{\text{tot}}}{k}} = e^{-\frac{\Delta G}{kT}} \approx e^{-\frac{1}{kT}\left\{E_{\min} + \cancel{\Delta S \frac{\partial E}{\partial S}}\Big|_V + \cancel{\Delta V \frac{\partial E}{\partial V}}\Big|_S + \frac{(\Delta S)^2}{2} \frac{\partial^2 E}{\partial V^2}\Big|_V + 2\Delta S \Delta V \frac{\partial^2 E}{\partial S \partial V} + \frac{(\Delta V)^2}{2} \frac{\partial^2 E}{\partial V^2}\Big|_T - \cancel{T \Delta S} + \cancel{p \Delta V}\right\}} \tag{23.12.7}$$

This exponent can be rewritten as, by employing $\Delta f(S, V) = \partial f / \partial S|_V \Delta V + \partial f / \partial V|_S \Delta S$ with f representing separately $\partial E / \partial S$ and $\partial E / \partial V$,

$$\frac{1}{2}\left(\Delta\left(\frac{\partial E}{\partial S}\bigg|_V\right) \Delta S + \Delta\left(\frac{\partial E}{\partial V}\bigg|_S\right) \Delta V\right) = \frac{\Delta T \Delta S - \Delta p \Delta V}{2} \tag{23.12.8}$$

Since, however, any pair of the variables $\{T, S, p, V\}$ are uniquely specified by the values assigned to the remaining two, the variation in the system must be specified as a function of only two of these quantities as independent variables; the remaining two are then termed *derived variables*. Setting

$$\Delta S = \left.\frac{\partial S}{\partial T}\right|_V \Delta T + \left.\frac{\partial S}{\partial V}\right|_T \Delta V = \frac{C_V}{T}\Delta T + \left.\frac{\partial p}{\partial T}\right|_V \Delta V$$
$$\Delta p = \left.\frac{\partial p}{\partial T}\right|_V \Delta T + \left.\frac{\partial p}{\partial V}\right|_T \Delta V = \left.\frac{\partial p}{\partial T}\right|_V \Delta T - \frac{1}{\kappa_{\text{isothermal}} V}\Delta V \tag{23.12.9}$$

yields the probability of a fluctuation in the form

$$p \propto e^{-\frac{1}{2}\left\{\frac{(\Delta T)^2}{\sigma_T^2} + \frac{(\Delta V)^2}{\sigma_V^2}\right\}} \tag{23.12.10}$$

with $\sigma_T = T_{\text{res}}\sqrt{k/C_V}$ and $\sigma_V = \sqrt{kT_{\text{res}} V \kappa_{\text{isothermal}}}$. Therefore, if both $c_v > 0$ and $\kappa_{\text{isothermal}} > 0$, the system is stable. That is, if increasing the pressure on a system at constant T decreases its volume, compressing the gas from equilibrium generates a restoring force. Similarly, if raising the temperature at constant V increases S, reversing the direction of entropy flow by transferring entropy to the reservoir lowers the system temperature.

23.13 PHASE TRANSFORMATIONS

Generally, materials can exist in several phases, such as solid, liquid, gas, or different crystalline structures. At fixed temperature and pressure, the global equilibrium state corresponds to the phase that minimizes the Gibbs free energy. However, if at some value of T and p the Gibbs free energies of two or more phases coincide, any ratio of these phases can be present. However, above a certain *critical temperature*, T_{critical}, only the gaseous phase exists. The value of T_{critical} together with the pressure at which both gaseous and liquid phases coexist just below T_{critical} defines the *critical point*. Additionally, if a temperature and pressure exist at which $G(T, p)$ is identical for solid, liquid, and gas phases, the corresponding coordinates T, p are termed the *triple point*.

A *phase diagram* graphs the pressure at which two or more phases coexist as a function of temperature. Since the coincident phases differ in volume and entropy per mole, $\partial G/\partial p|_T = V$ and $\partial G/\partial T|_p = -S$ are discontinuous across these lines. The energy difference of the phases per mole is termed the *latent heat* and is designated. Here the value of $l = T\Delta s$, where in this section l, s and p denote molar quantities, i.e., values per mole, is termed the latent heat. However, as opposed to its derivatives, G remains continuous as it can be obtained from integrals of the finite quantities V and S.

Since the Gibbs free energy accordingly coincides along both sides of a transition line in the phase diagram, for a small displacement (ΔT, Δp) along the line, the Gibbs free energy changes identically for each phase:

$$\Delta G = \left.\frac{\partial G}{\partial p}\right|_T \Delta T + \left.\frac{\partial G}{\partial T}\right|_V \Delta V = -s_1 \Delta T + v_1 \Delta p = -s_2 \Delta T + v_2 \Delta p \tag{23.13.1}$$

The *Clausius–Clapeyron equation* is derived by recasting the above equation as

$$\left.\frac{dp}{dT}\right|_{\substack{\text{along} \\ \text{transition line}}} = \frac{s_2 - s_1}{v_2 - v_1} = \frac{\Delta s}{\Delta v} = \frac{l}{T \Delta v} \tag{23.13.2}$$

Here, dp/dT coincides with the slope of the phase transition line, while Δv designates the difference in the molar volume between the two phases at two closely separated points on the transition line. Since melting (or boiling) increases the entropy, if a substance contracts in such a phase transition, the associated transition line exhibits a negative slope; otherwise, the slope is positive.

Example

For a transition between a liquid and an ideal gas with a far greater volume, $\Delta v \approx RT/p$ and

$$\frac{dp}{dT} = \frac{l}{R} \frac{p}{T^2} \tag{23.13.3}$$

Solving for p as a function of temperature from $dp/p = (l/R)dT/T^2$,

$$p(T) = p(T_0) e^{-\frac{l}{RT}} \tag{23.13.4}$$

23.14 THE CHEMICAL POTENTIAL AND CHEMICAL REACTIONS

The *chemical potential*

$$\mu = \left.\frac{\partial E(S, V, N)}{\partial N}\right|_{S,V} = -\frac{\left.\frac{\partial S}{\partial N}\right|_{V,E}}{\left.\frac{\partial S}{\partial E}\right|_{V,N}} = -T \left.\frac{\partial S}{\partial N}\right|_{V,E} \tag{23.14.1}$$

quantifies the energy gained by the system as a result of its chemical attractive forces when a particle is displaced from an infinite distance to a point at rest inside the system. To conceptualize this statement, suppose that a system chemically repels electrons. If an electron requires a minimum kinetic energy equal to μ at infinity in order to come to rest inside the system, the energy gained by the system by incorporating the additional particle equals the chemical potential μ. This is equivalent to the work performed by the system on the particle and hence the potential energy gained by the particle. Two systems that exchange particles possess equal chemical potentials in equilibrium since otherwise particles would move down the potential slope from the region of higher μ to that of lower μ and thus greater chemical attraction. Normally, either the variables $\{T, V\}$ or $\{T, p\}$ are controlled in which case from, e.g., $F = E - TS$ together with $dE = TdS - pdV + \mu dN$,

$$\mu = \left.\frac{\partial F(T,V,N)}{\partial N}\right|_{T,V} = \left.\frac{\partial G(T,p,N)}{\partial N}\right|_{T,p} \tag{23.14.2}$$

That is, depending on the boundary conditions, the chemical potential can represent the change in, e.g., the Helmholtz or Gibbs free energy per molecule.

A reaction among interacting species with chemical formulas X_j is described by a chemical equation

$$\sum_j m_j X_j = 0 \tag{23.14.3}$$

where the m_j are small integers such as in $2H_2 + O_2 - 2H_2O = 0$. At equilibrium, the energy is minimized implying that, e.g., for a system at fixed V and T, any rearrangement among the species consistent with Equation (23.14.3) does not affect F and consequently, since $\Delta N_j = m_j$ by definition,

$$\Delta F = \sum_j \left.\frac{\partial F}{\partial N}\right|_{T,V} \Delta N_j = \sum_j \mu_j \Delta N_j = \sum_j m_j \mu_j = 0 \tag{23.14.4}$$

For an ideal gas of indistinguishable particles, if z_j denotes the single particle canonical partition function, from $F = -kT \ln Z$ with $Z = \prod_i z_i^{N_i}/N_i!$ and $N! \approx N^N e^{-N}$,

$$\mu_j = \left.\frac{\partial F}{\partial N_j}\right|_{T,V} \approx \frac{\partial}{\partial N_j}\left(-kT \sum_l N_l \left(\ln \frac{z_l}{N_l} + 1\right)\right) \approx -kT\left(\ln z_j - \ln N_j\right) = -kT \ln \frac{z_j}{N_j} \tag{23.14.5}$$

Inserting into the last equation of Equation (23.14.4) and exponentiating yield the *law of mass action*

$$\prod_j N_j^{m_j} = \prod_j z_j^{m_j} \equiv K(T,V) \tag{23.14.6}$$

Example

For $H_2 \leftrightarrow 2H$, $N_{H_2}/(N_H)^2 = K(T,V)$.

23.15 THE FERMI GAS

Since each spatial electronic state can contain at most one two electrons with opposing spins, at $T = 0$, electrons fill all states with energies up to a value of the chemical potential termed the *Fermi energy*, designated ε_F. At finite temperatures, however, all states have a nonzero occupation probability. From the Fermi–Dirac distribution, this quantity is determined for an ideal *Fermi gas*, typified by a metal of volume V containing N noninteracting electrons by solving

$$\sum_m \frac{1}{e^{\beta(\varepsilon_m-\varepsilon_F)}+1} \equiv \sum_m f(\varepsilon_m) = N \tag{23.15.1}$$

The Fermi function $f(\varepsilon)$ obeys the relationships $f(\varepsilon_F)=1/2$ and $f(\varepsilon_F-\Delta\varepsilon)=1-f(\varepsilon_F+\Delta\varepsilon)$. While at finite temperature Equation (23.15.2) must be approximated or evaluated numerically, at $T=0$, Equation (23.15.1), rewritten as an integral over energies, becomes, after introducing the density of states $\mathrm{P}(\varepsilon)$ per unit volume given by the derivative of Equation (23.5.2),

$$\int_0^\infty \frac{1}{e^{\beta(\varepsilon-\varepsilon_F)}+1}\mathrm{P}(\varepsilon)d\varepsilon = \int_0^\infty \theta(\varepsilon_F-\varepsilon)\frac{1}{2\pi^2}\left(\frac{2m}{\hbar^2}\right)^{\frac{3}{2}}\varepsilon^{\frac{1}{2}}d\varepsilon = \frac{N}{V} \tag{23.15.2}$$

with $\theta(x)$ unity for $x>0$ and zero for $x<0$. Therefore, as can be obtained more directly by equating the number of states per unit volume within a sphere of radius k_F in wavevector space for a cube of side length L assuming periodic boundary conditions, namely, $(1/V)\times 2\left(4\pi k_F^3/3\right)/(2\pi/L)^3$ to N/V, and applying $\hbar^2k_F^2/2m=\varepsilon_F$,

$$\frac{2}{3}\frac{1}{2\pi^2}\left(\frac{2m\varepsilon_F}{\hbar^2}\right)^{\frac{3}{2}} = \frac{N}{V} \tag{23.15.3}$$

or

$$\varepsilon_F = \left(\frac{3\pi^2 N}{V}\right)^{\frac{2}{3}}\frac{\hbar^2}{2m} \tag{23.15.4}$$

The Fermi temperature, defined by $\varepsilon_F=kT_F$, attains values on the order of tens of thousands of degrees for most metals.

Example

In a magnetic field at small T, electrons with spins aligned opposite the magnetic field possess a magnetic potential energy $U_{\text{magnetic}}=-\vec{m}\cdot\vec{H}=\mu_m H$ and hence nonmagnetic mechanical energies up to $E_{\text{mechanical}}=\varepsilon_F-U_{\text{magnetic}}=\varepsilon_F-\mu_m H$, while the mechanical energies of electrons with aligned spins instead extend to $\varepsilon_F+\mu_m H$. Hence, more electrons possess aligned spins resulting in *paramagnetism*. Approximating the Fermi distribution function by unity for $\varepsilon_{\text{total}}=E_{\text{mechanical}}+U_{\text{magnetic}}<\varepsilon_F$ and the density of states per unit volume, $\mathrm{P}(\varepsilon_F)$, near the Fermi energy by its value at the Fermi energy, Equation (23.15.3), corrected for the participation of only a *single* unpaired spin state in each alignment process, for a material of volume V containing N participating electrons

$$\mathrm{P}(\varepsilon_F)=\frac{1}{2}\cdot\frac{1}{2\pi^2}\left(\frac{2m}{\hbar^2}\right)^{\frac{3}{2}}\varepsilon_F^{\frac{1}{2}}=\frac{3}{4\varepsilon_F}\frac{N}{V} \tag{23.15.5}$$

Multiplying the magnetic moment by the width of the energy region, $2\mu_m H$, over which a single spin is aligned parallel to the magnetic field and further by the density of states per unit volume, which corresponds to the number of such states per unit energy interval, yields a magnetization per unit volume of

$$M=\mu_m(2\mu_m H)\left(\frac{3}{4\varepsilon_F}\right)\frac{N}{V} \tag{23.15.6}$$

and a magnetic susceptibility $\chi=M/H=3N\mu_m^2/2\varepsilon_F V$.

23.16 BOSE–EINSTEIN CONDENSATION

In Bose–Einstein materials near $T=0$, electrons collect in the ground state rather than distribute evenly among states as in a Fermi gas. Dividing the electrons into those occupying the ground state and excited state electrons and replacing sums over the latter states by integrals over $\mathrm{P}(\varepsilon)Vd\varepsilon$ yields

$$N=N_{\text{ground}}+N_{\text{excited}}=\frac{1}{e^{-\beta\mu}-1}+V\int_0^\infty 2\pi\left(\frac{2m}{h^2}\right)^{\frac{3}{2}}\frac{\varepsilon^{\frac{1}{2}}}{e^{\beta(\varepsilon-\mu)}-1}d\varepsilon \tag{23.16.1}$$

Reversing the original derivation of the Fermi–Dirac occupation factor,

$$\frac{1}{e^{\beta(\varepsilon-\mu)}-1}=\frac{e^{-\beta(\varepsilon-\mu)}}{1-e^{-\beta(\varepsilon-\mu)}}=e^{\beta(\mu-\varepsilon)}\sum_{l=0}^{\infty}e^{l\beta(\mu-\varepsilon)} \tag{23.16.2}$$

from which

$$\begin{aligned}N_{\text{excited}}&=2\pi V\left(\frac{2m}{h^2}\right)^{\frac{3}{2}}\sum_{l=1}^{\infty}e^{l\beta\mu}\int_0^\infty e^{-l\beta\varepsilon}\varepsilon^{\frac{1}{2}}d\varepsilon\\&=2\pi V\left(\frac{2m}{h^2}\right)^{\frac{3}{2}}\sum_{l=1}^{\infty}e^{l\beta\mu}\frac{1}{(l\beta)^{\frac{3}{2}}}\int_0^\infty e^{-l\beta\varepsilon}(l\beta\varepsilon)^{\frac{1}{2}}d(l\beta\varepsilon)\end{aligned} \tag{23.16.3}$$

Since the integral evaluates to $\sqrt{\pi}/2$, as $T\to 0$, if the lowest energy level is associated with zero energy, the chemical potential μ must approach zero sufficiently rapidly that the first term in Equation (23.16.1) equals the number of nonexcited particles. This requires $\exp(l\beta\mu)\to 1$ for which the sum in Equation (23.16.3) can be approximated by $\sum_{l=1}^{\infty}1/l^{3/2}\approx 2.61$. When T exceeds the *condensation temperature* $T_{\text{condensation}}$ (approximately 3 K for helium) defined by

$$\frac{N}{V} = \frac{N_{\text{excited}}}{V} \equiv 2.61\left(\frac{2\pi mkT_{\text{condensation}}}{h^2}\right)^{\frac{3}{2}} \tag{23.16.4}$$

the excited states cannot accommodate the total number of electrons and at least $N - N_{\text{excited}}$ particles must accumulate in the ground state, an effect termed *Bose–Einstein condensation* occurs. The fraction of atoms in the ground state below $T_{\text{condensation}}$ is approximated by

$$\frac{N_{\text{ground}}}{N} = 1 - \left(\frac{T}{T_{\text{condensation}}}\right)^{\frac{3}{2}} \tag{23.16.5}$$

Since the number of particles in the excited state varies as $T^{3/2}$ and the energy per particle grows as kT, the total energy increases as $T^{5/2}$ while its derivative, the specific heat, varies as $T^{3/2}$ above $T_{\text{condensation}}$, which is termed a *second-order phase transition*. Realistic models that incorporate the details of the particle–particle interaction predict specific heat discontinuities with more involved properties.

23.17 PHYSICAL KINETICS AND TRANSPORT THEORY

The Maxwell distribution, while derived for a system in contact with a heat reservoir, also applies to single particles considered as the system of interest where the remainder of the actual system acts as a reservoir. The probability, Π, of observing a single particle in a region $d^3r\, d^3p$ of phase space is then

$$\Pi(\vec{r},\vec{p})d^3rd^3p = \text{constant}\cdot e^{-\beta\left(\frac{p^2}{2m}+V(\vec{r})\right)}d^3rd^3p \tag{23.17.1}$$

Example

In a spatially varying gravitational potential, the probability of finding a particle with a given momentum varies with height, z, as $ce^{-\beta gz}$.

For a constant potential, integrating over position yields for the mean number of particles per unit volume with a velocity in a region d^3v around $\vec{v}$, after normalizing to $n = N/V$,

$$\varpi(\vec{v})d^3v = n\left(\frac{m}{2\pi kT}\right)^{\frac{3}{2}} e^{-\frac{mv^2}{2kT}}d^3v \tag{23.17.2}$$

Integrating over two components of $\vec{v}$ leads to the distribution for a single velocity component, namely,

$$\varpi(v_j)dv_j = n\left(\frac{m}{2\pi kT}\right)^{\frac{1}{2}} e^{-\frac{mv_j^2}{2kT}}dv_j \tag{23.17.3}$$

while the distribution of the magnitude of the velocity (speed) is obtained from $d^3v = 4\pi v^2 dv$:

$$\varpi(v)dv = n\left(\frac{m}{2\pi kT}\right)^{\frac{3}{2}} e^{-\frac{mv^2}{2kT}} 4\pi v^2 dv \tag{23.17.4}$$

Accordingly, the most probable velocity is zero, while the most probable speed is at the maximum

$$\frac{d}{dv}\left(v^2 e^{-\frac{mv^2}{2kT}}\right) = \left(2v - 2v\frac{mv^2}{2kT}\right)e^{-\frac{mv^2}{2kT}} = 0 \tag{23.17.5}$$

or $v_{\max} = \sqrt{2kT/m}$. The mean velocity is however given by

$$\bar{v} = \left(\frac{m}{2\pi kT}\right)^{\frac{3}{2}} 4\pi \int_0^\infty v^3 e^{-\frac{mv^2}{2kT}} dv = \left(\frac{8kT}{m\pi}\right)^{\frac{1}{2}} \tag{23.17.6}$$

The root-mean-squared (r.m.s.) value of each velocity component can be calculated directly from the equipartition theorem since $m\overline{v_x^2}/2 = kT/2$ or $\overline{v_x^2} = kT/m$, while the r.m.s. velocity equals $m\overline{v^2}/2 = m\left(\overline{v_x^2} + \overline{v_y^2} + \overline{v_z^2}\right)/2 = 3kT/2$.

Recalling that the *particle flux* per unit area through a surface is defined as the number of particles per unit time crossing a plane of unit area perpendicular to the surface, for a surface perpendicular to the x-direction, all molecules within a distance v_x of the surface intersect the surface per unit time. Weighting by the probability of v_x and integrating over all $v_x > 0$ yield for the flux

$$F = n\left(\frac{m}{2\pi kT}\right)^{\frac{1}{2}} \int_0^\infty e^{-\frac{mv_x^2}{2kT}} \frac{1}{2} d(v_x^2) = n\left(\frac{m}{2\pi kT}\right)^{\frac{1}{2}} \frac{1}{2} \cdot \left(\frac{2kT}{m}\right) = \frac{n\bar{v}}{4} \tag{23.17.7}$$

Since the momentum transfer to a wall from a particle with velocity $\vec{v}$ is $2mv_x$, the ideal gas law can further be derived by computing the pressure, i.e., the momentum transfer per unit time per unit area

$$P = n\left(\frac{m}{2\pi kT}\right)^{\frac{1}{2}} \underbrace{\int_0^\infty 2mv_x^2 e^{-\frac{mv_x^2}{2kT}} dv_x}_{m\int_{-\infty}^{\infty} v_x^2 e^{-\frac{mv_x^2}{2kT}} dv_x} = 2n\left(\frac{m}{2}\overline{v_x^2}\right) = 2\frac{N}{V}\frac{kT}{2} \tag{23.17.8}$$

In a cavity at finite temperature, photons associated with the electromagnetic modes propagate in all directions with an energy density $u(T) = 4\sigma T^4/c$ with an energy flux within a solid angle $d\Omega$ of $cu(T)d\Omega/4\pi$. Therefore, the energy passing through the surface per unit time associated with photons propagating at an angle θ with respect to the normal to the surface is given by

$$dF_{\text{energy}} = \frac{cu(T)\cos\theta d\Omega}{4\pi} = \frac{c\cos\theta u(T)}{4\pi} 2\pi \sin\theta d\theta = \frac{cu(T)}{2} \sin 2\theta d\theta \tag{23.17.9}$$

Integrating between 0 and $\pi/2$ yields $F_{\text{energy}} = cu(T)/4 = \sigma T^4$.

In *transport processes* such as particle and heat transfer by *diffusion*, energy is transported through a medium by colliding particles. The relationship between microscopic behavior and macroscopic properties is quantified by the average time and therefore distance (the *mean time* and *free path*) between collisions. To estimate these quantities, note that in a system of particles with effective radii r, each particle collides with all particles located within a distance $2r$ of the center of its path yielding a *cross section* or effective area for collisions equal to $\sigma = \pi(2r)^2$. Additionally, the average squared relative velocity of two particles $\overline{\vec{v}_r^2} = \overline{(\vec{v}_1 - \vec{v}_2)^2} = \overline{(\vec{v}_1^2 + \vec{v}_2^2 - 2\vec{v}_1 \cdot \vec{v}_2)^2} = 2\overline{\vec{v}^2}$, so that $\left(\overline{\vec{v}_r^2}\right)^{1/2} \approx \sqrt{2}|\bar{v}|$. In unit time, a particle collides with all particles within a cylinder of volume $\sqrt{2}\bar{v}\sigma$ so that the mean number of collisions per unit time equals the average number, $\sqrt{2}\bar{v}\sigma n$, of particles in this volume. The average time per collision equals the reciprocal of this quantity, yielding a mean free path

$$\lambda_{\text{free}} = \frac{\bar{v}}{\sqrt{2}\bar{v}n\sigma} = \frac{1}{\sqrt{2}n\sigma} \tag{23.17.10}$$

Often, for simplicity, the velocity is employed in place of the relative velocity, leading to the above expression without the $\sqrt{2}$ factor. As expected, the velocity affects the number of collisions per second but not the average distance between collisions.

Consider a quantity, $Y(x)$, such as the transverse momentum, temperature, or particle number along a plane of constant x that is transported by particle motion. From Equation (23.17.7), $n\bar{v}/4$ particles cross this plane per unit area per second from each side. Assuming that the value of Y transported by each particle is given by Y at the average location of its previous collision, the *flux* of Y is determined by the mean distance $\overline{\Delta x}$ along x traveled by a particle between collisions. As a particle traveling at an incidence angle θ to a plane $x =$ constant travels an average distance $\Delta x = \lambda_{\text{free}} \cos\theta$ in the x-direction before colliding while the number of particles with a given value of v_x traveling in the $+\hat{e}_x$ direction that cross the plane per unit time is proportional to $v_x = v\cos\theta$,

$$\overline{\Delta x} = \frac{\int \lambda_{\text{free}} \cos\theta \cdot v \cos\theta \, d\Omega}{\int v \cos\theta \, d\Omega} = \frac{\int_1^0 \cos^2\theta \, d(\cos\theta)}{\int_1^0 \cos\theta \, d(\cos\theta)} \lambda_{\text{free}} = \frac{2}{3}\lambda_{\text{free}} \tag{23.17.11}$$

Consequently, the *flux* of Y in the $+\hat{e}_x$ direction along a plane at $x = a$ is given by $+n\bar{v}/4$ times $Y(a - 2\lambda_{\text{free}}/3)$ minus the flux in the $-\hat{e}_x$ direction from the particles at $x = a + 2\lambda_{\text{free}}/3$. Employing the Taylor series expansion $Y(x + \Delta x) \approx Y(x) + dY/dx|_x \Delta x + \ldots$, the net flux of Y can be approximated by

$$\frac{n\bar{v}}{4}\left(Y(a) - \frac{2\lambda_{\text{free}}}{3}\frac{\partial Y}{\partial x}\bigg|_a - \left\{Y(a) + \frac{2\lambda_{\text{free}}}{3}\frac{\partial Y}{\partial x}\bigg|_a\right\}\right) = -\frac{n\bar{v}\lambda_{\text{free}}}{3}\frac{\partial Y}{\partial x} \tag{23.17.12}$$

If Y is identified with the mean energy per molecule, ε, in a medium between two parallel plates held at different T, Equation (23.17.12) represents the heat flux across a plane parallel to the plates given by

$$\frac{1}{A}\frac{dQ}{dt} = -\frac{n\bar{v}\lambda_{\text{free}}}{3}\frac{d\varepsilon}{dT}\frac{dT}{dx} \equiv -\text{K}_{\text{thermal}}\frac{dT}{dx} \tag{23.17.13}$$

where the *thermal conductivity*, $\text{K}_{\text{thermal}}$, is given in terms of the specific heat per molecule, c_{specific}, by

$$\text{K}_{\text{thermal}} = \frac{n\bar{v}\lambda_{\text{free}}c_{\text{specific}}}{3} \tag{23.17.14}$$

If Y instead represents the transverse momentum mv_z of a particle of mass m between a stationary plate along $x = 0$ and a plate at $x = a$ moving in the z-direction, Equation (23.17.12) yields the shear force per unit area

$$\frac{F_z}{A} = -\frac{n\bar{v}\lambda_{\text{free}}}{3}\frac{d(mv_x)}{dx} \equiv -\eta\frac{dv_x}{dx} \tag{23.17.15}$$

with a *coefficient of viscosity* as follows, in which $mn = \rho$ corresponds to the medium density:

$$\eta = \frac{\rho\bar{v}\lambda_{\text{free}}}{3} \tag{23.17.16}$$

Finally, if Y is the fraction $n_{\text{impurity}}/n_{\text{total}}$ of impurities in a medium to the total number of particles in the media with $n_{\text{total}} \gg n_{\text{impurity}}$, Equation (23.17.12) describes the impurity transfer per unit area per unit time, i.e., the impurity current density,

$$J_x = -\frac{n_{\text{total}}\bar{v}\lambda_{\text{free}}}{3}\frac{1}{n_{\text{total}}}\frac{dn_{\text{impurity}}}{dx} = -\frac{\bar{v}\lambda_{\text{free}}}{3}\frac{dn_{\text{impurity}}}{dx} \equiv -D\frac{dn_{\text{impurity}}}{dx} \tag{23.17.17}$$

The *diffusion coefficient* D equals

$$D = \frac{\bar{v}\lambda_{\text{free}}}{3} \tag{23.17.18}$$

Quasi-one-dimensional impurity motion in, e.g., the $\hat{e}_x$ direction, J_x, obeys a continuity equation of the form

$$\frac{dJ_x}{dx} = -\frac{\partial\rho(x)}{\partial t} = -\frac{\partial n_{\text{impurity}}}{\partial t} \tag{23.17.19}$$

Inserting Equation (23.17.17) for the impurity current density leads to the *one-dimensional diffusion equation*

$$\frac{\partial n_{\text{impurity}}}{\partial t} = D\frac{\partial^2 n_{\text{impurity}}}{\partial x^2} \tag{23.17.20}$$

which generalizes in three dimensions to

$$\frac{\partial n}{\partial t} = D\nabla^2 n \tag{23.17.21}$$

Particle diffusion can be modeled with random number generators. The function `rand( )` in Octave, which returns a uniformly distributed number within [0, 1], is employed in the following code to generate a random walk with steps of unit length in the positive or negative direction. The final positions of 1000 random walks with 2000 steps each are displayed in a *histogram* that sums the number of steps within each 30 uniformly distributed intervals. Associating a time interval $\Delta t = 1$ with a single random walk step, *two* random walk steps implement $f(x, t+2) - f(x, t) = (1/4)[f(x-2, t) + 2f(x, t) + f(x+2, t)] - f(x, t)$ with an r.m.s. spread $\langle x^2 \rangle = 2$ after two steps. Dividing by $2\Delta t(\Delta x)^2$ with $\Delta t = \Delta x = 1$ approximates $\partial f/\partial t = (1/2)\partial^2 f/\partial x^2$ and therefore implies $D = 1/2$ for a large number of steps. The program writes the sum of the squares of the final positions divided by the total number of steps taken in all random walks. This result can be seen to coincide with the diffusion equation result $\langle x^2 \rangle = 2mDt$, where m represents the number of spatial dimensions:

```
numberOfRealizations = 5000;
numberOfTimeSteps = 1000;
sum = 0;

for outerLoop = 1 : numberOfRealizations
   position = 0;
   for innerLoop = 1 : numberOfTimeSteps
       position = position + sign( rand( ) - 0.5 );
   end
   finalPoint (outerLoop) = position;
   sum = sum + position^2;
end

sum / ( numberOfTimeSteps * numberOfRealizations )
hist( finalPoint, 30 );
```

24

CONDENSED MATTER PHYSICS

Condensed matter physics relates to solids and liquids with closely spaced, highly interacting atoms or molecules. The subfield of solid and in particular ordered solid matter is further termed solid-state physics.

24.1 CRYSTAL STRUCTURE

In contrast to liquids or "supercooled" solids (materials that solidify too rapidly for the atoms to relocate into energetically favorable ordered locations) for which the atomic positions are randomly oriented and polycrystalline materials formed from randomly oriented microscopic crystals, a crystalline substance consists of a single periodic arrangement of atoms. The atomic positions of a *simple cubic lattice* occupy the vertices, $(x, y, z) = (m_x, m_y, m_z)a$, of a periodically repeated set of cubes where m_x, m_y, m_z take on all integer values. In a *body-centered cubic* (bcc) *lattice*, an additional atom is situated at the center of each unit cell forming two simple cubic lattices with a relative displacement of $a/2(1, 1, 1)$. A *close-packed structure* is formed from a planar layer with eight atoms arranged in an octagon around every atom. The plane above this layer is formed from similar octagonal sets of atoms with a relative rotation of $\pi/8$ radians where, for a *hexagonal close-packed structure*, the atoms revert to their locations in the original layer of atoms, while the atomic positions are instead rotated by $2\pi/8 = \pi/4$ radians with respect to the first layer in a *face-centered cubic* (fcc) lattice.

Fundamental Math and Physics for Scientists and Engineers, First Edition.
David Yevick and Hannah Yevick.

The latter arrangement viewed from a different perspective forms a cubic lattice with atoms added to the center of each face of the cube. The *diamond structure* contains two fcc lattices with a relative displacement of $a/2(1, 1, 1)$. The resulting bonds form linked tetrahedrons, which yield an enhanced structural stability. The smallest periodic rectangular structure in a crystal is termed the *unit cell*, while the lengths of its sides are the *lattice constants*. The smallest crystal volume that when periodically repeated reproduces, the lattice is instead termed the *primitive cell*.

24.2 X-RAY DIFFRACTION

In an ideal crystal structure, numerous families of differently oriented identical scattering planes can be identified. A planar electromagnetic wave scatters coherently from a given set of planes if the Bragg condition (cf. Section 19.13) $2d\cos\theta_{\text{incidence}} = m\lambda$, where m is an integer and d is the minimum distance between two successive planes, is fulfilled. The Bragg condition can also be expressed as $\vec{k}_{\text{initial}} - \vec{k}_{\text{final}} = (2\pi m/d)\hat{e}_{\vec{A}}$, where $\vec{A}$ is the surface vector of the planes as taking the dot product of this equation with $\hat{e}_{\vec{A}}$ and employing the equality of the incoming and reflection angles yields $\Delta k_{\perp} = 2\pi m/d$ where $\Delta k_{\perp} \equiv 2k_{\text{initial}}\cos\theta_{\text{incidence}}$ is the difference between the components of the wavevectors of the incoming and outgoing beams perpendicular to the crystal planes (setting $k_{\text{inital}} = 2\pi/\lambda$ then reproduces $2d\cos\theta_{\text{incidence}} = m\lambda$). Accordingly, light with $\lambda > 2d$ cannot be Bragg scattered for any incoming angle. Thus, for diffraction to occur for a typical atomic spacing of 5 Å, the light wavelength must be less than 10 Å, corresponding to an X-ray photon energy of 1.24×10^3 eV.

As every plane of atoms in the lattice is associated with a diffraction condition, the values of $\Delta\vec{k}$ satisfying these Bragg conditions form a second lattice termed the *reciprocal lattice*. Mathematically, for an incoming plane wave $A\exp\left(i\left(\vec{k}_{\text{initial}}\cdot\vec{r} - \omega t\right)\right)$, the amplitude at a distant observation point $\vec{r}$ is given by

$$E = A' \sum_{\substack{\text{all lattice}\\ \text{sites } \vec{r}'}} e^{i\left(\vec{k}_{\text{initial}}\cdot\vec{r}' - \omega t\right)} e^{i\left(\vec{k}_{\text{final}}\cdot\left(\vec{r}-\vec{r}'\right)\right)} \equiv A' e^{i\left(\vec{k}_{\text{final}}\cdot\vec{r} - \omega t\right)} \int s\left(\vec{r}'\right) e^{i\left(\left(\vec{k}_{\text{initial}} - \vec{k}_{\text{final}}\right)\cdot\vec{r}'\right)} d^3 r' \tag{24.2.1}$$

where the periodic function $s(\vec{r})$ is unity at the position of each atom and zero elsewhere. The scattering amplitude therefore constitutes a Fourier transform of the particle distribution with respect to the difference between the incoming and outgoing wavevectors. For a simple rectangular lattice, if $\vec{a}$ is one of the three primitive vectors defining the primitive cell of the physical (direct) lattice, the scattered fields add coherently for reciprocal lattice vectors satisfying $\Delta k_i a_i = 2\pi m_i$ with $i = x, y, z$ so that

$$\vec{k}_{m_x, m_y, m_z} = 2\pi\left(\frac{m_x}{a_x}, \frac{m_y}{a_y}, \frac{m_z}{a_z}\right) \quad (24.2.2)$$

for all sets of integers m_i. In general, the components of the primitive vectors of the reciprocal lattice are given by $b_i = \varepsilon_{ijk}\, a_j\, a_k / \varepsilon_{ijk}\, a_i\, a_j\, a_k$ in terms of the primitive vectors $\vec{a}$ of the direct (physical) lattice.

In X-ray (or equivalently neutron or electron) scattering experiments, a crystal is typically rotated in a stationary beam. When the orientation of the crystal matches the Bragg scattering condition, a point is recorded on a photographic emulsion behind the crystal. Alternatively, the material can be converted into a powder for which a certain fraction of crystals satisfies each Bragg reflection condition. The Bragg angles are then recorded as rings. The diffraction pattern of extended objects such as molecular crystals again describes a set of dots corresponding to the angles of the reciprocal lattice vectors. However, the amplitude of each dot is proportional to that of the diffraction pattern of a single molecule at the corresponding angle. As well, for certain crystal structures, points in the diffraction pattern may be missing if the fields generated by certain planes of atoms internal to the unit cell interfere destructively with those of the remaining planes of atoms with the same orientation.

24.3 THERMAL PROPERTIES

According to kinetic theory, the thermal conductivity of a gas resulting from energy transfer through molecular motion equals $\kappa = n\lambda\bar{v}c_{\text{specific}}/3 = \lambda\bar{v}C/3$ where c_{specific} and C are the specific heat per molecule and per unit volume and λ is the mean free path. In a solid, energy is instead transferred through molecular vibrations. However, linear superpositions of vibrations termed *phonons* behave as localized quasiparticles. The corresponding thermal conductivity of a solid is $\kappa = \lambda\bar{v}C_{\text{phonon}}/3$, where C_{phonon} is the specific heat per unit volume of phonons. Long-wavelength acoustic phonons and short-wavelength optical phonons dominate at low and high temperatures, respectively, yielding a T^3 (cf. Section 23.7) and a constant temperature dependence in the two cases. The total phonon mean free path is given by

$$\frac{1}{\lambda_{\text{total}}} = \sum_{m=1}^{N} \frac{1}{\lambda_m} \quad (24.3.1)$$

where λ_m denotes the mean free path associated with the mth collision channel (e.g., the mth physical process that induces collisions) in isolation, since the number of collisions per unit time is the sum of those associated with each channel, each of which is in turn proportional to the inverse of its corresponding mean free path. Phonon scattering in a crystal occurs at point defects such as impurities or isotopes, crystal boundaries, and extended defects such as dislocations. Phonons can also interact through nonlinear interactions in which anharmonic terms in the atomic restoring force result in 1 or more than 2 phonons emerging from a two-phonon collision. Finally, phonons can be Bragg reflected by successive lattice planes. Typically, metals and near-perfect crystals such as diamond exhibit high thermal conductivities, while the disorder inherent in glasses and plastics yield small phonon mean free paths and therefore low thermal conductivities.

24.4 ELECTRON THEORY OF METALS

Since the electron wavefunctions of conduction electrons in different atoms in a metal overlap, the eigenfunctions extend throughout the lattice. In this manner, a metal resembles a single atom with a delocalized nuclear volume in which the electrons occupy all energy levels up to that of the highest filled state at the *Fermi energy*, E_F. The electronic wavefunction then resembles a sinusoidal wave confined to the crystal volume. Since sinusoidal functions are composed of equal amplitude, oppositely directed traveling waves, electrons are effectively distributed among all traveling waves with energies below E_F.

An electric field within an imperfect conductor accelerates its electrons with a force $q\vec{E} = -e\vec{E}$. However, as the velocity of an electron increases, its de Broglie wavelength contracts enhancing its scattering from lattice vibrations and imperfections. After an average distance termed the *mean free path*, these interactions effectively randomize the ensemble-averaged electron velocity, returning it to zero. The effective net average velocity that the electrons experience between randomizing scattering events is thus $\Delta v \approx F\Delta t/m = -eE\tau/m$, where τ is the mean time between these events. As the current density corresponds to the charge per unit time passing through a surface of unit, cross-sectional area is oriented perpendicularly to the direction of the charge flow $J = nq\Delta v = e^2 En\tau/m = \sigma E$, yielding a conductivity $\sigma = e^2 n\tau/m$. At low temperatures scattering from impurities and other localized defects and at high temperatures scattering from thermal lattice vibrations (phonons) dominate the free path. Empirically, the resistivity increases from a value that varies with sample purity first as $\sim T^5$ and then as T as T is raised from 0K.

In a conductor as in an atom, only electrons in the highest energy states normally contribute to electron transport. That is, while all electrons experience an electric force, lower-energy electrons with positive momentum are paired with reflected electrons with equal negative momentum, forming a standing wave with zero net transport. Therefore, only unpaired high-energy electrons with $E \approx E_F$ are scattered with a scattering time approximated by the scattering length divided by the velocity of electrons at the Fermi surface

$$\sigma = \frac{ne^2 l_{\text{free}}}{mv_{\text{Fermi}}} \tag{24.4.1}$$

where for copper $\sigma \approx 5 \times 10^7 (\Omega\text{m})^{-1}$, $v_{\text{Fermi}} = \sqrt{2m_e E_F} \approx 10^6 \text{m/s}$, and $n \approx 7 \times 10^{28}\ \text{m}^{-3}$, yielding $l_{\text{free}} \approx 2.5 \times 10^{-8}$ m or ≈ 50 atomic radii.

24.5 SUPERCONDUCTORS

The resistivity of certain materials vanishes below a critical temperature, T_{critical}, that does not vary with impurity concentration (although the extent of the temperature interval over which the resistivity drops to zero increases with the impurity density). This *superconducting* behavior results from a long-range attractive force between electrons mediated by phonons. Thus, T_{critical} is generally higher in poorly conducting metals with strong phonon coupling than in good conductors, while the T_{critical} of isotopes varies as the inverse square root of the atomic mass, replicating the $\sqrt{k/m}$ dependence of the phonon frequency. Effectively, an electron with momentum $\vec{k}$ and a certain spin propagating through the lattice alters the electron distribution of neighboring atoms, leading to an attractive potential for a second electron with momentum $-\vec{k}$ and opposite spin. As a result, *Cooper pairs* form at a critical temperature T_{critical} and increase in density for $T < T_{\text{critical}}$.

Some features of superconductors include the following:

1. Normal resistivity can be restored at $T = 0$ by applying a magnetic field greater than a specimen-dependent critical value $B_{\text{critical},T=0}$. For $0 < T < T_{\text{critical}}$, the critical magnetic field is

$$B_{\text{critical}} = B_{\text{critical},T=0}\left(1 - \left(\frac{T}{T_{\text{critical}}}\right)^2\right) \tag{24.5.1}$$

   Since current flowing through a superconductor generates a magnetic field, a superconducting sample becomes resistive at sufficiently large current densities.
2. Applying a magnetic field to a superconductor generates a persistent current opposing the magnetic field change, resulting in a zero internal magnetic field, which is termed the *Meissner effect*.
3. Superconductors can be *type I* or *type II*. Type I superconductors, which include almost all elements, transition directly from the normal to the superconducting state at B_{critical}. In type II superconductors, which comprise most alloys as well as Nb, Tc, and V, a *mixed state* forms above a *lower critical field* for which magnetic flux cylinders or *flux quanta* form microscopic filaments through normally conducting material. The entire sample only conducts normally above the *upper critical field*. If an electric field is applied to a type II superconductor between the two critical fields, the filament motion experiences an effective viscosity that dissipates energy resulting in a nonzero resistance. This can be counteracted through *flux pinning* by intentionally added defects that prevent filament migration.
4. The Gibbs free energy of a superconductor is continuous in a superconducting transition, but the specific heat, $T\partial S/\partial T$, is discontinuous, which is termed a *second-order phase transition*.

5. Heat is only transferred in a superconductor by normal, nonpaired electrons so that the thermal conductivity decreases rapidly below the critical temperature.
6. While the resistance of a superconductor is zero for DC currents, the fluctuating field of an AC current dissipates energy through the motion of the normal electrons, yielding optical absorption at all T.

24.6 SEMICONDUCTORS

The chemical bonding energies of valence electrons while negligibly small in alkali elements and metals on the bottom and left of the periodic table increase to several eV for atoms at the upper right-hand side. Since this range far exceeds the mean thermal fluctuation energy of $kT \approx 1/40$ eV at $T = 300$ K, the conductivities of insulators and conductors differ by ~ 25 orders of magnitude (more for ideal samples). The centrally located *elemental semiconducting* group IV elements, Si and Ge, possess intermediate binding energies. As well, in *compound semiconductors*, formed from alloys of group III and V elements or group II and VI elements, neighboring atomic electronic wavefunctions overlap, and hence, the electrons experience the average potential of the component atoms, yielding binding energies approximating that of the group IV elements. The intermediate binding strengths enable the modulation of the properties of such semiconductors by adding conducting or insulating impurities or applying, e.g., electrical, optical, or thermal forces. Consequently, mechanical switches can be replaced by rapid electrical or optical switches.

The atoms in a crystal bond chemically, possibly leading to different sequences of energy levels than those of the isolated atoms. Each modified energy level further broadens into an *energy band*. To illustrate, consider a one-dimensional crystal of N coupled atoms spaced a distance a apart along a line segment of length $L = Na$. The envelope of the eigenmode of an electron confined within the crystal must vanish at $x = 0, L$ and therefore varies as $\sin(m\pi x/L)$ for some integer $m < N$. The $\sin(m\pi x/L)$ envelope is composed of a sum of equal amplitude forward- and backward-traveling waves such that the phase of each wave differs by $\Delta\phi = \pm\, \pi m/N$ with $m = 1, 2, \ldots, N$ between adjacent atomic sites. Alternatively, artificial periodic boundary conditions can be introduced (corresponding to bending the line segment into a circle) in which case the eigenfunctions correspond to traveling waves with independent amplitudes and phase differences of $\Delta\phi = 2\pi m/N$ with $m = -N/2 + 1, \ldots, N/2 - 1, N/2$ between adjacent sites, yielding the same total number of modes as for zero boundary conditions. In either representation, the energy of each eigenfunction can be plotted as a function of the phase change $\Delta\phi$ written as the *"lattice" momentum* p through $\Delta\phi = \pi m/N = k\Delta x = ka$ with $p = \hbar k$, yielding an *energy band diagram*. For electrons that are strongly bound to their host atoms, the energy is nearly independent of $\Delta\phi$ and hence p. The ratio of applied force to acceleration or *effective mass*, $m^* = F/a$, which is given in analogy to the free electron mass by the reciprocal of the band curvature according to, $1/m^* = \partial^2 E/\partial p^2$ (from $E = p^2/2m^*$), is then large as expected.

In a metal, nearly degenerate or degenerate filled and empty levels bond or overlap to form a partly filled energy band. In contrast, as the electron level separations are larger in semiconductors and insulators, after bonding an energy difference E_g termed the *energy or bandgap* exists between the highest of the completely filled (at $T=0$) *valence bands* and the lowest energy level of the first unfilled *conduction band.* Hence, unlike a metal, applying a (moderate, nonionizing) electric field to a semiconductor or insulator at $T=0$ does not produce a current. Rather, since all bound states are filled, electrons propagate equally in all directions both in the presence and absence of an external field. For $T>0$, however, a bound electron near the top of the valence band can acquire an energy larger than E_g as a result of an interaction with a low-probability high-energy thermal lattice vibration, internally ionizing the electron from a valence to a conduction band state. An applied electric field accelerates the unpaired conduction band electron while displacing an electron from a neighboring atom into the unoccupied state of the ionized host atom. The latter current can be modeled as that of a positively charged *vacancy* or *hole.*

As the thermal generation of free electrons is characterized by a transition energy $\approx E_g$, its transition rate varies as $B_{\text{thermal}} \exp(-E_g/kT)$ in analogy with chemical reaction rates. The number of recombination events per second per unit volume of the conduction band electrons with ionized atoms is then given by a "bimolecular" rate $B_{\text{bimolecular}} np$ where n and p, which coincide for an *intrinsic* (pure) semiconductor, denote the density of electrons and holes per unit volume. That is, since the density of the background atoms far exceeds n and p, doubling both n and p yields four times as many collision events between the electrons and holes. Equating the recombination and generation rates yields with $n=p$

$$n \approx \left(B_{\text{thermal}}/B_{\text{bimolecular}}\right) \exp\left(-E_g/2kT\right) \tag{24.6.1}$$

While for room-temperature semiconductors E_g is on the order of an eV, in insulators, the energy gap is typically several eV, effectively suppressing the thermally induced free carrier density and hence the conductivity.

Since electrons possess a moderate ionization energy in semiconductors, the finite temperature conduction band electron density is greatly enhanced in *extrinsic (impure) semiconductors* through the addition of impurities. *Donor atoms* from the fifth column of the periodic table exhibit small ($\sim kT$) internal ionization energies when present in, e.g., group IV semiconductors such as Si or Ge. In particular, in a diamond (fcc) lattice of group IV atoms, four of the group V donor atom's valence electrons bond with the four neighboring atoms, while the fifth electron propagates in the potential of its host nucleus screened by the four bonded electrons. As these are easily polarized, the effective dielectric constant is large, reducing the effective atomic potential (which varies as the inverse of the dielectric constant). Hence, the fifth donor electron is only weakly bound to its host nucleus and hence is easily ionized by thermal fluctuations. Consequently, a large fraction of donor atoms are ionized at room temperature, leading to excess mobile free electrons accompanied by positive ions with *fixed lattice locations.* The conductivity of such an extrinsic *n-type semiconductor* far exceeds than that of the intrinsic semiconductor. Similarly, group III *acceptor* impurities form *p*-type extrinsic semiconductors as they readily capture electrons from neighboring atoms, forming four bonds with the neighboring group IV atoms. This leads to negatively ionized fixed *acceptor* sites together with mobile vacancies termed *holes* with the effective properties of *positively charged* particles.

That current is almost exclusively carried by the migration of negative carriers in *n*-type semiconductors and by positively charged holes in *p*-type materials enables a *p–n junction diode* formed by placing two samples of *p*- and *n*-type materials in contact to constrain or *rectify* electric current to flow in a single direction. That is, a positive voltage applied to a contact to the *p* section relative to an *n* contact produces an electric field from the *p* to the *n* contacts. Holes in the *p*-type material and electrons in the *n*-type material then both propagate toward the junction between the two materials where they recombine generating a current. Negatively biasing the *p* region relative to the *n* region reverses the electric field and therefore the direction of carrier motion, inhibiting recombination.

Free carriers can also be created directly by external forces as in the *field-effect transistor* (FET), which implements a voltage-controlled switch. In this device, the semiconductor conductivity varies between high and low values in response to an external electrical field in the same manner as a large electric field in a fluorescent light bulb ionizes a dilute gas of neutral mercury atoms, creating a conducting channel between two spatially separated contacts. To attain the required high Si ionization field $\sim 10^5$ V/cm, a thin ~ 1 μm insulating SiO_2 glass layer is grown on a Si substrate by exposure to oxygen at high T and a metal contact deposited over the glass. Patterning the layers results in a narrow capacitor plate over a segment of intrinsic Si separating two highly doped and therefore conducting *source* and *drain* regions (the source voltage is negative in normal operation relative to the drain forming a source of electrons). Applying a potential of a few volts to the metal contact yields an electric field $E = -\Delta V/\Delta x$ that ionizes the Si atoms directly under the glass plate. This forms a conducting channel between the source and drain. Once the applied voltage is removed, the free electrons recombine with the ionized atoms. The source–drain current is then eliminated, turning off the switch.

25

LABORATORY METHODS

25.1 INTERACTION OF PARTICLES WITH MATTER

When an energetic particle beam passes through matter, its energy and orientation is degraded. If the particles undergo single, destructive collisions, the beam preserves its angular distribution, but the number of particles decays according to $N(x) = N(x = 0) \exp(-\mu x)$, where μ is termed the *absorption coefficient* and $1/\mu$ the *mean free path*. If multiple collisions occur, the angular spread of the beam increases with distance. Its mean range, R_0, is defined as the distance at which half of the incoming particles dissipate their excess energy. The energy loss of a beam is quantified by the derivative $dE/d(\rho x)$, where E is the particle energy and ρ the medium density. This quantity varies as the square of the product of the particle charge with the mean atomic number of the absorbing material.

At low energies, electrons or heavy particles passing through matter collide with electrons and $dE/d(\rho x) \propto 1/v^2$ except at velocities sufficiently small that the incoming particle velocity is comparable to that of the atomic electrons. On the other hand, for energies higher than about 1 GeV, termed the *ionization minimum*, the derivative increases slowly with energy. At the minimum, $dE/d(\rho x)$ is approximately a few MeV/g-cm^2 for all materials although electrons undergo frequent high-angle scattering processes unlike heavy charged particles. Electron beams with energies larger than a critical energy of ≈ 600 MeV/Z_{target}, where Z_{target} is the charge number of the atoms in the target material, generate *bremsstrahlung* photons through heavy nuclei scattering. The resulting energy loss scales as $1/m^4$ where m is the incoming particle mass

Fundamental Math and Physics for Scientists and Engineers, First Edition.
David Yevick and Hannah Yevick.

and is therefore far more significant for electrons. The density of photons of a given energy is proportional to $1/E_\gamma$ up to energies that approximate that of the incoming electron. Energetic electrons produce high-energy photons that subsequently produce further electron–positron pairs, inducing electron showers.

For light beams at low energies, photons are principally absorbed through the *photoelectric effect* in which interactions with atomic electrons generate free electrons with an energy $E = E_\gamma - E_{\text{bound}}$, where E_{bound} represents the electron binding energy. *Compton scattering* of photons from electrons becomes significant at intermediate energies, while for $E_\gamma > 2m_e c^2$, electron–positron pairs are generated, becoming the dominant decay channel at high energies. The total absorption coefficient is obtained by summing the coefficients of all three processes.

25.2 RADIATION DETECTION AND COUNTING STATISTICS

Particles and particle trajectories are typically detected from their scattering or collisions in a background medium. If only the number of particles is of interest, then *scintillation counters* record photons produced by energetic charged particles in a transparent dielectric medium. For incoming gamma rays, these are typically high-energy secondary electrons created by electron–positron pair production, Compton scattering, or the photoeffect. In *avalanche photomultiplier* tubes, a fraction of this light is converted to photoelectrons at a photocathode. These are accelerated by an electric field onto a second metal *dynode* plate that produces additional secondary electrons from each collision. Subsequent stages yield a large amplitude output pulse with a height that is related to the energy initially transferred by the incoming particle to the photomultiplier. However, since the Compton effect produces electrons with a wide range of energies, low-amplitude pulses can be generated even by energetic incoming particles.

In an *ionization chamber*, a gas is placed in a high electric field between two capacitor plates. A particle that ionizes a gas molecule generates an electron and a positively charged ion that move in opposite directions, forming a current pulse when they arrive at the plates. Replacing the gas by a reverse-biased semiconductor *p–n* junction results in a *semiconductor detector*. Ionizing an atom in the junction region produces an electron and hole that propagate in opposite directions, creating an electric pulse. In a *semiconductor avalanche detector*, the field within the junction is sufficiently large that the accelerating electron and hole generate secondary electron–hole pairs. Large semiconductor detectors are however difficult to fabricate. Further, the probability, $P(n) = e^{-\bar{n}}\bar{n}^n/n!$, of initially generating a pulse with n electrons is Poisson distributed with standard deviation $\sqrt{\bar{n}}$. For small average numbers, $\bar{n}$, statistical fluctuations in pulse heights degrade the energy resolution of the measurement. With lithographic (photographic patterning) techniques, semiconductor detectors can be constructed that provide micron resolution, enabling the detection of very short-lived particles.

The actual particle trajectory can be observed in *bubble chambers* constructed of nearly boiling liquid in contact with a piston. Just before the passage of a stream of particles, the pressure is suddenly decreased, lowering the boiling point and causing the formation of bubble tracks in the resulting superheated liquid along the charged particle paths. A few milliseconds after the tracks are recorded, the piston is brought back to its original position. Normally, a high magnetic field is also present so that the charged particle energies can be deduced from the curvature of their paths.

While bubble chambers cannot be activated upon detection of specific events, *spark chambers* are surrounded by detectors such as scintillation counters that identify events such as a single particle entering the chamber and a certain number of particles leaving the chamber. When an event is detected by the outside counters, a voltage close to the breakdown voltage of the gas in the chamber is applied across alternating sets of metal plates, producing sparks at the locations where the particles previously ionized the gas. The location of these events can be recorded automatically by, e.g., replacing the plates with a wire mesh in which the electric signals are registered by ferrite cores at the wire intersections.

Calorimeters measure the energy of particles deposited in an absorbing material by monitoring its temperature. Particularly sensitive devices employ, e.g., the transition between superconducting and normal resistance induced by slight temperature changes.

25.3 LASERS

An excited carrier can decay through *spontaneous emission* that is unaffected by the presence of any background electromagnetic field and *stimulated emission* that is induced through field interactions. Simulated radiation retains the phase of the inducing field, resulting in long-range correlations of decay events, while spontaneous emission processes necessarily occur with random relative phase shifts.

Since stimulated emission results from an electron–photon interaction, its rate varies as the product of the photon density (or equivalently the electromagnetic energy since the energy is proportional to the number of photons through $E = \hbar\omega$) and the electron density. In a system with two energy levels in the presence of a photon density $\rho_\gamma(\hbar\omega)$, the absorption and emission rates must coincide in equilibrium. Denoting by N_1 and N_2 the number of electrons in the ground and excited state, respectively, and the *Einstein rate coefficients* for absorption, spontaneous, and stimulated emission by $A_{1\to2}$, $T_{2\to1}$, and $S_{2\to1}$, then, if the energy level difference equals $\hbar\omega_{21}$,

$$A_{1\to2}\rho_\gamma(\hbar\omega_{21})N_1 = T_{2\to1}N_2 + S_{2\to1}\rho_\gamma(\hbar\omega_{21})N_2 \tag{25.3.1}$$

Since $\rho_\gamma(E)$ and the ratio of occupation factors N_2/N_1 at thermal equilibrium are both known, the ratios of the various coefficients follows from Equation (25.3.1). For stimulated emission to surpass absorption and spontaneous emission, the density of electrons in the upper state, N_2, must exceed that of the density in the lower state N_1, which is termed *population inversion*. In a two-level system, however, as a result of spontaneous emission, the absorption rate is always greater than the simulated emission rate, as the number of transitions generated by absorption must equal the number of decays through both emission processes combined. Accordingly, one or more additional states are required at energies higher than E_2 that decay into state 2 more rapidly than the rate of spontaneous emission of state 2 into state 1. Exciting the electrons in state 1 into these higher levels can then result in population inversion such that an electromagnetic beam with frequency $\omega_{12} = (E_2 - E_1)/\hbar$ experiences net amplification. If the medium is placed between two parallel mirrors, a high-power directed beam is created at the frequency of greatest amplification.

While atomic lasers require an atomic system with three or more levels, in a *heterojunction semiconductor diode laser*, electrons and holes are instead injected from forward-biased p and n regions into a weakly doped or undoped high-index *active region* with a smaller bandgap. The carriers rapidly lose energy through lattice collisions until the free electrons and vacancies accumulate in low momentum states in the conduction and valence bands. The resulting population inversion enables lasing at photon energies slightly above the bandgap energy. The stimulated light is further confined by total internal reflection along the active region creating a directional beam.

INDEX

Fundamental Math and Physics for Scientists and Engineers, First Edition.
David Yevick and Hannah Yevick.